P. L. Huyskens · W. A. P. Luck
T. Zeegers-Huyskens (Eds.)

Intermolecular Forces

An Introduction
to Modern Methods and Results

With 251 Figures

Springer-Verlag

Berlin Heidelberg NewYork
London Paris Tokyo
Hong Kong Barcelona Budapest

Prof. Dr. Pierre L. Huyskens
Department Scheikunde
Katholieke Universiteit Leuven
B-3030 Leuven (Haverlee), Belgium

Prof. Dr. Werner A. P. Luck
Physikalische Chemie und
Zentrum für Materialwissenschaften,
Philipps Universität,
D-3550 Marburg, West-Germany

Dr. Therese Zeegers-Huyskens
Department Scheikunde
Katholieke Universiteit Leuven
B-3030 Leuven (Haverlee), Belgium

ISBN-13: 978-3-642-76262-8   e-ISBN-13: 978-3-642-76260-4
DOI: 10.1007/978-3-642-76260-4
Library of Congress Cataloging-in-Publication Data
Intermolecular forces : an introduction to modern methods and results /
P. L. Huyskens, W. A. P. Luck, T. Zeegers-Huyskens, eds.
Includes lectures presented at a meeting at the University of Leuven, Belgium in 1989.
Includes index.

1. Intermolecular forces--Congresses. I. Huyskens, P. L. (Pierre L.). 1927 -.
II. Luck, Werner A. P. III. Zeegers-Huyskens, T. (Therese), 1922 -
QD461.I56   1991
541.2'26--dc20

Typsetting: Thomson Press (India) Ltd., New Delhi

51/3020-543210 Printed on acid-free paper.

Preface

The study of intermolecular forces began over one hundred years ago in 1873 with the famous thesis of van der Waals. In recent decades, knowledge of this field has expanded due to intensive research into both its theoretical and the experimental aspects. This is particularly true for the type of very strong cohesive force stressed in 1920 by Latimer and Rodebush: the hydrogen bond, a phenomenon already outlined in 1912 by Moore and Winemill. Hydrogen bonds exert a profound influence on most of the physical and chemical properties of the materials in which they are formed. Not only do they govern viscosity and electrical conductivity, they also intervene in the chemical reaction path which determines the kinetics of chemical processes.

The properties of chemical substances depend to a large extent on intermolecular forces. In spite of this fundamental fact, too little attention is given to these properties both in research and in university teaching. For instance, in the field of pharmaceutical research, about 13 000 compounds need to be studied in order to find a single new product that can be successfully marketed. The recognition of the need to optimize industrial research efficiency has led to a growing interest in promoting the study of intermolecular forces. Rising salary costs in industry have encouraged an interest in theoretical ideas which will lead to tailor-made materials.

These developments require increased cooperation between fundamental research and industrial development and are facilitated by work in fields such as computer simulation of chemical and physical properties.

The research workers who in the future will bridge the gap between fundamental and applied research are the cream of today's university students. In order to bring the most recent fundamental and applied research in the field of intermolecular forces to these students, an intensive Erasmus course was organised at the University of Leuven, Belgium in 1989. Lectures on the many aspects of this topic were given by specialists from Belgium, Germany and the Netherlands (where van der Waals was born) and other countries. This book contains these lectures adapted for publication and extended for a larger public by several supplementary articles. We hope that this book will not only promote the status of intermolecular forces for advanced students and researchers but also influence university teachers to pay more attention in their courses to the fruitful field of intermolecular forces and their role in determining the properties of materials.

P.L. Huyskens
W.A.P. Luck
T. Zeegers-Huyskens

Table of Contents

List of Authors

Professor Josef Barthel,
Universität Regensburg, Institut für Physikalische Chemie,
Universitätsstr. 31, D-8400 Regensburg, Germany

Professor Phillippe Bopp,
RWTH Aachen, Institut für Physikalische Chemie,
Templergraben 59, D-5100 Aachen, Germany

Professor Udo Buck,
Max-Planck-Institut für Strömungsforschung,
Postfach 2853, D-3400 Göttingen, West-Germany

Professor Arnout Ceulemans,
Department of Chemistry, University of Leuven,
Celestijnenlaan 200F, B-3001 Heverlee, Belgium

Professor Yves Engelborghs,
Laboratory of Chemical and Biological Dynamics,
University of Leuven, Celestijnenlaan 200D,
B-3001 Leuven, Belgium

Professor Pierre Huyskens,
Department of Chemistry, Univesity of Leuven,
Celestijnenlaan 200F, B-3001 Heverlee, Belgium

Professor Thérèse Zeegers-Huyskens,
Department of Chemistry, University of Leuven,
Celestijnenlaan 200F, B-3001 Heverlee, Belgium

Professor Geoffrey S.D. King,
Laboratorium voor Kristallografie,
University of Leuven, Celestijnenlaan 200C,
B-3001 Leuven, Belgium

Dr. Hubertus Kleeberg,
Trifolio-M GmbH, Sonnenstr. 22,
D-6335 Lahnau 2, Germany

Professor Hans-Heinrich Limbach,
Universität Freiburg, Institut für Physikalische Chemie,
Albertstr. 21, D-7800 Freiburg, Germany

Professor Werner A.P. Luck,
Philipps-Universität Marburg,
Fachbereich 14—Physikalische Chemie,
Hans-Meerwein-Straße, D-3550 Marburg, Germany

Dr. Guido Maes,
Department of Chemistry,
University of Leuven, Celestijnenlaan 200F,
B-3001 Heverlee, Belgium

Professor Leo C.M. De Maeyer,
Max-Planck-Institut für Biophysikalische Chemie,
Nikolausberg am Fassberg, Postfach 2841,
D-3400 Göttingen, Germany

Professor Klaus Rademann,
Philipps-Universität Marburg,
Fachbereich Physikalische Chemie und
Wissenschaftliches Zentrum
für Materialwissenschaften, Hans-Meerwein-Straße,
D-3550 Marburg, Germany

Professor Georges G. Siegel,
Department of Chemistry,
University of Leuven, Celestijnenlaan 200F,
B-3001 Heverlee, Belgium

Professor G. Somsen,
Department of Chemistry,
Vrije Universiteit, De Boelelaan 1083,
NL-1080 HV Amsterdam, The Netherlands

Professor Luc G. Vanquickenborne,
University of Leuven, Department of Chemistry,
Celestijnenlaan 200F,
B-3001 Leuven, Belgium

CHAPTER I
Intermolecular Forces

Th. Zeegers-Huyskens and P. Huyskens

No fundamental difference exists between cohesion forces and chemical bonds. They chiefly originate from coulombic interactions between charged particles. Repulsion forces are only important when the distance between the atoms falls-below the sum of the van der Waals radii. *Van der Waals* forces are cohesive attractions between molecules that are already active at long interdistances. They result from interactions between permanent, induced or temporary electric dipoles. The last are called "dispersion forces". *Specific interactions* are cohesion forces that are only effective when so called specific sites of both molecules come into contact. In fact, specific interactions, as for instance hydrogen bonds, are short-range site-bounded cohesion forces that considerably weaken a given chemical bond of one of the partners. In the A—H...B hydrogen bond, the interdistance between the proton and the nearest nucleus of B is much shorter than the sum of the van der Waals radii, the distance A—H is larger than in the unperturbed molecule and the cohesion energy is intermediate between that of pure dispersion forces in the liquid state and the energy of normal chemical bonds. Hydrogen bonds appear as an intermediate step of the transfer of a proton from AH to B. In this transfer a new chemical bond BH^+ is formed. H-bonds already share two characteristics of the chemical bonds: the stoichiometry and the directionality. In contrast their lifetime is very short. These characteristics are also those of the so-called n—σ EDA bonds. The energy of a hydrogen bond is governed by the difference in proton affinity of B and the anion A^- A quantitative expression is proposed.

1 Cohesion Forces

1.1 Cohesion Forces and Chemical Bonds

Cohesion forces and chemical bonds govern the distances between the nuclei.

As a rule chemical bonds or valence forces can be defined as those that are preserved when low-molecular substances are brought in the gas phase. At normal pressures, cohesion forces in the gaseous state are very weak and their contribution to the total energy of the system is small. To a good approximation the cohesive energy of low-molecular substances in the liquid or in the solid state is therefore equal to the energy needed to vaporize them.[1]

The existence of cohesion forces was for the first time envisaged by Laplace in 1806 in order to explain the phenomena of capillarity.

In fact, no fundamental differences exist between cohesion forces and chemical bonds. They both chiefly originate from the coulombic interactions between charged particles, electrons and nuclei, present in the atoms. When interactions between charges of different sign predominate, the resulting effect is an attraction. Repulsion occurs in the opposite case.

---

[1] Carboxylic acids constitute an exception because they are still strongly H bonded in the saturated vapor phase

1.2 Repulsion Forces

In general, the repulsion forces become only important when the distances between the nuclei fall below a given limit. The order of magnitude of this limit is the Ångström ($1 Å = 0.1$ nm). For atoms belonging to different molecules this limit is the sum of the so-called Van der Waals radii of the atoms. The Van der Waals radius of a given atom is the halve of the shortest distance observed in crystals between the nuclei of atoms of the same nature belonging to different molecules.

Table 1. Van der Waals radii $r_{vdw}/Å$ of atoms

				H	1.20	He	1.30
N	1.50	O	1.40	F	1.35	Ne	1.40
P	1.90	S	1.85	Cl	1.80	Ar	1.70
As	2.00	Se	2.00	Br	1.95	Kr	1.80
Sb	2.20	Te	2.20	I	2.15	Xe	2.05

CH_3 and CH_2 groups: 2.00Å from the center of the carbon atom via the H-atoms.
Thickness of aromatic ring: 2×1.85Å.

In a chemical bond the internuclear distance is markedly lower than the sum of the Van der Waals radii. For instance, in H_2 the H...H distance is 0.74Å instead of 2.40 Å, in O_2 1.21 Å instead of 2.80 Å, and in N_2 1.10 Å instead of 3.00 Å.

1.3 Van der Waals Forces and Specific Interactions Between Molecules

When the interdistance between the nuclei is larger than the sum of the Van der Waals radii of the atoms, the interaction results as a rule in an attraction.

Such interactions were considered for the first time in 1873 by Van der Waals in order to explain the deviations of gases from the ideal behavior. According to Van der Waals, the pressure exerted by a gas on the walls of a vessel is lower than that predicted by the ideal law because the molecules that collide with the wall are somewhat retained by the attraction they undergo from the other molecules in the bulk of the gas:

$$p = \left(\frac{nRT}{V - nb} \right) - a\left(\frac{n^2}{V^2} \right) \tag{1}$$

The pressure correction and the "a" factor of Van der Waals are thus due to some remanent cohesion energy in the gas phase.[2]

The attraction forces that are at the origin of this cohesion are called "Van der Waals" forces. They are stronger in liquids or in solids owing to the shorter distances between the nuclei.

One can thus define the Van der Waals forces as cohesive attractions between molecules that are already active at long interdistances between atoms.

Besides the Van der Waals forces there exist cohesion forces acting at short distances, namely the hydrogen bonds and the charge transfer (EDA) bonds. These short-range forces are also called "specific" cohesion forces because they require the contact between given specific sites of the molecules of the partners.

1.4 Origin of the Van der Waals Cohesion Energies. Their Quantitative Expression in Gases

From a quantitative point of view, Van der Waals forces between gaseous molecules correspond to interactions between electric dipoles. One may distinguish three kinds of electric dipoles in molecules:

1.4.1 Permanent Dipoles μ^0

Molecules have a permanent dipole moment when in the unperturbed state, the center of charge of the nuclei does not correspond with that of the

Table 2. Dipole Moments of molecules μ^0 in the gaseous phase in Debye and SI units (Cm) [1]

	Debye	10^{-30} Cm		Debye	10^{-30} Cm
n-Butane	0.00	0.00	Pyridine	2.23	7.44
n-Pentane	0.00	0.00	Cyclohexanone	2.90	9.67
n-Hexane	0.00	0.00	Acetone	2.90	9.67
Cyclohexane	0.00	0.00	Nitroethane	3.20	10.7
Benzene	0.00	0.00	Propionitrile	3.50	11.7
Toluene	0.36	1.20	Dimethylacetamide	3.70	12.3
Triethylamine	0.78	2.60	Dimethylformamide	3.90	13.0
Hydrogen chloride	1.10	3.67	Dimethylsulfoxide	3.92	13.1
Diethylether	1.21	4.04	Propylenecarbonate	4.98	16.6
1,2-Dichloroethane	1.39	4.64	Lithium fluoride	6.40	21.3
Dichloromethane	1.58	5.27	Lithium chloride	7.09	23.6
Tetrahydrofurane	1.76	5.87	Natrium chloride	9.06	30.2
Methylacetate	1.80	6.00	Potassium chloride	10.70	35.7
Water	1.84	6.13			

[2] To a first approximation, for mono-atomic gasses the b parameter is related to the crystallographic Van der Waals radius and to Avogadro's number by the relation $b = (16\pi N/3)r_{vdW}^3$

electrons. Permanent moments of neutral molecules range from 0 to 15 Debye (1 Debye = 3.3356×10^{-30} Cm).

1.4.2 Induced Dipole Moments

These are caused by the displacements of electrons and nuclei under the influence of an external electric field E. The induced moment μ^{ind} is proportional to the applied field. The proportionality coefficient is the polarizability α of the molecule:

$$\mu^{ind} = \alpha E \qquad (2)$$

The polarizability can be calculated from the refractive index n and the density ρ of the substance according to the Lorenz–Lorentz equation:

$$\alpha = \frac{3}{4\pi N} \left(\frac{n^2 - 1}{n^2 + 2} \right) \frac{M_B}{\rho} \qquad (3)$$

M_B is the molecular weight of the molecule B and N Avogadro's number. Polarisabilities are of the order of 10^{-24} cm^3. They are rather unsensitive to changes in temperature or density. They depend on the frequency of the alternating field according to the equation:

$$\alpha_v = \frac{a_B}{v_B^2 - v^2} \qquad (4)$$

where a_B and v_B are characteristic constants of the substance. v_B is called "dispersion" frequency and is of the order of 10^{15} s^{-1}.

Table 3. Dispersion frequencies v_B/s^{-1} of various molecules

Gases		Liquids	
He	5.9×10^{15}	Water	3.17×10^{15}
Ne	5.2×10^{15}	Cyclohexane	3.24×10^{15}
Ar	3.5×10^{15}	Diethylether	2.37×10^{15}
Xe	2.8×10^{15}	Benzene	3.13×10^{15}
H_2	4.0×10^{15}	Methanol	3.34×10^{15}
N_2	4.1×10^{15}	Ethanol	3.54×10^{15}

To a good approximation the polarizability of a given molecule can be calculated from group contributions α_g given in Table 4.

$$\alpha = \Sigma \alpha_g \qquad (5)$$

Table 4. Group contributions $\alpha_{gD}/10^{-25}$ cm^3 for the polarizability (Na-line)

—H	3.9	—NO$_2$	19.7	≡CH	13.7
—F	4.5	—CH$_3$	22.4	—C≡C—	27.1
—Cl	22.1	—CH$_2$	18.4	—C≡C—C≡C—	57.7
—Br	33.1	—CO | H	22.0	≡N	10.4
—I	52.9			—S—S	60.8
—O—H	9.6	—CO | O	23.5	—C≡C— | |	25.6
—S—H	3.2			—C— |	9.8
		—O—	6.7		
—NH$_2$	16.4	—S—	30.0		
—CN	20.7				80.5
		=SO	32.8		
		=NH	13.7		

1.4.3 Temporary Moments

The dispersion frequency is directly related to the motions of the electrons. During a time shorter than the inverse of the dispersion frequency ν_B, an atom or a molecule (even apolar) exhibit temporary moments due to the motions of the electrons.

Equations were derived for the various interaction energies in the gas phase. Interactions between the permanent dipole moments μ_A^0 of the molecule A and μ_B^0 of the molecule B separated by a distance r in the gas phase provoke a lowering of the molar energy calculated by Keesom (1912):

$$\varepsilon_{\mu_A^0 \mu_B^0} = -\left(\frac{2N^2}{3RT}\right)\frac{\mu_A^{0^2}\mu_B^{0^2}}{r^6} \quad \text{(Keesom)} \tag{6}$$

R is the gas constant and T the absolute temperature. Debye (1920) calculated the molar energy of interaction between the permanent dipole μ_A^0 and the polarizable molecule B at a distance r:

$$\varepsilon_{\mu_A \alpha_B} = -\mu_A^{0^2}\alpha_B N/r^6 \quad \text{(Debye)} \tag{7}$$

A quantitative expression for the interaction between temporary moments ("dispersion forces") was derived by London (1930):

$$\varepsilon_{\alpha_A \alpha_B} = -\frac{3N}{2}\left(\frac{h\nu_A^0 \nu_B^0}{\nu_A^0 + \nu_B^0}\right)\frac{\alpha_A^0 \alpha_B^0}{r^6} \quad \text{(London)} \tag{8}$$

where α_A^0 and α_B^0 are the polarizabilities in a permanent field ($v = 0$). h is Planck's constant (6.6256×10^{-34} Js).

The dispersion frequencies of the various substances are all of the same order of magnitude. Furthermore, at a given temperature and under a given pressure the molar volume of gases (and, thus, the mean interdistances between the molecules) is independent of the nature of the gas. We may therefore conclude that the cohesion energy of gases is governed by the two molecular characteristics: the polarizability and the permanent dipole moment.

1.5 Dispersion Forces in Liquids

In apolar substances only dispersion forces are responsible for the cohesion. According to London, the cohesive energy per unit volume originating from the dispersion forces in the gas phase is directly proportional to the square of

Table 5. Molar energy of vaporization ΔU_{vap}, refractive index n_D^{25}, molar volume $\bar{V}$, polarizability α, ratio $k = \Delta U_{vap}/\alpha^2$ of apolar liquids at 25°C [2]

	ΔU_{vap} kJmol^{-1}	n_D^{25}	$\bar{V}$ cm^{3}mol^{-1}	α 10^{-24}cm^3	k 10^{52}Jcm^{-3}
2,2-Me$_2$Propane	19.37	1.3417	118.2	9.86	2.36
n-Pentane	23.95	1.3560	116.1	10.05	2.75
n-Hexane	29.07	1.3735	131.6	11.90	2.70
n-Heptane	34.07	1.3861	147.5	13.74	2.67
n-Octane	39.01	1.3957	163.5	15.56	2.64
2,2,4,4-Me$_4$-Pentane	35.68	1.4032	179.2	17.34	2.12
n-Nonane	43.96	1.4037	179.6	17.40	2.61
n-Decane	48.89	1.4085	195.9	19.18	2.61
2,2,3,4-Me$_4$-Pentane	38.36	1.4111	174.5	11.18	2.26
SiCl$_4$	27.60	1.4151	115.1	11.43	2.46
n-Undecane	53.86	1.4154	212.2	21.03	2.59
BCl$_3$	20.60	1.4173	88.1	8.79	2.36
n-Dodecane	58.81	1.4198	228.6	22.92	2.51
2,3,3,4-Me$_4$-Pentane	39.37	1.4204	170.8	17.15	2.29
2,2,3,3-Me$_4$-Pentane	38.69	1.4218	174.3	17.55	2.19
Cyclohexane	29.16	1.4248	108.8	11.08	2.62
n-Tridecane	63.75	1.4272	244.9	24.93	2.51
n-Tetradecane	68.69	1.4284	261.3	26.67	2.53
n-Pentadecane	73.67	1.4297	277.7	28.42	2.53
n-Hexadecane	78.61	1.4327	294.1	30.28	2.52
n-Heptadecane	83.78	1.4351	310.4	32.11	2.52
GeCl$_4$	31.40	1.4592	116.8	12.68	2.23
CCl$_4$	30.30	1.4610	97.8	10.64	2.63
1,4-Et$_2$-Benzene	50.00	1.4946	156.5	18.08	2.39
1,3,5-Me$_3$-Benzene	45.00	1.4973	139.2	16.15	2.39
Benzene	31.40	1.4990	89.4	10.40	2.60
SnCl$_4$	37.50	1.5128	117.6	14.01	2.25
AsCl$_3$	33.90	1.6006	84.2	11.43	2.19
CS$_2$	25.50	1.6253	60.4	8.47	2.14

the polarizability. As said before, the polarizabilities are little sensitive to changes in the density. On the other hand, in contrast to the gas phase, the molar volume of the liquid depends on its nature. Furthermore, in an homologous series we expect that the cohesive energy per unit volume will tend to some limiting value within series. These considerations lead to the following correlation between the molar energy of vaporization $\Delta \bar{U}_{vap}$ and the polarizability of apolar substances [2]:

$$\Delta \bar{U}_{vap} = k\alpha^2/\bar{V} \simeq k'\bar{V}\left(\frac{n_D^2 - 1}{n_D^2 + 2}\right)^2 \tag{9}$$

where $\bar{V}$ is the molar volume and n the refractive index for the sodium line (taken to estimate to a first approximation the static polarizability).

These correlations were tested for some thirty apolar liquids. One observes a reasonable constancy of $k = 2.45 \pm 0.18 \times 10^{49}$ kJcm^{-3} mol^{-1} (and $k' = 3.97 \pm 0.29$ kJcm^{-3} mol^{-1}).

1.6 Contribution of Cohesion Forces other than Dispersion Forces in the Molar Energy of Vaporization of Liquids

The contribution of the other cohesion forces (dipole–dipole interactions or H Bonds) in the total molar energy of cohesion $\Delta U_{residual}$ can be calculated from the molar energy of vaporization and from the refractive index and the molar volume by means of the expression derived from Eq. (9):

$$\Delta \bar{U}_{residual} = \Delta \bar{U}_{vap} - 3.97\,\text{kJ cm}^{-3}\left(\frac{n_D^2 - 1}{n_D^2 + 2}\right)^2 \bar{V} \tag{10}$$

Examples are given in Table 6.

The first part of the list contains polar liquids without H-bonds. The intervention of dipole–dipole interactions in the cohesion energy is evident. To a rough approximation $\Delta \bar{U}_{residual}$ is proportional to the dipole moment per unit volume. One finds:

$$\Delta \bar{U}_{residual} \simeq 622 \pm 102\,\text{kJD}^{-1}\,\text{cm}^3\,\mu^0/\bar{V} \tag{11}$$

However, a proportionality between $\Delta \bar{U}_{residual}$ and $\mu^{04}/\bar{V}^2$ that should be expected on the basis of the expression for the gas phase is not observed.

For the alcohols of Table 6 two important remarks have to be made: first, the residual molar cohesion energy is markedly larger than the values expected on the basis of Eq. (11). For methanol one finds 28.3 kJmole^{-1} instead of 21.8 kJmole^{-1} and for 1-hexanol one finds 28.2 kJmole^{-1} while·Eq. (11) yields

Table 6. Dipole moments μ^0, molar volume $\bar{V}$, refractive index n_D^{25}, polarisability α, molar energy of vaporization ΔU_{vap}, contribution of dispersion forces ΔU_{disp} and residual part ΔU_{res} of polar substances at 25°C [3]

	μ^0 Debye	$\bar{V}$ cm³mol⁻¹	n_D^{25}	α 10^{-24}cm³	ΔU_{vap} kJmol⁻¹	ΔU_{disp}	ΔU_{res}
Diethylether	1.21	104.4	1.3492	8.89	25.9	18.5	7.4
1,2-Dichloroethane	1.39	80.8	1.4452	8.46	30.8	21.7	9.1
Dichloromethane	1.58	64.3	1.4208	8.46	25.2	12.7	12.5
Tetrahydrofurane	1.76	81.7	1.4023	7.89	28.3	18.7	9.6
Methylacetate	1.80	79.7	1.3576	6.93	30.6	14.8	15.8
Pyridine	2.23	80.9	1.5062	9.53	38.8	27.5	11.3
Cyclohexanone	2.90	104.0	1.4478	11.03	42.9	28.7	14.2
Acetone	2.90	74.0	1.3567	6.42	29.3	13.6	15.7
Nitroethane	3.20	71.5	1.3888	6.70	36.8	15.4	21.4
Propionitrile	3.50	70.8	1.3624	6.23	33.3	13.4	19.9
N,N-Dimethylacetamide	3.70	92.5	1.4269	9.57	45.1	24.3	20.8
Dimethylformamide	3.90	77.0	1.4273	7.84	47.4	19.6	27.8
Dimethylsulfoxide	3.90	71.3	1.4739	7.94	42.8	21.7	27.1
Propylene carbonate	4.98	85.0	1.4158	8.45	62.9	20.6	42.3
Methanol	1.70	40.7	1.3265	3.26	34.9	6.6	28.3
Ethanol	1.70	58.7	1.3594	5.12	39.8	11.3	28.5
1-Propanol	1.70	75.1	1.3837	6.86	45.1	16.3	28.8
2-Propanol	1.70	76.9	1.3752	6.98	42.9	16.0	26.9
1-Butanol	1.70	92.0	1.3971	1.79	49.95	21.2	28.8
Isobutanol	1.70	92.9	1.3938	8.81	48.32	21.1	27.2
2-Butanol	1.70	92.4	1.3949	8.77	47.19	21.1	26.1
1-Pentanol	1.70	108.7	1.4080	10.62	54.49	26.3	28.2
2-Pentanol	1.70	109.5	1.4044	10.62	50.66	26.0	24.6
3-Pentanol	1.70	109.0	1.4079	10.56	51.08	26.3	24.7
1-Hexanol	1.70	125.2	1.4161	12.43	59.49	31.3	28.2

only 7.1 kJmole⁻¹. A second remark is that the values of $\Delta\bar{U}_{residual}$ are very similar for all the primary alcohols. The additional cohesion energy is thus practically the same for all these alcohols. In fact, this supplementary cohesion force that binds the alcohol molecules in the liquid phase is an hydrogen bond.

2 Specific Intermolecular Forces

2.1 Influence of the Cohesion Forces on the Strength of the Chemical Bonds

Cohesion forces can only modify the strength of chemical bonds when the molecules come into contact with each other and even then their influence is in general very weak. Such influence can be experimentally detected by the change $\Delta\tilde{\nu}$ of the wavenumber of the stretching band corresponding to the given valence bond in the infrared spectra. The results of Luck and his coworkers [4] show for instance that the O—H stretching wavenumber of ethanol drops from $3677 \, cm^{-1}$ in the gas phase to $3660 \, cm^{-1}$ (0.5%) in an Argon matrix and $3635 \, cm^{-1}$ (1.1%) in dilute CCl_4 solution. A similar observation is made for the first overtone at $7130 \, cm^{-1}$ in the vapor and in the vicinity of $7100 \, cm^{-1}$ (0.4%) in heptane [5]. Thus, *in general*, the stretching frequencies and also the interdistances between the nuclei bound by valence forces are modified only from one percent or less when the molecules are transferred from the gas phase to the liquid phase. As a rule, chemical forces are rather insensitive for the competition of cohesion forces.

However, there are remarkable exceptions to this rule. This is among others the case for O—H or N—H bonds for which in some cases the perturbations brought about by given cohesion forces are much more important. For instance, according to the results of Luck [5], the stretching frequency O—H of ethanol decreases by 5% when the molecule is transferred from the gas phase in acetone. In pyridine it decreases by 10%, in triethylamine by 13.5%, and in pure liquid ethanol by 8.5%.

2.2 Specific Interactions

What makes triethylamine, pyridine, acetone, or ethanol different from an Argon matrix, heptane, or CCl_4? It is the presence at the border of the molecules of electron pairs that are not directly involved in chemical bonds. These electron pairs display some directional character.

It is the contact of the hydrogen atom of the O—H group with these electron pairs that provokes the weakening of the O—H chemical bond.

Of course, the weakening of the O—H chemical bond will only occur when it is overcompensated by a strengthening of the H... cohesion bond. This additional cohesion can be called a "specific interaction" because it requires the contact between "specific" sites of the partners: the "proton donor site" OH and the "electron donor" or "proton acceptor" site, the lone pair of electrons. For instance, the H bond between ethanol and acetone is formed only when the hydroxylic oxygen atom enters into contact with one of the lone pairs of electrons of the carbonyl group of acetone:

No H bond is formed at the contact of the O—H group with a CH_3 group.

We may thus propose the following definition [6, 7]: "*Specific interactions are short-range site-bounded cohesion forces that considerably weaken a given chemical bond of one of the partners*".

According to this definition a specific intermolecular bond is a cohesion force that competes with a given chemical bond.

2.3 Hydrogen Bonds

Hydrogen bonding constitutes a particular case of specific interactions where the weakened chemical bond involves a hydrogen atom and a more electronegative one (in general O, N, S, halogens). The major characteristics of H bonding are the following ones:

—H bonding, in a spectacular way, shortens the distance between the proton and the nucleus to which the lone pair of electrons belongs. This interdistance becomes much shorter than the sum of the Van der Waals radii of the corresponding atoms. In the ethanol-pyridine H bonds the interdistance falls from 2.70 to 1.80 Å. Of course, this is not yet the distance of 1.00 Å encountered in a valence N—H bond but this shows that the hydrogen bond lies here approximately midway between the cohesive dispersion bonding and the true chemical bond.

—H bonding increases the interdistance between the nuclei in the weakened A—H bond. For O—H bonds this decrease can be deduced from the relative decrease of the stretching frequency with respect to the gas phase by means of the correlation of Novak [8]:

$$- \Delta r \text{ (in Å)} = 0.28 \text{ Å } \Delta v/v \qquad (12)$$

According to this correlation the interdistance in the ethanol-pyridine H bond is decreased by some 0.03 Å. This seems not so important at first sight but it is markedly more than the perturbations provoked by Van der Waals forces.

—H bonding weakens the energy of the chemical bond A—H but strengthens much more the energy of the cohesion bond between the proton and the other partner of the H bond. What is called the (negative) energy of the H bond is actually the sum of these two effects (and also of some additional effects resulting from the perturbations in the neighboring bonds). For most

H bonds this overall decrease of energy lies between 10 and 35 kJmole^{-1}. The "strength" of the H bonds is thus intermediate between that of pure dispersion forces in the liquid state (2 kJmole^{-1} per H atom and CH bond in pentane) and true chemical bonds (436 kJmole^{-1} in H$_2$).[3]

However, one should notice that in some highly polar liquids as propylene carbonate where no H bonds can be formed because of the lack of proton donor sites, the energy of the dipole–dipole interactons is quite comparable with that of H bonds.

—H-bonding between neutral molecules provokes an increase of the permanent dipole moment in the direction of the specific bond. For the complex between ethanol and acetone this increase is of the order of 0.4 D [10] and is larger for H bonds between phenols and triethylamine where the increments range from 1 to 8 D, depending on the acidity of the proton donors [9].

From the comparison of the characteristics listed under numbers 2 and 4, it can be anticipated that the weakening of the A—H bond brought about by the formation of a A—H...B bond results from some competition of the A$^-$ and B entities to bind to proton.

2.4 n—σ EDA Bonding

Chemical bonds other than A—H ones can be weakened by the action of an electron donor site. When an iodine molecule is brought into contact with the electron pair of triethylamine:

in the sense indicated by the figure, the following phenomena are observed.

1. The I...N distance falls to 2.77Å. This interdistance is much smaller than the sum of the Van der Waals radii of iodine and nitrogen: $(2.15 + 1.50)$Å $= 3.65$Å.
2. The distance between the iodine nuclei increases from 2.67 to 2.83Å. This indicates a marked weakening of the chemical bond. Another experimental proof for this weakening is the decrease of the v_{I-I} stretching wavenumber from 213 cm^{-1} to 185 cm^{-1}.
3. The enthalpy of the system decreases by 54.8 kJmole^{-1} (in cyclohexane).

---

[3] The energy of interaction between HF or HCl and molecules or rare gases is small. For Ar...HF one has calculated a value of 1 kJ mole^{-1} and the interaction is still weaker for Ne...HF (0.2 kJ mole^{-1}). The weakness of these interactions is corroborated by the small shift of the v_{HF} stretching vibration in an argon matrix, -40 cm^{-1}

4. The dipole moment increases by 5.15 D (in the same solvent). The formation of this "n—σ" EDA bond also requires the contact between given "specific" sites: the n-electrons of triethylamine and, on the other hand, one of the σ-sites of the I—I bond. The last ones are "electron acceptor" sites.

The features that we have considered as essential for the H bond, namely the reinforcement of the cohesion (as reflected by the enthalpy of formation) and the weakening of the chemical bond (demonstrated by the lowering of the stretching frequency) are thus also characteristic of n—σ EDA bonds. In both types of cohesion forces, the same sites act as proton-acceptor or electron-donor.

The comparison is still more striking from a quantitative point of view. When the experimental molar enthalpies of formation of H-bonds the EDA bonds formed by amines or pyridines with the O—H group of phenols or with the sites of I_2 or Br_2 are plotted against the change in the logarithm of the stretching frequencies of O—H, I—I, and Br—Br, the points lie on the same curve (Fig. 1) [7].

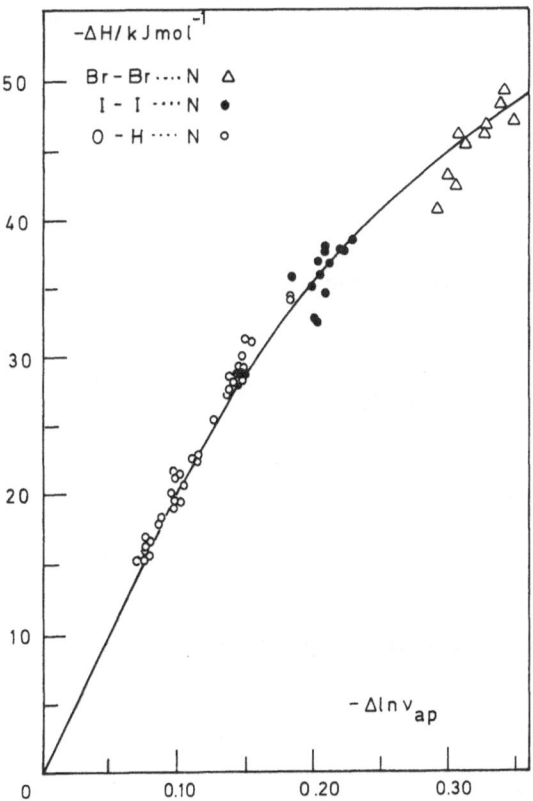

Fig. 1. Molar enthalpy of formation of H-bonds and n—σ EDA bonds between pyridines and triethylamine and substituted phenols, alcohols; I_2 and Br_2 as a function of the decrease of the logarithm of the O—H, I—I or Br—Br stretching vibration in CCl_4 at 25°C. (Huyskens P (1987)) Pure and Applied Chem 59: 1103. Reproduced with the permission of IUPAC)

2.5 Specific Sites for H-Bonding and n—σ EDA Bonding

2.5.1 Sites whose Contact Weakens a Chemical Bond in the other Partner

The principal electron donor (also proton acceptor) sites are the lone pairs of electrons of an oxygen atom or of a nitrogen atom in neutral molecules, and the external electron pairs in anions.

a) in neutral molecules:

N⃝   (Amines (strong); pyridines; anilines; nitriles (weak))

—N—O⃝   (Amine-Oxide (strong))

P=O⃝   (Phosphine-oxide (strong))

S=O⃝   (Sulphoxide (strong))

C=O⃝   (Ketones, aldehydes, carboxylic acids, amides, esters)

⃝O⃝   (Water, alcohols, ethers, phenols)

⃝S⃝   (Sulfides (weak))

b) in anions:
$$I^- < Br^- < Cl^- < F^-$$
—O⁻   (Anions of oxyacids, phenolates, hydroxide-ion)
—S⁻   (Sulfide anion)

2.5.2 Proton Donor Sites

The chemical bond of H atoms with a more electronegative atom weakens at the contact with a proton acceptor site. These sites are the "acidic" hydrogen atoms in chemical bonds with elements of stronger electronegativity.

a) in neutral molecules:

X—H	(X = halogen atom)
O—H	(oxo-acids, phenols, water, alcohols, silanols, oximes, hydro-peroxide or organophosphoric acids)
N—H	(pyrroles, amides, carbamates, peptides, anilines (weak), primary and secondary amines (very weak), NH_3
S—H	(hydrogen sulfide, thiols, thiophenols)
C—H	(only when the carbon atom is connected with strongly electronegative atoms: dinitroalkanes, HCN, $CHCl_3$ weak)
P—H	(phosphines, phosphonates (weak))

b) in cations:

N—H$^+$	(Ammonium and alkylammonium cations)
O—H$^+$	(Protonated amino oxides or phosphine oxides)

2.5.3 Electron Acceptor Sites for EDA Bonds

a) in neutral molecules:

Halogens	(I—I, Br—Br, Cl—Cl, F—F, ICl, IBr)
Lewis acids	(SO_2, SeO_2, $CdCl_2$, BF_3, SiF_4, $SnCl_4$, $TiCl_4$, $ZnCl_2$, $GeCl_4$, $SnBr_4$, SnI_4, $PbCl_4$, PF_5, $SbCl_5$ etc.)
Periodo, perbromo and perchloro alkanes (very weak)	CI_4, CBr_4, CCl_4, C_2Cl_6

b) in ions:

Iodo and bromo cations (I$^+$—N—, $\overset{+}{Br}$—N—)

2.6 Inversion of the Bond Strengths in Specific Interactions. Ionogenic Effect

It may happen that the perturbations brought about in a chemical bond by the interaction with an electron donor site lead to an inversion in the relative strengths of the bonds: the cohesion force becomes stronger than the weakened chemical bond.

For instance, in the crystalline complex between iodine and trimethylamine:

$$R_3N...I^+...I^-$$

the N...I bond is stronger than the I...I bond [10]. When dissolving the complex in a solvent of high dielectric permittivity that favors the separation of charges the weakest bond is broken and conducting ions are formed:

$$R_3N...I^+...I^- \longrightarrow R_3NI^+ + I^- \tag{13}$$

A similar observation can be made for the H bond between triethylamine and hydrogen iodide:

$$R_3N + HI \longrightarrow R_3N...H^+...I^- \longrightarrow R_3NH^+ + I^- \qquad (14)$$

In a polar medium an inversion of the bond strength occurs and separate ions can be formed. It must be emphasized that for this inversion of the bond strength and for the subsequent dissociation of ions, the medium plays a very important role.

One should observe that the overall ionogenic process which generates conducting ions through the intermediate formation of an H bond is not a transfer of *an electron* but that of *a proton* (There are as many electrons in HI as in I^-).

The hydrogen bond is an intermediate step in the transfer of a proton from HI to NR_3. In a similar manner we can state that the EDA complex between iodine and triethylamine is an intermediate step in the transfer of an I^+-ion from I_2 to R_3N:

$$R_3N + I\!-\!I \longrightarrow R_3N^+...I\!-\!I^- \longrightarrow R_3NI^+ + I^- \qquad (15)$$

One should notice here the symmetric character of the phenomenon. The process is reversible and can take place in the opposite direction. The entity R_3NI_2 can also be considered as an EDA complex between the cation R_3NI^+ and the anion I^-. A typical example where the two bonds become completely equivalent and where I^- is equally shared by the two partners is the I_3^- ion, formed by:

$$I\!-\!I + I^- \longrightarrow {}^{1/2-}I...I...I^{1/2-} \qquad (16)$$

The sum of the Van der Waals radii of both iodine atoms is equal to $4.36\,\text{Å}$, whereas the chemical bond in I_2 has a length of $2.67\,\text{Å}$. In the symmetric anion I_3^-, both $I\!-\!I$ distances are equal to $2.92\,\text{Å}$.

2.7 The Hydrogen Bond: Sharing of a Proton

This leads us to the real nature of H bonding: An H bond between neutral molecules AH and B consists in the sharing of a proton between the proton acceptor and the anion A^- derived from the proton donor and is therefore a situation intermediate between two extremes:

$$AH + B \rightleftharpoons A^-...H^+...B \rightleftharpoons A^- + H^+B \qquad (17)$$

This consideration leads to important conclusions about the characteristics of H bonding.

A first conclusion concerns the *stoichiometry*. The ion H^+B is of course stoichiometric. Normally, H bond are also stoichiometric because the number H bonds that B can form is governed by the number of lone pairs of electrons[4] (In the ion H^+B, this pair of electrons is involved in the valence bond H—B). The stoichiometric character of H bonds fundamentally differentiates them from strong dipole–dipole interactions for which the number of partners is never fixed.

A second is the *directionality*. In the ion H^+B, the direction of the H—B bond is fixed with respect to the neighboring valence bonds in the ion. In the preliminary state of H bonding, dipole measurements in the liquid phase show that the direction H...B is also more-or-less fixed with respect to the neighboring valence directions in B. Here also a marked difference appears with strong dipole–dipole interactions where always a whole fan of possible relative orientations has to be considered.

Strong specific interactions exhibit thus two characteristics of the true chemical bonds that strong dipole–dipole interactions do not possess: a) the stoichiometry and b) the directionality.

Let us illustrate this difference with an example:

The C=O group has a fairly large dipole moment (3 D) and can be surrounded by a *changing* number of other C=O groups, depending on their concentration in the liquid. The neighboring groups have a tendency to orient themselves so as to make the dipole–dipole interactions as exothermic as possible. This means a parallel orientation for the dipoles parallel to the interdistance and antiparallel for the dipoles perpendicular to the interdistance. This tendency is countered by the thermal agitation.

The interactions of the C=O group as proton acceptor in H bonding are different. One C=O group can form a maximum of two H bonds, because it disposes only of two free electron pairs. The direct consequence of the stoichiometric character of H bonds is that the concentrations of the various complexes is governed by the mass action law of Guldberg and Waage, in a similar way as that of chemical bonds.

One C=O group is thus characterized by a first equilibrium constant K_1, corresponding to the formation of a 1:1 complex:

$$\tag{18}$$

whereas the addition of a supplementary HA molecule is characterized by a second addition constant:

$$\tag{19}$$

---

[4] Only for very weak H-bonds, as for instance in liquid ammonia, a given lone pair can be involved in several H bonds

H bonds are thus stoichiometric, whereas dipole–dipole interactions are not.

The directional character of the H bonds is the consequence of the directionality of the lone pairs of electrons. For CO groups with a sp_2 hybridization, the two pairs from an angle of 60° with the C=O direction. As a consequence, an H bond with HCl, for instance, will try to adopt a structure where the HCl direction forms an angle of 60° with the C=O direction:

Accurate values of the angles can be obtained from the rotational spectra in the gas state. For the $R_2CO...HF$ complex, the θ angle determined experimentally by microwave spectroscopy is of the order of 70° [11]. Ab initio calculations performed on the $R_2CO...H_2O$ complexes reveal that the maximal stabilization occurs for a θ angle of the order of 60° [12–13]. This is confirmed by the experimental study in solution of the dipole moments of complexes involving carbonyl groups and hydroxylic proton donors [14].

It is worthy to note that this direction by no means corresponds to the most favorable orientation of the dipoles, namely:

$$\begin{matrix} C=O \\ H-Cl \end{matrix} \quad \text{or} \quad C=O...HCl$$

This constitutes a fundamental difference between strong H bonds and strong dipole–dipole interactions. The structure of an H-bonded complex is not determined by interactions between the overall dipoles but merely *by the direction of the covalent bond in the ion*:

This corresponds more-or-less to the direction of one of the lone pairs of electrons in the C=O group. The directional character weakens when the strength of the interaction decreases, and the directionality is much less strict in weak N—H...N hydrogen bonds than in strong O—H...N or O—H...O bonds. As a consequence, the stacking forces in crystals may make the angle in the H bond C=O   to differ markedly from 60° [15].
                                        HO

Moreover, in crystals, in order to enhance the lattice energy, the proton of an H bond can more-or-less deviate from the O...O line. Deviations from the linearity can also be caused by the existence of other H-bonds. This is the case in cyclic trimers of the type:

or in *intra*molecular H bonds as those observed in the enol form of β-diketones:

All these deviations do not prevent that the directional character of H-bonds and n—σ bonds is much more important than in other cohesion forces and that, from this point of view, they are very similar to the valence forces with which they compete. *Therefore an H bond or a n—σ EDA bond corresponds to what is called in German a "Nebenvalenz"* [5].

The sharing of the proton between A$^-$ and B, realized in the H bond, will be a priori the more possible the fewer the difference between the energy of protonation of the two entities. From this consideration one can already conclude that one of the leading factors for H-bond formation will be the difference in the proton affinity (PA) of the partners. This will be discussed in detail later.

2.8 Difference Between Specific Intermolecular Forces and Valence Forces: Lifetime of the Bonds

Specific intermolecular forces and chemical bonds share the properties of stoichiometry and of directionality but differ from each other by their spontaneous lifetime in the liquid and in the vapor state. In a liquid all the cohesion forces between nuclei that belong to different molecules are constantly broken and re-established. The partners of the cohesion bonds perpetually change and the duration of the contact between two hydrogen atoms belonging to two different pentane molecules in the pure liquid is very small: of the order of 10^{-12} s. In contrast, the partners of chemical bonds in one pentane molecule can remain the same for geological periods, as long as the temperature does not strongly rise or chemical reactions do not occur.

From the point of view of the lifetime, the H bond in liquids resembles much more the dispersion forces than the chemical bonds. Their spontaneous lifetime is perhaps longer than that of the dispersion forces but, anyway they are very low, as discussed elsewhere.

However, specific interactions can reduce the life-time of the chemical bonds that they weaken. This is due to the possibility of inversion of the roles between these chemical bonds and the H bonds, that was discussed before. This occurs for instance in cyclic H-Bond chains:

2.9 Differences Between Hydrogen Bonds and n—σ EDA Bonds. Electron Transfer

Although H bonds and n—σ EDA bonds are similar in the four properties described above, they exhibit differences that are observed in the ultraviolet and sometimes in the visible spectrum.

The formation of a n—σ EDA bond is accompanied by changes in the electronic absorption bands that are particularly spectacular. When iodine interacts with triethylamine one observes a new very intense band in the ultraviolet region at a wavelength of 266 nm. This band is called the "charge transfer" (CT) band. Moreover, the solution of iodine in cyclohexane undergoes a marked change in the color when triethylamine is added to the solution. This is due to the displacement of the absorption band of iodine towards the higher frequencies (blue shift) when the complex is formed.

The phenomena were explained by Mulliken on the basis of a partial transfer of the electronic charge from triethylamine to iodine.

In contrast, the formation of an H bond between neutral partners is only accompanied by a marked change in the electronic absorption spectrum when the inversion of bonding strengths previously described occurs and, thus, when ion pairs are formed. There does exist some transfer of the electronic charge in normal H bonds, but, as already said, this transfer is small.

Furthermore, one should observe that, in a normal H bond between a phenol and triethylamine, the electron transfer takes place form triethylamine towards the oxygen atom, whereas in the H bond between the phenolate ion and the triethylammonium ion, the electron transfer occurs in the opposite direction from the oxygen atom towards the nitrogen atom:

These differences are actually due to the fact that *in solution*, there exists for the potential energy of the proton a double minimum.

This leads to a tautomerism of the H-bonded complexes in solution. Such tautomerism does not exist in a n—σ EDA complex because in this case the curve of the potential energy of the central cation has only one minimum.

At the origin of this difference one finds the fact that in a H bond supplementary repulsion forces are acting because the interdistance between the oxygen and the nitrogen nucleus becomes of the order of magnitude of the sum of the Van der Waals radii of these two atoms ($(1.40 + 1.50)$ Å $= 2.90$ Å). In contrast, in the iodine-triethylamine complex the interdistance between the iodine nucleus at the left and the nitrogen nucleus is equal to 5.10 Å, which is much larger than the sum of the Van der Waals radii ($2.15 + 1.50 = 3.65$ Å).

2.10 Cooperativity in Hydrogen Bonding

Water has two hydroxylic protons and two electron pairs:

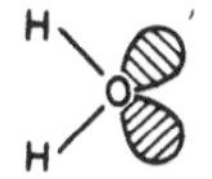

It can be involved in a first H-bonded chain of the type:

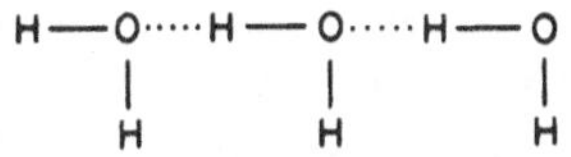

Furthermore, the second proton and the second lone pair can be involved in a second chain:

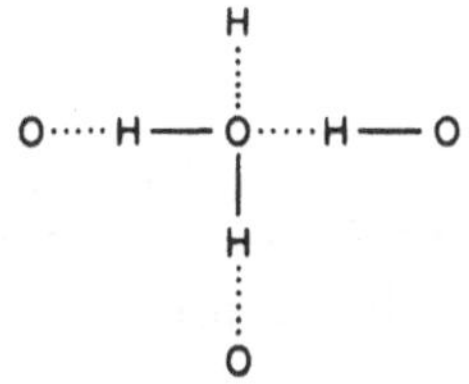

Let us consider the formation of a dimer.

The polarization of the bonds in the proton acceptor molecule at the left, and also that of the proton donor at the right, change according to the following figure:

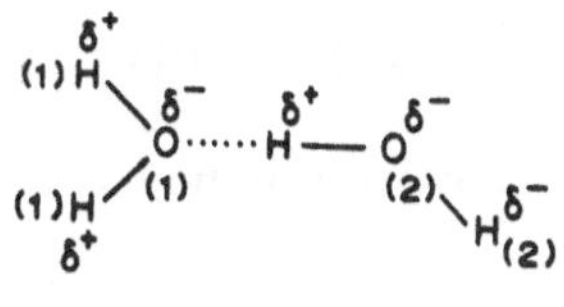

Numerous quantum-mechanical calculations on the water dimer can be found in the literature. The values of the changes in the charges depend on the chosen basis set (Chap. II by L. Vanquickenborne). However, from a qualitative point of view, the changes in the polarity are clear and show the direction in which cooperativity or anticooperativity is expected. It is clear that the two O_1-H_1 sites acquire a stronger proton donor character than the monomer. The opposite happens for the O_2-H_2 bond. On the other hand, the O_2 atom becomes a better proton acceptor.

The cooperativity effect explains the stronger acidity of the HX dimer. When we compare the two complexes $H_3N...HF$ and $H_3N...HF...HF$, the strength of the N...H interaction is higher in the trimer than in the dimer. This corresponds to a more important stretching of the HF bond and a larger shift of the ν_{H-F} vibration in the trimer than in the dimer (as observed in an Argon matrix). Ab initio calculations and infrared data show that the trimer displays the structure:

$$H_3N...HF...HF$$

and not the proton transfer one:

$$H_3NH^+...(FHF)^-$$

Ab initio calculations and spectroscopic measurements are in general limited to simple systems (dimers or trimers). Dynamic simulation calculations allow to study more complex systems, for instance, systems with 200 water molecules, 8 cations, and 8 anions. Such calculations lead to the radial and to the angular distribution of the entities in these systems (Chap. XIV by P. Bopp).

2.11 H Bonding and Proton Affinity

It was already mentioned that a leading factor in the formation of an H bond between a neutral proton donor HA and a neutral electron donor B would be the difference in proton affinity PA between B and the anion A^-.

This consideration can be extended, mutatis mutandis, to H bonds involving ions. For instance, the H bond between the ion AH^+ and the neutral molecule B, $AH^+...B$, is governed by the difference in proton affinity between B and the neutral A.

The proton affinity of a particle B is the decrease of molar enthalpy $-\Delta H$ accompanying the protonation in the gas phase:

$$B + H^+ \longrightarrow BH^+ \tag{20}$$

The gas phase basicity is the corresponding change in standard free energy $-\Delta G^0$.

A large amount of PA data for molecules, anions, and atoms is now available in the literature [11–16]. The experimental values of the PA, still usually expressed in kcal·mol^{-1}, have been obtained by mass spectrometric techniques at high pressure or by ion cyclotron resonance experiments. Most of the published values refer to NH_3 (PA = 205 kcal·mol^{-1}; 857.7 kJmol^{-1}) as reference.

The PA of some important bases, anions, or atoms are listed in Table 7. It can be observed that in homologous series such as the hydrogen halides, when the PA of the neutral molecule increases, its value decreases in the conjugated anion. The PA of molecules without free electron pair(s) such as methane is much lower than that of typical organic bases such as amines, ethers, ketones, pyridines, and so on. 1,8-bis-dimethylamine naphthalene which is considered a "proton sponge", is characterized by a PA of 241.8 kcal·mol^{-1}, the highest value up to now measured for a neutral organic base. The PA of rare gases also listed in Table 7 is very low and ranges between 42.5 (He) and 118.6 (Xe) kcal·mol^{-1}.

Table 7. PA of some molecules, anion, and atoms

PA	kcal·mol^{-1}	(kJmol^{-1})	PA	kcal·mol^{-1}	(kJmol^{-1})
HF	112	(468.0)	F^-	372	(1555)
HCl	134.8	(563.5)	Cl^-	333	(1392)
HBr	136	(475)	Br^-	324	(1354.5)
HI	150	(627)	I^-	314	(1312.5)
H_2O	166.5	(696)	OH^-	389.5	(1628)
CH_3OH	181.9	(760)	CH_3O^-	379.2	(1627)
C_2H_5OH	188.3	(787)	$C_2H_5O^-$	376.1	(1549)
CH_3COOH	190.2	(795)	CH_3COO^-	348.5	(1457)
C_6H_5OH	196.3	(820.5)	$C_6H_5O^-$	351.4	(1469)
H_2S	170.2	(711.5)	SH^-	353	(1475.5)
			NH_2^-	400	(1674)
			CH_3^-	417	(1745)
CH_3CN	190.9	(798)	CO_2	130.9	(547)
C_2H_5CN	192.8	(806)	$(CH_3)_2O$	192.1	(803)
n-C_3H_7CN	193.7	(797.5)	$(C_2H_5)_2O$	200.2	(837)
n-C_4H_9CN	195.8	(818.5)	CH_3COCH_3	197.2	(824)
pyrazine	206.4	(862.5)	CH_3COOCH_3	198.3	(829)
pyrimidine	208.6	(872)	1,4-dioxane	193.8	(810)
pyridine	220.5	(921.5)	pyridine N-oxide	220.3	(921)
$C_6H_5NH_2$	209.5	(875.5)	$(CH_3)_2SO$	211.3	(883)
NH_3	205	(857)	$CH_3CON(CH_3)_2$	218.1	(911.5)
CH_3NH_2	214.1	(895)			
$(CH_3)_2NH$	220.6	(922)	CS_2	167.1	(698.5)
$(CH_3)_3N$	237.2	(991.5)	$(CH_3)_2S$	200.6	(838.5)
H_2N-$(CH_2)_3$-NH_2	234.1	(979)	$(C_2H_5)_2S$	205	(857)
1,8-bisdimethyl-aminonaphtalene	241.8	(1012)			
CH_4	132	(551.5)	He	42.5	(177.5)
CH_3F	150	(627)	Ne	48.1	(201)
CH_3Br	165.7	(692.5)	Ar	88.6	(370)
			Kr	101.6	(424.5)
			Xe	118.6	(495)

As previously discussed, an AH...B hydrogen bond results from the share of a proton between the proton acceptor B and the A-anion derived from the proton donor:

$$AH + B \longrightarrow A^- ... H^+ ... B \qquad (21)$$

One can thus expect that the energy of a given bond will depend on the difference in proton affinity of the proton acceptor B and the A^- anion. For hydrogen bonds between cations and neutral molecules:

$$AH^+ + B \longrightarrow A ... H^+ ... B \qquad (22)$$

the energy will depend on the difference in proton affinity of A and B.

The PA obtained in the gas phase are free from solvation effects and, as a consequence, are expected to correlate better with the thermodynamic parameters of hydrogen bond formation usually obtained in aprotic organic solvents, than the aqueous pK_a values.

The same remarks also hold for the frequency shift of the stretching or out-of-plane deformation vibrations of the AH bond which are observed in organic solvents, in the gas phase, in low-temprature matrices, in solution of rare gases at high pressure or in the solid state.

Solvation effects strongly influence the pK_a values and this has been extensively discussed in Ref. [16] and others. The classical example is that of the aliphatic amines whose PA is ordered according $R_3N > R_2NH > RNH_2$ (Table 7) in agreement with the inductive effect of the R group but whose pK_a values are ordered according $R_2NH > R_3N > RNH_2$ owing to competition between inductive effects and solvation effects of water. The comparison of pyridine (Pyr), N,N-dimethylacetamide (DMA) and NH_3 is also interesting. The PA are ordered according to:

$$Pyr > DMA > NH_3$$

while the aqueous pK_a values follow a complete different order:

$$NH_3 > Pyr > DMA$$
$$(9.3) \quad (5.17) \quad (\sim 0.5)$$

The PA scale is also very interesting for the comparison of hydrogen bonds involving organic bases such as ketones, esters, and nitriles whose aqueous pK_a are too low to be determined with a great accuracy. The thermodynamic data for complex formation between a given hydroxylic derivative and aliphatic nitriles can be compared with the PA of the nitrile (see Table 7) which is known with an accuracy of about $0.5 \, kcal \cdot mol^{-1}$. The pK_a values of aliphatic nitriles are very low and uncertain. Up to now, the PA scale has been chiefly used for the comparison of the energetics of strong hydrogen bonds in the gas phase and for the comparison of the frequency shifts of the H—X stretching vibration in low-temperature Argon (or nitrogen) matrices. Relatively little effort has been devoted to the comparison of the thermodynamic data of hydrogen bond in solution and the PA of the base or the anion conjugated to the acid.

The concept of normalized proton affinity, which is defined by the expression:

$$\Delta PA_n = \frac{PA(A^-) - PA(B)}{PA(A^-) + PA(B)} \tag{23}$$

has been introduced by Ault and Pimentel [24] in order to predict the proton transfer structure *in low-temperature N_2 matrices*. These authors have investigated the hydrogen bond complexes between HCl or HBr and organic bases. The plot of the wavenumbers of the stretching vibration v_{HX} versus ΔPA_n shows a minimum value for $\Delta PA_n = -0.23$. This value corresponds to a maximum sharing of the proton between the A^- and B entities. In Argon matrices, this minimum value is somewhat lower (-0.21) [25] and is observed for the DMA·HCl complex characterized by a very low v_{HCl} frequency ($740\,cm^{-1}$) shifted by $2130\,cm^{-1}$ from the value in the free HCl molecule [26]. The v_s HCl vibration in the ND_3 complex having a ΔPA_n value of -0.238 [27] is observed at $1350\,cm^{-1}$. This suggests that in this complex, the elongation of the HCl bond is important but that the proton stays preferentially nearer to the chlorine atom. The pyridine complex ($\Delta PA_n = -0.204$, $v_s HCl = 841\,cm^{-1}$) [28] has a very small tendency to show the proton transfer structure. This tendency is more accentuated in the $(CH_3)_3N...HCl$ complex characterized by ΔPA_n and $v_s HCl$ values equal to -0.195 and $1485\,cm^{-1}$ [29]. Interestingly, there is no discontinuity between and N and O bases and this would not be the case if the aqueous pK_a where used instead of the PA.

Several correlations between the stretching frequencies and the PA have been discussed for the HF complexes in Argon matrices. For the complexes of 1–1 stoichiometry, no minimum in the vibrational correlation diagram has been observed and this means that even in the strong complexes involving aliphatic amines, the proton stays preferentially nearer the F atom.

In the last decade, the thermochemistry of numerous strong hydrogen bonds involving cations or anions and neutral molecules has been investigated in the *gas phase*. The interaction energy of substituted pyridinium ions and water:

first investigated by Kebarle et al. [32] ranges between 10 and 15 kcal·mol^{-1} and decreases with increasing basicity of the pyridines. The largest hydrogen bonding stabilization is generally obtained when the basicity of B_1 and B_2 are equal ($\Delta PA = 0$) in agreement with theoretical predictions [33].

The energies of several $NH^+...O$, $NH^+...N$, $OH^+...O$ systems involving alcohols, ethers, ketones, esters, amines, have been measured and for structurally simple dimers, linear correlations between the energies of hydrogen bond formation ($- \Delta H_{HB}$) and ΔPA, the difference in proton affinities of the proton donor BH^+ and the proton acceptor A are observed [32].

For $NH^+ \ldots O$ systems, involving ammonium and pyridinium ions complexed with water characterized by energies ranging between 12 and 27 kcal mol^{-1}, the correlation can be written as:

$$- \Delta H_{HB} = 30 - 0.26 \, \Delta PA \quad \text{(correlation coefficient} = 0.923) \qquad (24)$$

The slope and the intercept depend on the nature of the hydrogen bond.

For $NH \ldots H^+$ systems, involving ammonium, pyridinium, trialkylammonium ions and their conjugated bases, the hydrogen bond energies range between 16 and 30 kcal·mol^{-1}, the slope and intercept of the correlation being somewhat lower [34].

$$- \Delta H_{HB} = 23.2 - 0.25 \, \Delta PA \text{ (correlation coefficient} = 0.897) \qquad (25)$$

Similar correlations have been found for $RS^- \ldots HOR$ [35], $SH^+ \ldots O$ [36], $NH^+ \ldots S$ [36], $OH \ldots O^-$ [37] hydrogen bonds.

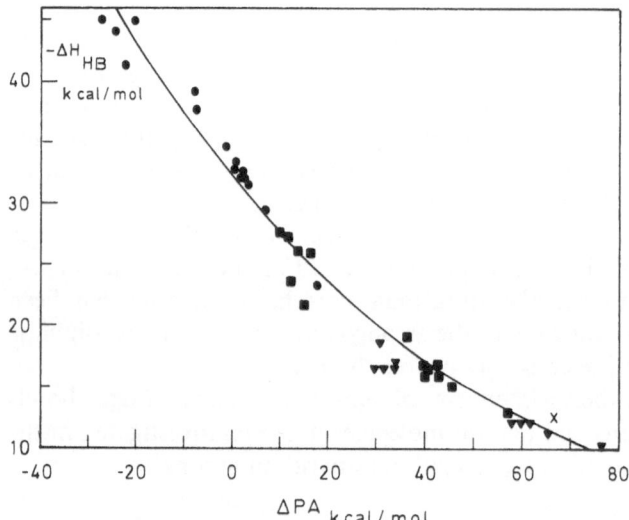

Fig. 2. $- \Delta H_{HB}$ as a function of ΔPA
- OH...F$^-$ bonds
▾ OH...Cl$^-$ bonds
X OH...Br$^-$ bonds
▲ OH...I$^-$ bonds
——curve calculated by the equation

$$- \Delta H_{HB} = 32 \exp(- 0.0156 |\Delta PA|)$$

(Zeegers-Huyskens T (1986) Chem Phys Lett 129: 172. Reproduced with the permission of Elsevier Sci Publ BV)

Ion cyclotron resonance equilibrium measurements have also been carried out to obtain accurate binding energies of the fluoride ion with Brønsted acids [38]. For $OH...F^-$ hydrogen bonds, the energies range from $-45\,kcal\cdot mol^{-1}$ ($HCO_2H...F^-$) to $23.3\,kcal\cdot mol^{-1}$ ($HOD...F^-$). The slope of the $-\Delta H_{HB}$ vs ΔPA correlation of about -0.5 presumably indicates a nearly equal sharing of the proton. These results are in sharp contrast to the situation for complexes of substituted ions with water [32] which shows a slope of only 0.12. The smaller slope may be indicative of more localized protons.

The energies of hydrogen bond formation between hydroxyl derivatives and Cl^- ions range between 13 and $33\,kcal\cdot mol^{-1}$ [38] and the $-\Delta H_{HB}$ vs ΔPA correlation has a slope of -0.33; for the $OH...I^-$ hydrogen bonds, the energies are lower (10 to $19\,kcal\cdot mol^{-1}$) [39] and the slope of the $-\Delta H_{HB}$ vs ΔPA correlation is -0.17.

Figure 2 where $-\Delta H_{HB}$ has been plotted against ΔPA for the $OH...F^-$, $OH...Cl^-$, $OH...I^-$ and $OH...Br^-$ complexes, shows that there is in fact no discontinuity between the data relative to the four halide ions. The curve of this figure can be expressed by the following exponential equation [40]:

$$-\Delta H_{HB} = 32e^{-0.0156|\Delta PA|} \qquad \Delta PA > 0$$
$$(kcalmol^{-1})$$
$$ (26)$$
$$-\Delta H_{HB} = -\Delta PA + 32e^{-0.0156|\Delta PA|} \quad \Delta PA < 0$$
$$(kcalmol^{-1})$$

Whatever the sign of ΔPA, the derivative of this equation is continuous for $\Delta PA = 0$ and is equal to -0.5. It can thus be concluded that in a broad $-\Delta H_{HB}$ ($10–45\,kcal\cdot mol^{-1}$) ΔPA ($-20–80\,kcal\cdot mol^{-1}$) range the correlation between the energetics of the hydrogen bond and the difference of the proton affinity of the two partners is not linear but takes an experimental form. The coefficients x and x' of the exponential equation:

$$-\Delta H_{HB} = xe^{-x'\Delta PA} \qquad\qquad (27)$$

have been calculated for other hydrogen bonds [41].

The results show that the values of the derivatives strongly depend on the nature of the interaction. The derivative seems to be related to the shape of the potential curve for the proton motion and to the barrier for the proton transfer. For the $H_5O_2^+$ cation, the proton potential function at the minimum energy configuration ($R = 2.37–2.39\,Å$) is a single minimum corresponding to a symmetric position for the central proton [42–44]. For the $(OHO)^+$ systems, the derivative of the exponential function is very close to -0.5. This is in complete contrast with the $(NHN)^+$ systems for which most of the ab initio calculations predict an equilibrium structure with a non-symmetrically bound proton, the $R_{N...N}$ distance is between 2.716 and $2.817\,Å$, the r_{NH} distance between 1.060 and $1.090\,Å$ and the barrier to proton transfer between 1 and $2\,kcal\cdot mol^{-1}$

[45–47]. The derivative of the exponential function is equal to -0.14 and indicates a strongly localized proton in the $(NHN)^+$ bonds.

Within a given system, this value represents the energy of the hydrogen bond when $\Delta PA = 0$. The exponential equation predicts, in a series of closely related dimers, the same energy for homoconjugated cations (or anions). The experimental energies are not exactly the same; for the $(OHO)^+$ bonds, for example, the reported energies are 31.6, 30.7, 29.8, and 27 kcal·mol^{-1} when the oxygen bases are H_2O, $(CH_3)_2O$, $(C_2H_5)_2O$, and $(i\text{-}C_3H_7)_2O$, respectively [27]. As a consequence, the x represents the mean value of $-\Delta H_{HB}$ when $\Delta PA = 0$. For these symmetric systems, the identity of the proton donor seems to be the key factor in the binding energies because an increase in the acidity of the OH^+ bond exerts a greater influence on the stability of the $(OHO)^+$ bond than the decrease in the basicity of the oxygen base. The same bonding trends have also been observed when a proton is bound between two identical nitrogen bases [48], although in these systems the proton does not occupy a central position.

Interestingly, the bond energy between a proton and σ-electron molecules is very low; for the $CH_4\ldots H^+\ldots CH_4$ system, a value of 7.4 kcal·mol^{-1} has been reported in the literature [49].

The thermodynamic data of hydrogen-bond formation *in solution* have been compared with the pK_a values of the proton donor or the proton acceptor in numerous works. The correlations are valuable within a given series of complexes. The phenol derivatives, for example, are often taken as reference acids towards a given base and the relations between $-\Delta H_{HB}$, $-\Delta G^0_{HB}$ or the frequency shift of the ν_{OH} vibration are still valuable. As a matter of fact, in a given series of proton donors or proton acceptors, the difference of the free energies of ionization in the gas phase is linearly related to the corresponding quantities in aqueous solutions. The slope of these correlations depends strongly on the nature of the interacting molecules, being, for example, -6.6 for phenol derivatives [50], about 3 for pyridine derivatives [51], and 4 for substituted anilines [52].

If ΔG_A^{0i} and ΔG_B^{0i} represent the free energies of ionization of the proton donor and proton acceptor in aqueous solution, the $-\Delta G^0_{HB}$ measured in an inert solvent can be expressed by the relation:

$$-\Delta G^0_{HB} = A - a\Delta G_A^{0i} + b\Delta G_B^{0i} \tag{28}$$

where the a and b coefficients are usually different.

When considering the gas phase acidities of the proton donor (GA) and the proton acceptor (GB), the free energy of hydrogen-bond formation takes the simple form [53]:

$$-\Delta G^0_{HB} = A' - a'(GA - GB) \tag{29}$$

the A' and a' coefficients depending on the nature of the hydrogen bond and within a given series of complexes, on the hybridization of the proton acceptor atom. In our earlier investigations on the factors governing the proton transfer equilibrium, we have proposed a correlation between the proton transfer constant and the aqueous pK_a values of the proton donor and of the proton acceptor molecules [10, 54]. It should be very interesting to predict this constant from the gas phase protonic acidities and basicities. Unfortunately, the GA values for the stronger acids such as the dinitrophenols which generally protonate the organic bases like amines, pyridines...have not yet been experimentally measured.

3 References

1. Selected values from McClellan Al (1974) Tables of Experimental Dipole Moments. Rahara Enterprises, El Cerrito
2. Huyskens PL (1989) J Mol Struct 198: 123
3. Selected experimental data from Weast RC, Astle MJ (1980) CRC Handbook of Chemistry and Physics, CRC Press Boca Raton Fl; Zwolinski BJ, Wilhoit RC (1971) Vapor pressures and heats of vaporization of hydrocarbons and related compounds Thermodynamics Research Center College Station; WIlhoit RC, Zwolinski BJ (1973) J Phys Chem Ref Data 2: 23; Roth W, Scheel K (1936) Landolt-Börnstein Physikalische Chemische Tabellen, Springer Berlin (1984)
4. Schrems O, Oberhoffer H, Luck W (1984) J Phys Chem 88: 4335
5. Luck WAP (1986) Acta Chim Hungar 121: 119
6. Huyskens PL (1986) J Mol Struct 135: 67
7. Huyskens PL (1987) Pure and Applied Chem 59: 1103
8. Novak A (1974) Structure and Bonding 18: 177
9. Ratajczak H, Sobczyk L (1969) J Chem Phys 50: 556
10. Zeegers-Huyskens T, Huyskens P (1980) in: Ratajczak H, Orville-Thomas WJ (eds) Molecular Interactions, vol 2, J Wiley; and references herein
11. Legon AC, Millen DJ (1987) Chem Soc Rev 16: 467
12. Morokuma K (1971) J Chem Phys 55: 1236
13. Del Bene JE (1975) J Chem Phys 62: 1314
14. Huyskens PL, Cleuren W, Van Brabant-Govaerts HM, Vuylsteke MA (1980) 84: 2740
15. Olovsson I. Jönsson, PG (1976) in Schuster P, Zundel G, Sandorfy C (eds) The Hydrogen Bond II, North Holland, Amsterdam
16. Johnson GL, Andrews L (1982) J Am Chem Soc 104: 3043
17. Kunnig IJ, Szczesniak MM, Scheiner S (1986) J Phys Chem 90: 4253
18. Aue DH, Bowers MT in Bowers MT (ed) (1979) Gas-Phase Ion Chemisry, Academic Press, New York, p 1
19. Lias SG, Liebman JF, Levin RD (1984) J Phys chem Ref Data, 13: 695
20. Bartness JE, Scott JA, McIver RT (1876) J Am Chem Soc 101: 6046
21. Larson JW, McMahon TB (1984) J Am Chem Soc 106: 517
22. Davis DW (1985) J Mol Struct 127: 337
23. Taft RW (1983) Prog Phys Org Chem 14: 247 (and references herein)
24. Ault BS, PImentel GC (1975) J Phys Chem 79: 615
25. Zeegers-Huyskens Th (1990) J Mol Struct 217: 239
26. Mielke Z, Barnes AJ (1986) J Chem Soc Faraday Trans 2 82: 437
27. Barnes AJ, Beeck TR, Mielke Z (1984) J Chem Soc Faraday Trans 2 80: 455
28. Barnes AJ, Szczepaniak K. Orville-Thoms WJ (1980) J Mol Struct 59: 39
29. Barnes AJ, Kuzniarski JN, Mielke Z (1984) J Chem Soc Faraday Trans 2 80: 466
30. Johnson GL, Andrews L (1982) J Am Chem Soc 104: 3043
31. Andrews L, Davis SR, Johnson GL (1986) J Phys Chem 90: 4273
32. Davidson WR, Sunner J, Kebarle P (1979) J Am Chem Soc 101: 1675

33. Desmeules DJ, Allen LC (1980) J Chem Phys 72: 4731
34. Meot-Ner (Mautner) M (1984) J Am Chem Soc 106: 1257
35. Sieck LW, Meot Ner (Mautner) M (1989) J Phys Chem 93: 1586
36. Meot-Ner (Mautner) M, Sieck LW (1985) J Phys Chem 89: 5222
37. Meot-Ner (Mautner) M, Sieck LW (1986) J Am Chem Soc 108: 7525
38. Yamdagni R, Kebarle P (1971) J Am Chem Soc 93: 7139
39. Caldwell G, Kebarle P (1984) J Am Chem Soc 106: 967
40. Zeegers-Huyskens T (1986) Chem Phys Lett 129: 172
41. Zeegers-Huyskens T (1988) J Mol Struct 177: 125
42. Kollman PA, Allen LC (1970) J Am Chem Soc 92: 6101
43. Del Bene JE, Frisch MJ, Pople JA (1985) J Phys Chem 89: 3669
44. Scheiner S, Bigham LD (1985) J Chem Phys 82: 3316
45. Scheiner S, Redfern P (1986) J Phys Chem 90: 2969
46. Delpuech J, Serratrice G, Streck A, Veillard A (1975) Mol Phys 29: 849
47. Merlet P; Peyerimhoff SD, Buenker RJ (1972) J Am Chem Soc 94: 8301
48. Hiraoka K, Takimoto H, Yamabe S (1986) J Phys Chem 90: 5910
49. Hiraoka K, Kebarle P (1975) J Am Chem Soc 97: 4179
50. Fujio M, McIver RT, Taft RW (1981) J Am Chem Soc 103: 4017
51. Arnett EM, Chawla B, Bell L, Taagepera M, Hehre WJ, Taft RW (1977) J Am Chem Soc 99: 5729
52. Law YK, Nishizawa K, Tse A, Brown RS, Kebarle P (1981) J Am Chem Soc 103: 6291
53. Zeegers-Huyskens T (1986) J Mol Liquids 32: 191
54. Huyskens P, Zeegers-Huyskens T (1964) J Chim Phys 61: 81

CHAPTER II
Quantum Chemistry of the Hydrogen Bond

L.G. Vanquickenborne

This chapter reviews the quantum-chemical approach to the hydrogen bond, and compares this approach with the theoretical description of other bond types. Morokuma's partitioning of the hydrogen bond into electrostatic, exchange, polarization, and charge transfer contribution is discussed in some detail. Special attention is given to the construction of density difference maps, as these maps are able to illustrate the bond formation in a rather illuminating way.

1 Introduction

The focus of this chapter will not be on the details of quantum-chemical calculations, but rather on certain general questions of the following type: can the existence of the hydrogen bond be understood from first principles? What makes one hydrogen bond stronger than another hydrogen bond? What are the factors that determine the specific structure and geometry of a hydrogen bond?

Intermolecular Forces
© Springer-Verlag Berlin Heidelberg 1991

At the present time, the answer to most of these questions is known. Or at least, by carrying out quantum-chemical calculations, it is possible to find a reasonably correct, and a reasonably complete answer to this kind of questions. The approximate solution of the appropriate Schrödinger equations generates rather reliable values of the structural parameters and the energy of the hydrogen bonds.

In many cases, however, scientific curiosity is not completely satisfied by a simple comparison of an experimental bond energy and a calculated bond energy, or between a theoretical and a measured bond angle. Usually, one also likes to have a more qualitative understanding of the phenomenon; one wants to know in what way the more important properties of the hydrogen bond can be related to the main features of the wavefunction. In the case of the hydrogen bond, this qualitative interpretational aspect of quantum chemistry does not always appear to be available in a completely satisfactory way. Indeed, the hydrogen bond energy E_b is usually a rather small quantity (of the order of a few kcal/mole), but it has to be calculated as the difference between two huge quantities: $E_b = E_2 - E_1$, where E_1 is the energy of the bound dimer, and E_2 is the sum of the energies of the two individual monomers. Considering that E_1 and E_2 are themselves sums of a large number of physically different effects, it is obvious that a simple interpretation may not be easy to find, especially not when this simple interpretation is also supposed to be valid for a large category of different hydrogen bonds.

In this chapter, we will try to present a few general results and ideas that may be useful in a variety of different cases. We will limit ourselves to those systems that are theoretically best characterized, that is, small isolated dimers. This means that we will not discuss solvent effects or cooperative effects, or anything involving more than the interaction between two simple monomers.

2 Hydrogen Bonds versus other Bond Types

Figure 1 shows a schematic representation of a number of different types of chemical bonds. The scheme is definitely an oversimplification, but it allows us to cast everything in terms of molecular orbital theory. In Figs. 1a and b, we represent a conventional covalent bond for two simple diatomic molecules H_2 and Li_2. In both cases, the actual bond is carried by one bonding electron pair; in H_2 this is the only electron pair; for Li_2 we have an additional set of electron pairs, where bonding and antibonding interactions tend to cancel each others contributions, so that the only net bonding contribution can be associated with the valence orbital.

In Fig. 1c we have a heteronuclear covalent bond, such as in carbon monoxide, where we show only the doubly degenerate π-orbitals. In this case, we have two net bonding π-electron pairs, since the corresponding π^* antibonding orbitals are much higher in energy and remain vacant.

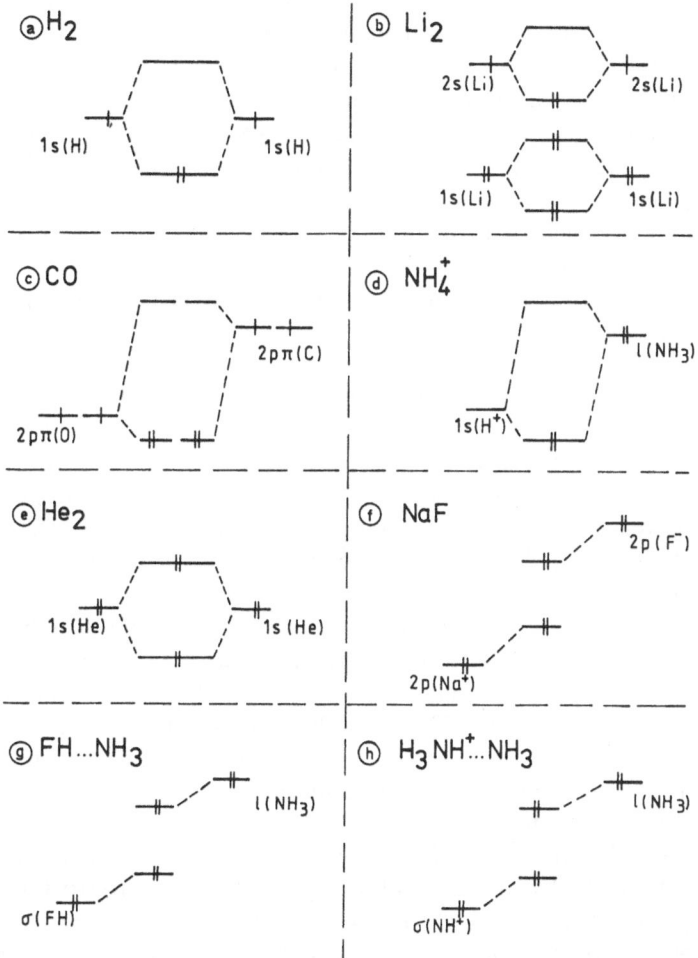

Fig. 1. Schematic representation of different bonding types within the framework of molecular orbital theory

In Fig. 1d we show a somewhat similar picture describing the formation of the ammonium ion from NH_3 and a bare proton. Both electrons in the bonding orbital were originally situated in the lone pair of ammonia; the resulting bond energy—which is the proton affinity of ammonia—is of the same order of magnitude as a regular chemical bond (205 kcal/mole).

We obtain a completely different picture in the description of the He—He interaction (Fig. 1e) where bonding and antibonding orbitals are equally occupied, and consequently no net bonding results.

In the case of sodium fluoride (Fig. 1f), it does not make much sense to set up a covalent description between neutral sodium and neutral fluorine, because the ionization potentials of these two entities are so vastly different. Also, the singly occupied 3s-orbital of sodium does not play a predominant role in the bonding phenomenon. Therefore, Fig. 1f shows a bond description where the atomic ions Na^+ and F^- are taken as a starting point. The dominant contribution to the bond energy is the ionic contribution, and the figure shows only what happens after a hypothetical electron transfer has taken place. The fluoride molecular orbitals get stabilized in the field of the positive sodium ion, but the sodium orbitals are destabilized in the field of the negative fluoride ion. The net balance of the different energy shifts is, of course, stabilizing, since it represents the quantum-chemical equivalent of the classical electrostatic attraction between two oppositely charged shells.

Finally, the figure also shows what we were initially interested in, namely the hydrogen bond. At the right-hand side, we show a very strong hydrogen bond where two ammonia molecules are linked by a proton, and at the left-hand side, we show the much weaker bond between ammonia and hydrogen fluoride. Both figures are qualitatively very similar, and their difference is essentially concerned with the magnitude of the interactions. If we make a comparison with all the previous bonding types, it is obvious that the hydrogen bond is most closely related to the ionic bond as represented in Fig. 1f. This similarity suggests that the predominant contribution to the H-bond energy will be of an electrostatic nature.

Of course, the electrostatic interactions in the hydrogen bond will be smaller than in sodium fluoride, where the individual partners were characterized by a net positive charge and a net negative charge. In Fig. 1h we have a net charge on only one of the two partners, and in Fig. 1g we have no net charge on either of the individual partners. Therefore, in the latter case, the electrostatic interactions will be of the multipole–multipole type, and the bond energy of Fig. 1h may be expected to be intermediate between Fig. 1g and f.

More sophisticated ab initio calculations are generally in agreement with this qualitative picture of the hydrogen bond as a predominantly electrostatic bond of two charge distributions (which may or may not be developed as a multipole expansion). Nevertheless, in quite a few cases, it is necessary to incorporate rather important corrections to this electrostatic picture and for more quantitative work, a number of additional interactions have to be taken into account.

3 Ab initio Methods

The most important numerical calculations of the hydrogen bond have been carried out during the last 15 years [1–10]. In general, ab initio methods can be subdivided into two categories, depending on whether or not they take electron correlation into consideration. If they do not, the methods are of the Hartree–Fock or the self-consistent field (SCF) type. The resulting Hartree–Fock equations describe the electrons in a molecule or in a dimer as if each electron did not feel the instantaneous repulsion of the other electrons, but only their average repulsion. This approximation is called the independent model: it is obviously not exact, but it yields the best possible solutions of the orbital type.

In order to obtain numerically reliable results, it is usually necessary to go beyond the SCF approximation, because correlation effects are often responsible for very important corrections. In the case of the hydrogen bond, however, the most important parameters, such as the geometry of the dimer and the evolution of the bond energy, can usually be described rather reliably already at the Hartree–Fock level. One of the reasons is that the number of paired electrons is identical in the monomers $AH + B$ and in the dimer $AH \dots B$. Therefore, to a first approximation, the correlation error can be assumed to be of comparable magnitude in the "reactants" and in the "product".

Yet, we should keep in mind that the independent model necessarily remains an approximation. For instance, within the framework of SCF theory, it is in principle impossible to account for the dispersion interactions between the monomers. The reason is that dispersion interactions are by definition correlation interactions. They result from the fact that the instantaneous positions of the electrons in the one monomer adapt themselves to the instantaneous position of the electrons in the other monomer. Dispersion interactions could therefore be considered as inter-monomer correlation effects. It is true that they are small (of the order of 1 kcal/mole) but in principle they are not within the reach of Hartree–Fock calculations, and if we want to carry out calculations of very high precision, we will have to introduce electron correlation effects [9, 10].

An important point in carrying out SCF calculations is the choice of the basis set. It will be necessary to use a basis of sufficiently high quality, so as to obtain results that are reasonably close to the Hartree–Fock limit. In practice, this means that we have to select a basis set of so-called double zeta quality for the valence orbitals, supplemented by a number of additional polarization functions.

In this context, we should mention a problem that is of particular interest in the study of the hydrogen bond, namely the basis set superposition error (BSSE). In the calculation of the energy of the dimer, one usually works in the vector space, spanned by the basis sets of the two monomers together, the so-called sum basis set. But this means that the description of the dimer is carried out in a better basis set than the basis set used for the individual partners. This will always tend to increase (artifically) the value of the hydrogen bond energy.

There are a variety of ways to minimize the BSSE, the most obvious one being to work with a still larger basis set (by including more diffuse functions) so that the addition of even more basis functions does not make much difference.

4 The Hydrogen Bond at the Hartree–Fock Level

We wish to obtain a refinement of the very approximate MO schemes that were shown in Fig. 1. We will follow the analysis of Morokuma and his coworkers [1] as represented in Fig. 2. Separate Hartree–Fock calculations can be carried out for each one of the two monomers AH and B, and for the dimer,

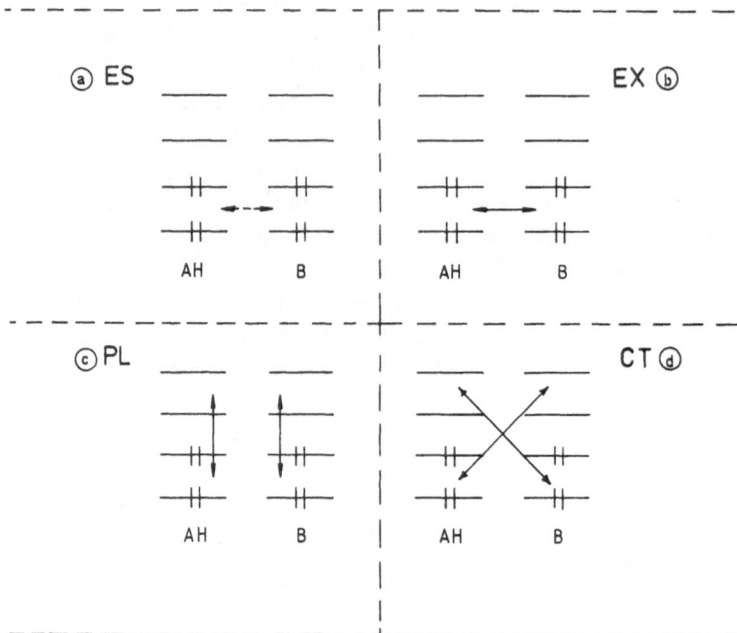

Fig. 2. Schematic representation of four different contributions to the total hydrogen bond energy: the electrostatic (ES), exchange (EX), polarization (PL), and charge transfer (CT) contributions

by using the sum basis set. Now, it is possible to calculate the dimer in a different, but entirely equivalent way by not using the sum basis set as such, but by using the MOs of the two monomers. If one takes both the occupied and the vacant MOs of the monomers, one uses in fact exactly the same vector space as before, but in a rotated coordinate system. From Fig. 2, we see that there are four types of orbitals: occupied orbitals of AH and B, and vacant orbitals of AH and B. If we allow the complete interaction of the four sets of orbitals with each other, one obtains the optimal dimer-energy. If one does not allow a complete interaction of the four sets of orbitals, but if instead one allows gradually more and more interaction possibilities, it becomes possible to dissect the total hydrogen bond energy into different components, each one being assignable to a physically different origin.

The first contribution is schematically represented in Fig. 2a: it is essentially a purely electrostatic (ES) interaction between two unmodified monomer charge densities; this contribution corresponds to a classical Coulombic interaction between two frozen density distributions.

In addition to this classical—and obviously stabilizing—ES interaction, we also have EX, accounting for the changes in the wavefunction due to the mixing of the occupied orbitals of both monomers and the corresponding action of the Pauli exclusion principle. The Pauli principle keeps electrons of the same spin away from each other; therefore, electron density is removed from the interaction zone, and the EX interaction is repulsive and destabilizing: it prevents the two monomers from approaching each other too closely.

A third interaction is related to the polarization effect (PL): it means that the approach of the partner molecule, say B, modifies the physical environment, and therefore also the electronic structure of AH. Within the framework of the present description, this can only happen by mixing the occupied and the vacant orbitals of AH. Similarly, the approach of AH modifies the electron density at B by mixing the vacant and the occupied orbitals of B. The polarization energy PL is obviously a stabilizing interaction.

Finally, we have the charge transfer (CT) contribution illustrated in Fig. 2d, where we also allow the mixing of the vacant orbitals of one partner and the occupied orbitals of the other partner. The result is an electron transfer between the two monomers—of course essentially from proton acceptor (B) to proton donor (HA).

The SCF-value of E_b is then given by:

$$- E_b = \Delta E = ES + EX + PL + CT + MIX$$

where the last term would be zero, if the first four contributions were strictly independent of each other. The magnitude of MIX is therefore a measure of the separability of the four components. As a rule, MIX is not very large, but it is an increasing function of ΔE.

If one calculates the different contributions to the hydrogen bond energy for a number of specific hydrogen bond complexes, one finds the results of Table 1. The first seven complexes of the table are neutral species and they are characterized by typical hydrogen bond strengths, whose value is small with respect to a regular chemical bond. From the calculations it also appears that the geometry of the monomers is not very much affected by the formation of the bridge. Within this category of seven complexes, E_b varies significantly, but the fact that all energy components are increasing functions of ΔE itself, suggests

Table 1. Decomposition of the hydrogen bond energy into its components; all energies are in kcal/mole and are taken from Morokuma [1]

Proton acceptor	Proton donor	$\Delta E = -E_b$	ES	EX	PL	CT	MIX
H_3N	HF	−16.3	−25.6	16.0	−2.0	−4.1	−0.7
H_2O	HF	−13.4	−18.9	10.5	−1.6	−3.1	−0.4
HF	HF	−7.6	−8.2	4.5	−0.4	−3.2	−0.3
H_3N	HOH	−9.0	−14.0	9.0	−1.1	−2.4	−0.4
H_2O	HOH	−7.8	−10.5	6.2	−0.6	−2.4	−0.5
H_3N	HNH_2	−4.1	−5.7	3.6	−0.6	−1.3	−0.2
H_2O	HNH_2	−4.1	−4.6	2.5	−0.3	−1.5	−0.2
H_2O	HNH_3^+	−27.3	−34.1	5.9	−4.1	−5.0	
F^-	HF	−62.7	−86.1	67.5	−5.9	−27.9	−10.3

that essentially the same bonding type prevails in all cases. However, if one tries to fit ΔE as a linear function of any one of its components (ES, EX, ...), the scattering of the points is definitely too large to be satisfactory. This fact only illustrates that nothing guarantees that the trends in hydrogen bonding are supposed to be simple. It should be stressed however, that the agreement with experiment is generally acceptable, although quantitatively satisfactory results can only be obtained if electron correlation and zero-point vibrational corrections are included [9, 10].

The last two examples of Table 2 are HF_2^- and $H_2O-H_4N^+$. It is obvious that these strong, ionic complexes are markedly different from the first series of seven complexes. On the one hand, they are characterized by bond energies that are significantly larger; and on the other hand, the relative value of the different contributions does not follow the same pattern as before: for HF_2^- the charge transfer contribution is unusually large, and for the $H_2O-H_4N^+$ positive ion the exchange destabilization is unusually small. Moreover, in these strong complexes, the geometrical structure of the monomers is definitely affected more thoroughly strongly than in the neutral complexes.

5 The Hydrogen Bond in Neutral Molecules

5.1 Bonding Type

Table 1 clearly shows that the most important stabilizing contribution to the bond energy is due to the classical electrostatic interaction ES. And since the other components are roughly increasing functions of ES, it is not surprising that it is possible to obtain a reasonably good understanding of the H-bridge by focusing on simple electrostatic considerations.

Before we do so, however, it may be useful to point to the small but non-negligible bonding contribution of the CT term. In all cases, we are concerned essentially with an electron transfer from B to HA through σ-interaction. As stressed before, the CT contribution is calculated in MOT by allowing the mixing of the vacant orbitals of one partner and the occupied orbitals on the other partner. In VBT the same phenomenon would be expressed by writing the wavefunction as:

$$\Psi = c_1 \Psi_1(B, HA) + c_2 \Psi_2(B^+, HA^-)$$

where c_2 is small, but not negligible.

5.2 Structural Predictions

Since the CT mixing is small and c_1 is much larger than c_2, the classical ES interactions are clearly predominant, and one might be inclined to try to make structural predictions on the basis of pure electrostatics. To some extent this procedure can lead to qualitatively correct results, but in many cases, it does not. One possible, albeit extremely simplified procedure, consists in replacing the monomers by purely classical dipoles. If one follows this line of thinking, one does not even try to make predictions on the distance between the two monomers in the dimer, but one might perhaps hope to be able to predict certain qualitative geometrical features of the dimer. Even this modest ambition is generally not realized by the simple dipole model. For instance, for the water dimer, one might a priori consider three different plausible structures: the linear, the bifurcated, and the cyclic structure.

If the geometry were determined by the most favorable dipole–dipole orientation $\oplus\ominus\oplus\ominus$, then one would expect the bifurcated structure to be the most stable. Now, from ab initio calculations, it is beyond doubt that the linear structure is the more stable of the three. Therefore, the dipole model does not appear to be very adequate, but in this particular case —for the water dimer—it is possible to arrive at a more acceptable answer by going only slightly beyond the dipole model, namely by associating a point charge to each atom, that is giving each hydrogen atom a charge of $+\delta$, and each oxygen atom a charge of -2δ. If a classical electrostatic calculation is carried out in this way, one finds that the linear structure is the more stable one, although it has only one H bond versus two H bonds in the two alternative structures.

The basis for the stability in Fig. 3a is due to the very favorable linear structure of the OHO group where the two negative charges are separated by the positive hydrogen atom. So, already from classical electrostatic interactions, it can be shown that one linear bridge is more stable than two nonlinear bridges.

As a matter of fact, the linearity of the B...HA structure is one of the most general properties of H-bonding, although it is true that secundary interactions can result in small deviations from linearity. These deviations can be due to the repulsion between the bridging hydrogen and the other hydrogen atoms, but they do not alter significantly the global picture of the H-bridge with a quasi-linear structure.

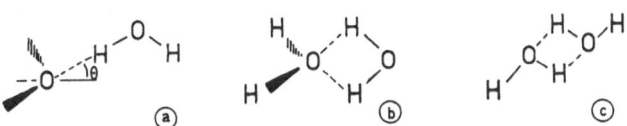

Fig. 3. Three different plausible structures for the water dimer (a) linear; (b) bifurcated; (c) cyclic structure

From the above considerations, one might get the impression that one can obtain already a reasonable picture of dimer-structures from an atomic charge model. This is only very partly true. For instance, the angle θ between the plane of the proton-acceptor and the linear OHO axis is very poorly predicted from this atomic charge model, where one finds about 20°, whereas the correct value is much closer to 60°.

The reason is that the charge density of an atom in a molecule is not spherically symmetrical, and therefore, it cannot adequately be described by a point charge (only a free, isolated atom has a spherical charge density). In the water molecule, the electron density at oxygen can approximately be described by four sp^3 hybrids. Therefore, at the right-hand side of the oxygen atom in Fig. 4a, the proton acceptor is characterized by a lower electron density in the plane of the molecule than at both sides of this plane. On purely electrostatic grounds, it is therefore favorable to have a bond dipole of the type H^+A^- approach along the axis of maximal density; this axis is at about 55° with the molecular plane.

Similar considerations can be invoked to understand the structure of the $(HF)_2$ dimer, which is again quasi-linear (Fig. 4b), with the angle φ of $\sim 6°$ and an angle θ of $\sim 57°$. Again this structure can be rationalized to some extent by considering the density pattern and by realizing that the electron density has

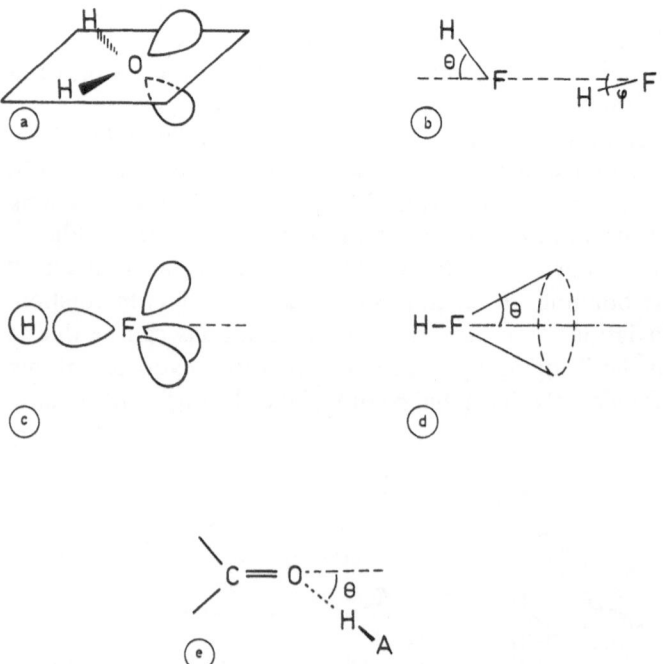

Fig. 4. Schematic representation of the relevant orbitals as related to the structure of the hydrogen-bound dimers (a) H_2O; (b) $(HF)_2$; (c) and (d) HF; (e) carbonyl group

been decreased at the backside of the FH bond axis (Fig. 4c and 4d). Therefore, the most favorable line of approach is not what a simple electrostatic model would suggest.

In a similar way, it is possible to discuss the carbonyl group (Fig. 4e). Here too, the O...HA bonding is quasi-linear, and the angle is close to 60°, in agreement with what one expects on the basis of conventional sp^2 hybrids. For a general review of directional bonding to sp^2- and sp^3-hybridized atoms, the reader is referred to the work of Murray–Rust et al. [11]. In many cases, however, the ab initio results cannot be summarized as simply as suggested by the previous considerations. As a matter of fact, there has been a recent paper [12] where the structure of the water dimer was reconsidered, and where Dannenberg claims that the most stable structure is not one of the three alternatives that were shown in Fig. 3, but rather a fourth alternative with *three* rather than one or two H-bridges. It is true though that the additional stabilization obtained by Dannenberg was very small, and basically the picture is not changed dramatically, but the example illustrates once more that quantitative predictions generally require elaborate calculations.

5.3 Population Analysis and Density Shifts

Although it was stressed in the previous section that the association of formal charges to each individual atom is only partly satisfactory, a dissection of the electron density into individual atom contributions is often used as one of the first steps in discussing the properties of the wavefunction. One of the most commonly used procedures is the Mulliken population analysis. As an example, Fig. 5 shows the results of a population analysis for water and the water dimer-molecule based on a 4-31G basis set calculation. The left-hand side displays the Mulliken atomic charges on the water monomers—which are of course identical for both molecules. At the right-hand side, we show the charge *shifts* upon H-bonding (not the atomic charges themselves). The specific numbers are of course only valid for the water dimer, but the global trends are rather general, and typical for H-bridging in neutral species. As expected, the overall electron transfer takes place from the proton acceptor to the proton donor. It is important to observe, however, that this electron transfer is extremely small: of the order of 1/100 of an electron.

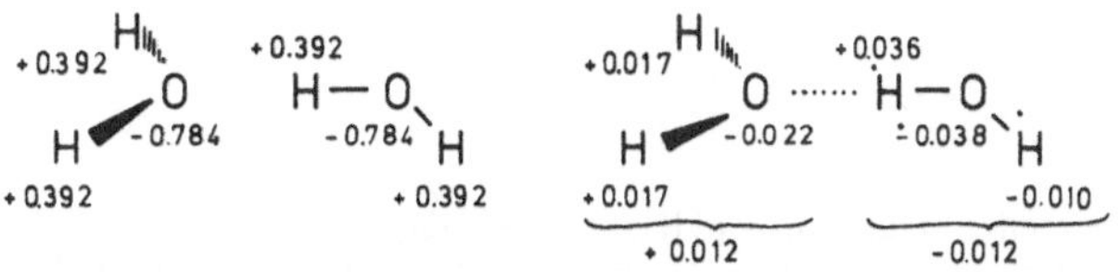

Fig. 5. Results of a Mulliken population analysis for H_2O and $(H_2O)_2$. At the left-hand side, partial charges on the atoms in isolated H_2O; at the right-hand side, density shifts upon H-bond formation

Other procedures of charge assignment (alternatives to the Mulliken population analysis) yield similarly very small values. In fact, the intramolecular charge shift, that is the polarization effect of both molecules on each other, are actually larger than the global CT. These polarization effects are seen to amplify the existing polarities even further: the dipole moment of the dimer is larger than the sum of the dipoles of the monomers. As a consequence, we obtain the rather paradoxical situation where an $O^{\delta-}$ and an $H^{\delta+}$ are brought together, and where the negative entity becomes even more negative, whereas the positive entity becomes more positive. The case of $(H_2O)_2$ illustrates a general trend, schematically represented as follows:

$$Y - B \ldots H - A - X$$

Upon the formation of a hydrogen bridge between the proton donor HAX and the proton acceptor YB, the electron density increases at B and A, while it decreases at Y and H.

It is possible to study the density shifts in much more detail by simply calculating the difference, as a continuous function, of the dimer density minus the sum of the two monomer densities. The formation of a H-bond corresponds

Fig. 6. Hartree–Fock density difference map $\Delta\rho = \rho[(H_2O)_2] - 2\rho(H_2O)$ in the yz-plane, containing the nuclei of the proton donor (right) and the oxygen atom of the proton acceptor. The densities were calculated using a 6–31G** basis set; the numerical values of $\Delta\rho = \pm 0.00125$, ± 0.0025, $\pm 0.005\,au^{-3}$, etc. The O—O distance amounts to 3 Å (SCF-optimum)

to electron density flowing into those regions of space where $\Delta\rho > 0$. $\Delta\rho$ is a tridimensional function, and in order to obtain a graphical representation we might choose a section through the molecule in a particularly relevant plane, usually containing as many nuclei as possible.

Figures 6 and 7 show the example of the water dimer. The plane of intersection (yz) contains the nuclei of the proton donor and the oxygen atom of the proton acceptor; the other two H atoms are sticking out at both sides of the plane. The curves shown are isodensity curves, somewhat similar to the contour lines of a geographical map. The full lines correspond to contours where $\Delta\rho$ is positive, the dashed lines to contours where $\Delta\rho$ is negative and the dotted lines to contours where $\Delta\rho = 0$. In its general outline, we see that the density map is compatible with the results of the population analysis. More specifically, the region around the bridging H atom is depopulated, which means that the positive H atom becomes even more positive upon bridging.

Now, it is possible, as we did before with the H-bond energy, to subdivide also $\Delta\rho$ in its different constituent parts. If one does so, one usually finds that $\Delta\rho \simeq \Delta\rho_{PL}$, although in some cases the exchange and the CT contributions are not completely negligible. But the dominant part of $\Delta\rho$ is definitely due to internal polarization effects. This is of course in sharp contrast to the energy decomposition; indeed Table 1 showed that PL was systematically the smallest contribution to the H-bond energy!

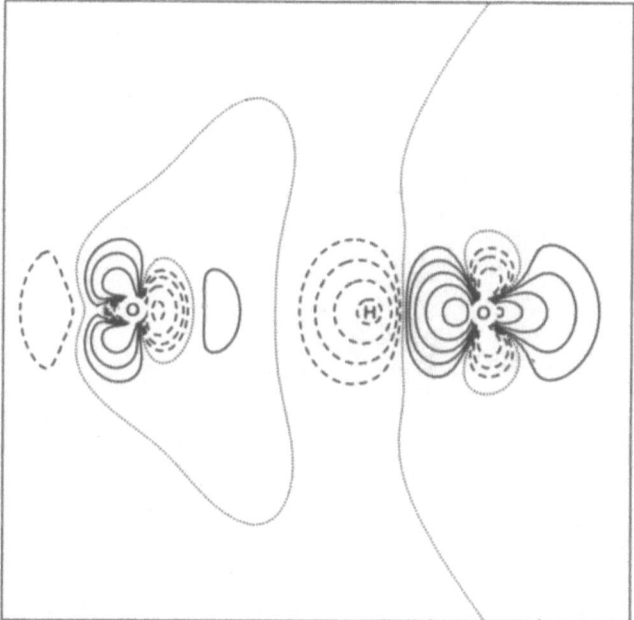

Fig. 7. Density difference map for the formation of $(H_2O)_2$ in the xy-plane; the other relevant elements of information are identical as in Fig. 6

As a matter of fact, the electrostatic contribution illustrates even more spectacularly that there is not a simple connection between $\Delta\rho$ and ΔE. Indeed: $\Delta\rho_{ES} \equiv 0$ by definition, because the electrostatic interaction results from an approach of two frozen monomers, where the density does not change. Yet, from Table 1 we know that ES was always by far the largest contribution to ΔE.

For the sake of comparison, Figs. 8–11 show a few density difference maps for a number of other bonding types. Figure 8 shows the interaction between two noble gas atoms Ne_2. The interaction is non-bonding and if the two atoms are forced too close together, the region between them becomes depopulated and the electrons flow outward.

On the contrary, in Fig. 9 we have a bonding interaction, where two fluorine atoms come together so as to form the stable F_2 molecule. The situation has been completely reversed with respect to Fig. 8, and now, the electrons flow into the central bonding zone.

Figure 10 shows the formation of sodium-fluoride from a sodium atom and a fluorine atom. The contour lines show in a very striking way how electron density is removed from sodium at the left and transferred to fluorine at the right; this looks like a purely ionic bond.

The next figure (Fig. 11) shows again the formation of sodium-fluoride, but this time from an Na^+ ion and an F^- ion (that is, *after* the electron had been transferred). And now we see that part of the electron density flows back into

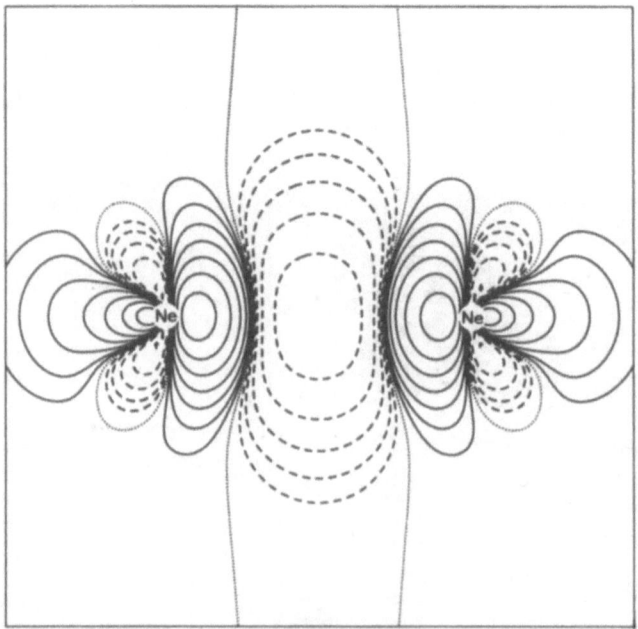

Fig. 8. $\Delta\rho = \rho(Ne_2) - 2\rho(Ne)$. The densities were calculated using a 6–31G* basis set; the Ne—Ne distance is 1.59 Å

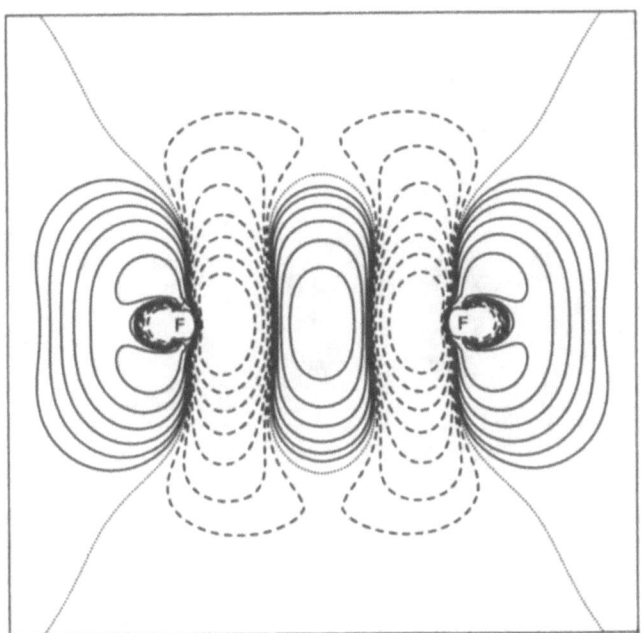

Fig. 9. $\Delta\rho = \rho(F_2) - 2\rho(F; 2p_z^1)$, calculated with 6–31G* basis sets; the z-axis is the internuclear axis; the F—F distance is 1.43 Å

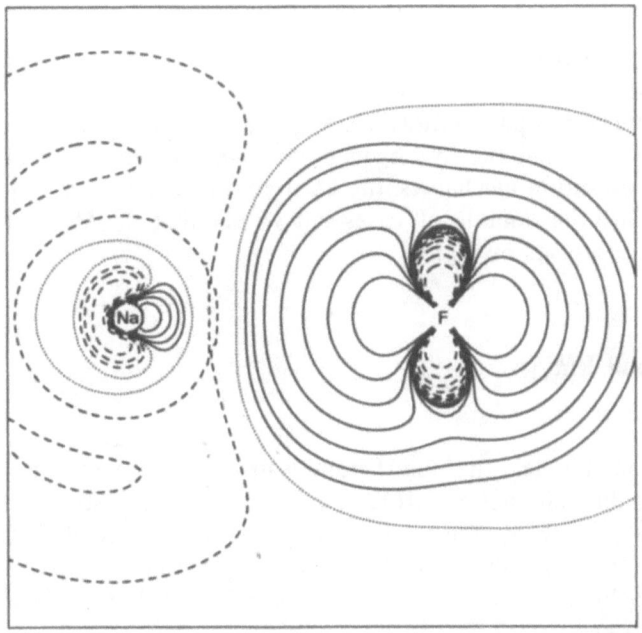

Fig. 10. $\Delta\rho = \rho(NaF) - [\rho(Na) + \rho(F; 2p_z^1)]$, using 6–31G* basis sets; the z-axis is the internuclear axis; the internuclear distance is 2.36 Å with the F-atom at the right-hand side

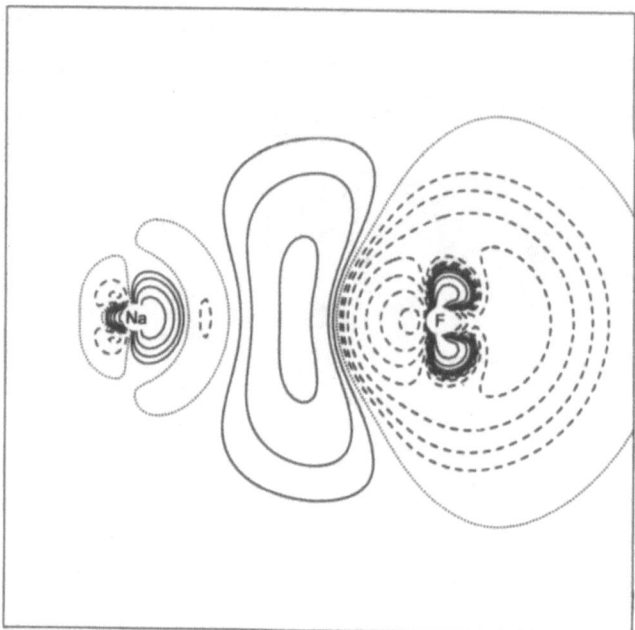

Fig. 11. $\Delta\rho = \rho(NaF) - [\rho(Na^+) + \rho(F^-)]$, using 6–31G* basis sets; the internuclear distance is 2.36 Å; the fluoride ion is at the right-hand side

the central bonding zone—which illustrates that even in NaF there is a certain amount of covalent character.

If we compare all these bonding patterns (Figs. 8–11) to the map of the water dimer (Figs. 6 and 7), we see that the hydrogen bond is different from all of them. The basic laws of quantum mechanics, the Hartree–Fock simplification, and even the basis sets are the same in all cases but we see that the H-bond represents a density pattern of its own [13].

5.4 The Hydrogen Bond Energy

From the previous section, it is clear that the H-bond is made up from a (subtle) interplay of a number of different factors. Moreover, since the bond is so weak, even a small effect might destroy, or at least, seriously jeopardize any regularity that one might have hypothesized. Therefore, it seems probable that any model, that is simpler than the Schrödinger equation itself, is bound to be characterized by quite a few imperfections.

The most obvious attempts at simplification will of course focus on the largest component of the H-bond, that is the electrostatic component ES. As

Table 2. Mulliken population analysis (4-31G basis set) of a number of simple hydrides

Molecule H_nA	q_A	q_H
HF	−0.479	+0.479
H_2O	−0.784	+0.392
H_3N	−0.896	+0.299
HCl	−0.230	+0.230
H_2S	−0.182	+0.091
H_3P	+0.063	−0.021

stressed before, within the framework of an electrostatic theory, we cannot have much hope for success if we do not reach—at the very least—the level of sophistication of an atomic point charge model.

Table 2 shows the results of a Mulliken population analysis for a number of simple hydrides. From this table, one might try to draw a number of conclusions. For instance, if one considers a dimer of ammonia and hydrogen fluoride, which one of the two monomers is going to be the proton donor? In this case, the answer is obvious from electrostatic arguments: HF is the proton donor and NH_3 the proton acceptor because in this case the neighboring charges are − 0.896 and + 0.479; if HF were the proton acceptor, the neighboring charges would only be − 0.479 and 0.299. But in general, the applicability of this type of population analysis is rather limited, certainly if one is interested in a semi-quantitative prediction of the H-bond energy.

It has been argued by a number of authors like Kollman [2] that molecular electrostatic potentials provide a better guide in the prediction of the H-bond energy.

6 The Strong (Ionic) Hydrogen Bond

If one of the partners is an ion, the resulting H-bond is usually much stronger. We have already briefly considered two examples (Table 1), the first one being a positive ion (a dimer between water and a positive ammonium ion), and the other one being a negative ion (a dimer between HF and the F^- ion).

6.1 Positive Ions

Quite a few ions are known of the general type $[H_mB—HAH_n]^+$. In general, their bond strength is rather high, and in the neighborhood of some 15–30 kcal/mol. At the same time, the length of the bridge (the B—H distance) is some 20 or

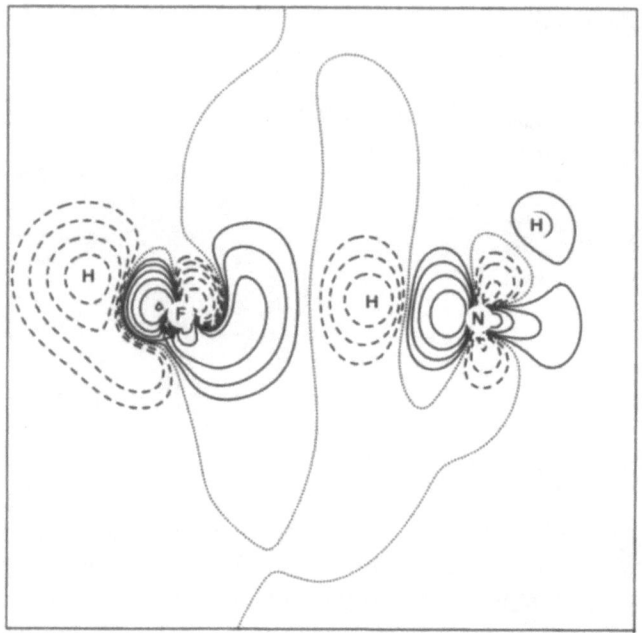

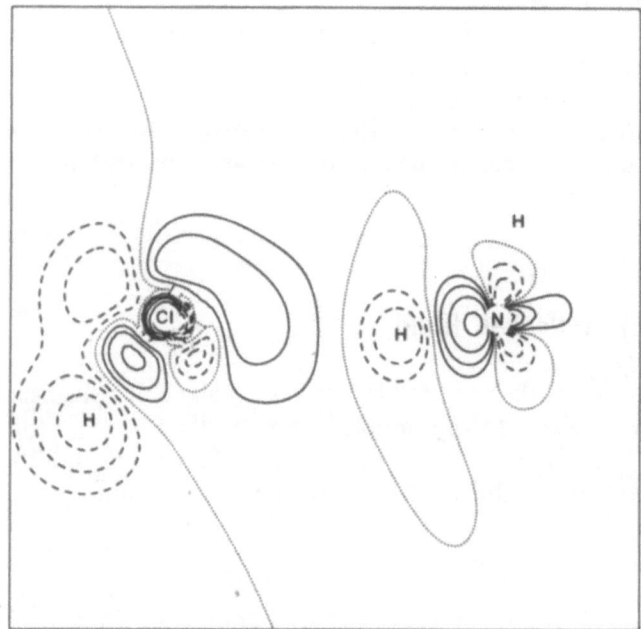

Fig. 12. Density difference maps for three strong ionic hydrogen bonds using 6–31G* basis sets
(a) $\Delta\rho = \rho(HF\ldots HNH_3^+) - [\rho(HF) + \rho(NH_4^+)]$
(b) $\Delta\rho = \rho(HCl\ldots HNH_3^+) - [\rho(HCl) + \rho(NH_4^+)]$
(c) $\Delta\rho = \rho(H_3N\ldots HNH_3^+) - [\rho(NH_3) + \rho(NH_4^+)]$

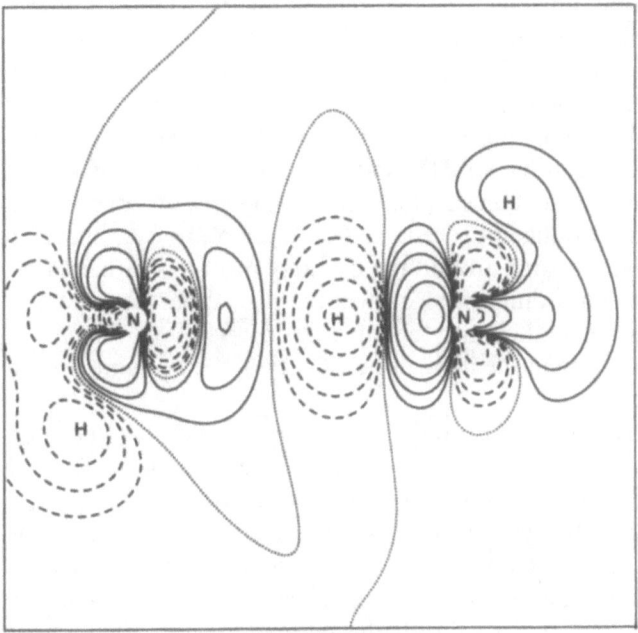

Fig. 12. (*Continued*)

30% smaller than in the neutral dimers. Correspondingly, the AH distance is increased significantly (up to 5%) with respect to the positive monomer ion AH_{n+1}^{+}. From the numerical calculations (Table 1), it appears that the decomposition of the bond energy in the positive ions is rather similar to the pattern for the neutral dimers, except for one thing: the exchange contribution is comparatively much smaller. This feature can easily be understood from the fact that the wavefunctions in the positive ions are more compact, and therefore the overlap of the dimers is smaller. But the main feature remains that the electrostatic interaction is the dominant contribution, just as in the neutral dimers.

Still there are a few characteristics of these strong H-bonds that require special attention. Figure 12 shows three density difference maps; in each case the proton donor is the ammonium ion NH_4^{+}; the three proton acceptors are HF, HCl, and NH_3. In all cases, we observe the same typical alternating pattern that was observed for the neutral complexes: we have a zone of electron depopulation, population, etc. Apparently this pattern is quite general and even in the third complex $N_2H_7^{+}$, where the central zone of electron population is absent, one might say that there is still an attenuated presence of this population zone, since the depopulation is clearly less pronounced in the middle. It has to be stressed that the gradients of $\Delta\rho$ are much larger in the charged complexes, sometimes up to three times larger than in the neutral species [7].

Figure 12 also illustrates how there can be an apparent discrepancy between the density difference plots and the results of a Mulliken population analysis. Indeed, in the three complexes of Fig. 12, it is clear that the bridging H-atom is situated in a zone of electron depopulation—as usual. Therefore, we expect the positive charge on this bridging proton to increase. A Mulliken population analysis confirms this expectation for HF and NH_3, but not for HCl. This illustrates a phenomenon that is often encountered in carrying out a population analysis. By definition the Mulliken procedure is an integration over space: in the HCl case, the negative $\Delta\rho$-zone around the bridging proton is rather small, and surrounded by two important zones of positive $\Delta\rho$. Apparently, part of these positive zones have been taken into the integration, and they overshadow the depopulation phenomenon in the immediate vicinity of the bridging proton [14].

Finally, we should stress that the overall CT from proton acceptor to proton donor remains very small: $\sim 0.1\,e^-$, but this is still an order of magnitude more than in the neutral complexes, where we had a CT of only $0.01\,e^-$.

Since the interaction is stronger in these positive ions, this means that the bridging H atom has become relatively more independent of the original positive monomer ion and it has become bound more intensively to its new partner. In this respect, it may be interesting to sketch the potential energy of the bare proton in the field of two hypothetical neutral partners.

Figure 13a shows the qualitative features of this potential energy curve for the two complexes $HF—NH_4^+$ or $HCl—NH_3$ because in those two cases the proton does not find a local minimum in the neighborhood of F or Cl.

On the other hand, for $NH_3—H—NH_3^+$ the curve has to be symmetrical. For a symmetrical curve, there can be two possibilities: either they are characterized by a central maximum or they are characterized by a central minimum. If two equal partners are pulled far enough apart, one will always end up with a local maximum, as in Fig. 13b, and if one pushes them close enough together, one will always end up with a single minimum as in Fig. 13c.

Now it turns out that for the equilibrium position of $N_2H_7^+$ we have the situation of Fig. 13b, whereas for the equilibrium position of $O_2H_5^+$ or $F_2H_3^+$, we have the situation of Fig. 13c. In the latter case, it is obvious that the distinction disappears between proton donor and proton acceptor. But so far we have been basing our molecular orbital description precisely on the possibility of this distinction. Therefore, for the case of Fig. 13c, it is indicated to use a new qualitative MO scheme, as shown in Fig. 14.

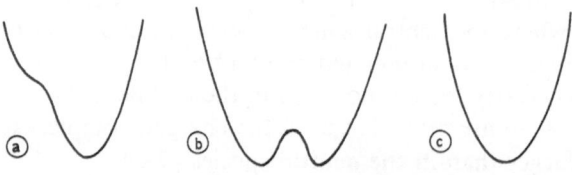

Fig. 13. Potential energy curves for a bare proton in the field of two hypothetical neutral partners

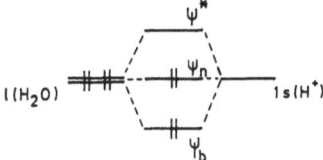

Fig. 14. Simplified molecular orbital scheme for a symmetrical ionic complex, such as $H_5O_2^+$. The (H_2O) orbital represents a lone pair on H_2O

In this case the H-bridge might be described as a 3 center-4 electron bond, which is an electron-excess compound, like for instance XeF_2. Essentially the bonding is carried by the lowest occupied orbital; the middle orbital is virtually non-bonding, so that the bond order for the two OH bonds is 1/2. This conclusion is compatible with the bond strength of this interaction, which is intermediate between a weak H-bond and a normal covalent bond.

6.2 Negative Ions

Let us focus on the simplest entity of this kind, mainly the linear, symmetric and very stable HF_2^-. The corresponding MO-diagram is completely similar to Fig. 14.

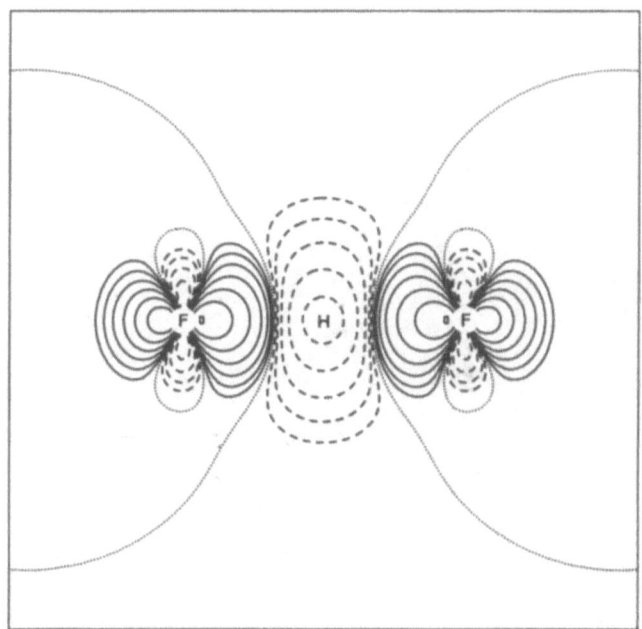

Fig. 15. $\Delta\rho = \rho(FHF^-) - \frac{1}{2}[\rho(HF)_{left} + \rho(F^-)_{left} + \rho(HF)_{right} + \rho(F^-)_{right}]$, based on 6–31G* basis sets, with an F—F internuclear distance of 2.36 Å

A population analysis of HF_2^- yields the charges of $+0.58$ on the bridging H-atom, and -0.79 for F. Although the overall charge of the complex is -1, the bridging atom has definitely a positive charge. This fact can be rationalized rather easily by using the MOs of Fig. 14: the non-bonding ψ_n orbital has no density at H. The population of this orbital shifts the electron density towards the outsides of the molecule. If we want to verify this picture by means of density difference maps, we have to find a suitable reference state. Obviously, we cannot compare FHF^- with $F^- + H^+ + F^-$, because if we do, we know beforehand that charge has to flow toward H. A more relevant description compares FHF^- with the average of $HF + F^-$ and $F^- + HF$ (Fig. 15). Here too, the proton zone gets depopulated, and this time not with respect to a free H atom but even with respect to a bound H-atom. Figure 16 shows—for the sake of comparison—an asymmetric bond formation where one has F^- at the left approach HF at the right. Even here, the approach of the electron-rich F^- ion pushes electrons away from the bridging proton, which was already positive in the first place!

In summary, it seems fair to say that density difference maps offer one of the most simple, and at the same time, one of the most illuminating ways to describe the detailed electronic rearrangements accompanying the hydrogen bond.

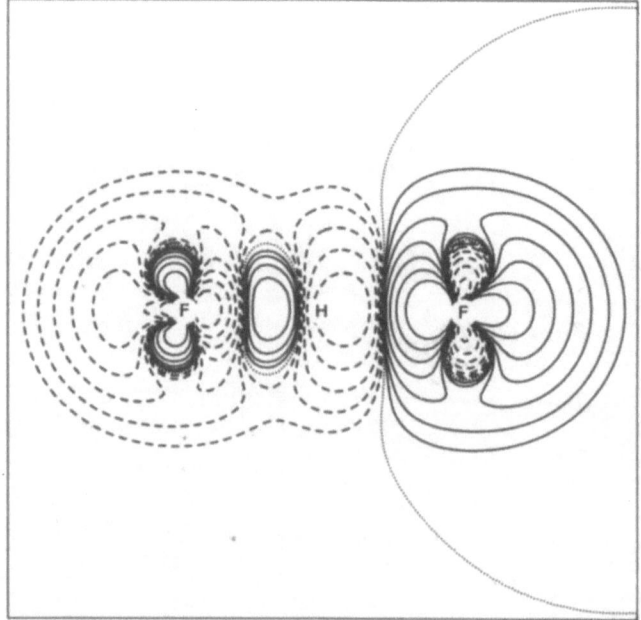

Fig. 16. Asymmetrical $\Delta\rho = \rho(FHF^-) - \rho(HF)_{right} - \rho(F^-)_{left}$ based on 6–31G* basis sets and an F—F internuclear distance of 2.36 Å

Acknowledgement

The author is indebted to Ms B. Coussens for preparing the density difference maps.

7 References

1. Morokuma K, Kitaura K (1980) in: Ratajczak H, Orville-Thomas WJ (eds) Molecular Interactions, Wiley, Chichester, Vol 1, p 21
2. Kollman PA (1977) in: Schaefer HF (ed), III, Applications of Electronic Structure Theory, Modern Theoretical Chemistry, Vol 4, Chap 3, Plenum, New York
3. Kollman PA, Allen LC (1970) J Am Chem Soc 92: 6101
4. Noble PN, Kortzeborn RN (1970) J Chem Phys 52: 5375
5. Kollman PA (1972) J Am Chem Soc 94: 1837
6. Umeyama H, Morokuma K (1977) J Am Chem Soc 99: 1316
7. Desmeules PJ, Allen LC (1980) J Chem Phys 72: 4731
8. Beyer A, Karpfen A, Schuster P (1984) Topics in Current Chemistry 120: 1
9. Del Bene JE, Frisch MJ, Pople JA (1985) J Phys Chem 89:3669; Frisch MJ, Pople JA, Del Bene JE (1985) J Phys Chem 89: 3664
10. Brédas JL, Street GB (1988) J Am Chem Soc 110: 7001; Vanquickenborne LG, Coussens B, Verlinde C, De Ranter C (1989) J Mol Struct (Theochem) 201: 1
11. Murray-Rust P, Glusker JP (1984) J Am Chem Soc 106: 1018
12. Dannenberg J (1988) J Phys Chem 92: 6869
13. Carroll MT, Bader RFW (1988) Mol Phys 65: 695
14. In a few exceptional cases—as $H_3F_2^+$ and $H_3Cl_2^+$, the bridging proton is situated in a region of positive $\Delta\rho$

CHAPTER III
How to Understand Liquids?

W.A.P. Luck

A simple model is given for the understanding of the liquid state. This model is based on the intermolecular pair potential and the linear density—temperature (T) relation in an extended T-region. This linear relation is estimated with the population of the intermolecular vibration levels based on Boltzmann statistics. An ideal liquid is defined. The deviations of the linear density—T relation give the real corrections of the ideal liquid. The usefulness of the idealized liquid model is demonstrated: with a common equation for the heat of vaporization as measure of the inter-molecular energies and with the estimation of the surface energy. The advantage of the surface energy in comparison with the surface tension is demonstrated.

1 Introduction

For a proper understanding of specific interactions and their influence on the structure of liquids we need some knowledge of normal non-specific interactions. Non-polar and spherical molecules are the best systems for their description. This paper will be restricted to these systems. It is always necessary to take the normal van der Waals forces, which are always present, into account for a full understanding of specific interactions.

Even though the liquid state is extremely important in chemistry it is rather neglected in teaching. Probably this is because complicated distribution functions are important for it. Recently Hertz [1] has stressed: "The 'correct' description of the structure of a liquid is in general not practicable... one needs

Intermolecular Forces
© Springer-Verlag Berlin Heidelberg 1991

an infinite number of words". The importance of liquids forces us to try to build approximate models of them. We should not forget that the complicated Maxwell–Boltzmann distribution of thermal energies is important even for ideal gases and can be efficiently described by its average value RT for many purposes. We can try to do the same for liquids if we restrict our description to the calorific properties of non-polar or weak polar systems. In this chapter we give the fundamentals of liquids with non-specific interactions.

2 The Intermolecular Interaction Potential

One of the first attempts to describe intermolecular interactions was given by van der Waals as a correction to the ideal gas equation. He corrected the pressure p by an "inner pressure" a/V^2 as a consequence of attraction forces and the volume V by the "self volume" b of the molecules, where b takes into account the repulsion forces. During a contact between two molecules the intermolecular potential is related via a/V and a/b to the pair potential.

The second indication concerning intermolecular forces has been found by the temperature dependence of the intermolecular collision diameter σ, determined by gas viscosities, heat-conductance or diffusion. This effect can be described to a good approximation by:

$$\sigma_T^2 = \sigma_\infty^2 (1 + C/T) \tag{1}$$

This relation has been described by Sutherland [2] as the increase of the probability of a collision due to intermolecular attractions. If we assume one particle as the centre of a coordination system and (Fig. 1) discuss the relative movement of a second particle and we describe the first particle as having a radius $\sigma = 2r$ (with r: radius of one particle) and the second colliding particle as if it were a point, the orbit of the second particle will then depend on the

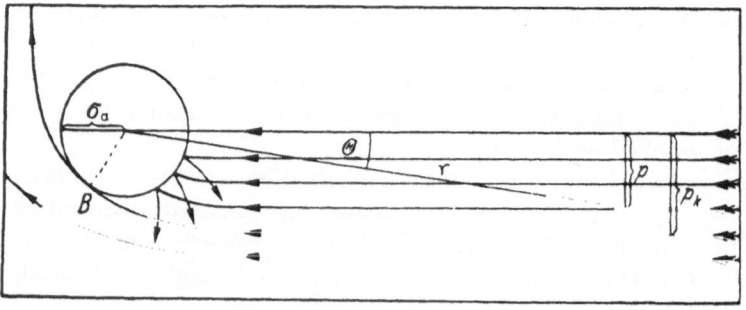

Fig. 1. Scheme of a gas kinetic collision with Sutherland correction, indicating that attraction forces will enlarge the collision diameters σ to p_k (depending on E_{pot}/E_{kin})

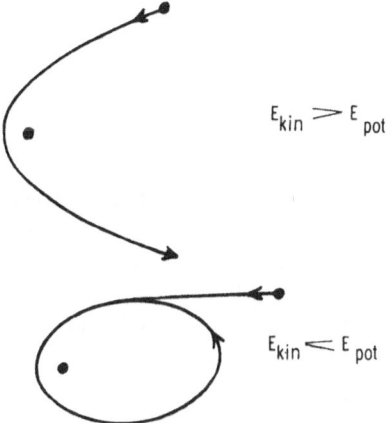

$E_{kin} > E_{pot}$

$E_{kin} < E_{pot}$

Fig. 2. Kepler's rule for a two-particle collision; at the top: model of gas collision, below: model of liquefaction

ratio of the potential energy E_{pot} and the kinetic energy $3RT/2$. As Kepler showed, the orbit of two particles will be either a hyperbola or an ellipse if this ratio-E_{pot}: $3RT/2 = C/T$ is less or greater than unity respectively (Fig. 2). The discussion of the laws of energy and momentum [2, 3] shows that Eq. (1) could be understood with $3RC/2 = -E_{pot}(\sigma_\infty)$ if E_{pot} is the potential energy per mole at a collision with the largest possible collision parameter p_K in Fig. 1.

The so-called Sutherland constant C therefore provides an easy way of determining the intermolecular pair potential (attraction part). For molecules with specific interactions, larger deviations from Eq. (1) have been found.

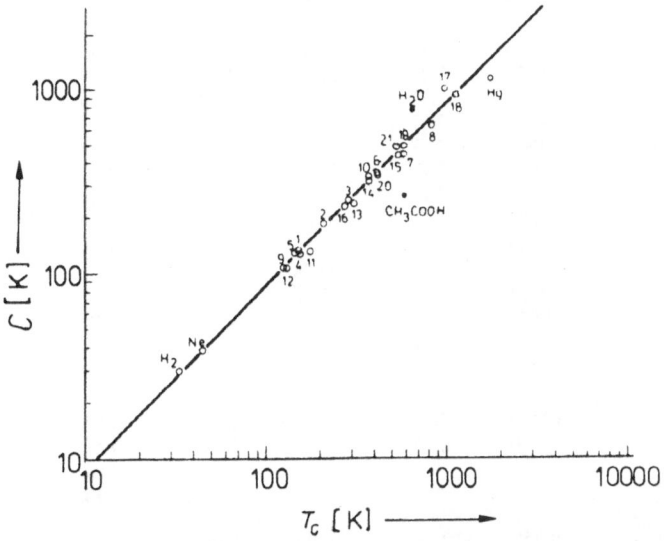

Fig. 3. Proportionality between Sutherland constant C and the critical temperatur T_C, indication on the proportionality between T_C and the intermolecular potential. 1: Ar, 2: Kr, 3: Xe, 4: O_2, 5: F_2, 6: Cl_2, 7: Br_2, 8: I_2, 9: N_2, 10: H_2S, 11: NO, 12: CO, 13: CO_2, 14: COS, 15: CS_2, 16: SiH_4, 17: $HgCl_2$, 18: $HgBr_2$, 19: $SnCl_4$, 20: NH_3, 21: CH_3OH

Rankine [4] was the first to find for small spheroidal like molecules an empirical relation with the critical temperature T_C:

$$C = 0.89T_C \tag{2}$$

The number 0.89 is an average for molecules with known values of C. Equation (1) is an approximation in that it assumes collisions between only two molecules. Experimental C values are somewhat temperature dependent; it therefore seems reasonable to compare C values extrapolated to T_C. In this case we obtained [4, 5] a mean factor of about $0.833 = 1/1.2$ in Eq. (2). Figure 3 demonstrates this proportionality of Eq. (2) for non-polar compounds with deviations for specific interactions of water or carbon acids.

As a consequence we can conclude as an approximation for the pair potential for small and non-polar molecules at direct contact σ_∞:

$$-E_{pot} = 3RC/2 = 0.833 \times 3RT_C/2 = f' 3RT_C/2 \tag{3}$$

with $f' = 0.833$.

Table 1. $T_C = T_k$ and estimated pair potentials of non- or weak polar molecules, molecular radius r calculated by the van der Waals b and gas kinetic diameter σ_∞ at $T \to \infty$

Gas	T_C [K]	$\frac{3}{2}RT_C$ [Joule/Mol]	$\frac{a}{b}\left[\dfrac{\text{Joule}}{\text{Mol}}\right]$	$\dfrac{a/b}{\frac{3}{2}RT_C}$	$b\left[\dfrac{\text{cm}^3}{\text{Mol}}\right]$	r $[10^{-8}\text{cm}]$	$\dfrac{\sigma_\infty}{2}$ $[10^{-8}\text{cm}]$
He	5.2	64.9	176	2.71	23.7	1.32	1.13
H_2	33.3	415	1130	2.72	26.6	1.38	1.32
Ne	44.2	551	1528	2.77	17.09	1.19	1.24
N_2	126	1570	4187	2.67	39.13	1.57	1.62
CO	133	1660	4605	2.77	39.85	1.58	1.61
F_2	144	1800	4940	2.7	26.78	1.38	1.58
Ar	150.6	1880	5150	2.73	32.19	1.47	1.49
O_2	155	1933	5275	2.73	31.83	1.47	1.50
NO	177.2	2210	5945	2.69	27.89	1.40	1.54
Kr	209.2	2609	7201	2.76	39.78	1.58	1.6
SiH_4	272.7	3401	9211	2.71	57.86	1.79	1.78
Xe	289.8	3615	9881	2.73	51.05	1.72	1.77
CO_2	304	3790	10380	2.74	42.67	1.62	1.7
H_2S	373.6	4660	12980	2.78	42.87	1.62	1.59
COS	378	4715			58.17	1.79	1.87
Cl_2	417	5201	14240	2.73	56.22	1.77	1.83
CS_2	546	6810	15320	2.25	76.85	1.97	1.89
Br_2	575	7172					1.89
J_2	826	10300					2.14
$HgCl_2$	976	12170					
Hg	1732	21600	48350	2.23	16.96	1.19	1.24
ethylene	282.1	3600	7913	2.2	57.14	1.78	1.85
ethane	305.3	3808	10430	2.74	63.8	1.85	1.93
propane	370.1	4616	12610	2.73	84.45	2.03	2.23
benzene	561	6997	19260	2.75	115.4	2.25	2.28

This equation corresponds to the experience that many intermolecular properties are proportional [5] to T_C or—with the empirical relations: $T_B \approx 2T_C/3$ or $T_M \approx 2T_B/3$ (with T_B: boiling point and T_M: the melting point)—proportional to T_B or roughly to T_M. The van der Waals equation demands:

$$a/V_C = a/3b = 9RT_C/8 \quad \text{or} \quad \frac{a/b}{3RT_C/2} = \frac{9}{4} = 2.25 \tag{4}$$

Table 1 demonstrates that experiments give a factor of about 2.7 for the ratio of Eq. (4) indicating the approximate character of the van der Waals equation especially near T_C.

An intuitive understanding of the meaning of the critical point can be given based on Eq. (3) and Fig. 2. Below T_C the relation $E_{pot} > E_{kin} = 3RT/2$ is valid and all collisions at sufficient density will give rise to associations and possibly to liquefaction; for $T > T_C$ stable large intermolecular aggregates are excluded. The total pair potential of van der Waals forces (attraction + respulsion) can be described by the so-called Lennard–Jones potential as a function of the pair distance R (dashed line in Fig. 4):

$$U_{11} = 4\varepsilon[[\sigma/R]^{12} - [\sigma/R]^6] \tag{5}$$

ε is the depth of the potential minimum; σ is the value of R for which U_{11} is zero.

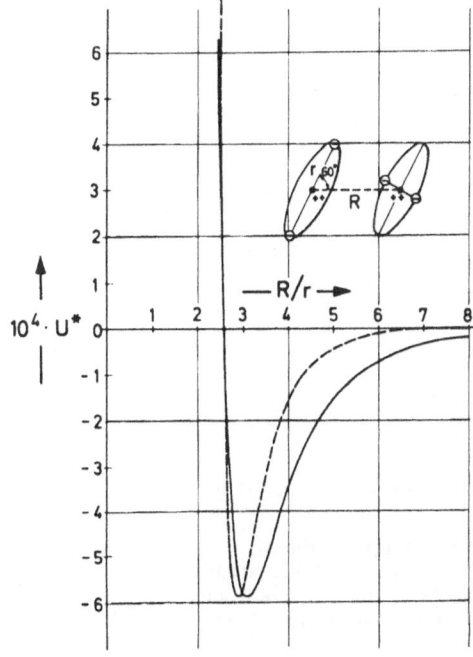

Fig. 4. *Dotted line*: Lennard–Jones (LJ) potential
Continuous line: Quadrupole–quadrupole interaction as model of the LJ potential (model for He—He)
Distance R and energy U* in units of the electron radius r

The quantum theory of the R^{-6} attraction term was given by London. He described [6] this result, contrary to the statistical orbital description, as being due to the fact that molecules recognize the point charges of neighbours which induce small short-lived dipole–dipole interactions. To a first approximation these disturbances do not influence interactions with third particles because of the severe distance dependence. The dotted line demonstrates that the Lennard–Jones potential is similar to the Coulomb interaction energy of two quadrupoles (full line in Fig. 4 as a model of two He atom–atom interaction).

As Fig. 5 shows, T_C and therefore the pair potential is proportional to the sum Z_e of the electron numbers of non-polar molecules. The proportionality is nearly linear for rare gases and a little weaker for organic molecules (about $\sqrt{Z_e}$). Figure 5 includes the deviations from this rule for specific H-bond

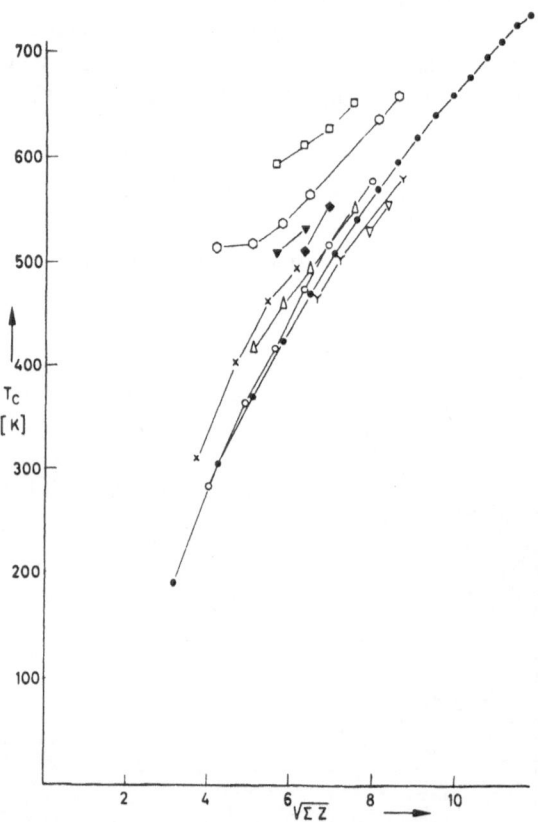

Fig. 5. The critical temperature T_C ($T_K = T_C$) in dependence on the square root of the sum Σ of all electrons Z

●: n-parafins   ○: primary n-alcohols   ◆: cyclic hydrocarbons
○: n-olefins   □: saturated n-carbonic acids   △: mono-chloro hydrocarbons
×: alcines   ▼: ketones   Y: mono-bromo hydrocarbons
                                             ▽: mono-iodo hydrocarbons

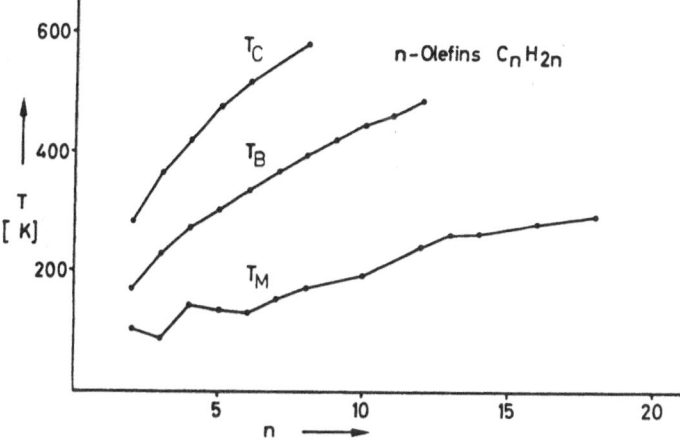

Fig. 6. Critical T_C, boiling T_B and melting point T_M of olefins in dependence on the number n of carbon atoms

interactions of alcohols or acids. For C_nH_{2n} compounds with one double bond Fig. 6 gives the relations between the critical temperature T_C, the boiling point T_B and the melting point T_M.

3 The Density of Condensed Media

The sum of all pair interactions U_{11} with the Z next nearest neighbours $(Z \times U_{11})$ gives an approximate value for the total intermolecular interactions in condensed media. In the case of dispersion interactions the influence of higher (second etc...) neighbours increases the interaction by about a factor of 1.2.

A plane model is calculated [5] in Fig. 7 of a central particle no. 7 surrounded by 6 others at a distance r_0/σ from the coordination centre (compare left corner in Fig. 7). The ordinate is the sum of 6 Lennard–Jones potentials of Eq. (5) as a function of the movement of particle 7 in the direction of the x-axis (dotted lines: movements in the y-axis). The curve for $(r_0/\sigma) = 1.122$ corresponds to the temperature $T = 0$ K; all particles are in the pair-potential minimum resulting in a minimum of 6ε (left model in Fig. 8). At higher temperatures the quantized vibration levels in the sum potential of Fig. 7 will be excited, corresponding to the Boltzmann e-law. At finite temperature, the vibration at these levels will induce a distance partition and a vibration volume (Fig. 8 on the right). As an approximation we can describe these vibrations in Fig. 7 with an average distance r_0 to all six nearest neighbours. This model gives the other potential curves of Fig. 6 with different r_0-values. For $r_0/\sigma = 1.214 = 1.08 \times 1.122$ we get a lower flat potential minimum, which does not change much near the central position of particle 7. This assumed average distance corresponds to the increased volume V_M at the melting point T_M. The so-called empirical Lorenz

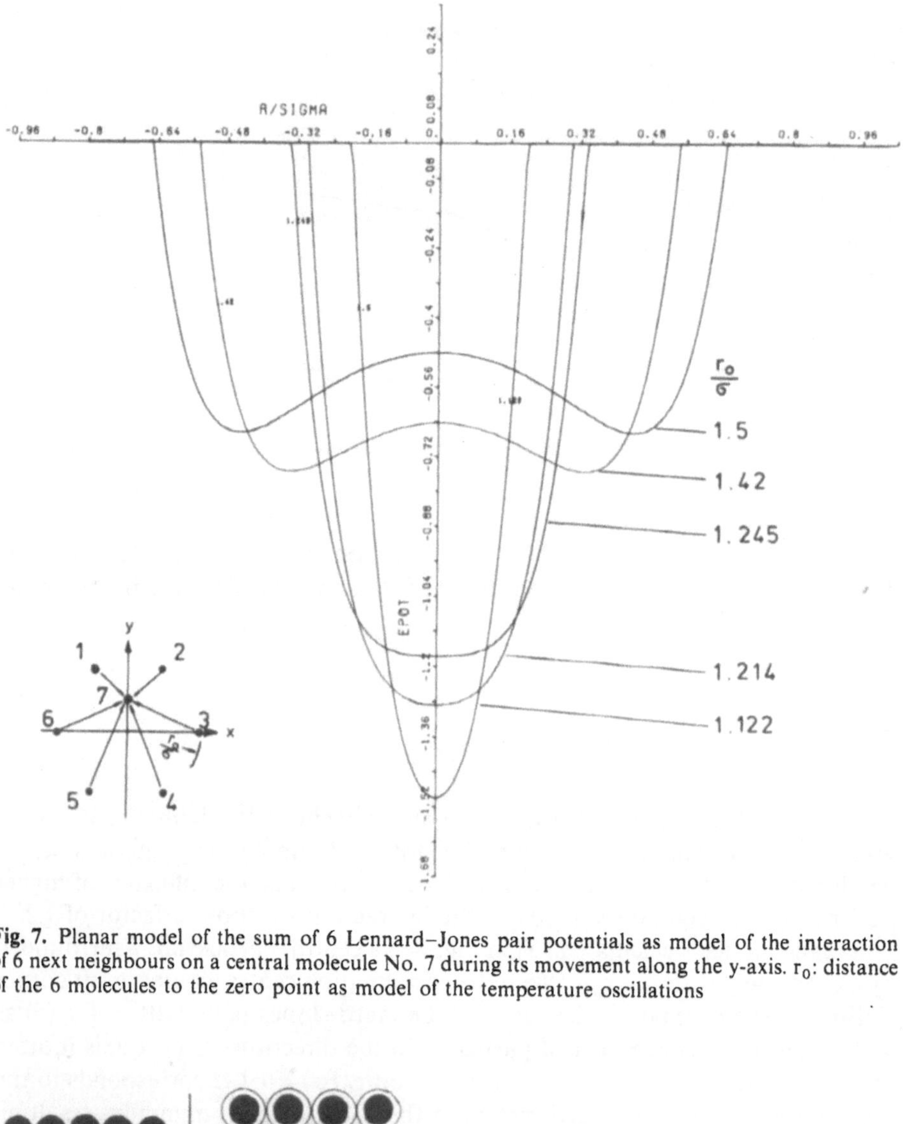

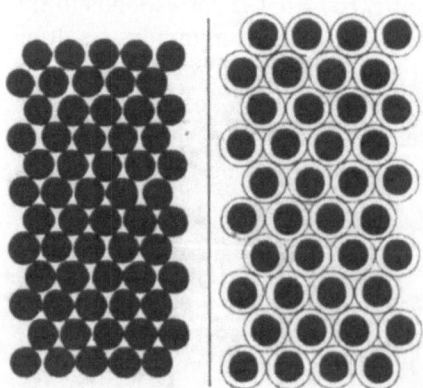

R/SIGMA

EPOT

$\frac{r_0}{\sigma}$

1.5
1.42
1.245
1.214
1.122

Fig. 7. Planar model of the sum of 6 Lennard–Jones pair potentials as model of the interaction of 6 next neighbours on a central molecule No. 7 during its movement along the y-axis. r_0: distance of the 6 molecules to the zero point as model of the temperature oscillations

Fig. 8. Left model of a crystal; **right:** model of averaged thermal moments in a liquid

positions of neighbours

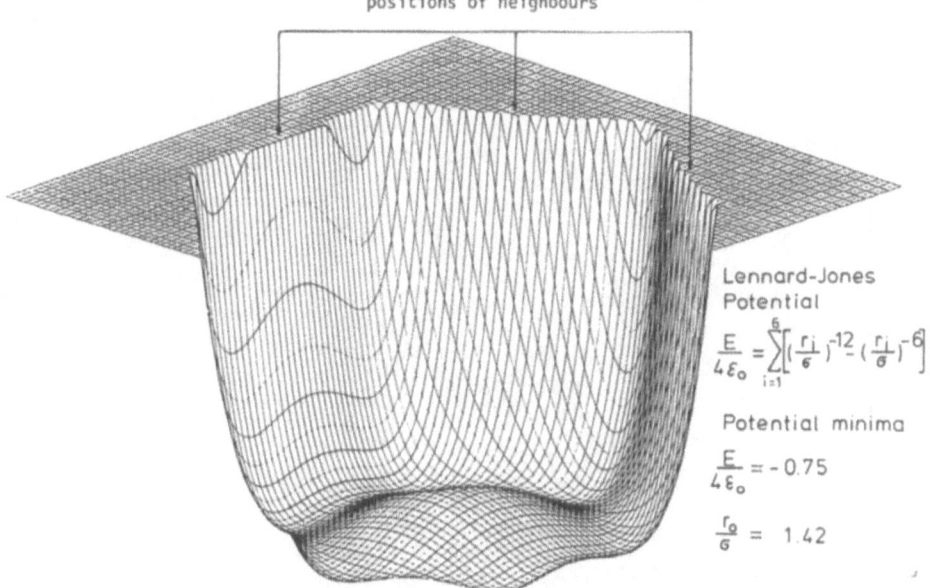

Lennard-Jones
Potential

$$\frac{E}{4\varepsilon_0} = \sum_{i=1}^{6}\left[(\frac{r_i}{\sigma})^{-12} - (\frac{r_i}{\sigma})^{-6}\right]$$

Potential minima

$$\frac{E}{4\varepsilon_0} = -0.75$$

$$\frac{r_0}{\sigma} = 1.42$$

Fig. 9. Planar model of the intermolecular interaction energies of 6 molecules on a central one. Perpendicular to the particles plain is given the sum of energies.
Average distances of the 6 molecules to the origin: $r = 1.42\sigma$.
Model of the liquid state with oscillations between the 6 minima

rule stresses that for many crystals at T_M in the liquid, $V_M(T_M) = 1.21 V_0$ ($T = 0$ K) or $r_M = (1.21)^{1/3} r_0$ ($T = 0$ K) $= 1.07 r_0$. At higher $r_0 > 1.214\sigma$, the potential splits into six minima (two of them are plotted in the plane of Fig. 7). Figure 9 shows all 6 minima. The three-dimensional calculation with 12 nearest neighbours gives a similar result, but 12 minima appear and their relative depths are slightly lower. These minima seem to characterize the liquid state. The central particle of our model could oscillate from one flat minimum to the next. If we plot the curves of Fig. 7 as a function of the distance of the central particle from one of the outer ones (Fig. 10) the positions of the minima correspond nearly to the pair-potential minimum. This may be related to the experimental result that the X-ray scattering pair distribution curves of nearly all liquids show a first maximum (position of the nearest neighbours) nearly independent of temperature. As the example of liquid argon (Fig. 11) shows, only the area of the first peak—corresponding to the coordination number—decreases but the peak maximum hardly changes.

The volume of a crystal increases and the density ρ decreases with increasing temperature as a result of the increasing thermal vibration volume (Fig. 8). This ρ decrease is nearly linear with temperature (the exception is water see page 217, chapter IX).

$$\rho = \rho_0 - aT \quad \text{for } T < T_B \tag{6}$$

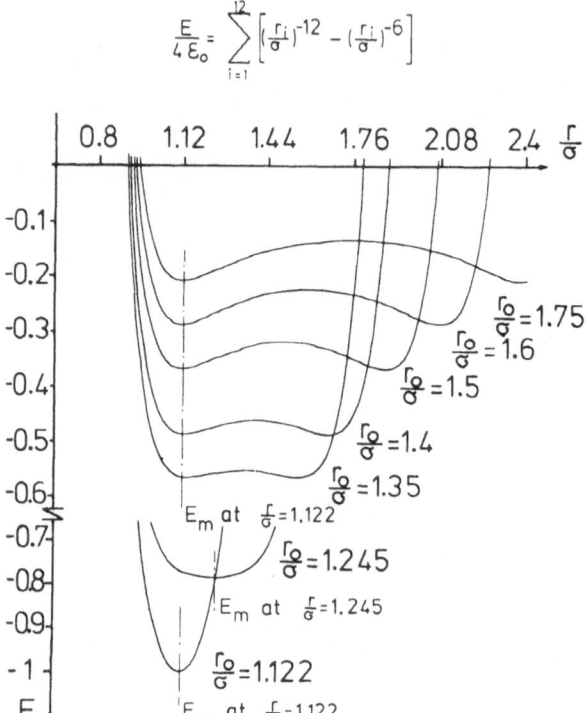

LENNARD JONES POTENTIAL

$$\frac{E}{4\,\varepsilon_o} = \sum_{i=1}^{12}\left[\left(\frac{r_i}{\sigma}\right)^{-12} - \left(\frac{r_i}{\sigma}\right)^{-6}\right]$$

Fig. 10. Sum of 12 next neighbours LJ energies along the y-axis in dependence of the distance from one of the outer molecules as origin. (Model for the T-independent maximum of the scattering pair distribution)

The linear $\rho - T$ function could be established with the model of the Lennard–Jones potential sum of the 12 nearest or the 24 first and second neighbours [7]. If we calculate the quantized levels of the intermolecular vibrations of this sum potential with the so-called semi-classical phase integral ($\oint p\,dr = h$ where p is the momentum), we can estimate the average volume of these thermal excited oscillations. For the rare gases (for one example see Fig. 12) we obtain near $T = 0\,K$ a temperature constancy of ρ and then a nearly linear decrease of ρ with increasing temperature in agreement with experiments. The ρ-constancy is a direct experimental macroscopic test of the quantized intermolecular vibrations as was also established by far-IR spectra. This model could explain why all known $\rho - T$ functions are nearly linear; this is also valid for liquids with the exceptions of those like water, methanol and ethanol which show specific interactions. The semi-classical method for the crystal density gives poorer results for liquids. We can conclude that the model of averaged neighbour positions is nearer to the truth for solids because the thermal movements in small areas could be nearly in phase (phonons) while in liquids

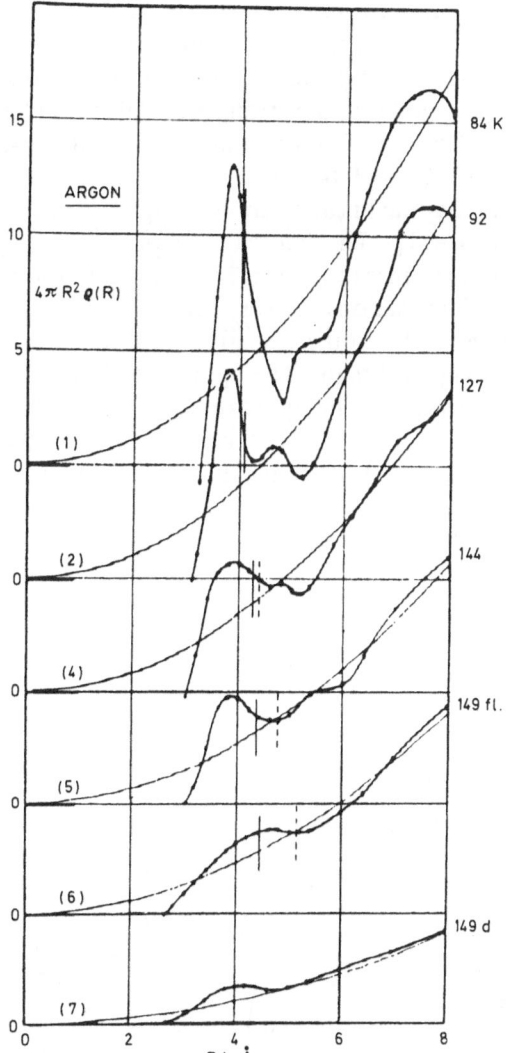

Fig. 11. X-ray scattering pair distribution around one central atom in liquid argon indicating the T-constancy of the first maximum position

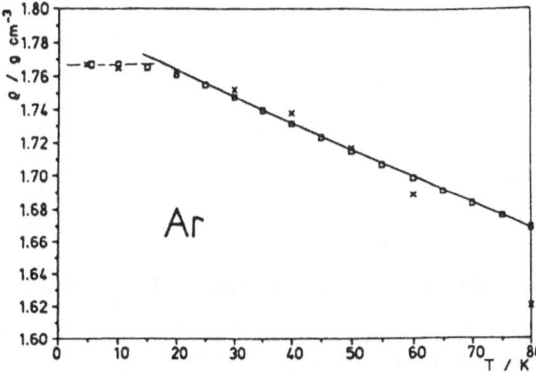

Fig. 12. Densities in solid argon.
×: experimental data
○: calculated with the thermal oscillations of 12 LJ potentials

the thermal oscillations of single molecules are more independent. At T_M the volume increases or ρ decreases by about 10% for many compounds (second Lorenz rule). This means that the distances between molecules change by about 3% during melting. These experiments suggest that the arrangement of molecules in a liquid are crystal-like near T_M. The liquid volume increases to at most $V_C = 3b$ in the van der Waals model ($\sqrt[3]{3}r_0 = 1.445r_0$).

The curves of Fig. 13 demonstrate that the linear relation of Eq. (6) is nearly valid, too, for liquids in the region $T < T_B$. The linearity disappears in the temperature region where the vapour pressure is finite (Fig. 13). The ideal vapour law is still very useful even though its idealisations are very rough. It is defined and tested with the pressure–volume (p–V) isotherms. Figure 13 allows the definition of an ideal liquid as one obeying Eq. (6). The approximation of an ideal gas is better the higher the temperature. The approximation of an ideal liquid is better at lower temperatures. Figure 13 also demonstrates that the linear ρ-relation also has a meaning in a region where real liquids deviate from

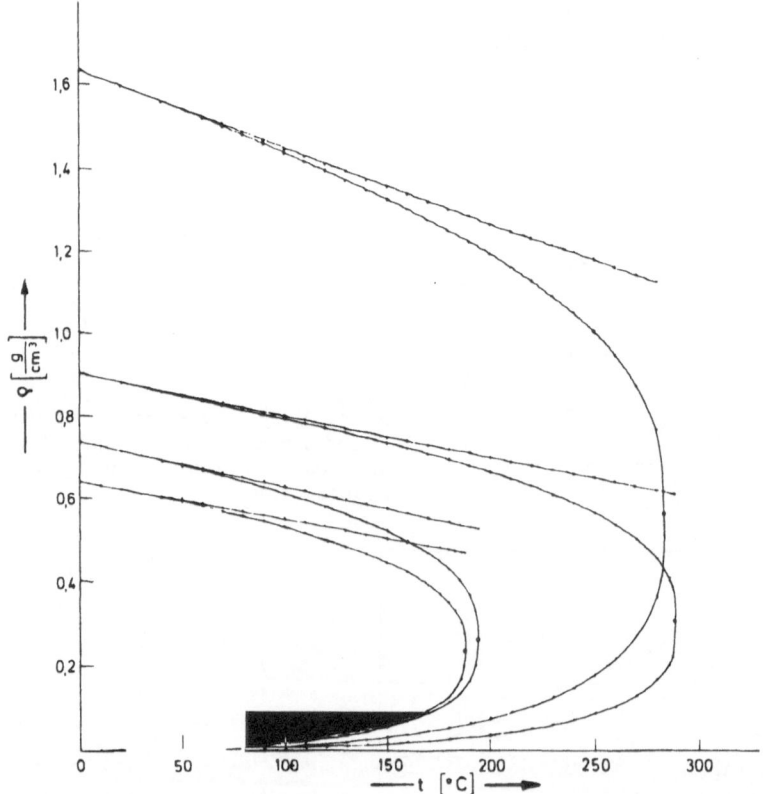

Fig. 13. Densities ρ of liquids (**upper part of the curves**) and saturated vapours (**lower part of the curves**).
The straight lines are: (at low T): ρ(liquid); (at high T): [ρ(liquid) + ρ(vapour)].
From the top to the bottom: CCl_4, benzene, diethylether and butane-dimethyl

the ideal Eq. (6). It gives, at higher temperature, the so-called empirical Cailletet–Mathias rule:

$$\rho_{liq} + \rho_{vap} = \rho_0 - aT \qquad (7)$$

ρ_{liq}: ρ of liquids in equilibrium with its vapour; ρ_{vap}: ρ of saturated vapours.

Equation (7) is valid for most liquids with the exception of H-bonded liquids with specific interactions [5]. This means that the differences between an ideal and a real liquid are easier to correct than are the van der Waals corrections of a real vapour. The linearity of Eq. (6) depends on the high degree of symmetry of the potential curves in Fig. 7 when only lower vibration levels are excited at low temperatures. At higher $T > T_B$ the anharmonicity of the Lennard–Jones potential induces deviations from Eq. 6. This deviation is proportional to ρ_{vap}. We could therefore correct the percentage deviation of Eq. (6) to ρ_{real} of a real liquid to:

$$\rho_{real} = [\rho_0 - aT][1 - x_F] \qquad (8)$$

with $x_F = \rho_{vap}/(\rho_{liq} + \rho_{vap})$.

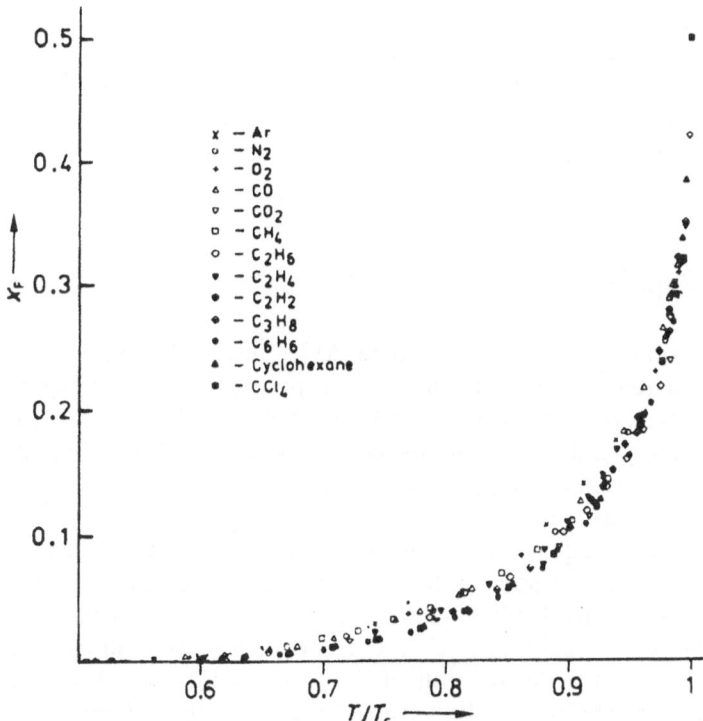

Fig. 14. The nonlinearity x_F (hole concentrations in Eyring's model) of different liquids in the reduced scale T/T_C

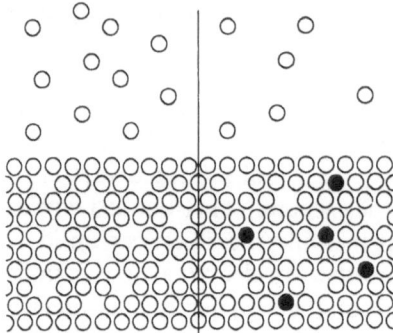

Fig. 15. Left: Eyrings hole model of a liquid, in addition to the thermal vibration-volume some holes appear; **right:** Solutes with negligible vapour pressures p_s reduce p_s proportional to its molar concentration (model of Raoult law)

As Fig. 14 demonstrates, these "defect" concentrations x_F are similar for normal liquids in the reduced temperature scale T/T_C.

The Eyring [8]–Cremer [9] hole model of liquids assumes that, in addition to the increase of the thermal vibration volume, holes appear in a liquid at higher temperature. This model is shown in Fig. 15 on the left. This model assumes that the number of holes is exactly the same as the number of vapour molecules per unit volume. This model is able to demonstrate Raoult's law for the reduction of the vapour pressure proportional to the mole concentration of a solute with negligible vapour pressure. Such a solute would be expected to be dissolved in the holes and this effect should reduce the vapour pressure proportional to the number of holes filled [5, 6]. The model of Fig. 15 also describes the meaning of the critical temperature T_C as the temperature at which the concentrations of holes in the liquid is the same as the concentration of vapour molecules. Both are 50% and the two phases become identical.

4 The Heat of Vaporization

The heat of vaporization at constant volume ΔU_{vap} gives the total inter-molecular energies at low temperatures and low vapour pressures. At higher temperatures and larger vapour pressures and intermolecular forces, ΔU_{vap} is the intermolecular energy in the liquid minus the rest energies in the real vapour phase. The attempt to estimate ΔU_{vap} therefore suggests an interesting model for estimating the temperature dependence of the intermolecular interactions. This attempt is made in the equations of Fig. 16: The left column gives the heat of vaporization in the ideal vapour state: the potential energy of Z nearest neighbours has to be corrected for $Z(1 - x_F)$ neighbours at $T > T_B$, in the non-linear region (region with "holes" in Eyring–Cremer model). The influence of higher neighbours and f' for the relation of Eqs. (2) and (3) are taken into account by the factor f. The second line gives the heat content of the inter-molecular degrees of freedom. Our experience thus gives a good result for molecules if we assume that every molecule-pair gives one intermolecular degree

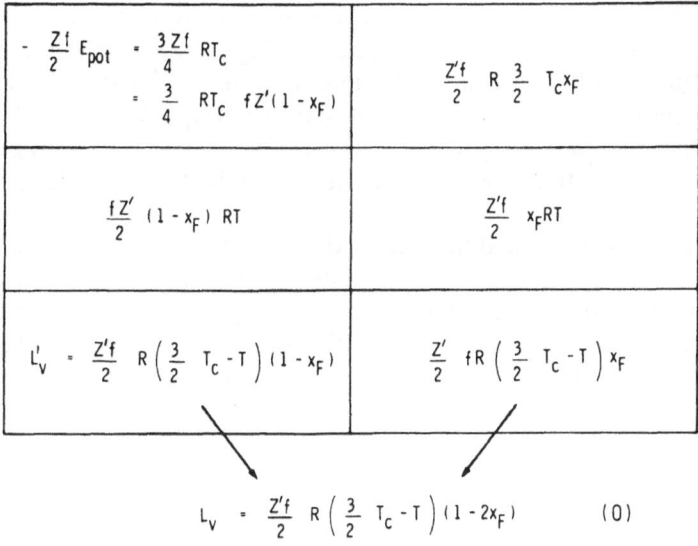

$$-\frac{Zf}{2} E_{pot} = \frac{3Zf}{4} RT_c$$
$$= \frac{3}{4} RT_c \; fZ'(1-x_F)$$

$$\frac{Z'f}{2} R \frac{3}{2} T_c x_F$$

$$\frac{fZ'}{2} (1-x_F) RT$$

$$\frac{Z'f}{2} x_F RT$$

$$L'_v = \frac{Z'f}{2} R\left(\frac{3}{2} T_c - T\right)(1-x_F)$$

$$\frac{Z'}{2} fR\left(\frac{3}{2} T_c - T\right) x_F$$

$$L_v = \frac{Z'f}{2} R\left(\frac{3}{2} T_c - T\right)(1-2x_F) \qquad (0)$$

Fig. 16.

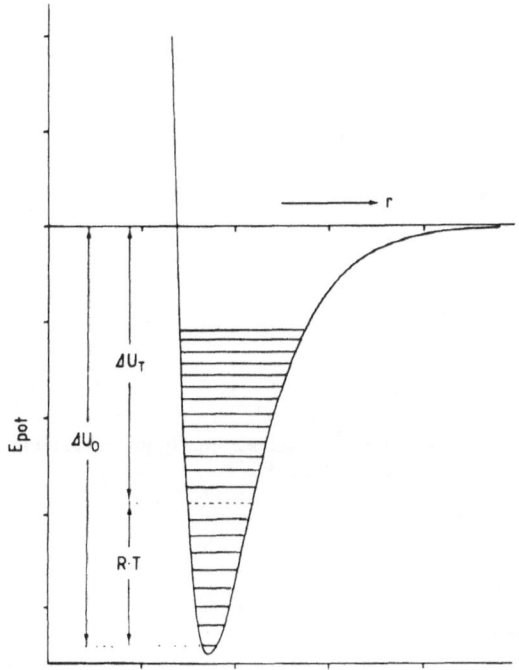

Fig. 17. The distribution on the inter-molecular vibration levels symbolised by an averaged vibration level of the height RT

of freedom by oscillations in the intermolecular potential curve with a specific heat of R. The pair numbers are given by $fZ/2$ and must be corrected by $(1 - x_F)$ in temperature regions with finite x_F. This intermolecular heat content is liberated during evaporation and has to be substracted from E_{pot}. This approximation takes the distribution of distances into account by assuming an average (dotted) intermolecular vibration level in Fig. 17 instead of the thermal partition on the different levels.

The third line at the left as the difference of the first two lines gives L_v for evaporation into a vacuum. The real $\Delta U_v = L_v$ into the real vapour state is L'_v minus the result of the right column, which gives step by step the energy of the van der Waals forces in the real vapour state with the same assumptions. The end result L_v describes experimental heats of vaporization surprisingly well. The

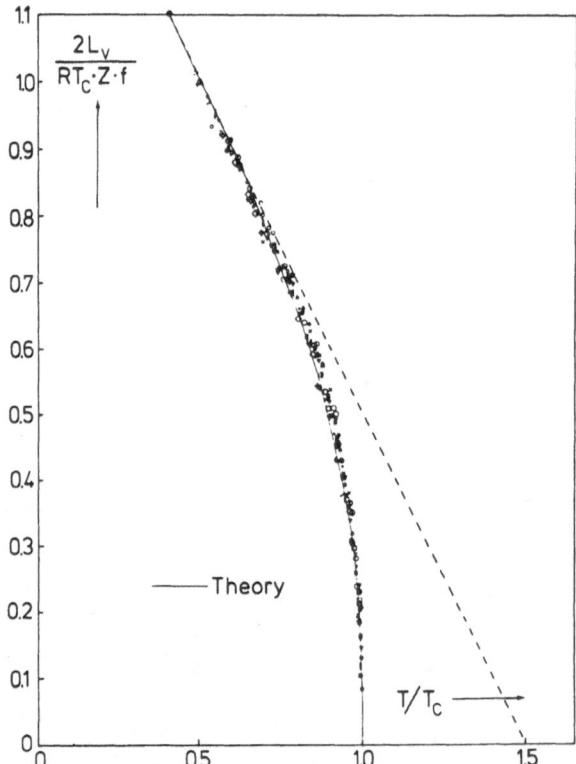

Fig. 18. Experimental heats of vaporization L_v multiplied by $2/RT_CZf$ (see Eq. (9)) give a common curve as function of T/T_C. *Continuous line*: theoretical values of Eq. (9).

× CH$_4$	· Ar	◇ C$_2$H$_5$—O—C$_2$H$_5$
o N$_2$	⊙ O$_2$	☆ naphthalene
○ CO	* CO$_2$	D C$_2$H$_6$
□ SF$_6$	○ hexane	> CCl$_4$
▽ Kr	▲ heptane	○ HCl
▼ Xe	+ benzene	● chloro-benzene
			NH$_3$

experimental values of L_v multiplied by the factor $2/RT_CZf$ (corresponding to the equation in Fig. 16) are given in Fig. 18 in a reduced T-scale T/T_C. As we should expect from this equation the values at temperatures below T_B give a straight line of slope $3/2$ which crosses the ordinate axis at $3/2 = 1.5$. If we multiply all ordinates of this straight line by the factor $(1 - 2x_F)$, all experimental values are really given by the derived equation for L_V. It is necessary to adjust at a single temperature, the factor Zf which expresses the type of package (Z) and the action of the higher neighbours (f). This factor f may also express the type of the distance dependence of the intermolecular potential. This simple factor Zf seems to rectify the so-called invalidity of the theorem of correspondence states. In Fig. 18 this theorem seems to be valid. The excellent and unexpected agreement of the equation of ΔU_v with experimental data demonstrates the efficiency of this model of non-polar or weakly polar liquids and should encourage its use in teaching and in daily research work.

The equation for L_v of our model also gives the interaction energy in a solution. In this case we can adjust the pair interaction energy U_{12} between the solvent 1 and the solute 2 by the Berthelot Rule: $U_{12} = (U_{11}U_{22})^{1/2} \approx 3R(T_{C1}T_{C2})^{1/2}/2$. This interaction could also be studied by measuring the shift of the OH stretching vibration frequency Δv, which is proportional to the van der Waals interaction energy [12]. We have also found a good linear relation for $\Delta v = \Delta v(T = 0\,K) - aT$ for alcoholic OH in different non-polar solvents [13, 14, 15] in the region $100\,K < T < T_B$. In an analogous way, for $T > T_B$ we obtain [16] parallel to equation for L'_v [Fig. 16]:

$$\Delta v = \{\Delta v(0\,K) - aT\}(1 - x_F) \tag{10}$$

as expected. At low temperature Δv is nearly temperature-independent. This may be related to fairly large intermolecular zero point energies by the large coordination numbers [15]. Table 2 gives the assumed values for Zf. If we apply a close-packing with $Z = 12$ the f-values of normal liquids are in the range: $1.03 < f < 1.2$. For HCl with specific interactions we get $f = 0.84$.

Table 2. Values of the effective coordination numbers Zf in Eq. (9)

Substance	Zf	f
$C_2H_5-O-C_2H_5$	14.5	1.2
CCl_4	12.4	1.03
C_6H_6	12.82	1.06
C_6H_5Cl	14.4	1.2
C_6H_5Br	12.88	1.07
C_6H_5J	13.8	1.15
naphtalene	14.36	1.2
HCl	10.1	0.84
NH_3	13.8	1.15

Our model also gives the so-called Trouton rule: $(\Delta H_{vap}/T_B) = (\Delta U_{vap}/T_B + R) = 20\,cal/mol\,K$ or $84\,J/mol\,K$ and could cover [5] the scatter of this value by assuming different values of Zf. The model gives with: $12 < Zf < 14.4$ the Trouton rule as:

$$17 < \Delta H_{vap}/T[cal/mol\,K] < 20 \tag{11}$$

$$71 < \Delta H_{vap}/T[J/mol\,K] < 84$$

The model also gives [5] Nernst's [17] empirical vapour Eq. (12) and describes the empirical specific constant as $3Zf/4$:

$$\ln(p/p_C) = -[T/T_C - 1]3Zf/4 \tag{12}$$

Last but not least, the enthalpy of the liquid ΔH_{liq}, ΔH_{vap} (enthalpy of the saturated vapour) or the specific heat C_σ of a liquid at saturation conditions are also given with sufficient accuracy [5]. As an example Fig. 19 gives the comparison between experimental values (full line) of ΔH_{liq} and ΔH_{vap} for CO and the theoretical values (o) of our model [5]. The agreement could be improved if we added one additional degree of freedom in the liquid state. (Δ in Fig. 19). This could be a translation degree of freedom. $-C_\sigma$ contains a factor dx_F/dT which dominates near T_C (compare the slope of the curve in Fig. 14). Figure 20 gives the comparison between the experimental values of $(C_\sigma - C_{vid})/R$ for benzene (x) in equilibrium with its vapour and the theoretical values (o) given by our model [5]. C_{vid} values are calculated assuming harmonic vibration

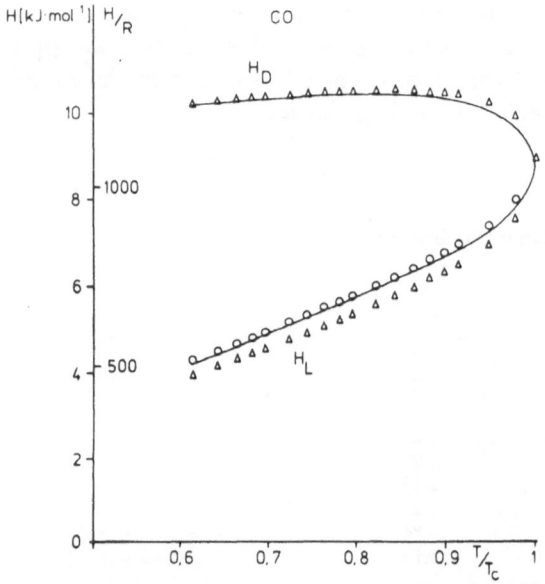

Fig. 19. *Continuous line*: enthalpy of CO under saturation conditions, in vapour: H_D and in liquid: H_L. *circles* and *triangles* calculated by the simplified liquid model. $\triangle$: calculated H; O: calculated: $H + 1f_{trans}$; —— literature

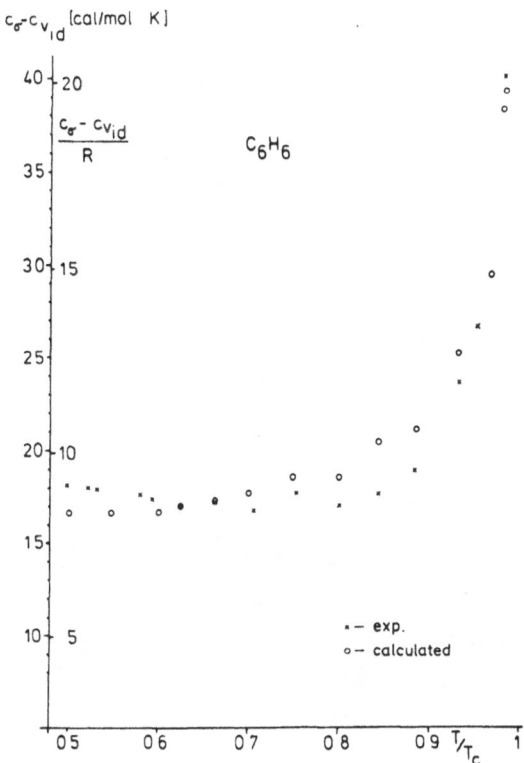

Fig. 20. ×: Experimental values of the specific heat C_σ of saturated liquid benzene
o: calculated by the simplified liquid model.
C_σ in reduced units

frequencies [18] of ideal benzene vapour. The agreement is sufficient and describes the large increase of C_σ by the large values of dx_F/dT.

5 Surface Tension and Surface Energy

The surface tension σ and the surface energy U_σ gives the isothermic work or the energy necessary to enlarge the surface. Both illustrate a difference between the bulk and the surface, both are sensitive tests of models. Figure 21 shows some experimental values of σ given in erg/cm² in the reduced temperature scale T/T_C. The large water values are often discussed as an indication of the specially large intermolecular forces of water. However a structure discussion of σ is not relevant, because 1 cm² contains different numbers of molecules. Even during the temperature changes in Fig. 21 these numbers change. Therefore we have to calculate [5, 6, 19] the molar surface tension

$$\sigma_M = \sigma N_L^{1/3} V^{2/3} \tag{13}$$

[N_L is Loschmidt number and V the molar volume].

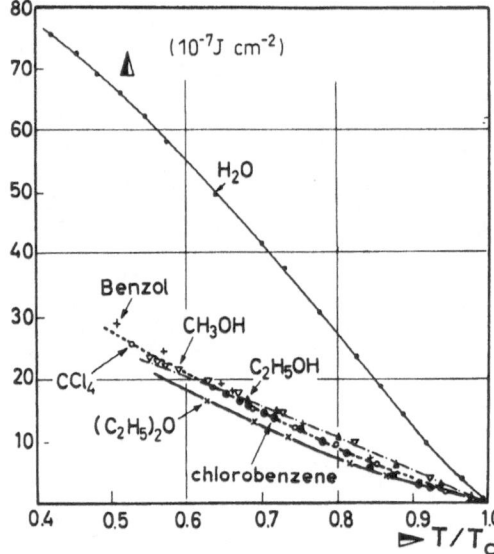

Fig. 21. Surface tension σ of some liquids in reduced T/T_C scale as energy (in 10^{-7} Joule) to enlarge the surface by 1 cm²

To obtain the isothermal work required to bring one mole to the surface (see Fig. 22). In this diagram water loses its dominant role but still gives larger σ_M-values than methanol or ethanol because of its two H-bonds compared with the one in alcohols.

σ_M gives the isothermally measured work required to enlarge the surface. During this process, energy is exchanged with the heat of the environment. To

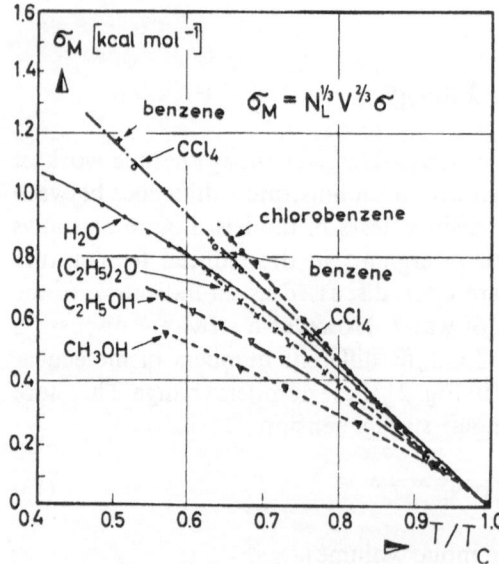

Fig. 22. Molar surface tension σ_M as work (kcal/mol) needed to bring one mole to the surface. (1 kcal mol^{-1} = 4.187 kJ mol^{-1})

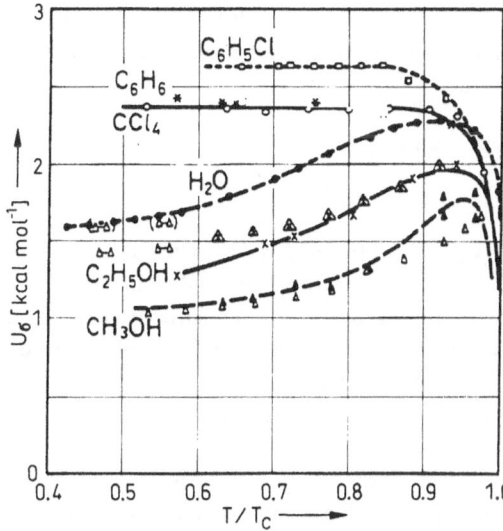

Fig. 23. Surface energy U_σ calculated by the Gibbs–Helmholtz equation from σ_M (free energy)

obtain the energy U_σ to bring one mole to the surface we have to apply the Gibbs–Helmholtz equation:

$$U_\sigma = \sigma_M - T d\sigma_M/dT \qquad (14)$$

As Fig. 23 shows, the surface energy U_σ is temperature independent for normal liquids for T/T_C less than about 0.9, meaning that the specific heats in the surface and the bulk are the same. (Further examples are given in Ref. [5]). For liquids with specific interactions, a temperature maximum of U_σ is typical.

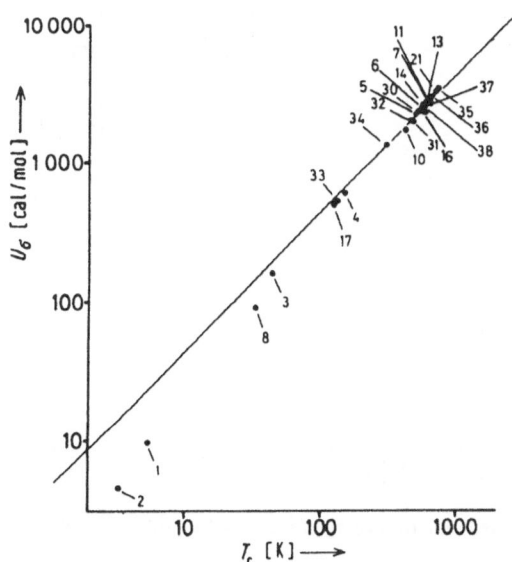

Fig. 24. The proportionality between the surface energy U_σ and T_C as predicted by Eq. (15). 1: ^{4}He, 2: ^{3}He, 3: Ne, 4: Ar, 5: n-Hexane, 6: Benzene, 7: Toluene, 8: H_2, 10: Cl_2, 11: o-Xylene, 12: m-Xylene, 13: p-Xylene, 14: Cyclohexane, 16: CCl_4, 17: N_2, 21: Aniline, 30: Ethylacetate, 31: Methylformiate, 32: Diethylether, 33: CO, 34: CO_2, 35: N,N-Dimethylaniline

In our simplified model of non-polar spherical molecules U_σ corresponds to the difference of intermolecular interactions between the bulk and the surface. Based on Eq. (9) we get from our model:

$$U_\sigma = f(Z - X)3RT_C/4 \qquad (15)$$

X: coordination number at the surface.

As Fig. 24 shows, the predicted proportionality of Eq. (15) with T_C is verified for a large variety of liquids. With the idealised assumption of $Z = 12$ and $X = 9$ we would expect: $(Z - X) = 3$. Experiments give about 2.85 with the exception of molecules with low T_C and relatively large zero point energies of the intermolecular forces [25]. The small deviation between 3 and the experimental value 2.85 may be caused by a reduction of Z during the melting process with its volume increase.

The U_σ-maximum of H-bonded liquids indicates a temperature-dependent change of the H-bond content $(1 - O_F)$ (see Chap. IX) and agrees with the spectroscopic determinations [20]. At low temperature, $(1 - O_F)$ does not change much so that U_σ would not be expected to change with temperature. This prediction has been confirmed experimentally [5]. Our simplified model is also able to describe the complicated surface properties. The temperature constancy of U_σ indicates that the loss of intermolecular degrees of freedom is fully compensated by the possibility of a translation movement of the surface layer.

One of the first to claim the necessity of describing surface phenomena per mole and not per surface was Eötvös [21]. His idea seems to be too often

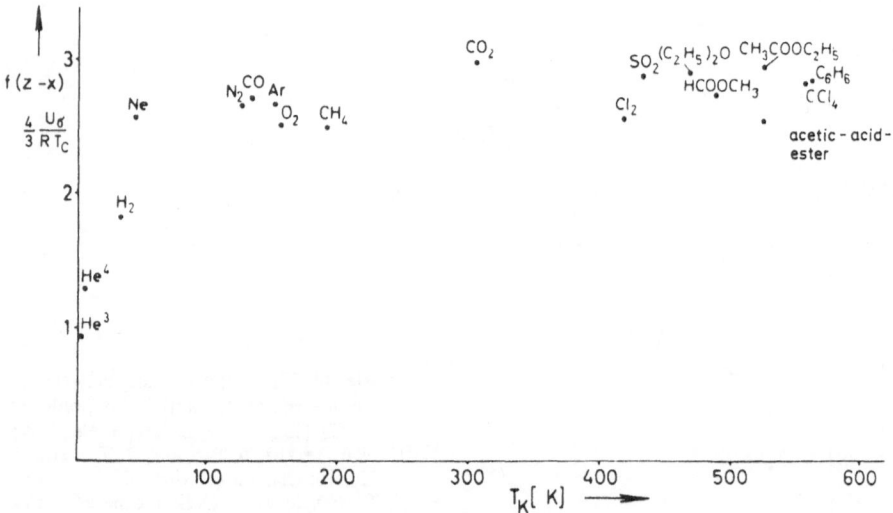

Fig. 25. Surface energy U_σ over $3RT_C/4$ corresponds on average for normal liquids to a value of about 2.85

ignored. In addition he proposed what later became knows as the Eötvös rule [21], that for not too large molecules:

$$d\sigma_M/dT = -2.13R = -2.1\,10^{-7}(J\,K^{-1}\,mol^{-2/3}) \tag{16}$$

This rule can be understood and quantitatively given by our model: Because of the temperature constancy of Eq. (15):

$$\Delta\sigma_M/\Delta T = [\sigma_M(T_2) - \sigma_M(T_1)]/[T_2 - T_1] = -2.13R$$

and with $T_2 = T_C$; $T_1 = T$ and $\sigma_M(T_C) = 0$:

$$[0 - \sigma_M(T)]/[T_C - T] = -2.13R \tag{17}$$

or $\sigma_M(T) = 2.13R(T_C - T)$.

It follows from Eqs. (14), (15) and (17) that:

$$\sigma_M - Td\sigma_M/dT = 2.13RT_C = U_C = f(Z - X)3RT_C/4 \tag{18}$$

or with $f(Z - X) = 2.85 \times 0.75 = 2.137$.

This result gives the value 2.13 of the Eötvös constant nearly quantitatively, predicts that this rule is only approximately valid and has to vary with: $f(Z - X)$, as the experimental data confirm.

Our model also corrects the so-called Stefan rule [22], which claims theoretically that $\Delta H_{vap}/U_\sigma = 2$. Our model predicts [5, 6] a temperature dependency

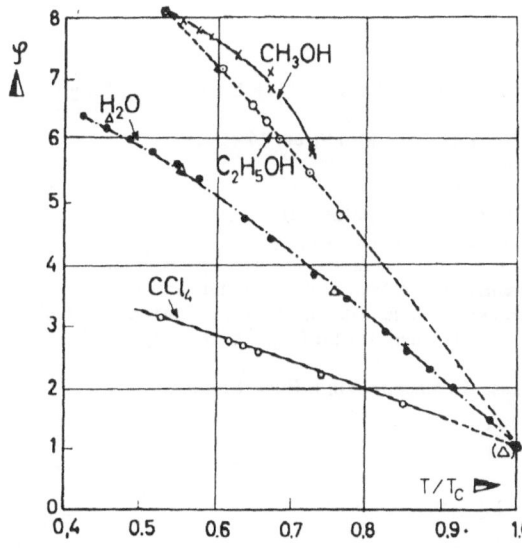

Fig. 26. T-dependence of the Stefan rule: $\varphi = \Delta U_{vap}/U_\sigma$ with limit value of about 4 at $T = 0$ for CCl_4 (predicted by Eq. (20)). H-bonded liquids give much higher φ-values

of the Stefan rule:

$$\Delta H_{vap}/U_\sigma = Z[(3T_C/2) - T]/[Z - X]3T_C/2 \tag{19}$$

and for T against O:

$$\lim_{r \to 0} [\Delta H_{vap}/U_\sigma] = Z/[Z - X] = 12/3 = 4 \tag{20}$$

Both predictions are well verified by the experimental data [6].

This ability of our simplified model to describe the surface properties and especially to give a better understanding of the Eötvös and the Stefan rules increases the confidence in the possibility of using an approximate model to describe the heat required for understanding normal liquids. It seems illogical on one hand to use the rough ideal gas law and on the other to avoid approximate models for important liquids.

6 References

1. Hertz G (1978) Prog Colloid & Polymer Sci 65: 60
2. Sutherland W (1893) Phil Mag 36: 507
3. Eucken A (1948) Lehrbuch der chemischen Physik, Vol II, 1. Akademische Verlagsgesellschaft, Leipzig, p 308
4. Rankine AO (1910) Proc Roy Doc, London 84: 190
5. Luck WAP (1979) Angew Chem 91: 408; Angew Chem Int Ed Engl 18: 350
6. Ditter W, Luck WAP (1971) Tetrahedron 27: 201
7. London F (1933) Trans Farad Soc 33: 8
8. Lorenz R (1916) Z Anorgan Allgem Chem 94: 240
9. Wess T, Luck WAP (1991) Chem Phys Lett
10. Eyring E (1936) Z Phys Chem 4: 283
11. Cremer E (1944) Z Phys Chem 193: 287
12. Luck WAP (1990) J Mol Struct 217: 281
13. Luck WAP, Zheng HY (1984) J Chem Soc Faraday Trans 80: 1253; Luck WAP, Zheng HY (1984) Z Naturforsch 39A: 888
14. Mentel Th, Peil St, Schiöberg D, Luck WAP (1986) J Mol Struct 143: 321
15. Peil S (1990) Thesis, University of Marburg
16. Kümmerle I (1990) Thesis, University of Marburg
17. Nernst W (1926) Theoretische Chemie, 15th edn, Enke, Stuttgart
18. Zeise H (1954) Thermodynamik, Bd.III/1, Tabellen Hirzel, Leipzig
19. Luck WAP (1973) In: Die Bedeutung der Oberflächenenergie. Chemie, Physikalische Chemie und Anwendungstechnik der grenzflächenaktiven Stoffe, Kongreßband IV. Internationaler Kongreß für grenzflächenaktive Stoffe, Zürich, Carl Hanser, München, p 83
20. Luck WAP (1980) Angew Chem 92: 29; Angew Chem Int Ed Engl 19: 28
21. Eötvös Rv (1886) Ann Phys 27: 452
22. Jellinek K (1928) Lehrbuch der Physikalischen Chemie, Vol I. Enke, Stuttgart, p 909

CHAPTER IV
Dynamic Aspects of Intermolecular Interactions

L.C.M. De Maeyer

Molecules fluctuate constantly between different states. Their lifetimes in a particular state depend on the mechanisms of energy exchange during interactions with other molecules. These mechanisms also determine the dynamic behaviour of a system on the macroscopic scale. Some physical or chemical properties, e.g. the boiling point, or a chemical equilibrium constant, may inform us about the energies of interaction states. Quantities that do not include the time scale explicitly in their dimensional definition, however, cannot tell us anything about the molecular mechanism of the transitions, which are essentially processes evolving in time.

The study of properties that depend on the time scale involves the construction of models describing these processes. This is exemplified by simple descriptions for elementary steps in transport processes in gases and liquids. Transport is a process taking time, whether it involves the transposition of real particles in real space, or of a representative point in an abstract configuration space (which is only a more convenient way of treating the former). Viscosity and diffusion are treated as examples. Their limiting influence on chemical rate processes is discussed. The chemical lifetime of a particular species in a particular state is related to, but distinguished from the relaxation times of macroscopically observable normal modes of fluctuation, in which many interactions are coupled.

Intermolecular Forces
© Springer-Verlag Berlin Heidelberg 1991

1 Introduction

Molecular interactions are the origin of the formation of recognizable structures in the spatial arrangement of the particles. Unordered localization of particles exists only in the abstraction of the ideal gas where, by definition, they exert no interaction forces upon each other. The probability of finding a particle of an ideal gas in an arbitrary volume element of the available space is completely independent of the position of the other particles. Particles of real matter, however, exert interaction forces of differing nature. These are attractive as well as repulsive, and may depend in very different ways on distance and mutual orientation. The spatial distribution of the positions and movements of the particles is governed by the ordering forces, but a substantial lack of information on the actual distribution of energy over the available modes of motion may remain. The entropy of the system quantifies this lack of information. The smaller the available space for a given number of material particles is, the stronger the ordering forces will become and the smaller the entropy of the arrangement will be, when it is in equilibrium with a constant thermal environment. An individual molecule takes part in the disorder by energy exchange with the other molecules. It spends on the average a certain time in each of the distinct recognizable states. This time depends upon the stability of the structure and upon the dynamic properties of the acting intermolecular forces. Information on the nature of the forces and the formation of structures can be obtained by observing the lifetimes of the states. Time-dependent processes also greatly influence the behaviour of structured matter subjected to external forces.

The primary interaction forces leading to structure formation in the systems of interest here are of electrodynamic nature. The strong electrostatic repulsion of negative electron charges that are not localized pairwise in a common molecular orbital is the origin of the mutual impenetrability of atoms. The mechanical behaviour of colliding gaseous particles and the low compressibility of liquids and solids are a consequence of the small range of these repulsive forces. The experimentally observed phenomena in inert gases can be explained satisfactorily by a repulsive potential $E_{rep}(r) = B/r^n$ with $n \geqq 10$. Here r is the distance between two atoms and B and n are atom-specific empirical parameters. Other forms of the repulsive potentials are used as well, e.g. the Born–Mayer potential a exp(br), also depending upon two parameters. Quantum-mechanically calculated potentials can be approximated rather well by an appropriate choice of these parameters.

The attractive forces between neutral particles (van der Waals forces) also have an electric origin. They derive from the reduction of the field energy of the continuously fluctuating charge distribution in an atom, when electrically polarizable matter is present in its vicinity. The distance law is that of electrical dipoles:

$$E_{dip}(r) = -\frac{2\vec{p}_1\vec{p}_2}{4\pi\varepsilon_0 r^3} = -\frac{2\alpha\bar{p}_1^2}{4\pi\varepsilon_0 r^6}$$

$\vec{p}_1$ is the dipole moment of an instantaneous charge distribution in an atom and $\vec{p}_2 = \alpha \vec{p}_1 / r^3$ the induced moment in the neighbouring atom; α is the polarizability and ε_0 is the dielectric permittivity of free space.

The repulsive and attractive contributions to the energy function of two identical rare gas atoms are summarized by the Lennard–Jones potential:

$$E(r) = 4\varepsilon \left\{ \left(\frac{\sigma}{r} \right)^{12} - \left(\frac{\sigma}{r} \right)^{6} \right\}$$

The atom-specific constants B and $\langle \alpha p_1^2 \rangle$ are expressed here by two parameters σ and ε, which have a simple physical interpretation. σ is a measure for the linear dimension (diameter) of the atom and $-\varepsilon$ is the depth of the energy minimum, located at $r = 2^{1/6} \sigma \simeq 1.12 \sigma$.

The Lennard–Jones potential is a good approximation for the ubiquitous interactions between the electron shells of particles in contact. Other interactions can be active in addition, e.g. electrostatic interactions between ions or molecules with permanent dipole moments, or chemical interactions such as the formation of coordinatively bonding molecular orbitals. These can lead to specific locally interacting structures.

For numerical simulations of processes in gases and atomic liquids (molecular dynamics calculations) a very much simplified hard-sphere model is often used. This model neglects attractive forces and replaces the repulsive ones by $E = \infty$; $r < \sigma$, $E = 0$; $r \geq \sigma$.

In experiments with crossed molecular beams, on the other hand, the interaction potentials must be taken into account with great precision. Apart from the already mentioned dipole interaction, higher order terms (C_8/r^8, C_{10}/r^{10}) of the multipole expansion of the interaction between two charge distributions are taken into account as well.

There is a limit to the description of interaction potentials by generalized equations containing more and more parameters, especially when the assumption of spherical symmetry cannot be upheld, as in the case of most polyatomic molecules. Certain molecules share properties that are specific for a particular part of their chemical structure. A great systematic advantage is often gained by treating the interaction forces resulting from such characteristic features as a specific category of a unique chemical nature. The formation of hydrogen bonds is a specific interaction of this kind. In many cases the specific interaction forces, that make hydrogen-bond formation between certain molecules possible, exceed by far the general van der Waals forces, present for any type of molecule.

We will be occupied in this chapter with the lifetime of interaction processes and interaction states. If one is looking for a theory of macroscopic matter on the basis of elementary corpuscular elements and wants to include statements about the temporal evolution of changes, the approach must include dynamics. Neither the thermodynamic, nor the statistical method are sufficient to handle dynamic questions. The *thermodynamic* method attempts to derive the behaviour of certain properties of matter (e.g. thermal expansion, compressibility, chemical equilibrium) from appropriate functions of macroscopic state variables under given external conditions. The *statistical* method starts from the corpuscular properties of the interacting elements, but limits itself to stable distributions and averages over large numbers. Only in the *kinetic* method is time explicitly included in the treatment of the elementary processes. It is then necessary to follow the movement of individual particles under the influence of acting forces and to propose models that enable predictions about the evolution of variables of motion.

Some measurable physical properties related to time-dependent phenomena enable us to estimate the average lifetime that a particle or a structure spends in a particular state. All measurable properties, explicitly containing the dimension of time in their definition, are candidates for learning something about the dynamic processes in which the elementary constituents of the system are involved.

2 The Collision Process

When two molecules collide, their kinetic energy is temporarily converted into potential energy of repulsion forces. It is possible that some of the potential energy is redistributed over other intramolecular motions (rotation, vibration, electronic excitation). It depends on the dynamics of a collision and its duration, whether energy is exchanged evenly with all the coupled atomic motions in the molecule, or whether only a few vibration modes, or perhaps none at all, are perturbed by the temporary change in the electronic wavefunctions of colliding particles.

The dynamics of a collision process in which two particles are involved depend on their mutual interaction potential $E(r)$ and on their initial relative kinetic energy $\frac{1}{2}mu_0^2$. u_0 is the initial relative speed and m is the reduced mass of the particles with masses m_1 and m_2, given by $m^{-1} = m_1^{-1} + m_2^{-1}$. A further important factor is the offset between the initial trajectories, known as the impact parameter b. b would be the distance of closest approach if the interaction potential were inactive. As a consequence of the conservation of angular momentum and total energy, the initial kinetic energy is divided at any time over three terms:

$$\tfrac{1}{2}mu_0^2 = \tfrac{1}{2}m\left(\frac{dr}{dt}\right)^2 + \tfrac{1}{2}mu_0^2\left(\frac{b^2}{r^2}\right) + E(r)$$

$r(t)$ is the linear distance between the colliding particles. The middle term represents the kinetic energy of rotation. It vanishes for a head-on collision (b = 0). We can include it together with the last term in a new effective interaction potential $E'(r) = E(r) + \tfrac{1}{2}mu_0^2(b^2/r^2)$, where it represents a centrifugal part of the repulsive potential.

In the interaction equation the variables r and t are separable. The duration of a collision can therefore be estimated by calculating the integral

$$\tau(u_0, b, r_0) = 2 \int_{t_0}^{t_{min}} dt = 2 \int_{r_0}^{r_{min}} \frac{dr}{\sqrt{u_0^2 - \frac{2}{m} E'(r)}}$$

The upper integration limit r_{min} is the closest encounter distance during the collision. At this distance the initial kinetic energy is completely absorbed by the effective interaction potential:

$$E'(r_{min}) = \tfrac{1}{2} m u_0^2$$

The lower integration limit r_0 may be a somewhat arbitrarily chosen distance at which a collision is assumed to begin, e.g. when the colliding atoms are separated by a few atomic diameters, $r_0 = n\sigma$, say $2 \le n \le 4$. The integral is multiplied by a factor two, because it covers only the deceleration period r_0 to r_{min}, which must be followed by an equal period of reacceleration.

In view of the arbitrariness of the "beginning" of a collision and its dependence on u_0 and b, the "lifetime" of a collision-interaction is not a well defined quantity. Moreover, there is no simple procedure for evaluating the integral. Even numerical integrations must be based on approximations since the integrand diverges at $r = r_{min}$ and contains high powers of r when realistic interaction potentials like the Lennard–Jones potential are substituted into $E'(r)$.

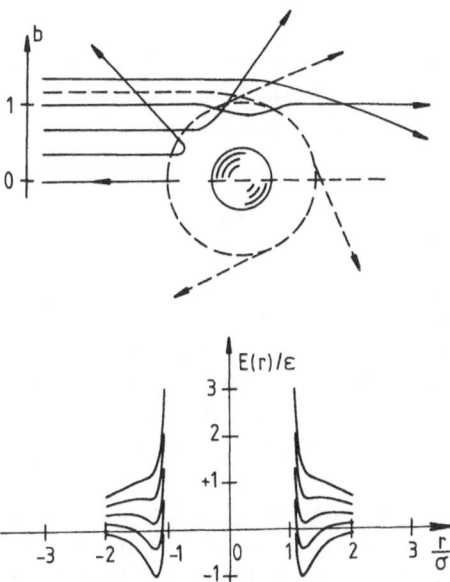

Fig. 1. Collision trajectories for particles with a Lennard–Jones interaction potential for different impact parameters and effective potential functions including different centrifugal contributions

Depending upon the initial kinetic energy $mu_0^2/2$ in relation to ε, and the impact parameter b in relation to the diameter σ, two colliding atoms may show quite different behaviour. This is illustrated schematically in Fig. 1. $b < 2\sigma$ is the range of "head-on" collisions, with deflections between zero and 180 degrees (when $b = 0$). For b between 2σ and 4σ, the interaction depends on the initial kinetic energy. If this is much larger than the attractive potential ε, the centrifugal repulsion dominates during the fast passing and the deflection remains small. For smaller initial velocities a grazing collision takes place in which the attractive force is able to deflect the trajectories during the slow passing. At some particular combinations of (small) initial kinetic energy and impact parameter b, the attractive force may be sufficient to compensate the centrifugal force. The collision trajectories can encircle each other several times before separating again. Two colliding atoms may then spend an appreciable time in each others vicinity, although separated by more than an atomic diameter. This "orbiting" collision must be distinguished from a "bound" state in which a pair of atoms is trapped in the minimum of the Lennard–Jones potential well. A bound pair cannot be formed during a binary collision but requires a third collision partner to take away the excess energy.

The average kinetic energy of a gaseous particle is $m\langle u^2 \rangle/2 = 3kT/2$. This amount of energy must be divided over three possible directions of motion. The relative kinetic energy of two colliding particles is therefore on the average $kT/2$. The time needed for deceleration and reacceleration during a bimolecular collision in the gas phase is of the order of 10^{-12} s.

An estimate for the average velocity of a particle can be obtained from

$$\sqrt{\langle u^2 \rangle} = \sqrt{\frac{3kT}{m}} = \sqrt{\frac{3RT}{M}} \quad (M = \text{molar mass})$$

The arithmetic average velocity $\langle u \rangle$ is somewhat smaller, because higher velocities are counted with higher weight in quadratic averaging. Using the Maxwell–Boltzmann velocity probability distribution $\rho(u)$ one gets

$$\langle u \rangle = \int_0^\infty u\rho(u)du = \sqrt{\frac{8kT}{\pi m}} = 0.921\sqrt{\langle u^2 \rangle}$$

The sound velocity in gases is a measurable quantity giving us a hint about the average velocity of the particles in these media. When the particles do not meet each other very often, a perturbation of their velocities by a sound wave cannot progress faster than the particles themselves. In fact, the experimentally determined sound velocity is of the order of 2/3 of the average particle velocity (cf. Table 1).

The frequency of collisions is obtained from the average velocity $\langle u \rangle$ and the mean collision-free path length λ. The mean free path length should not be confused with the mean distance between the particles. This distance is

Table 1. Sound velocity c_0 in gases (273 K)

	M(g)	$\sqrt{\langle u^2 \rangle}$ (m s^{-1})	c_0 (m s^{-1})
H_2	2.01	1840	1261
He	4.00	1304	971
D_2	4.03	1299	890
NH_3	17.03	627	415
Ne	20.18	580	433.4
N_2	28.01	493	337.5
O_2	32	461	315.5
Ar	39.95	412	308.5
CO_2	44.01	393	258.3
SO_2	64.06	326	209

proportional to the cube root of the particle density. The flight path of a particle, whose size is small compared to the mean distance, can pass many neighbouring particles without touching them.

To calculate the mean free path we consider a cylindrical volume part of the gas with the diameter σ of a molecule. This volume is passed through by a moving molecule in the course of its flight. The length of the cylindrical volume, traversed on the average without colliding, is the mean free path λ. Whenever another molecule projects into this volume, there will be a collision. Its center may be separated at most by the molecular radius from the boundary of the cylinder. The center of the second molecule must therefore be inside a cylinder with diameter 2σ to produce a collision.

In a cylinder with this diameter, length λ and volume $\pi\sigma^2\lambda$ just one molecule will be found if the particle density is $n = 1/\pi\sigma^2\lambda$. Between λ and n we therefore have the relation $\lambda = 1/\pi\sigma^2 n$. In reality the mean free path is slightly smaller. The attractive interaction forces deviate the flight path, leading also to collisions with molecules that are within a marginally larger distance than $\sigma/2$ from the traversed cylinder. This correction leads to the expression

$$\lambda = \frac{1}{\pi\sigma^2 n \left(1 + \dfrac{C}{T} \right)}$$

The constant C represents a substance-specific temperature, known as Sutherland's constant. The quantity $\sigma\sqrt{1 + C/T}$ is often combined into a (temperature-dependent) effective molecular diameter σ_{eff} leading to $\lambda = 1/\pi n \sigma_{eff}^2$.[1]

The duration between consecutive collisions is $\tau = \lambda/\langle u \rangle$. A molecule, e.g. nitromethane, with an effective diameter of 0.6 nm has a mean free path of about 33 nm at normal pressure 1 bar $= 0.1$ MPa and 273 K. Its molar mass is

---

[1] C could be related with the intermolecular potential see Chap. III, page 57.

$61\,\mathrm{g\,mol^{-1}}$ and its average velocity $330\,\mathrm{m\,s^{-1}}$. This molecule spends about 1% of its time in interaction with other molecules of the gas.

3 Diffusion

Almost every collision will change the direction in which·a free path length is traversed. After j collisions the distance from the starting point therefore is not $j\lambda$, but a shorter stretch. To calculate this, we consider first the one-dimensional case, where the motion of the particle is constrained to a straight line. The position of the particle is indicated by the coordinate value x. From the starting position x_0 the particle moves with equal probabilities to $x_0 + \lambda$ or to $x_0 - \lambda$. The distance after the first collision is then $\Delta x_1 = \pm\lambda$, and the mean square value is $\langle \Delta x_1^2 \rangle = \lambda^2$. After a second collision the squared distance will be $\Delta x_2^2 = (\Delta x_1 + \lambda)^2 = \Delta x_1^2 + 2\Delta x_1 \lambda + \lambda^2$, or $\Delta x_2^2 = (\Delta x_1 - \lambda)^2 = \Delta x_1^2 - 2\Delta x_1 \lambda + \lambda^2$. The average value is $\langle \Delta x_2^2 \rangle = \Delta x_1^2 + \lambda^2 = 2\lambda^2$. Continuing this reasoning, after j collisions $\langle \Delta x_j^2 \rangle = j\lambda^2$. The mean distance after j collisions is only $\sqrt{j}\lambda$. The total time needed is $\Delta t_j = j\tau = j\lambda/\langle u \rangle$. It is the average *square* of the distance that increases proportionally with the number of collisions, or linearly with time. In the three-dimensional case only one third of the collisions will take place in x-, y- or z-direction, respectively. On the other hand one has $\langle \Delta r^2 \rangle = \langle \Delta x^2 \rangle + \langle \Delta y^2 \rangle + \langle \Delta z^2 \rangle$. Eliminating j leads to the expression

$$\langle \Delta r^2 \rangle = \lambda \langle u \rangle \Delta t$$

The ratio $\langle \Delta r^2 \rangle/\Delta t = \lambda \langle u \rangle$ is a characteristic quantity for the kind of gaseous molecules considered.

The direction, in which a molecule departs from its initial origin, is no longer random if there exists a concentration gradient for its species. More particles diffuse away from the more concentrated side as there arrive from the less concentrated one. This leads to a particle current, proportional to the concentration gradient:

$$\phi = - D\frac{\partial n}{\partial x}$$

n is the number of particles per unit volume, ϕ is the excess of particles flowing per unit time and unit area through a surface, perpendicular to the gradient. This linear expression (first Fick's law) defines the diffusion coefficient as a proportionality constant with dimension $l^2 t^{-1}$, the same as that of $\langle \Delta r^2 \rangle/\Delta t$. Between the macroscopic quantity D, which is also related to the mechanical mobility μ of a particle, and the microscopic (molecular) quantity $\langle \Delta r^2 \rangle/\Delta t$ there exists a general relation

$$\frac{\langle \Delta r^2 \rangle}{\Delta t} = 2kT\mu = 2D$$

which was derived for the first time by Einstein, from considerations about Brownian motion.

The quantity D discussed so far is the coefficient of self-diffusion, related to the transport of a particle between particles of the same kind. In a measurement one must be able to distinguish the diffusing particle from the rest. This can be done by special marking (e.g. radioactive isotopes, spin-orientation).

In gaseous mixtures the diffusing particles have different mean velocities and effective diameters. The diffusion coefficients D_{ij} of a gas i in a gas j depend on the composition of the mixture. For every mixing ratio one must have $D_{ij} = D_{ji}$.

The effective diameter of a particle is determined essentially by the spatial extension of the electron shell (repulsive interaction). The polarizability of the electron shell (attractive interaction) introduces only a correction. The finite diameter and its consequence, the molecular collisions, are also responsible for the viscosity of fluid media. The viscosity η is a measure for the lateral transport of momentum (through a surface perpendicular to a velocity gradient). It can be expressed by momentum per unit time, per unit area, per unit of velocity gradient. Its dimension is then

$$\left(m\frac{l}{t}\right)\left(\frac{1}{t}\right)\left(\frac{1}{l^2}\right)\left(\frac{d}{dl}\left(\frac{l}{t}\right)\right)^{-1} = m\,l^{-1}\,t^{-1}$$

The cgs-unit (Poise) is one gram $s^{-1}\,cm^{-1}$, the SI-unit (mks) is Newton $s\,m^{-2}$. There is a quantity $v = \eta/\varrho$ known as *kinematic* viscosity, which is the viscosity defined here, divided by the density. Its cgs-unit (Stokes) is one $cm^2\,s^{-1}$.

In dilute gases the viscosity is a transport quantity (momentum transport), related to the mean free path. The transport of momentum from a region S with linear velocity $w_x + dw_x$ (this macroscopic velocity is superposed for all particles onto their randomly oriented thermal velocity $\langle u \rangle$) to a region Q with velocity w_x, separated from S by a distance dy (orthogonal to x), can be effected only by particles that have not meanwhile collided with others and changed their momentum. The total transport $d\phi$ per unit area and time over a free path λ is proportional to three quantities. The first one is the density n of particles in S and Q. The second one is the velocity $\langle u \rangle$, with which they traverse the path. The third one is the momentum difference $m(dw_x/dy)\lambda$ brought along. Consequently one has

$$\eta = \frac{d\phi}{dw_x}\,dy = \frac{1}{3}n\langle u \rangle m\lambda = \frac{2\sqrt{mkT}}{3\pi^{3/2}\sigma_{eff}^2}$$

The proportionality factor $\frac{1}{3}$ takes account of the fact, that $\frac{1}{6}$ of the particles run in the positive and $\frac{1}{6}$ in the negative direction along the y-axis, and that particles running from S to Q bring along a positive momentum excess, whereas particles running from Q to S transport negative momentum excess. The equation shows that the viscosity of a gas depends only upon its chemical nature (m and σ) and upon temperature, but it is independent of pressure, since the particle density has disappeared from the final expression.

It follows from the derivations for $D = \lambda \langle u \rangle /2$ and $\eta = \varrho \lambda \langle u \rangle /3$, obtained so far for dilute gases ($\lambda \gg \sigma$), that diffusion coefficient and viscosity are proportional to each other in these media, according to the relation $D = 1.5 \, \eta/\varrho$.

In liquids this proportionality must be replaced by a completely different relation. Here these quantities are in inverse relation to each other. In liquids the mean free path is smaller than or of the same order as the average particle distance. It is no longer the mean free path but the size of the particles that plays a dominant role.

4 Liquids

Gases condense when the average kinetic energy per particle is less than the depth of the attractive potential well. Rather than giving each particle a very small amount of kinetic energy and keeping them all freely moving in the available space, the energy distribution of maximum entropy is one in which many particles form a liquid or solid phase, and a few of them remain gaseous. Each of the liquid as well as the gaseous molecules still retain the same *average* translational energy $3kT/2$, but their velocity distributions are different. The molecules of the liquid move in the force fields of their neighbours. The *instantaneous* direction of motion of a particle is still mostly independent of the direction of motion of particles in its neighbourhood and one can use the notion of collisions. These occur, of course, much more often than in the gaseous phase. The average collision frequency is the ratio between average velocity $\langle u \rangle$ to the distance $\bar{s}$ travelled between collisions. Due to the proper extension of the particles the latter is only a fraction of the average distance between the particle centers. Taking $\bar{s} \simeq 0.1$ nm as the order of magnitude, collision frequencies as large as 10^{13} Hz are obtained at normal temperature.

The mean free path is no longer larger, but smaller than the average particle distance. In solids, the instantaneous motion of a particle is strongly correlated with the motion of neighbouring particles. The kinetic energy of translation has been transformed in vibration modes of the ordered solid-state structure, in which each particle has its own place. The displacement of a particle from one position to another in a liquid occurs as a sequence of mutual exchange processes. A change of place can happen when the neighbours of two molecules create a space large enough for them to rotate together around its center. This means that the diameter of the required space must be about 2σ.

During this process the two molecules occupy a volume $4\pi\sigma^3/3$. Normally the space available for two molecules is only $2M/N_A\varrho$ (M: molar mass, N_A: Avogadro's number, ϱ: density). To enlarge the available volume, the work $E_P = P(4\pi\sigma^3/3 - 2M/N_A\varrho)$ must be done against the external pressure P. The relative probability of this state being reached is given by the Boltzmann equation:

$$\frac{p}{p_0} = \exp\left\{-\frac{E_P}{kT}\right\}$$

For this fraction of the time a particle is able to change place. It moves with average velocity $\langle u \rangle$ and travels over a distance σ. In analogy to the case of gases, where $2D = \lambda \langle u \rangle$, the diffusion coefficient in a liquid should be given

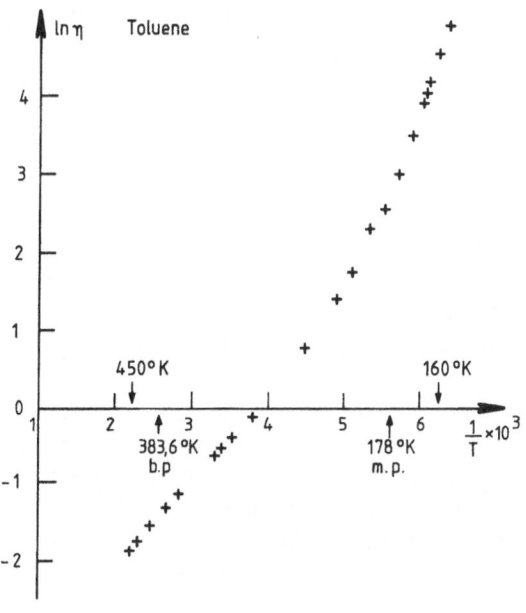

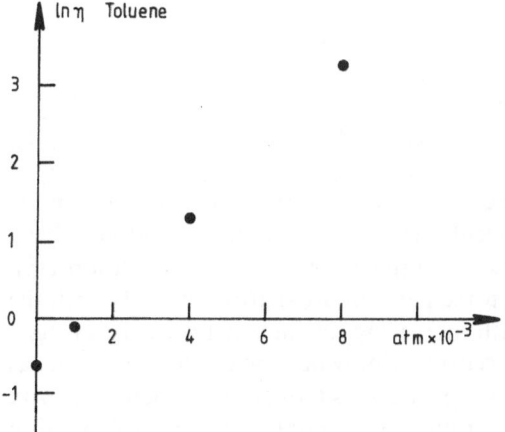

Fig. 2. a) Measured viscosities of toluene at different temperatures (Barlow AJ, Lamb J, Matheson AJ (1966) Proc Roy Soc A292: 322). **b)** Measured viscosities of toluene at different pressures (Bridgman PW (1925) The physics of high pressure, Bell, London)

by an expression of the form

$$2D = \sigma \langle u \rangle \exp \left\{ -\frac{E_P}{kT} \right\}$$

In this simple derivation it has been neglected that consecutive exchanges of place are not always independent of each other. In associated liquids an additional activation energy may be required to overcome specific chemical interactions keeping the molecules together.

It follows from these still primitive considerations that the diffusion coefficient, and also the viscosity, of liquid systems should show an exponential dependence upon pressure and inverse temperature. This is confirmed by experiments, as shown in Fig. 2.

5 Viscosity

The viscosity of liquid media is much larger than that of gases, where the effect of increasing density on viscosity during compression is compensated by the decreasing mean free path. The viscosity is low enough, however, for a liquid to be able to adapt its shape without much expense of energy when subjected to outer forces. This can only be explained by a relatively large amount of free space, allowing sufficient mobility to the individual particles. By compression or cooling this amount of free space decreases, viscosity increases until a glassy or crystalline state is formed. Some of the free space remains in the crystal under the form of lattice vacancies. The excess free space V_f available to a molecule in the liquid is given by the difference between the molar volume in the liquid (M/ϱ) and the solid (V_s) state:

$$V_f = \frac{\dfrac{M}{\varrho} - V_s}{N_A}$$

Transport of momentum is not effected in the liquid by molecules carrying along excess momentum over long distances large compared to the average molecular distances, as it is in the gas phase. In the condensed state one must consider neighbouring layers of molecules with a cohesion force between them. During relative motion the layers exert a tangential shearing force upon each other. The shearing stress (shearing force per unit area) grows linearly with the velocity difference and decreases with increasing distance between the layers.

The proportionality factor is viscosity. Equivalent with the definition as measure for the rate of momentum transport across a velocity gradient, viscosity also represents the ratio between a deforming force and the resulting deformation velocity. A rate of change of momentum corresponds to a force (Newton's first law). The deformation velocity corresponds to a difference in speed of motion of neighbouring layers of an object. It is possible only in a fluid medium to maintain a continuous deformation; this process corresponds to fluid flow, in which parts of the fluid move relatively to each other. Their velocity differences are small compared to the thermal speed of the molecules. The behaviour of a fluid is called *newtonian* when its viscosity does not depend on the shearing stress. The shearing stress between neighbouring molecular layers is then given by

$$f = \eta \frac{\delta v}{\sigma}$$

where f represents the shearing stress, δv is the velocity difference and the distance between the layers is taken to be the thickness of a contact layer of molecular diameter σ.

In order to maintain a velocity gradient, molecules travelling in opposite directions in the contact plane must exchange places. Those belonging to the faster side of the layer move more often in the direction of macroscopic motion whereas those of the slower side move more often in the opposite direction. We now must distinguish between the two directions in the calculation of the frequency factor p/p_0. Since the exchange of place takes place in a volume of 2σ, and f is the force per unit area, the total shearing force on the cross-section of the volume containing the two exchanging particles is $\pi\sigma^2 f$. On each particle a force of magnitude $\pi\sigma^2 f/2$ is acting. When a particle travels a distance σ in the direction of the force, the work done is $A = \pi\sigma^3 f/2$. For the other direction the work done is $- A$. The work done by the shearing force must be included in the Boltzmann expressions:

$$p^+ = p_0 \exp\left\{ -\frac{E_P - A}{kT} \right\}$$

$$p^- = p_0 \exp\left\{ -\frac{E_P + A}{kT} \right\}$$

or, since $A \ll kT$, with $p = p_0 \exp\{-(E_P/kT)\}$

$$p^+ = p\left(1 + \frac{A}{kT} \right)$$

$$p^- = p\left(1 - \frac{A}{kT} \right)$$

Now, a particle moves on average with velocity $\langle u \rangle$ during an exchange. The velocity difference δv corresponding to the shearing f is then:

$$\delta v = \frac{p^+ - p^-}{p_0} \langle u \rangle = 2 \langle u \rangle \frac{A}{kT} \exp \left\{ -\frac{E_P}{kT} \right\}$$

With $\eta = f\sigma/\delta v$ and $A = \pi\sigma^3 f/2$ we obtain the following expression for the viscosity:

$$\eta = \frac{kT}{\langle u \rangle \pi\sigma^2} \exp \left\{ \frac{E_P}{kT} \right\}$$

Comparing this with the expression for D, one finds

$$D = \frac{kT}{2\pi\eta\sigma}$$

A relation of this kind exists not only for the coefficient of self-diffusion, but also for the diffusion coefficient of spherical particles with arbitrary diameter in a fluid medium with viscosity η. This is known as the Stokes–Einstein-equation. According to Stokes' equation in hydrodynamics, the force required to move a sphere with radius r with velocity v through a viscous fluid is given by

$$F = 6\pi\eta rv$$

The ratio $F/v = R_{tr}$ defines the mechanical translational friction coefficient, its inverse is the mechanical mobility μ. For spherical particles $R_{tr} = 6\pi\eta r$. Einstein showed that the ratio kT/D must also correspond to the friction coefficient. This leads to the often encountered relation for the diffusion coefficient D_i of dissolved spherical particles with radius r_i:

$$D_i = \frac{kT}{6\pi\eta r_i}$$

Small deviations from this equation are found when the size of the dissolved particles becomes comparable to the size of the solvent molecules. Most real molecules also do not have a spherical shape. All derivations given above are, of course, quite approximate because of this fact, and because distributions of velocities, mean free path etc. have not been considered exactly. They do describe, nevertheless, the experimentally observable behaviour of simple systems rather well.

Values of D, η and other quantities depending on them, measured experimentally as functions of pressure, temperature, chemical species and composition of the medium, allow us to obtain values for the molecular parameters, such as particle mass, size, interaction potentials etc. by fitting these using the theoretically derived models. One needs these fundamental kinetic models of interaction in order to get insight in microscopic events, starting from the macroscopically observable time course of a phenomenon, e.g. diffusion.

The quantity E_P, appearing in the equations for D and η as an activation energy, represents the energy requirement for the formation of a particular configuration making possible an exchange of position. Apart from the estimated term for volume work, it may include other contributions. Mainly in this quantity the influence of molecular shape and of specific interactions with neighbouring molecules will be observable as additional contributions. The preexponential factors, describing processes in the place-exchange configuration, have been treated in a simplified manner for spherical particles. They should be valid also for particles with approximately spherical shape, but will not apply in the case of polymers or very anisotropic molecules.

6 Rotational Diffusion

Most molecules are anisotropic. They can have distinct orientations in space. Under the influence of collisions a given orientation will persist only for a short time. In liquids, rotational states of different energy cannot exist for a sufficiently long time to define sharp energy levels. The rotational structure of IR-vibration spectra, characteristic for gases, disappears for this reason in liquids and liquid solutions.

Different orientations can be distinguished by specific anisotropic features, e.g. electric moment of the molecular charge distribution, transition moment of chromophores etc. This anisotropy also influences the interaction with other molecules, which becomes dependent on mutual orientation.

A molecule with a moment of inertia respective to an axis possesses a thermal energy $kT/2$ in this degree of rotational freedom. Since free rotation is not possible, the molecule executes torsional vibrations around this axis. These are called librations. The mean orientation then changes with the position of the neighbours hindering the free rotation. The presence of torsional motion is noticeable by a contribution to the internal energy and to the entropy. Torsional vibration levels of increasing energy are excited with increasing temperature, contributing to the molar heat capacity $C_v = (\partial U/\partial T)_v$. Not only the kinetic energy of the torsional motion ($kT/2$ per principal axis of rotation), but also the increasing potential energy of the constraint resistance must be considered. In liquids, therefore, the rotational degrees of freedom of a molecule usually contribute more than $R/2$ per degree of freedom to the molar heat capacity. The spatial positions of neighbouring molecules change, due to the fluctuation of the free volume. A particle can thereby change the orientation in which it executes torsional vibrations. The velocity of this "rotation", or change of orientation direction, is less determined by the moment of inertia than by the inner friction, or the viscosity, of the surrounding medium.

For a macroscopic sphere, with radius r, in a medium of viscosity η, the frictional resistance coefficient to rotation is the ratio of applied torque M to the angular velocity $\omega = d\psi/dt$ around an axis through the center. According to the Stokes formula one has

$$R_{rot} = M/\omega = 8\pi\eta r^3$$

R_{rot} represents the energy dissipation per radian of revolution per unit time. Its inverse $\mu_{rot} = \omega/M$ is the rotational mobility.

A general relation, derived by Einstein, states that the average squared fluctuation $\langle \Delta\alpha^2 \rangle$ of an observable parameter α in a time Δt is proportional to the mobility of a system with respect to this parameter. The proportionality factor is 2kT. We then have

$$\frac{\langle \Delta\psi^2 \rangle}{\Delta t} = 2kT\mu_{rot} = \frac{2kT}{R_{rot}} = 2D_{rot}$$

Here, D_{rot} is the rotational diffusion coefficient, introduced in analogy with the case of translational motion.

Rotations correspond to diffusive displacements of a representative point in a configuration space of orientations. Every possible distinct orientation is defined by the coordinates of a point in this space. Neighbouring points correspond to neighbouring orientations.

As an example, the points in the surface of a sphere with unit radius form a suitable configuration space for the possible orientations of a pointer, drawn from the sphere's center. The location on the sphere's surface, hit by the pointer, is determined by two independent coordinates, e.g. its colatitude ($0 \leq \theta \leq \pi$) and its azimuth ($0 \leq \varphi \leq 2\pi$). θ and φ are defined with respect to the euclidean three-dimensional space in which the orientations take place and in which the curved, two-dimensional, configuration space is embedded. θ is the angle between the z-axis and the pointer, φ is the angle between the x-axis and the projection of the pointer on the xy-plane.

This configuration space is sufficient for dealing with the orientations of simple uniaxional polar molecules, as long as they are of spherical shape, or truly uniaxial, like CO, HCl and other diatomics.

The number of particles with a given orientation can be represented by a density function $n(\theta, \varphi)$ on the spherical surface. Changes of orientation by thermal motion are described by a diffusion equation in configuration space:

$$\phi = -D_{rot}\,\text{grad}\,n; \quad \frac{\partial n}{\partial t} = -\text{div}\,\phi = \text{div}(D_{rot}\,\text{grad}\,n)$$

For the general case of a 2- and 3-dimensional rigid object, the orientational configuration space requires three linearly independent coordinates, because the object has also a degree of freedom of rotation around the direction indicated by θ and φ. Moreover, the friction coefficients for the rotations around the principal axes are different if these axes have a different length. The rotational diffusion coefficient is then a tensorial quantity.

The orientational configuration space is a mathematical tool for deriving the influence of external forces on the distribution of orientations. Macroscopically measurable quantities (electrical polarization, optical properties) can be calculated by a convolution of the relevant molecular property with the density function over orientational space.

The average "lifetime" of a state of orientation is an important quantity for the study of molecular interactions. Specific interactions usually can occur only at certain mutual orientations. The distribution of relative molecular orientations contributes to the entropy of the system. This must be taken into consideration in accounting for the free energy of interaction.

In simple cases the molecule considered will have a distinguished uniaxial anisotropy and an approximate rotational symmetry around this axis. The distribution $n(\theta, \varphi)$ then reduces to a function of a single angular variable, namely the angle between the axis of anisotropy and an externally determined reference direction. This direction could be that of an external static or periodic electric field, or the direction of polarization of a beam of light, etc. For the description of dynamic phenomena, time will be a second explicit variable. The number of particles oriented within a solid angle $d\Omega = d\theta \sin\theta \, d\varphi$, directed at (θ, φ), is $n(\theta, t) \sin\theta \, d\theta \, d\varphi$.

A separation of the variables t and θ, however, is possible. An arbitrary orientational distribution $n(\theta, t)$ can always be represented by a linear superposition of Legendre polynomials:

$$n(\theta, t) = \sum_k a_k(t) P_k(\cos\theta)$$

Legendre polynomials P_k form a series of orthogonal functions ($u = \cos\theta$):

$$P_0(u) = 1; \quad P_1(u) = u; \quad P_2(u) = \tfrac{1}{2}(3u^2 - 1); \quad P_3 = \tfrac{1}{2}(5u^3 - 3u) \cdots$$

Some of these functions correspond to particular cases. P_0 represents the (normalized) random distribution, where every orientation has equal probability. P_1 corresponds to the distribution impressed by a (weak) static electric field upon molecules with a permanent dipole moment. P_2 is the distribution resulting from an induced, instead of a permanent moment. An induced moment produces an orienting force only when the induced moment does not have the same direction as the field. This is the case for spherical particles with an internal anisotropy of polarizability. Ellipsoidal or rod-like particles with an internally isotropic dielectric constant ε that is different from that of the surrounding medium, are also oriented according to P_2 by an electric field. In strong electric fields the higher order terms ($k > 1$) cannot be neglected. A weak field is one in which the maximal potential energy of a molecule in an orienting force field is still small compared to kT.

The time-dependent coefficients $a_k(t)$ reflect the temporal change and the nature of the orienting force. They depend upon the dynamic properties of the particles (rotational diffusion coefficient, friction factor) and upon the factors (field strength, dipole moment, polarizability) determining the orienting force field. After switching off an electric field, in which a distribution was established, the coefficients $a_k(t)$ take the following form:

$$a_k(t) = c_k \exp\{-k(k+1)D_{rot}t\} = c_k \exp\{-t/\tau_k\}$$

The time constants τ_k are the *relaxation times*. Each polynomial term in the expansion of $n(\theta, t)$ has a different relaxation time. The experimentally observed relaxation time for the return to the equilibrium distribution depends on the contribution of each P_i to the convolution with the molecular property relevant for the measurement. The dielectric relaxation time for particles with a permanent dipole moment is $\tau_1 = 1/(2D_{rot})$. The relaxation time measured by the decay of optical birefringence is $\tau_2 = 1/(6D_{rot})$.

Particles with ellipsoidal shape, oriented in an electric field, still behave like spherical particles, as long as the direction of the dipole moment, or the anisotropy of polarizability, coincides with one of the principal axes of the ellipsoid. The orientational distribution of ellipsoidal particles is not characterized by a single angular variable, but in this particular case the directions of the two other principal axes remain random. The rotational diffusion coefficient appearing in the equation for the relaxation time, is then the average of the rotational diffusion coefficients around these two axes:

$$2D_{rot} = D_1 + D_2 = \frac{kT}{R_1} + \frac{kT}{R_2} = \frac{R_1 R_2 kT}{R_1 + R_2}$$

7 Diffusion-Controlled Processes

Molecular interactions are called *specific*, when the energy of molecular orbitals in one or in both interacting partners is substantially lowered. The lowering of energy is the cause of the stability of the association complex. This implies that the molecular partners are in spatial contact with each other. When the interacting partners are dissolved in a not too high concentration in a solvent, they first must come together by diffusion. Not all chemical interactions are so fast, that they occur at the first encounter of the reacting molecules. But if they do, the rate at which the association complex can be formed, is limited by the diffusion process. This includes rotational diffusion if there are strict sterical constraints on the interaction. The association rate is then dependent on the viscosity of the solvent. When this is the case, then the dissociation rate must also be limited by diffusion, since the thermodynamic stability (lowering of free energy) and the kinetic stability (ratio of association to dissociation rate constant) must be identical. We can always consider the reaction as two consecutive processes:

$$A + B \underset{k_{-E}}{\overset{k_E}{\rightleftharpoons}} (A, B) \underset{k_{-I}}{\overset{k_I}{\rightleftharpoons}} AB$$

$$\text{Encounter} \qquad \text{Interaction}$$

(A, B) symbolizes the encounter complex, in which A and B are in contact, but the change in binding energy has not yet taken place. AB represents the association complex. If $k_I \gg k_{-E}$, AB will be formed before A and B can diffuse apart. Each encounter then leads to the formation of AB. The reaction is diffusion-controlled. k_E, k_{-E}, k_I, k_{-I} are chemical rate constants in the kinetic equations:

$$\frac{dc_A}{dt} = \frac{dc_B}{dt} = -k_E c_A c_B + k_{-E} c_{(A, B)}$$

$$\frac{dc_{AB}}{dt} = k_I c_{(A, B)} - k_{-I} c_{AB}$$

In usual chemical kinetics, chemical reactions are considered to be scalar processes. In this sense $c_{(A,B)}$ is the concentration of a quasi-substance. Together with the equations for mass conservation $c_{(A,B)} + c_{AB} = c_A^\circ - c_A = c_B^\circ - c_B$, the kinetic equations can be solved for the unknown, time-dependent quantities A, B, (A, B) and AB. c_A° and c_B° are the total numbers of particles A resp. B, divided by total volume, which is supposed to be constant. For $t \to \infty$ the $c_i(t)$ become the equilibrium concentration $\bar{c}_i, d\bar{c}_i/dt = 0$. The kinetic condition $d\bar{c}_i/dt = 0$ defines the equilibrium state, equivalent to the thermodynamic one corresponding to the minimum of free energy. In order to obtain an assertion about the duration of time for the transitions between different states, we consider small fluctuations around the equilibrium, putting $c_i = \bar{c}_i + \delta c_i$. When terms quadratic in δc_i are neglected, the non-linear differential equation system can be linearized. With $\delta c_A = \delta c_B = x_1$ and $\delta c_{AB} = x_2, \delta c_{(A,B)} = -(x_1 + x_2)$ one obtains:

$$\frac{dx_1}{dt} = -(k_E(\bar{c}_A + \bar{c}_B) + k_{-E})x_1 - k_{-E}x_2$$

$$\frac{dx_2}{dt} = -k_I x_1 - (k_I + k_{-I})x_2$$

The general solution for this system of coupled equations can be written:

$$x_1 = g_{11} \exp\{-\lambda_1 t\} + g_{12} \exp\{-\lambda_2 t\}$$
$$x_2 = g_{21} \exp\{-\lambda_1 t\} + g_{22} \exp\{-\lambda_2 t\}$$

λ_1 and λ_2 are the characteristic relaxation rates (inverse relaxation times $\lambda_i = 1/\tau_i$) and g_{ji} the relaxation amplitudes. These quantities can be determined experimentally by analytically observing, with the required time-resolution, the time course of reestablishing equilibrium after an external perturbation (e.g. a temperature change that is much faster than τ_1 and τ_2). The λ_i have the dimension t^{-1}. They are functions of the rate constants k_i. The following general relations hold:

$$\lambda_1 = \frac{S + \sqrt{S^2 - 4P}}{2}$$

$$\lambda_2 = \frac{S - \sqrt{S^2 - 4P}}{2}$$

with

$$S = k_E(\bar{c}_A + \bar{c}_B) + k_{-E} + k_1 + k_{-1}$$

$$P = k_E(\bar{c}_A + \bar{c}_B)(k_1 + k_{-1}) + k_{-E}k_{-1}$$

The two relaxation times are not to be assigned to the diffusion step and the reaction step. They belong to two dynamically different *normal modes* of the coupled reactions. Normal modes of reactions are distinguished by a difference in the symmetry of reaction flow, just as in the case of the normal vibration modes of coupled resonators. The faster normal mode with relaxation rate λ_1 corresponds to a fast change of the intermediate product $c_{(A,B)}$, compensated by simultaneous changes in c_A, c_B and c_{AB}. The slower normal mode represents the (positive or negative) reaction flow from A and B to AB via (A, B), the concentration of which can thereby change as well. The faster reaction mode is always the one in which a perturbation tries to equalize over the accessible parallel pathways, while the slower modes encompass consecutive pathways for adjustment.

Although we will see that the concentration of encounter complexes can be substantial, its experimental determination is quite difficult. This is understandable, since we have presumed that energetic changes have not yet occurred. There is then no relevant molecular property that distinguishes an isolated molecule from one in an encounter complex. But since two relaxation processes can be observed if the (A, B)-complex is present in amounts comparable to the other reactants, it is possible to deduce information about all rate constants from the measured relaxation times. When a reaction becomes diffusion controlled, the equilibrium concentration of encounter complexes becomes very small, since they are converted very rapidly in the association complex. The amplitude of the fast relaxation process becomes too small to be measurable. The remaining relaxation rate corresponds to a simple association—dissociation:

$$A + B \underset{k_D}{\overset{k_A}{\rightleftharpoons}} AB$$

$$\lambda = k_A(\bar{c}_A + \bar{c}_B) + k_D$$

where

$$k_A = \frac{k_E k_I}{k_{-E} + k_I}; \quad k_D = \frac{k_{-E} k_{-I}}{k_{-E} + k_I}$$

With $k_I \gg k_{-E}$ this simplifies to:

$$k_A \simeq k_E; \quad k_D = k_{-E}\frac{k_I}{k_{-I}} = k_{-E}K_I$$

When the transition from the encounter complex to the interaction complex is a rare event compared to the two partners diffusing apart, then k_I and k_{-I} contribute almost nothing to the fast relaxation process. The amplitude g_{21} then approaches zero. Practically, we have an association with a fast pre-equilibrium. One now has $k_I \ll k_{-E}$, leading to

$$k_A = k_I\frac{k_E}{k_{-E}}; \quad k_D = k_{-I}$$

One will often meet this type of reaction when the process of interaction requires substantial changes in molecular structure, needing activation energies that are seldom available to the reaction partners in the encounter complex. The rate constants k_E and k_{-E} can be derived from the diffusion equation under appropriate boundary conditions. The following simple model may be used.

We designate by $n_i(r_{ij})$ the density of particles i, distant from a particle j by r_{ij}. The distance of the particles i and j in the encounter complex is $a_{ij} = a_{ji}$. The concentration $n_i(a_{ij})$ is zero, because here i changes into the chemical quasi-substance (i, j). On the other hand, $N_i(r_{ij} \to \infty) = \bar{n}_i$ is the concentration of i in the system. A diffusion current ϕ arises, due to the concentration gradient:

$$\phi_i = - D_{ij} \, \text{grad} \, n_i$$

The relative diffusion coefficient D_{ij} is equal to $D_i + D_j$. r_{ij} is expressed in polar coordinates:

$$\phi_i(r, \theta, \varphi) = - D_{ij}\left(\frac{\partial n_i}{\partial r} + \frac{1}{r}\frac{\partial n_i}{\partial \varphi} + \frac{1}{r \sin \theta}\frac{\partial n_i}{\partial \theta}\right)$$

Since n_i obviously depends only on r and not on θ and φ, the total flux through a spherical surface with radius r is

$$\Phi_i(r) = \int_0^\pi \int_0^{2\pi} \phi_i(r, \theta, \varphi)r^2 \sin \theta \, d\theta \, d\varphi = - 4\pi r^2 D_{ij}\frac{dn_i}{dr}$$

Hereby we obtain an expression for the concentration gradient required to maintain the diffusion current Φ:

$$\frac{dn_i}{dr} = -\frac{\Phi_i(r)}{4\pi D_{ij}r^2}$$

In the stationary case the total flux does not depend on r, otherwise local enrichments or reductions in concentration would be the consequence. Φ_i must, in the stationary case, correspond to the encounter frequency of particles i with particles j. For each particle j the number of particles i that disappear in unit time is $k_E \bar{n}_i/N_A$. With $\Phi_i(r) = \text{const} = -k_E\bar{n}_i/N_A$ one obtains a differential equation that can be integrated immediately from a_{ij} to ∞:

$$\int\limits_{a_{ij}}^{\infty} dn_i = \bar{n}_i = -\frac{\Phi_i}{4\pi D_{ij}}\int\limits_{a_{ij}}^{\infty}\frac{dr}{r^2} = \frac{k_E\bar{n}_i}{4\pi N_A D_{ij}a_{ij}}$$

In this way one obtains the often used expression for the rate constant of encounter of neutral particles

$$k_E = 4\pi N_A(D_i + D_j)a_{ij}$$

When k_E is to be expressed in the usual units (lit $mol^{-1} s^{-1}$), the right side of this equation must be multiplied by 1000, since D is usually given in $cm^2 s^{-1}$ and a_{ij} in cm.

In this derivation it has been neglected that mutual attraction or repulsion forces can already be active during the diffusion process. This is the case with electrically charged ions. The right side of the equation must then be corrected by the factor

$$\frac{f_{ij}}{\exp\{f_{ij}\} - 1}; \quad f_{ij} = \frac{z_i z_j e_0^2}{4\pi \varepsilon \varepsilon_0 kT}$$

z_i and z_j are the (signed) ionic valencies. In media of low dielectric permittivity the long-range electrostatic interaction can easily increase or decrease the rate constant of encounter by an order of magnitude when the reaction distance is smaller than 1 nm. In thin liquids (H_2O, organic solvents) the rate constant k_E attains values of 10^{10} lit $mol^{-1} s^{-1}$ and more.

It is also possible to derive k_{-E}, using the diffusion equation with different boundary conditions. It is, however, simpler to estimate the equilibrium constant of the formation of encounter complexes $K_E = k_E/k_{-E}$. For neutral particles the density of encounter pairs $n_{(i,j)}$ is given by the density of particles j, multiplied by the probability that there is a particle i inside a spherical shell of thickness a_{ij} around the center of j. The total volume of these shells is $V_j = N_A n_j 4\pi a_{ij}^3/3$.

The fraction of particles i inside this volume is $n_i V_j$, yielding

$$K_E = \frac{n_{(i,j)}}{n_i n_j} = \frac{4\pi N_A}{3} a_{ij}^3$$

The right-hand side of this equation is to be corrected by the factor $\exp\{-f_{ij}\}$ for electrically charged particles. Even for small molecules with $a_{ij} \simeq 0.5$ nm K_E attains a value of 1 lit mol^{-1}.

8 Reaction Mechanisms in the Encounter Complex

Diffusion controlled reactions often appear in acid-base equilibria with participation of H^+ and OH^- ions, especially in aqueous media. In this and other H-bond associated media H^+ and OH^- are often characterized by a special mechanism of diffusion transport that should not be generalized. In neutralization reactions the diffusing entity is not an individual identifiable proton, but a solvating "structure", transporting a solvated proton charge. This charge does not remain fixed to a specified hydrogen nucleus or water molecule, but is confined to a local structure of constantly exchanged solvent molecules. The reaction distance a_{ij} for the solvated charge (excess-proton in case of H^+, defect proton in case of OH^-) and its also solvated reaction partner is reached when both solvation structures come into contact. The charge inside the solvation structure can be transferred rapidly by proton shifts in the hydrogen bonds. One finds this special mechanism of proton mobility in H_2O and some other solvents also for proton exchange reactions between proton donor and proton acceptor molecules. Both are initially associated with the solvent, which can act as donor as well as acceptor.

Solvent-assisted transport during diffusion and reaction is only possible if the solvent molecule has a proton donor and a proton acceptor site located on the same atom. Then a proton can be taken up by the acceptor site while simultaneously delivering a proton from the donor site. Place-exchange processes are not required. The transport of a proton by this mechanism, however, causes an effective rotation of the assisting solvent molecule. The original acceptor and donor sites have been substituted for each other. A back-rotation of the molecule is required before a second proton can be transported. In solids with an extended H-bond lattice the rate of reorientation can limit the mobility of protons.

The conditions for the participation of solvent molecules in the transport of protons are not fulfilled in all solvents. If the solvent molecule is strongly solvating (having, for instance, an acceptor site, without being a donor) the formation of a hydrogen bond between a solvated molecule A_{solv} and a partner B can only occur after breaking the bond between A and the solvent molecule. If this does not take place within the encounter time of about 10^{-10} s, the reaction between A and B is no longer diffusion controlled. The energy of the H-bond with the solvent molecule appears as activation energy in k_l. This is even the case when the interaction between A and B is much stronger than that between A and the solvent. As an example, the free proton in N-methylformamide is strongly solvated. It is enclosed in a homonuclear hydrogen bond between two solvent molecules:

$$CH_3NHCHO—H^+—OCHNHCH_3$$

About hundred encounters are required before the proton is transferred to a nitrophenolate-anion, although the free energy of the nitrophenol formed is $36\,kJ\,mol^{-1}$ lower than that of the encounter complex. The free energy of the hydrogen bond in the solvation structure must therefore be at least $-11.5\,kJ\,mol^{-1}$.

In solvents that do not enter into specific interactions with the reacting molecules, the rate constants for intermolecular hydrogen bond formation can be diffusion controlled. Sometimes, however, the presence of intramolecular hydrogen bonds in one of the reaction partners can prevent this, for instance, as in acetylacetone-enol:

$$CH_3—C{=}C—C—CH_3$$
$$\begin{array}{ccc} | & & \| \\ OH & \cdots & O \end{array}$$

The intramolecular bond must be broken before the molecule can bind to an acceptor molecule.

9 Chain Formation and Meshed Association

The possibility of a specific (chemical) interaction is not limited to two reaction partners. Many molecular species offer binding sites at different places, e.g. for the formation of hydrogen bonds. Bivalent molecules with donor as well as acceptor sites can form chain-like associations. Sometimes these can build closed rings. Molecules with appropriate geometry can form multiple bonds between two partners. This is the case with carboxylic acids, lactams, pyridones and other heterocyclic compounds. The particles cannot diffuse apart before both hydrogen bonds are broken simultaneously. The stability of multiply bonded association complex can become very high.

Chain-like associations require a consecutive mechanism

$$m_1 + m_1 \rightleftharpoons m_2; \quad K_{2,1}$$
$$m_1 + m_2 \rightleftharpoons m_3; \quad K_{3,1}$$
$$\vdots$$
$$m_1 + m_i \underset{z_{i+1,1}}{\overset{w_{i,1}}{\rightleftharpoons}} m_{i+1}; \quad K_{i+1,1}$$

The reactions

$$m_p + m_q \underset{z_{p+q,p}}{\overset{w_{q,p}}{\rightleftharpoons}} m_{p+q}; \quad K_{p+q,p}$$

are often neglected. These reactions have no influence on the equilibrium distribution of the chain length. They do, however, strongly influence the kinetics of the reaction.

The general reaction scheme has many parameters. In a theoretical interpretation of experimentally observable kinetic processes it is important to use models with the smallest possible number of parameters.

The simplest model does not consider the multitude of possible chain associates. It uses only a single equilibrium between two states: binding sites are either free; or they are occupied (two-state model):

$$\text{donor} + \text{acceptor} \overset{w}{\underset{z}{\rightleftharpoons}} \text{bond}; \quad K = \frac{w}{z}$$

This model is equivalent with the consecutive mechanism, if one assumes the following:

1. all $K_{n,m}$ are equal, independent of n und m;
2. all $w_{n,m}$ have the same value, w, for $n \neq m$. For $n = m$ one has $w_{n,m} = w/2$;
3. the same applies to $z_{n,m} = z$; $z_{n,n} = z/2$.
4. The number of donors is equal to the number of acceptors.

This strong simplification is often satisfactory for considering measurable quantities that only depend on the number of bonds formed, but that do not distinguish between chains of different length, e.g. the intensity of the long-wave-shifted IR-absorption of the OH-bond in hydrogen bonding, or the heat liberated during bonding.

The rate at which a chain-length distribution adapts itself to external conditions (e.g. temperature or pressure change) can be characterized by a single or by several relaxation times. The two-state model is characterized by a single relaxation time. A consecutive model to which the above assumptions apply has been called a random isodesmic model. It also shows a single relaxation time. The sequential isodesmic model, in which $w_{p,q} = z_{p,q} = 0$ for $q \neq 1$, shows a spectrum of relaxation times. In fact, the random isodesmic model is also characterized by many relaxation times, relating to the redistribution of the aggregate sizes. The amplitudes, however, of the size-redistribution relaxations are not observable by a method that only depends on the number of bonds formed.

The relaxation times for the bond formation reaction can be interpreted as the lifetime of the specific interaction involved in the bond. In the random isodesmic model the lifetime of the bond is independent of its position in the chain. In the sequential model, a bond in the middle of a long chain will have a longer lifetime than one near the ends or one in a short chain. For non- or weakly-cooperative interactions (where interaction energies are independent of the presence of neighbouring interactions) the random model is more realistic. When interactions become highly cooperative or interdependent, a sequential model may sometimes be more appropriate. The decision must be made on the basis of experimental evidence.

In many cases the relaxation times of fast specific interactions can be studied by ultrasonic techniques. In an ultrasonic wave the volume elements of the system are subjected to a very rapid succession of compression and dilatations. The compression is practically adiabatic. The phase of pressure increase is then accompanied by a temperature increase and the phase of dilatation by a temperature decrease. Viscosity and heat conductivity lead to absorption of energy from the sound field, since part of the energy transferred to the medium during compression is not transferred back reversibly during decompression.

The sound wave interacts also chemically with the system when the pressure or temperature increase influences the chemical association equilibrium. During the temperature rise a part of the associates is broken up, they must form again during the temperature decrease. If the sound frequency is comparable to the relaxation rate, the chemical energy taken up, or given off, also is not exchanged reversibly with the second wave, leading to an additional absorption. One finds these additional absorptions in solutions of alcohols, amides, carboxylic acids etc. in inert solvents, usually in the frequency range between 10 and 1000 MHz.

The concentration dependence of the frequency and amount of the excess ultrasonic absorption compared to the pure solvent, in combination with other (e.g. caloric, optical, dielectric) measurements delivering information on the amount of association, allows detailed interpretations in terms of one or another model of the reaction mechanism. The two-parameter mechanism (two-state approximation) is not always able to account for all observations, even if it reproduces rather well the kinetic behaviour. One is then forced to introduce a more complicated model.

As a first step one is often led to the conclusion that the equilibrium $K_{2,1}$ (dimer formation) is different from the following equilibria. If $K_{2,1} > K_{n,m}$, then the interaction in which a molecule is already involved lessens the formation of a further interaction. If $K_{2,1} < K_{n,m}$, then the first interaction "catalyzes" the formation of a larger aggregate. This is a kind of nucleation.

We are meeting here the first appearance of cooperative effects. Strong effects of this kind can happen with multiple hydrogen bonds within the same molecule (e.g. α-helix formation and folding of protein molecules) or with multiple interactions between two molecules (e.g. formation of the DNA double helix). The state of a segment of a chain is influenced by the state of the other segments, in the simplest case the neighbouring segments. $K_{2,1}$ can be viewed as a nucleation constant and $K_{n,m}$ as a growth constant.

Many molecules have several donor and acceptor sites. It is then possible to build a meshed network of associations. H_2O is the most prominent example. This molecule consists only of two donor and two acceptor sites on a single central atom. In the solid and in the liquid state the association structure is strongly meshed. The ability of the liquid to orient its dipoles in an electric field is, nevertheless, very high. The dielectric relaxation frequency is 10^{10} Hz in water. In ice, on the other hand, it is about 10^4 Hz.

The strong association of the water molecules with each other is responsible for the crowding together of other dissolved molecules that do not assimilate into the association structure. The crowded molecules show an apparent tendency to interact with each other. They appear as dimers, multimers, micelles, vesicles and other aggregation forms. This apparent interaction, however, must be traced back to the strong specific interactions between the solvent molecules. It is their interaction, finally, that is responsible for the so-called hydrophobic or solvophobic interactions of the dissolved molecules.

10 Lifetime of Molecular States

The average lifetime of a molecular state is the inverse of the sum of all velocities leading to states that are not equivalent to the state under consideration:

$$\bar{t} = \left(\sum_i k_i \right)^{-1}$$

The distinction between transitions that do not change the state in an *essential* way, and those that terminate its existence is often a matter of definition. Let us take, as an example, the dissociation reaction of water.

$$H_2O \underset{k_R}{\overset{k_D}{\rightleftharpoons}} H^+ + OH^-$$

The rate constants are

$$k_R = 1.4 \times 10^{11} \, \text{lit mol}^{-1} \text{s}^{-1}$$
$$k_D = 2.5 \times 10^{-5} \, \text{s}^{-1}$$

With $\bar{t} = k_D^{-1}$ the average lifetime of a water molecule in water is more than 40000 seconds or 11 hours, before it dissociates in an H^+ and an OH^- ion. This, however, is not the time during which an oxygen atom remains bound to the same hydrogen atoms. The hydrogen atoms are exchanged between the water molecules by the reactions

$$H_3O^+ + H_2O \overset{k_{A_1}}{\rightleftharpoons} H_2O + H_3O^+$$

$$OH^- + H_2O \overset{k_{A_2}}{\rightleftharpoons} H_2O + OH^-$$

The rate constants are the same in both directions:

$$k_{A_1} = 1 \times 10^{10} \, \text{lit mol}^{-1} \text{s}^{-1}$$
$$k_{A_2} = 5 \times 10^9 \, \text{lit mol}^{-1} \text{s}^{-1}$$

In the purest water $(c_{H_3O^+} = c_{OH^-} = 10^{-7}\,mol\,lit^{-1})$, a water molecule exchanges one of its hydrogen atoms already after 0.7 milliseconds:

$$\bar{t} = (k_{A_1}c_{H_3O^+} + k_{A_2}c_{OH^-})^{-1} = 6.66 \times 10^{-3}\,s$$

The lifetime of an H_3O^+ resp. OH^--ion is only

$$\bar{t} = (k_R c_{OH^-})^{-1} = (k_R c_{H_3O^+})^{-1} = 7.1 \times 10^{-5}\,s$$

This, again, is not the time during which an excess or defect proton charge remains associated with a particular water molecule, but the time between creation of a solvated excess change by H_2O dissociation and its disappearance by recombination. The water molecules participating in the solvation structure are constantly exchanged. The excess charge fluctuates between the participating molecules by shifting the hydrogen atoms in the hydrogen bonds. The solvation structure moves when new hydrogen bonds at its periphery are made and others are broken.

The lifetime of a hydrogen bond between two neighbouring H_2O-molecules in liquid water is of the order of 10^{-11} seconds. This is estimated from the relaxation frequency of dielectric reorientation. The duration of the state in which a water molecule is not hydrogen bonded must be even shorter. The short duration of an individual hydrogen bond in liquid water is an indication that the donor–acceptor interaction can be transferred with low activation energy when the breaking of one bond is simultaneously accompanied by the formation of another one on the same atom. An intramolecular H-bond with the same thermodynamic stability is kinetically much less labile. The relative orientation and the distance between donor and acceptor in an intramolecular interaction is less variable than in one between individual particles with their own energy of translation and libration.

The average lifetime of a state of interaction is rarely accessible to direct experimental observation, especially when the lifetime is short. Lifetimes are determined by the dynamics of several processes and influence these in turn. Many dynamic processes are coupled. Experimentally observable time constants (relaxation times) then do not reflect the lifetimes of individual states, but they must be assigned to the normal modes of the dynamic system.

11 References

1. Hirschfelder JO, Curtiss CF, Byron R (1954) Molecular theory of gases and liquids, Wiley, New York
2. Einstein A (1906) Zur Theorie der Brownschen Bewegung, Ann Physik, 371
3. Perrin F (1934) Mouvement Brownien d'un Ellipsoide (I). Dispersion Diélectrique pour des Molécules Ellipsoidales, J de Physique et le Radium 5: 497
4. Caldin EF (1964) Fast reactions in solution, Blackwell, Oxford

CHAPTER V
Vibration Aspects of the Hydrogen Bond

A. Ceulemans

This chapter offers an introductory theoretical treatment of the vibration aspects of the hydrogen bond. A hydrogen bonded system is characterized by the interaction of intra- and intermolecular forces. As a result a proper description of the corresponding vibrations can only be obtained from a quantummechanical treatment. This is illustrated in a pictorial way for the case of the stretching vibrations in a linear A—H—B system. Such a system exhibits a double-well potential, which deviates considerably from the classical two-dimensional harmonic oscillator. Solutions of the wave equation for nuclear motion in this potential are discussed, with special attention to the splitting effect of the double well and the coupling of the v_s and v_σ vibrations.

1 Introduction

Quantum-mechanical studies of the hydrogen bond mainly consist of electronic structure calculations, as is discussed in Chap. II on the Quantum Chemistry of the Hydrogen Bond in this volume. It must be kept in mind though that there are also important vibrational aspects to the problem, which can only be described properly in a quantum-mechanical perspective. In principle a separate treatment of electronic and vibrational aspects even seems artificial, since the motions of electrons and nuclei are coupled phenomena. Nonetheless, as a result of the much smaller electron mass, the electron cloud will adapt itself much more rapidly to the slowly varying nuclear positions,

Intermolecular Forces
© Springer-Verlag Berlin Heidelberg 1991

than vice versa. Hence, in practice it is possible to describe the electronic structure by means of a wave equation, containing the nuclear positions as fixed, i.e., immobile, parameters:

$$[T_e(q) + V(q; Q)]\psi(q; Q) = E(Q)\psi(q; Q) \qquad (1)$$

Equation 1 represents the electronic Schrödinger equation; lower case q refers to electronic coordinates and capital Q to nuclear coordinates. T_e and V are resp. the kinetic energy operator of the electrons, and the potential energy operator of electrons and nuclei. Since the kinetic energy operator of the nuclei, $T_N(Q)$, is not included in Eq. (1), there is no differentiation with respect to the Q coordinates. Consequently, the nuclear positions merely act as parameters. The resulting electronic eigenvalue, E(Q), which is a function of these para- meters, yields the potential in the equation which describes the motion of the nuclei:

$$[T_N(Q) + E(Q)]\chi(Q) = W\chi(Q) \qquad (2)$$

Here $\chi(Q)$ represents the nuclear wavefunction; W is the vibrational energy.

The separation of electronic and nuclear motions, the eigenvalues of the former being inserted as potential operators for the latter, is known as the Born–Oppenheimer approximation. In this approximation the electronic energy, E(Q), is also referred to as the adiabatic potential energy surface. The term 'adiabatic' means that the electronic structure follows the nuclear motions instantaneously, so that the electronic energy always remains in equilibrium with the nuclear positions.

The Born–Oppenheimer approximation undoubtedly is at the basis of almost all quantum-mechanical studies of molecular systems; the present study of the hydrogen bond will be no exception to this rule. More specifically we will be concerned with the vibration aspects that emerge from Eq. (2).

2 The Need for a Quantum-Mechanical Treatment

The simplest vibration model of the H-bond is based on a linear tri-atomic entity containing two donor atoms and a proton:

$$A ---H --- B \xrightarrow{z}$$

A particular feature of this entity is the coupling between the two stretching vibrations along the bond axis. Of course there are also interactions with the bending mode, as well as with the internal modes of the molecular environment. However, by way of example, our treatment will be limited to the stretchings, which certainly constitute the most prominent characteristics of the infrared spectrum. Identifying the bond axis as the cartesian z-axis, one obtains the following expression for the internal bond distances:

$$R_{B-H} = z_B - z_H$$
$$R_{H-A} = z_H - z_A \tag{3}$$

In order to solve the nuclear equations of motion, the use of mass-weighted cartesian displacement coordinates is indicated. These coordinates are given by:

$$\zeta_A = \sqrt{m_A} z_A \quad \zeta_B = \sqrt{m_B} z_B \quad \zeta_H = \sqrt{m_H} z_H \tag{4}$$

The z-coordinate of the center of mass corresponds to:

$$Z = (m_A z_A + m_B z_B + m_H z_H)/M$$

with

$$M = m_A + m_B + m_H \tag{5}$$

The mass-weighted translational coordinate, Q_t, thus becomes:

$$Q_t = (\sqrt{m_A}\zeta_A + \sqrt{m_B}\zeta_B + \sqrt{m_H}\zeta_H)/\sqrt{M} \tag{6}$$

The two remaining degrees of freedom along the z direction correspond to internal modes. In coordinate space these modes define a plane, which is perpendicular to Q_t. Some suitable criterion is needed to fix the unit vectors of this plane. A convenient choice defines the internal coordinates in such a way that they reduce to symmetry coordinates in the case of equivalent donor atoms. In this way an orthonormal transformation matrix can be defined, up to the usual phase factors:

$$(Q_x, Q_y, Q_t) = (\zeta_A, \zeta_B, \zeta_H)(\mathbf{C})$$

with:

$$\mathbf{C} = \begin{vmatrix} -\sqrt{\dfrac{m_A m_H}{M(m_A + m_B)}} & -\sqrt{\dfrac{m_B}{m_A + m_B}} & \sqrt{\dfrac{m_A}{M}} \\[2em] -\sqrt{\dfrac{m_B m_H}{M(m_A + m_B)}} & \sqrt{\dfrac{m_A}{m_A + m_B}} & \sqrt{\dfrac{m_B}{M}} \\[2em] \sqrt{\dfrac{m_A + m_B}{M}} & 0 & \sqrt{\dfrac{m_H}{M}} \end{vmatrix} \tag{7}$$

The relation between Q_x, Q_y, and the interatomic distances can easily be obtained:

$$Q_x = \sqrt{\frac{m_H}{M(m_A + m_B)}}[m_A R_{H-A} - m_B R_{B-H}]$$

$$Q_y = \sqrt{\frac{m_A m_B}{(m_A + m_B)}}[R_{B-H} + R_{H-A}] \qquad (8a)$$

Hence Q_x measures the position of the proton with respect to the center of mass of A and B, while Q_y is proportional to the distance between the two donor atoms. Clearly for equivalent donor atoms ($m_A = m_B$), Q_x and Q_y correspond resp. to the antisymmetric and symmetric stretching modes. In this

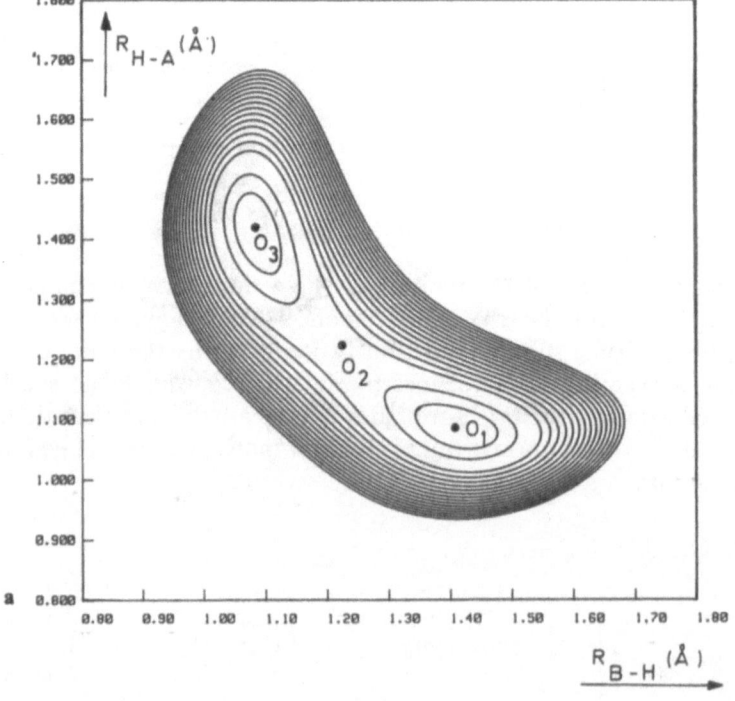

Fig. 1. Adiabatic potential energy surface for a linear symmetric A—H—B bridge, represented in three different coordinate systems. Adapted from Ref. [1]. **a.** Internal coordinates [R_{B-H}, R_{H-A}; see Eq. (8a)]. **b.** Symmetry coordinates [x, y; see Eq. (8b)]. **c.** Mass-weighted symmetry coordinates (Q_x, Q_y; see Eq. (7), with: $m_H = 1$, $m_A = m_B = 16$).

The potential is given by: $E(x,y) = -22x^2 + 583x^4 - 150x^2(y-y_0) + 31(y-y_0)^2$, with $y_0 = 2.4333$ Å (x and y in Å, E in units of kilokaiser; $1\,kK = 1000\,cm^{-1}$). E is null in saddle point O_2 ($x = 0$, $y = y_0$). There are two equivalent minima (O_1 and O_3), with $x = \pm 0.1655$ Å, $y = 2.50$ Å. The height of the barrier between the minima is $300\,cm^{-1}$

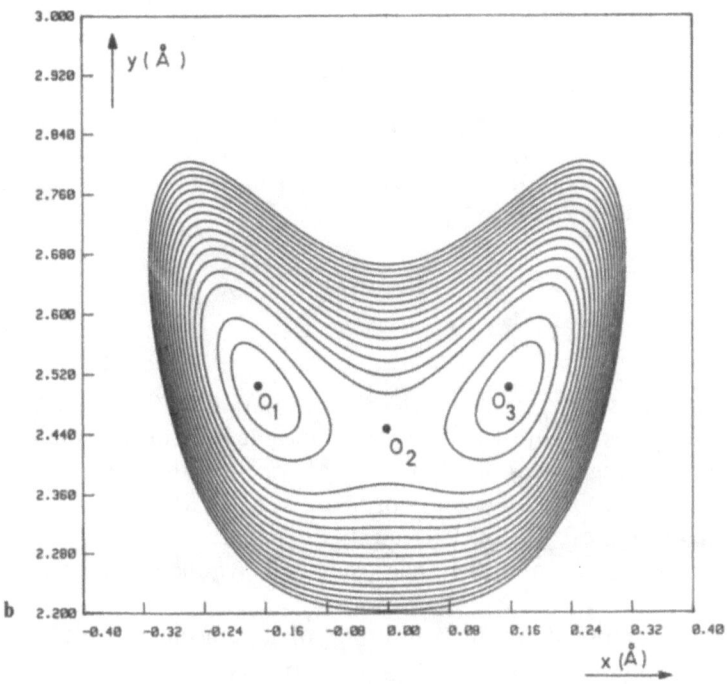

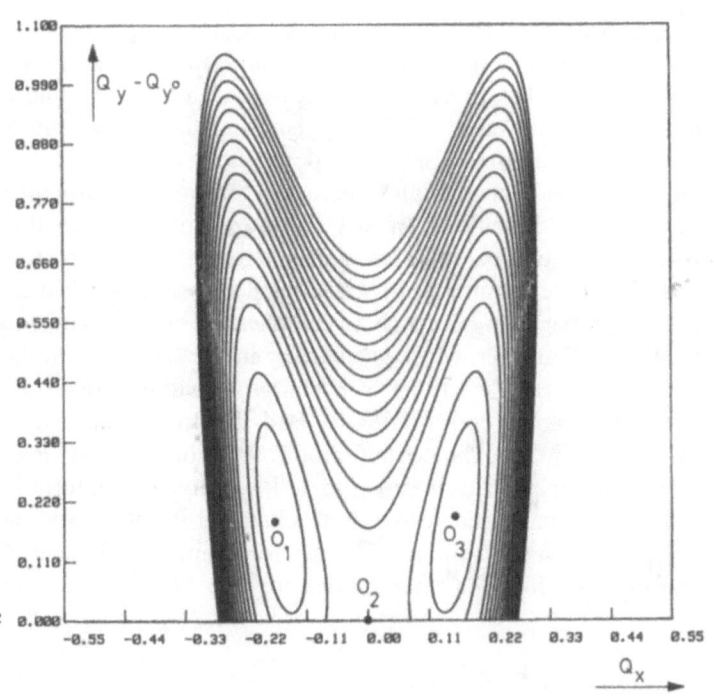

case, substitution of Eq. (3) into Eq. (8a) yields:

$$Q_x = \left[\frac{1}{2m_A} + \frac{1}{m_H}\right]^{-1/2} x$$

$$Q_y = \left[\frac{2}{m_A}\right]^{-1/2} y \qquad (8b)$$

with:

$$x = z_H - 1/2(z_A + z_B)$$

$$y = z_B - z_A$$

Here the expressions in square brackets refer to the reduced masses, associated with x and y coordinates. Figure 1 displays a simple example of the adiabatic potential for the stretching modes of a symmetrical system with two equivalent minima [1]. In the figure three different coordinate systems are compared: the usual 'chemical coordinates', i.e., the interatomic distances (R_{B-H}, R_{H-A}), the symmetry coordinates (x, y), and the mass-weighted symmetry coordinates (Q_x, Q_y).

In the *classical* treatment of the vibrations of a molecular equilibrium structure, one uses simplified potential energy expressions to construct the normal modes of vibrations. These modes are directed along the principal axes of curvature of the potential surface, in a system of mass-weighted orthonormal displacement coordinates. Notice that this definition depends explicitly on a particular choice of coordinate system, viz. the (Q_x, Q_y) system in Fig. 1c. Associating normal coordinates with the directions of curvature on the surfaces in Figs. 1a and 1b thus may be utterly misleading since the corresponding coordinates do not include the effect of the nuclear masses. Hence, unlike extrema, normal modes do not correspond to 'topological invariants'.

Even so, Fig. 1c makes quite clear that a classical vibrational treatment of the H-bonds soon runs into difficulties. In fact there are not less than three extremal points, O_1, O_2, O_3, two of which are genuine minima. The directions of principal curvature in these minima are rotated with respect to each other. Hence if the system hops back and forth between the two wells, its normal modes continually change character. More specifically in the saddle point, O_2, in between the two minima, the directions of principal curvature are seen to correspond to the two symmetry coordinates Q_x and Q_y. In contrast, at the bottom of the wells, in O_1 and O_3, the normal modes are reoriented to yield a high-frequency proton-donor vibration (v_s) and a low-frequency intermolecular vibration (v_σ). This situation corresponds to chemical intuition and can therefore most easily be grasped from the surface representation in Fig. 1a. To first approximation, the local normal modes for the O_3 minimum are given by:

$$(Q_s, Q_\sigma, Q_t) = (\zeta_A, \zeta_B, \zeta_H)(\mathbb{C}')$$

with:

$$\mathbb{C}' = \begin{vmatrix} 0 & -\sqrt{\dfrac{m_H + m_B}{M}} & \sqrt{\dfrac{m_A}{M}} \\[2ex] \sqrt{\dfrac{m_H}{m_H + m_B}} & \sqrt{\dfrac{m_A m_B}{M(m_H + m_B)}} & \sqrt{\dfrac{m_B}{M}} \\[2ex] -\sqrt{\dfrac{m_B}{m_H + m_B}} & \sqrt{\dfrac{m_A m_H}{M(m_H + m_B)}} & \sqrt{\dfrac{m_H}{M}} \end{vmatrix} \qquad (9)$$

The rotation between the local set (Q_s, Q_σ) and the symmetry set (Q_x, Q_y) can easily be expressed by the following matrix transformation:

$$(Q_s, Q_\sigma) = (Q_x, Q_y)(\mathbb{R})$$

with:

$$\sqrt{(m_A + m_B)(m_H + m_B)} \ \ \mathbb{R} = \begin{pmatrix} -\sqrt{m_B M} & \sqrt{m_A m_H} \\ \sqrt{m_A m_H} & \sqrt{m_B M} \end{pmatrix} \qquad (10)$$

For $m_A = m_B = 16 m_H$ the angle of rotation between Q_σ and Q_y is calculated [2] to be approximately $10°$, which apparently is in good agreement with the diagram in Fig. 1c. These results illustrate that the concept of a 'normal mode' breaks down when dealing with potential energy surfaces that allow large amplitude motions. Even in the case of a highly asymmetric H-bond, with only one well, the form of the potential will considerably deviate from the classical paraboloid function, describing the two-dimensional harmonic oscillator. For this reason a more detailed description of the vibrational spectrum is impossible without recurrence to the quantum-mechanical equations of nuclear motion on the full potential energy surface.

3 Solutions of the Wave Equations

For a linear A—H—B vibrator the nuclear wave equation, Eq. (2), becomes:

$$\left[-\frac{\hbar^2}{2m_A}\frac{\partial^2}{\partial z_A^2} - \frac{\hbar^2}{2m_B}\frac{\partial^2}{\partial z_B^2} - \frac{\hbar^2}{2m_H}\frac{\partial^2}{\partial z_H^2} + E(R_{B-H}, R_{H-A}) \right] \chi = W\chi$$

or, in mass-weighted coordinates:

$$\left[-\frac{\hbar}{2}\left(\frac{\partial^2}{\partial \zeta_A^2} + \frac{\partial^2}{\partial \zeta_B^2} + \frac{\partial^2}{\partial \zeta_H^2} \right) + E(R_{B-H}, R_{H-A}) \right] \chi = W\chi \qquad (11)$$

Note that the kinetic energy operator in this expression is invariant under the orthonormal transformation $\mathbb{C}$, specified in Eq. (7). Moreover, since the potential energy operator only depends on internal coordinates, it is possible to remove the translational degree of freedom from the equation, yielding:

$$\left[-\frac{\hbar^2}{2}\left(\frac{\partial^2}{\partial Q_x^2} + \frac{\partial^2}{\partial Q_y^2} \right) + E(Q_x, Q_y) \right]\chi = W\chi \qquad (12)$$

Equation (12) is the master equation for the stretching vibrations. As noticed before, the potential $E(Q_x, Q_y)$ can be generated in a pointwise way, by integrating the electronic Schrödinger equation for varying nuclear positions. In practice though one often uses model potentials, which considerably simplify the calculations. As an example, the potential surface in Fig. 1 corresponds to a straightforward polynomial in Q_x and Q_y (see also Ref. [1]).

$$E(Q_x, Q_y) = A_{00} + A_{20}Q_x^2 + A_{40}Q_x^4 + A_{21}Q_x^2Q_y + A_{02}Q_y^2 \qquad (13)$$

It should be noted that in this expression, odd powers of the antisymmetric stretching coordinate, Q_x, are lacking, because of obvious symmetry reasons. Of course if the H-bridge is asymmetric, such terms will no longer be forbidden. The term $A_{21}Q_x^2Q_y$ in Eq. (13) constitutes the simplest symmetry-allowed coupling term between Q_x and Q_y. If this term is neglected, Eq. (12) can be factored into two one-dimensional problems, one of which corresponds to a simple harmonic oscillation in Q_y direction, the other being a more complex oscillation in a double-well potential along Q_x.

From a perturbation point of view, the absence of coupling can be identified as the zeroth-order starting point. This situation would be represented by a potential energy surface, on which the directions of principal curvature are aligned along the Q_x and Q_y coordinates. The term in A_{21} then corresponds to the perturbation operator, which mixes the two modes and rotates the directions of principal curvature. Both aspects of this problem, i.e., the nuclear motion in a double well and the coupling of Q_x and Q_y, will now be studied in some detail.

3.1 The Double-Well Potential in Q_x

Extensive studies have been devoted to the quantum-mechanical description of a particle in a well with two minima. This potential is of wide interest, since it not only describes the tunneling of a proton, but also is relevant for the study of conformational changes, such as the inversion of amines. A generalized polynomial description of a double-well potential with the saddle point in the coordinate origin is given by:

$$E(Q_x) = A_{00} + A_{20}Q_x^2 + A_{30}Q_x^3 + A_{40}Q_x^4 \qquad (14)$$

The vibration spectrum for this potential has been calculated by Somorjai and Hornig [3]. These authors expressed the eigenfunctions as a series of classical oscillator functions:

$$\chi(Q_x) = \sum_n c_n \phi_n(Q_x)$$

with

$$\phi_n(Q_x) = \left(\frac{\alpha}{2^n n! \sqrt{\pi}}\right)^{1/2} H_n(\alpha Q_x) \exp(-\alpha^2 Q_x^2 / 2) \tag{15}$$

H_n is the Hermite polynome of rank n. The expansion coefficients c_n can be obtained in the usual way by diagonalizing the energy matrix over the oscillator basis. The factor α in Eq. (15) is a scaling factor, which takes into account the curvature of the potential walls. It should be remarked that the outer walls of the model potential in Eq. (14) are determined by the fourth power term in A_{40}. Of course, this term raises much more rapidly than the usual parabolic Hooke-type law, which is characteristic of the harmonic oscillator. This implies that the asymptotic behavior of the trial function in Eq. (15) is not well adapted to the form of the potential wall. Accordingly, the definition of the scaling factor in the trial functions is largely arbitrary.

Of course, alternative potential functions have been tried as well. An example is the potential in Eq. (16). In this case a parabolic well is combined with a gaussian-type barrier [4, 5].

$$E(Q_x) = B_{00} + B_{20}Q_x^2 + C \exp[-\beta(Q_x - Q_x^0)^2] \tag{16}$$

Of peculiar interest is the so-called Manning potential, which dates from as early as 1935 [6]. In this potential, the roles of the second and fourth powers in the polynomial expansion are reversed, by taking hyperbolic secant functions:

$$E(Q_x) = D_{00} + D_{20} \operatorname{sech}^2(\gamma Q_x) + D_{40} \operatorname{sech}^4(\gamma Q_x) \tag{17}$$

In Eq. (17) the outer walls are determined by the quadrate sech^2 term (with $D_{20} < 0$), while the quartic term provides the barrier in between the two minima. The Manning-potential can be integrated numerically. The bottom of this potential very much ressembles the function in Eq. (16).

For a quantitative evaluation of these various model potentials, a detailed comparison with *ab initio* results seems imperative. However, if one is merely interested in the qualitative characteristics of the vibration spectrum, the actual form of the double well is of secondary importance. Indeed all solutions seem to follow a similar pattern, which is illustrated in Fig. 2 [8]. The potential in this figure is characteristic of the inversion of the NH_3 molecule. Table 1 lists the observed energy levels, and compares these with an eigenvalue calculation, based on a Manning potential [9].

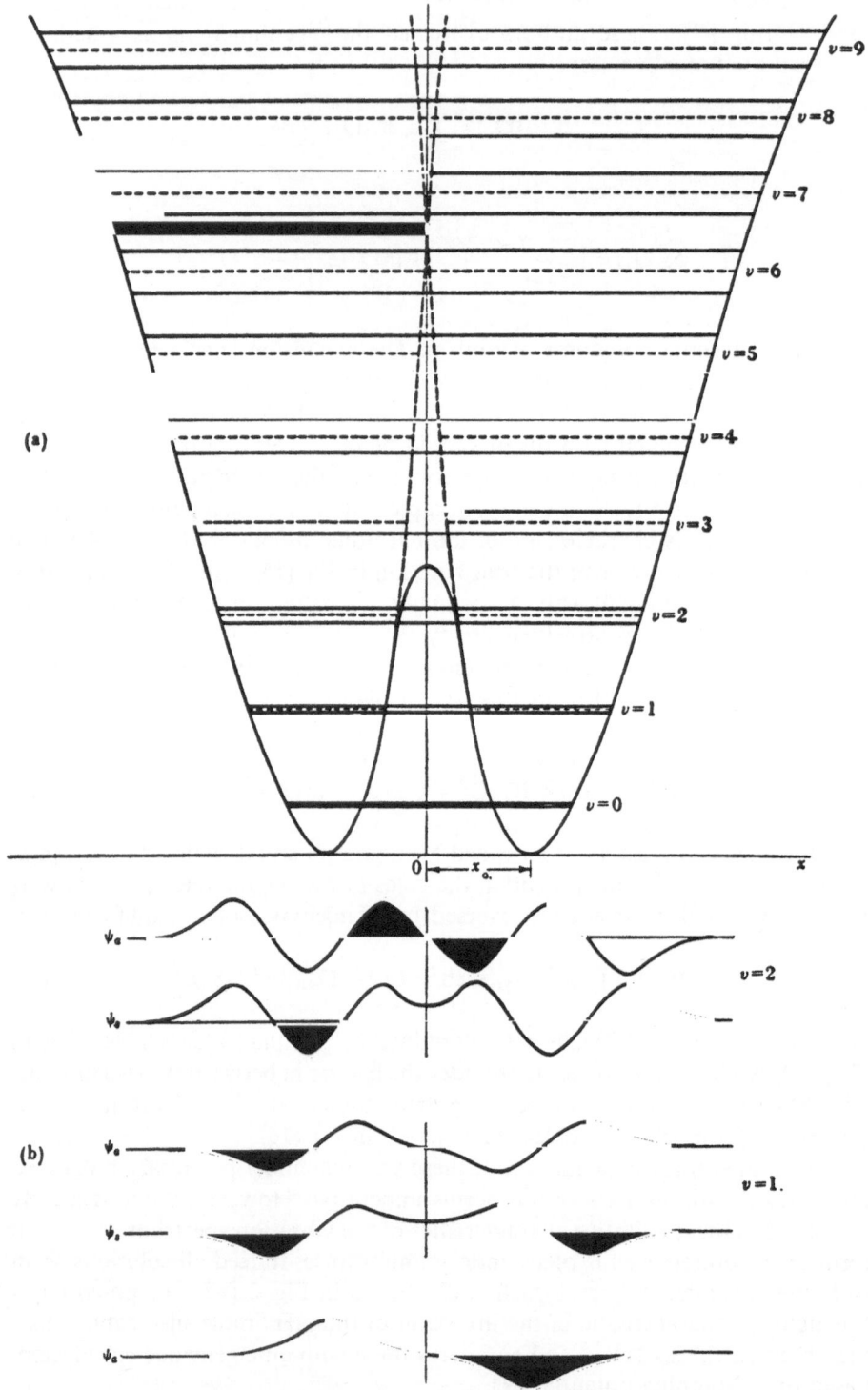

Table 1. Energy levels[a] for the inversion of NH_3

	Manning potential[b]	Observed[c]	
0^+	0	0 ⎫	
0^-	0.83	0.7935 ⎭	0.8
1^+	935	932.51 ⎫	
1^-	961	968.32 ⎭	35.8
2^+	1610	1602 ⎫	
2^-	1870	1882.2 ⎭	280.2
3^+	2360	2383.46 ⎫	
3^-	2840	2895.48 ⎭	512

[a] All values in cm^{-1}
[b] The barrier is $2000\,cm^{-1}$ above the minimum of the potential well
[c] Taken from Ref. [8]

The solutions of a double-well potential have the following properties:

i) The wavefunctions for a symmetric double well must be either symmetric (ψ_s) or antisymmetric (ψ_a) with respect to inversion through the origin. Both symmetries alternate, giving rise to a spectral series of the following type:

$$|0^+\rangle, |0^-\rangle, |1^+\rangle, |1^-\rangle, |2^+\rangle, |2^-\rangle, \ldots$$

Here the $+$ and $-$ superscripts denote resp. symmetric and antisymmetric levels. The lower part of the figure clearly shows that the eigenfunctions with lowest energy nearly correspond to sum and difference combinations of localized oscillator functions in left and right wells.

ii) The spectrum of energy eigenvalues consists of a convolution of two different spectral series. At high energy one notices a regular series of almost equidistant levels, with alternating symmetries. This pattern reflects the oscillator characteristics of the outer walls. In the low energy region, at energies below the saddle point, s and a terms associate in pairs, forming doublets. The energy gaps between these doublets becomes larger towards the bottom of the well. This of course reflects the narrowing of the potential well itself.

The doublet splitting itself is a measure for the penetration of the potential barrier (tunneling effect). If the barrier is high, tunneling between the two minima is very inefficient, and this leads to a quasi-degeneracy of the $|0^+\rangle$ and $|0^-\rangle$ levels [10]. As an example for the inversion of NH_3 the tunneling splitting of the ground state is found to be equal to approximately $0.8\,cm^{-1}$ ([11] compare Table 1).

In contrast, if the potential barrier is lower than one vibrational quantum, the pairwise clustering of the lowest vibrational levels cannot occur.

Fig. 2. Energies and six eigenfunctions ($v = 0, 1, 2$) of a double-well potential. The potential parameters concur with the inversion mode of NH_3, described in Table 1. The splitting of the $v = 0$ and $v = 1$ doublets is exaggerated. Taken from G. Herzberg, Ref. [8]

iii) Electronic selection rules: only s↔a transitions are allowed, in accordance
with the Laporte selection rule. On the other hand, the $\Delta v = \pm 1$ rule for
the harmonic oscillator is no longer valid.

iv) Asymmetrization of the potential gives rise to mixing of ψ_s and ψ_a
components. The resulting wavefunctions rapidly loose their delocalized
character [5].

3.2 The Coupling of Q_x and Q_y Vibrations

We recall that the zeroth-order description of the linear A—H—B vibrator is
a sum of a double-well potential in Q_x and a simple harmonic potential in Q_y;
usually the former potential gives rise to high frequency vibrations, while the
latter corresponds to a much softer mode at low frequency. The resulting states
are products of the two vibrational components, and can be written as follows:

$$|0^+0\rangle, |0^+1\rangle, |0^+2\rangle, |0^+3\rangle \cdots$$

$$|0^-0\rangle, |0^-1\rangle, |0^-2\rangle, |0^-3\rangle \cdots$$

$$|1^+0\rangle, |1^+1\rangle, |1^+2\rangle \cdots$$

$$|1^-0\rangle, |1^-1\rangle, |1^-2\rangle \cdots$$

$$|2^+0\rangle, |2^+1\rangle \cdots$$

$$|2^-0\rangle, |2^-1\rangle \cdots$$

or in general:

$$|m^+n\rangle, |m^-n\rangle$$

The corresponding energy diagram is very much dependent on the height
of the potential barrier along the Q_x coordinate. This can best be represented
in a correlation diagram, as indicated in Fig. 3 [1, 12, 13]. The left-hand side
of the figure shows the energy levels for a typical O—H—O bond, with a low-
energy barrier. A case in point is the system which is depicted in Fig. 1, with
a barrier of $300 \, \mathrm{cm}^{-1}$ and a O—O distance of 2.5 Å. The right-hand side is a
schematic representation of a system with a substantial barrier in the range of
$5000 \, \mathrm{cm}^{-1}$. Such would be the case for a O—O distance of 2.7 Å. Notice that
the fundamental transition on the left $(0^+0 \rightarrow 0^-0)$ correlates with the tunneling
transition between quasi-degenerate levels on the right. In contrast, the fundamental excitation on the right $(0^+0 \rightarrow 1^-0)$ is seen to correlate with an overtone
vibration on the left.

Now, following perturbation theory, one must study the action of the
coupling operator $A_{21}Q_x^2Q_y$ of Eq. (13) in the space of the zeroth-order basis
functions $|m^\pm n\rangle$. This action is described by a perturbation matrix, with

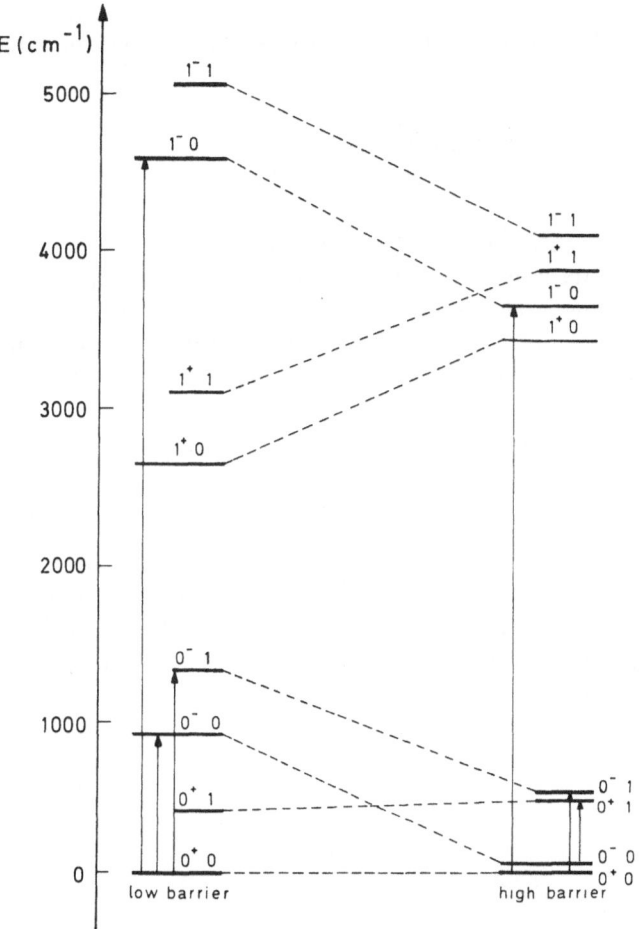

Fig. 3. Correlation diagram for the lowest vibrational levels of a symmetric A—H—A system as a function of the energy of the potential barrier. The vertical lines indicate allowed transitions

elements:

$$\langle m^{\pm}n|A_{21}Q_x^2Q_y|m'^{\pm}n'\rangle = A_{21}\langle m^{\pm}|Q_x^2|m'^{\pm}\rangle\langle n|Q_y|n'\rangle \qquad (18)$$

Evaluation of these matrix elements is rather straightforward. It should be noted that many elements will be zero. Indeed there can be no mixing between + and − functions, and, moreover, allowed values of n' are limited to $n+1$ and $n-1$. The result of the perturbation is a *mechanical* coupling between the Q_x and Q_y modes. This leads to a breakdown of the quantum numbers m and n for the individual modes. In this process both frequencies and intensities will be affected. Allowed transitions are as follows.

i) On the one hand, transitions may occur inside the m series, following the previously specified selection rules. These transitions are of two possible types:
   a) $|m^+n\rangle \leftrightarrow |m^-n\rangle$
      These are the so-called *tunneling* transitions.
   b) $|m^+n\rangle \rightarrow |(m+k)^-n\rangle$
      $|m^-n\rangle \rightarrow |(m+k)^+n\rangle$
      These are the proper excitations of the Q_x mode.
ii) On the other hand, excitations of the Q_y mode may gain intensity as well. In symmetric systems this mode is infrared-inactive, since it corresponds to a symmetric stretch. Nonetheless it can be observed in combination with excitations of the active Q_x vibration. Again two types of transitions are possible here.
   a) Combination of Q_y with a tunneling transition:
      e.g.: $|m^+n\rangle \rightarrow |m^-(n+1)\rangle$
   b) Combination of Q_y with a Q_x excitation:
      e.g.: $|m^+n\rangle \rightarrow |(m+1)^-(n+1)\rangle$

The intensity of these combination bands stems from two possible mechanisms: the previously mentioned mechanical coupling between Q_x and Q_y, and a similar so-called *electronic* coupling. The latter mechanism is due to the appearance of nonlinear terms in the nuclear coordinate dependence of the dipole moment [12, 14].

4 Conclusion

The principal vibration aspect of the hydrogen bond is the pronounced anharmonicity of the potential energy. This gives rise to a number of specific spectroscopic effects:

i) Spectacular shifts of the stretching frequencies, as illustrated in Fig. 3.
ii) Coupling of the A—H stretching mode with various other vibrations. A case in point is the $v_s - v_\sigma$ interaction.
iii) Band splitting due to a tunneling effect. Direct observation of this splitting is difficult though, due to the width and complex structure of the v_s band.

For a detailed discussion of the spectra the reader is referred to Chap. VI on the spectroscopic characteristics of hydrogen bonds in this volume.

Finally, it should be mentioned that all aspects discussed at present derive from the usual time-independent Schrödinger equation for the nuclear motion. It is clear that a double-well potential, with a high energy barrier in-between the two minima, might give rise to dynamic properties that can only be studied in a time-dependent framework. One of the results of such a treatment is that a wave packet, which was initially localized in one of the two wells, will oscillate back and forth between the two minima, with a period τ which is inversely proportional to the splitting of the $v = 0$ level [15].

$$\tau = \hbar / [E(0^-) - E(0^+)] \tag{19}$$

As an example for the tunneling of NH_3—with a splitting of $0.8\,\text{cm}^{-1}$—this period of fluctuation is found to be $10^{-12}\,\text{s}$. If the barrier increases, the tunneling process will of course slow down. In this way the inversion time for arsines amounts to almost one year! Obviously in such metastable systems the localized states acquire a quasi-stationary character, which implies that they can conveniently be described by time-independent equations, using however a partial potential operator for a single potential well.

References

1. Janoschek R (1976) In: Schuster P, Zundel G, Sandorfy C (eds) The hydrogen bond Part I: Theory. North-Holland, Amsterdam, chap 3
2. The lower-diagonal element of the $\mathbb{R}$ matrix in equation 10 is equal to the scalar product $\langle Q_\sigma | Q_y \rangle$. The inverse cosine of this term thus yields the desired angle.
3. Somorjai RL, Hornig DF (1962) J Chem Phys 36: 1980
4. Busch JH, de la Vega JR (1977) J Am Chem Soc 99: 2397
5. Flanigan MC, de la Vega JR (1974) J Chem Phys 61: 1882
6. Manning MF (1935) J Chem Phys 3: 136
7. The outer sech^2 well is also known as the Pöschl–Teller hole; see Poschl G, Teller E (1933) Z Phys 83: 143
8. Herzberg G (1945) Molecular Spectra and Molecular Structure II Infrared and Raman Spectra of Polyatomic Molecules. Van Nostrand, Princeton NJ, p 222
9. Swalen JD, Ibers JA (1962) J Chem Phys 36: 1914; Papousek D, Stone JMR, Spirko V (1973) J Mol Spectrosc 48: 17
10. Notice that the symmetric level $|0^+\rangle$ is always at lower energy than the antisymmetric $|0^-\rangle$ level. This is because the latter level has a nodal point in the origin, which keeps the function away from the saddle region. As a result it hits the outer walls at higher energies.
11. The experimental observation of this tunneling splitting constituted an important test for the validity of the wave-mechanical treatment
12. Ginn SGW, Wood JL (1967) J Chem Phys 46: 2735
13. Romanowski H, Sobczyk L (1977) Chem Phys 19: 361
14. Sandorfy C (1984) Top Curr Chem 120: 41
15. Brickmann J (1976) In: Schuster P, Zundel G, Sandorfy C (eds) The hydrogen bond Part I: Theory. North-Holland, Amsterdam, chap 4

CHAPTER VI
Experimental Vibrational Characteristics
of the Hydrogen Bond

Th. Zeegers-Huyskens

As shown in several books and reviews, vibrational spectroscopy, especially infrared spectroscopy is one of the powerful techniques for hydrogen bond investigations. The aim of this chapter is not to discuss all the vibrational characteristics of the hydrogen bonds, but to make a bridge between Chap. V (A. Ceulemans, vibrational aspects of the hydrogen bond) and the experimental characteristics, more specifically the structure of the v_s stretching band.

Some applications of infrared spectroscopy to the study of molecular interactions in biological systems where several sites are available for hydrogen bond formation, are briefly mentioned and discussed.

1 Introduction

Electronic spectroscopy (ultraviolet or visible) gives very valuable information on the charge redistribution accompanying charge transfer complexation or hydrogen bond formation, particularly in conjugated systems.

The rotational spectra measured in the gas phase can provide information on the geometry of the complexes. A quantity that has been in the forefront of

Intermolecular Forces
© Springer-Verlag Berlin Heidelberg 1991

discussions of hydrogen bonding in B...HA complexes is the lengthening of the HA bond ∂r which accompanies dimer formation. This elongation has been obtained recently for HF complexes from the ground-state hyperfine coupling constants determined by pulsed-nozzle Fourier-transform microwave spectroscopy [1]. The uncertainty on ∂r is at worst ± 0.003 Å. The ∂r value for the Van der Waals complex Ar.HF is zero; for HF complexed with N_2, H_2S and HCN, the ∂r values are 0.001, 0.010 and 0.014 Å respectively. The ∂r(HF) value increases monotonically with the hydrogen bond stretching constant [1]. These results are worth mentioning. In this chapter however, we will restrict ourself to the vibrational spectroscopy, particularly infrared spectroscopy, which is one of the most powerful tools for hydrogen bond investigations. Many characteristic properties of intermolecular complexes can be deduced from vibrational spectra in which bands due to the free, non-hydrogen bonded species are in most cases distinguished from those due to associated hydrogen bonded species. These properties allow us to determine the thermodynamic parameters in solution and possibly in the gas-phase, without a deep knowledge of the theory of the vibrational spectroscopy.

The vibrational spectra of hydrogen bonded systems have been extensively reviewed and discussed in several books or chapters in books dealing with intermolecular interactions [2–9]. The scope of this chapter which was originally a lecture of $1\frac{1}{2}$ hours in the Erasmus programm "Specific molecular interactions" is not to discuss all the aspects of the vibrational spectroscopy of these interactions but to construct a bridge between the theory expounded in Chap. V (A. Ceulemans; Vibrational aspects of the hydrogen bond) and the experimental vibrational properties. This chapter will also deal with some recent applications of infrared spectroscopy to the study of hydrogen bonded complexes involving polyfunctional bases of biological interest.

As a consequence, some important aspects of the vibrational spectroscopy of hydrogen bonds such as the solvent effects, the effect of hydrogen bond formation on the Raman intensities, the overtone spectra will not be discussed here.

Owing to the growing interest of the molecular interactions in biological systems, the last part of this chapter will deal with some recent applications of infrared spectroscopy to the study of hydrogen bonds in polyfunctional bases of biological interest.

2 Vibrational Degrees of Freedom and Symmetry in Hydrogen-Bonded Complexes

Hydrogen bond formation brings about new vibrational degrees of freedom. We shall consider the $CH_3C\equiv N...HF$ complex as an example. The $CH_3\equiv CN$ is a non linear molecule and has 6 translational and rotational degrees of freedom while HF has only 5 such degrees of freedom. When the complex is formed

$(6 + 5 - 6)$ degrees will be transformed into vibrational degrees of freedom. These intermolecular vibrations can be described as follows:

Description	Assignment	Symmetry	$\tilde{v}(\text{cm}^{-1})$ gas phase
$CH_3C\equiv\bar{N}...\overrightarrow{HF}$	intermolecular stretching vibration (v_σ)	A_1	170
$CH_3C\equiv N...HF$ $CH_3C\equiv N...HF$	N...HF deformation (v_d)	E	600
$CH_3C\equiv N...HF$ $CH_3C\equiv N...HF$	HF wagging (v_{wag})	E	~ 40

These 5 modes correspond to the intermolecular stretching vibration v_σ and to two degenerate modes (E). The molecules constituting the complex conserve, however, some memory of their old freedom because the N—HF deformation and the HF wagging can be considered as the rotational and translational modes with respect to the two directions perpendicular to the internuclear N...F axis. The v_d mode can be considered as a librational motion. In other hydrogen halide complexes, this mode is observed at much lower frequencies.

The v_s mode corresponding to the asymmetrical stretching vibration

$$H_3CC\equiv\bar{N}...\overrightarrow{HF}$$

is observed at $3630\,\text{cm}^{-1}$ and is shifted by about $300\,\text{cm}^{-1}$ from the value observed in the free molecule. The internal modes of the base (as for example the v_{C-C} and $v_{C\equiv N}$ vibrations) are shifted by a few wave numbers.

It must be noted here that the bending vibration is a degenerate vibration for the linear triatomic unit. Under lower symmetry, it will split into an in-plane (v_∂) and an out-of-plane (v_γ) bending motion. Those motions can be represented as follows

$$A—H...B \quad (v_\partial \quad \text{or} \quad \partial_{AH})$$
$$A—H...B \quad (v_\gamma \quad \text{or} \quad \gamma_{AH})$$

The v_s stretching vibration is by far the most studied among the vibrations of hydrogen-bonded complexes. The v_σ vibration is for most of the interactions involving organic acid and bases, observed between 200 and $100\,\text{cm}^{-1}$ [10]. In the AH...B (A $\neq$ B) complex, the v_σ mode is infrared active. Interestingly, the van der Waals complexes between hydrogen chloride and Argon, Krypton or Xenon have characteristic v_σ values between 30 and $45\,\text{cm}^{-1}$; these values are obviously much lower than for specific interactions [10].

In the $CH_3C\equiv N\ldots HF$ complex, two vibrational motions are degenerate. There are interesting examples where some vibrations are degenerate in the free molecule and where complexation brings about a removal of this degeneracy. As a consequence, the two components of the E mode(s) are observed in the complexes.

A typical example is ammonia which has the C_{3v} symmetry in the free state. The v_{NH}^{as} stretching vibration is degenerate and is observed at $3416\,cm^{-1}$ in carbontetrachloride. This mode can be described as follows:

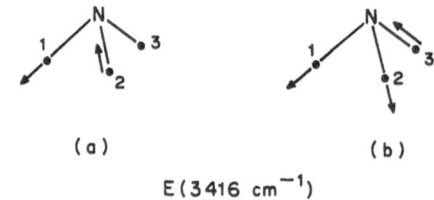

(a)   (b)

$E(3416\ cm^{-1})$

When NH_3 is complexed with a proton acceptor B (pyridine as for example), there is a lowering of the symmetry which becomes C_s in the 1–1 complex. As a consequence, the E mode is split into two components a and b which are

(a) $3415\ cm^{-1}$ (b) $3399\ cm^{-1}$

symmetric (a′) and antisymmetric (a″) with respect to the symmetry plane. Note also that for the a mode, the amplitude of the H(3) atom is zero. When this atom is complexed with B, the frequency of the a component is rigorously the same as in the free molecule.

In the 1–3 ($1\,NH_3\cdot 3\,B$) complex where the three NH bonds are equivalent, the complex recovers the C_{3v} symmetry and only one band is observed at $3378\,cm^{-1}$ [11].

Infrared spectroscopy is also a powerful tool for investigating the structure of aqueous systems [12, 13]. Water has C_{2v} symmetry in the free state and in 1–2 complexes where the intermolecular interaction is symmetrical to both OH bonds. In this case the frequency difference between the symmetric (v_1) and antisymmetric (v_3) stretching vibrations is nearly constant. (In carbon tetrachloride these two modes are observed at 3709 and $3614\,cm^{-1}$ respectively). In the 1–1 complexes characterized by the C_s symmetry, the band separation of v_3 and v_1 increases markedly when the complex becomes stronger. The degree of coupling (DC_s) can be defined as follows [12]:

$$DC_s = \frac{v_3(HOH\ldots B) - v_{OH}(HOH)}{v_3(HOH) - v_{OH}(HOH)} \times 100\%$$

where $v_3(HOH...B)$ and $v_3(HOH)$ are the experimental frequencies of the v_3 vibration in the complex and in the free molecule and v_{OH} is the decoupled frequency

$$v_{OH} = \tfrac{1}{2}(v_3 + v_1)$$

In Fig. 1, the DC_s values have been plotted in function of Δv_1, the frequency shift of the v_1 vibration, for different complexes and for HOD. The coupling is about 60% for a weak base such as acetonitrile and 25% for the stronger base triethylamine. The degree of coupling for HOD is 3%.

The harmonic potential of water in 1:1 complexes can be written as [12, 3]

$$2\mathscr{V} = k_{OH...B}(\Delta r)^2 + k_{OH}(\Delta r)^2 + k_\alpha^2(\Delta\alpha)^2 + 2k_{OH...B,OH}(\Delta r)^2$$

where $k_{OH...B}$ is the force constant of the bonded OH bond, k_α the bending force constant and $k_{OH...B,OH}$ the stretching-stretching interaction force constant. The stretching-bending force constant has been neglected. The normal vibrations of water in different 1:1 complexes calculated from the frequencies and force constants are presented in Fig. 2 [12, 13]. As can be seen from this figure, a strong change of the vibrational amplitudes with increasing asymmetry of the water molecules can be observed. For the complex with the strong base triethylamine, the v_3 vibration shows, to a good approximation, only a motion of the non-hydrogen-bonded OH group. The v_1 vibration describes, to the same

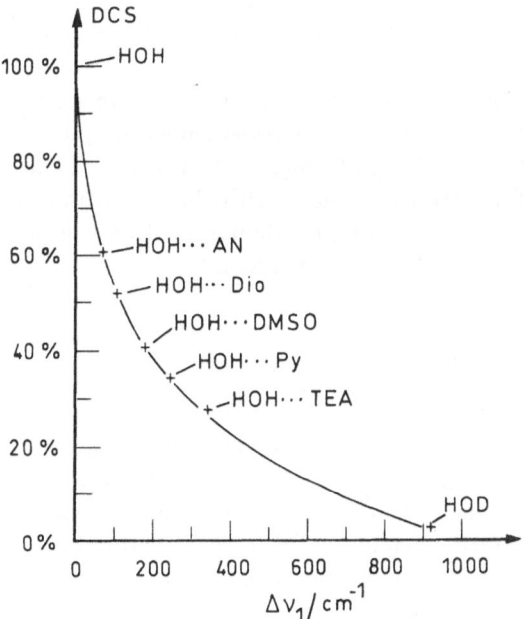

Fig. 1. Degree of coupling of water in different 1:1 complexes and of HOD versus Δv_1. Solvent = CCl$_4$. AN = acetonitrile, DIO = dioxane, DMSO = dimethylsulfoxide,   PY = pyridine, TEA = triethylamine. (Luck WAP, Schiöberg D (1979) Adv Mol Relaxation Interaction Proc 14: 277; Reproduced with the permission of Elsevier Science Publ. and of WAP Luck)

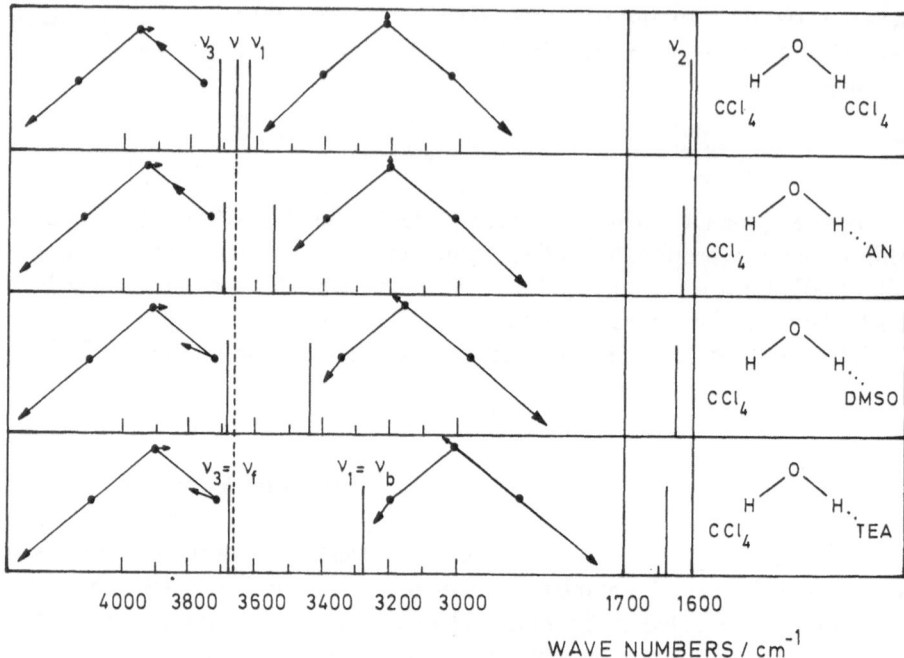

WAVE NUMBERS / cm^{-1}

Fig. 2. Wavenumbers of v_3 and v_1 and normal vibrations of HOH in CCl_4 and in 1 − 1 complexes. An = acetonitrile, DMSO = dimethylsulfoxide, TEA = triethylamine. (Luck WAP, Schiöberg D (1978) Adv Mol Relaxation Interaction Proc 14: 277. Reproduced with the permission of Elsevier Science Publ. and of WAP Luck)

approximation, only a vibration of the bonded OH group. But according to DC_s calculations, the vibrations of the two OH-bonds are always somewhat coupled even in this strong complex.

The value of the decoupled OH frequency of H_2O dissolved in carbon tetrachloride is 3661 cm^{-1}. Interestingly, the computed decoupled frequencies in the complex with 4-C_2H_5 pyridine are 3364 (v_1) and 3669 (v_3) cm^{-1} (These values have also been corrected for Fermi resonance with the $2v_2$ vibration) [14]. The v_3 value appears to be higher in the complex than in the free molecule and this suggests that the proton donor ability of the second

OH bond of the water molecule has decreased by hydrogen bond formation on the first one. This agrees well with the theory of Huyskens who has shown hydrogen bond formation between an acid and a base weakens the electron acceptor power of the other neighbouring acidic sites of the proton donor [15]. This result also agrees with the slight increase of charge density on the terminal H atom (Chap. II by L. Vanquickenborne).

3 General Considerations

The ν_s stretching vibration is by far the most studied among the vibrations of hydrogen bonded complexes. As outlined in Chap. V, the hydrogen bond is characterized by a very strong anharmonic potential which is at the origin of characteristic spectroscopic effects:

—spectacular shift of the ν_s stretching vibration to lower frequencies
—coupling of the ν_s vibration with the ν_σ or other vibrational modes
—splitting of the ν_s band by a tunneling mechanism.

Other bands also undergo spectacular shifts. This is the case for the γ_{AH} (out-of-plane deformation) vibration. The changes of the δ_{AH} (in-plane deformation) mode is less characteristic owing to its strong mixed character in many organic molecules. In aliphatic carboxylic acids the δ_{OH} vibration is coupled with the δ_{CH_3} and δ_{C-O} vibrations, in the phenol derivatives, the δ_{OH} mode is strongly mixed with the ν_{C-O} and ν_{C-C} ring modes, in aliphatic amides, the δ_{NH} vibration coupled with the ν_{C-N} vibration etc.

The relative frequency shifts $\Delta\nu/\nu$ and $\Delta\gamma/\gamma$ can reach values as high as 0.8 and 0.98 respectively. These values can be correlated to the $R_{A...B}$ or r_{AH} distances [16] determined by X-ray or neutron diffraction [17].

Table 1 indicates some spectroscopic and crystallograhic data for OH...O bonds incorporating the CH_3COO_- units. In this table, the frequencies of the ν_{OH} and γ_{OH} vibrations along with their relative shifts $\Delta\nu/\nu_0$ where ν_0 (or γ_0) corresponds to the frequency of the free OH group in the gas phase, are reported. This table also indicates the $R_{O...O}$ and r_{OH} distances [16]. The infrared spectra are schematically described in Fig. 3 for three different $R_{O...O}$ distances, showing that the changes in frequency, breadth and intensity are very sensitive to the strength of the hydrogen bond.

In spectrum a) where the $R_{O...O}$ distance is 2.65Å and r_{OH} distance is 1.02 Å, the ν_{OH} band is observed at 2900 cm^{-1} and the band breadth is about 300 cm^{-1}.

Table 1. Spectroscopic and crystallographic data for OH...O hydrogen bonds[a]

Compound	ν_{OH} (cm^{-1})	$\Delta\nu/\nu_0^b$	γ_{OH} (cm^{-1})	$\Delta\gamma/\gamma_0^b$	r_{OH} (Å)	$R_{O...O}$ (Å)
CH_3COOH (gas)	3585	—	650	—	0.97	—
$(CH_3COOH)_2$ (gas)	3020	0.16	940	0.45	1.03	2.68
CH_3COOH (crystal at 90 K)	2875	0.20	925	0.42	1.01	2.63
$KH(CH_3COO)_2$ (crystal at 90 K)	1400	0.61	1125	0.77	—	2.48
$NaH(CH_3COO)_2$ (crystal at 50 K)	720	0.80	1285	0.98	1.22	2.44

[a] Values quoted from Ref. [16]
[b] Computed from the ν_0 value in the gas phase

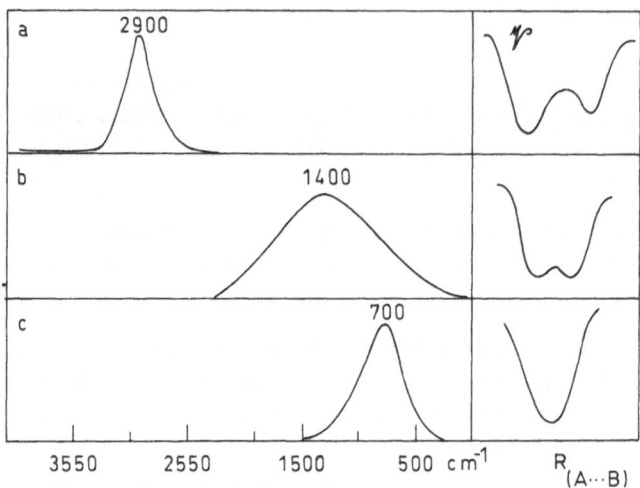

Fig. 3. ν_s(OH...O) band for different $R_{O...O}$ distances and potential curve for the proton motion
a) $R_{O...O} = 2.65$ Å
b) $R_{O...O} = 2.47$ Å
c) $R_{O...O} = 2.44$ Å

These data correspond to a hydrogen bond of medium strength where the proton is localized nearer one of the hydrogen atoms and where the elongation of the r_{OH} bond remains moderate (about 0.06 Å). In spectrum b), the ν_{OH} band is observed at $1400 \, cm^{-1}$ with a breadth of about $1000 \, cm^{-1}$, the $R_{O...O}$ corresponding to that observed in $KH(CH_3COO)_2$ where the proton is not exactly midway between the two oxygen atoms. This corresponds to a double minimum potential with a low barrier and a strongly delocalized proton. In spectrum c), the ν_{OH} is observed at $700 \, cm^{-1}$ with a breadth of about $300 \, cm^{-1}$. This corresponds to a symmetrical hydrogen bond as in $NaH(CH_3COO)_2$ where the proton is exactly midway between the two oxygen atoms. The approximate potential curve for the proton motion as a function of the $R_{O...O}$ distance is also indicated in Fig. 3.

It must be mentioned that the $\nu_s AH$ band is characterized in many complexes in the gas, liquid or solid state by a complex structure. This will be discussed in the next section.

There have been several attempts to correlate the frequency shift of the ν_{AH} stretching vibration with the charge transfer properties of the hydrogen bond [18–21]. The ν_{AH} values have been correlated with the weight of the dative structure defined by Mulliken as $b^2 + abS$, a and b being the coefficient of the wave function in the non bonded and dative structure and S the overlap integral. Recent second-order Møller-Plesset calculations have shown that the dispersion term contributing to the total hydrogen bond energy also contributes to the lengthening of the AH bond [22]. Some results of these calculations for complexes involving H—X(X=F, Cl) and simple bases are indicated in Table 2.

Table 2. Contribution of the dispersion term to calculated properties of some hydrogen bonded systems[a]

System	$-\Delta E^{tot}$ (kcal mol^{-1})	$\dfrac{-\Delta E^D}{\Delta E^{tot}}$ (%)	Δr_{HX} (%)
$H_3N...HF$	15.13	22	21
$H_3N...HCl$	11.03	40	43
$H_2O...HCl$	6.59	42	40
$H_2S...HCl$	4.95	69	55

[a] Values cited in Ref. [22]

This table also lists the calculated total energy $(-\Delta E^{tot})$, the contribution of the dispersion energy to the total energy $(\Delta E/\Delta E^{tot})$ along with the elongation of the HX bond (Δr) arising from the dispersion energy. These results show that the contribution of the dispersion energy makes up 1/4 of the total hydrogen bond energy when two first-row atoms are paired; this contribution rises to about 2/3 when second-row atoms are involved. The dispersion term also contributes to the elongation of the bond, amounting to over 50% of the total Δr value in some cases.

Even in weak hydrogen bond complexes there is usually no overlapping between the free and complexed v_{AH} band. This property can be used to calculate the thermodynamic parameters. The determination of the formation constants is based on the measurement of the intensity of a free band, usually the v_{AH} band of the proton donor and the variation of its intensity upon addition of a proton acceptor to the solution. The formation constants are usually obtained with a precision of about 5% and the $-\Delta H$ values with an accuracy of $\pm 1.5\,kJmol^{-1}$. The thermodynamic data obtained in this way are related to the proton donor or proton acceptor ability of the components or, with other words, to their pK_a or proton affinity values.

There have been numerous attempts to correlate $-\Delta H$, the experimental value of complex formation with Δv_{AH} (Badger–Bauer correlation) [23–24]. The general relation

$$-\Delta H = a + b\Delta v_{AH}$$

applied to several hydrogen bonded systems, shows that the slope and intercept of the equation depend on the nature of the proton donor and proton acceptor molecules. This has been discussed for OH...O and OH...N systems, the experimental results showing that at constant $-\Delta H$, Δv_{OH} is greater for OH...N than for OH...O hydrogen bonds [25]. These findings can be explained at least qualitatively by considering all the terms contributing to the total hydrogen bond energy: the electrostatic, repulsion, charge transfer, polarisation and dispersion energy and are illustrated by comparing the OH...O and OH...S systems.

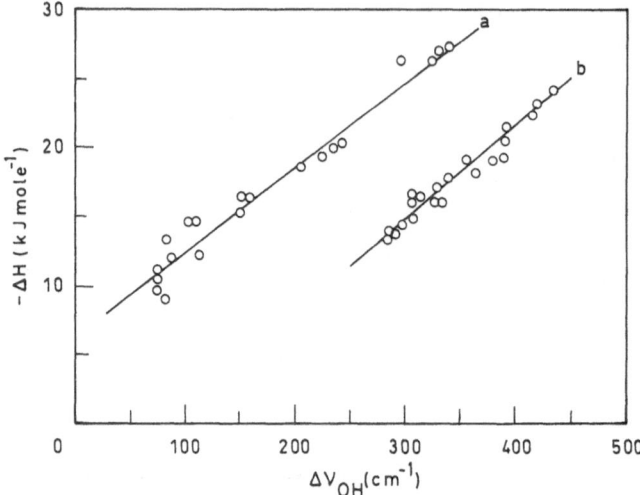

Fig. 4. $-\Delta H$ (kJmol^{-1}) vs $\Delta\nu_{OH}$ (cm^{-1})
a) OH...O=C systems
b) OH...S=C systems
(Reyntjens-Van Damme D, Zeegers-Huyskens Th (1979) Bull Soc Chim Belg 88 1003. Reproduced with the permission of the editor)

The Badger–Bauer correlation is shown in Fig. 4 for OH...O=C bonds involving hydroxylic proton donors and aliphatic ketones or amides and OH...S=C bonds with the same proton donors and aliphatic thioamides as proton acceptors. The experimental $-\Delta H$ and $\Delta\nu_{OH}$ have been obtained in carbontetrachloride [26], $\Delta\nu_{OH}$ being the difference between the wavenumber of the free and the complex band in the solvent.

The following equations characterize these systems (Fig. 4):

$$\text{OH...O=C systems:} - \Delta H = 0.062 \, (\text{kJ/mol cm}^{-1})\Delta\nu_{OH} + 6.83 \, (\text{kJmol}^{-1}) \quad (3)$$

$$\text{OH...S=C systems:} - \Delta H = 0.068 \, (\text{kJ/mol cm}^{-1})\Delta\nu_{OH} - 5.35 \, (\text{kJmol}^{-1}) \quad (4)$$

These correlations show that for the same $-\Delta H$ value, $\Delta\nu_{OH}$ is about 150 cm^{-1} higher for OH...S=C than for OH...O=C bonds. The $\Delta\nu_{OH}$ values are related to the elongation of the OH bond (or the r_{OH} distances). These distances have been calculated for different $\Delta\nu_{OH}$ values by the Lippincott–Schroeder function where the different parameters are available for OH...O [27] and OH...S hydrogen bonds [28]. The results of the calculations are presented in Table 3. The comparison of the two curves depicted in Fig. 5, shows that for the same $-\Delta H$ value (33 kJ mol^{-1}), r_{OH} is approximately equal to 0.986 Å for OH...O=C bonds and to 1.004 Å for OH...S=C bonds. The greater r_{OH} distance in sulfur bonds can be explained by the greater disperson and charge transfer effects in OH...S than in OH...O hydrogen bonds. The data of Table

Table 3. Δv_{OH}, $R_{O...O(S)}$ and $-\Delta H$ values for OH...O=C and OH...S=C hydrogen bonds

Δv (cm^{-1})	OH...O			OH...S		
	$R_{O...O}$ (Å)	r_{OH} (Å)	$-\Delta H$ (kJmol^{-1})	$R_{O...S}$ (Å)	r_{OH} (Å)	$-\Delta H$ (kJmol^{-1})
200	2.945	0.982	18.3	3.318	0.980	9.7
250	2.900	0.986	21.9	3.280	0.983	12.5
300	2.865	0.989	24.9	3.241	0.990	15.4
350	2.835	0.994	28	3.202	0.996	18.3
400	2.805	0.998	31.1	3.164	1.003	21.3
450	2.770	1.004	34.3	3.126	1.006	24.3

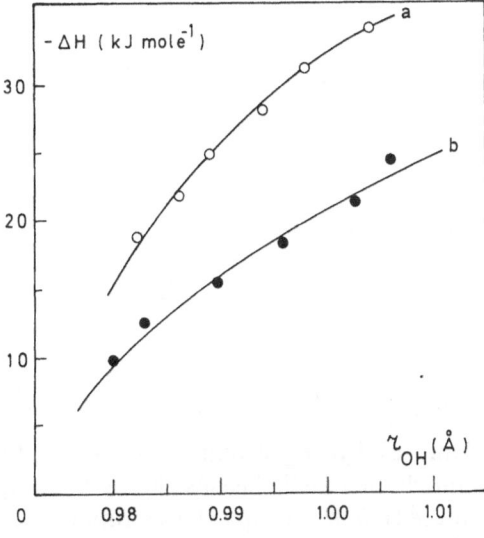

Fig. 5. $-\Delta H$ (kJmol^{-1}) vs r_{OH} (Å)
a) OH...O=C systems
b) OH...S=C systems
(Reyntjens-Van Damme D, Zeegers-Huyskens Th (1979) Bull Soc Belg 88: 1003. Reproduced with the permission of the editor)

2 show indeed that, for a given proton donor, the contribution of the dispersion energy to the total energy is greater for hydrogen bonds involving sulfur atoms. According to the theory of Allen [27] the charge transfer depends on the lone pair extent which is greater for sulfur (2.13 Å) than for oxygen (1.58 Å). According to previous considerations, the greater slope of Eq. (4) is related to the greater importance of covalent character for OH...S bond [30]. This is in agreement with CNDO calculations that have shown that the overlap population is greater for OH...S=C than for OH...O=C bonds (31). The negative intercept of Eq. (4) is associated with the greater repulsion energy for the sulfur bonds [30].

The intercepts of the Badger–Bauer correlation have also been related to the Van der Waals interaction between the proton donor and the solvent [32]. A plot of the heat of evaporation of water against $\Delta v_{OH...B}$ shows a linear region for $\Delta v_{OH...B} > 100 \, \text{cm}^{-1}$ which indicates the validity of the Badger-Bauer rule

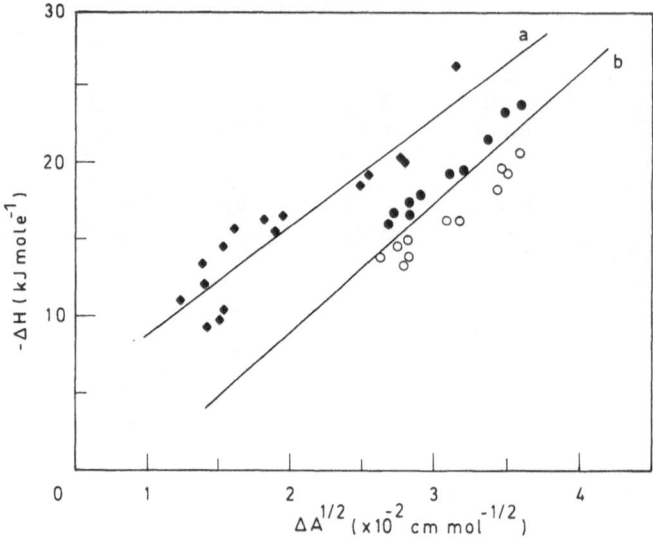

Fig. 6. $-\Delta H$ (kJmol^{-1}) against $\Delta A^{1/2}$ ($\times 10^{-2}$ cm mol$^{-1/2}$)
a) OH...O=C systems
b) OH...S=C systems
(Reyntjens-Van Damme D, Zeegers-Huyskens Th (1979) Bull Soc Chim Belg 88: 1003. Reproduced with the permission of the editor)

in this region. The change of slope for non hydrogen bonding solvents such as benzene and carbontetrachloride is related to the Van der Waals interaction which is of the order of magnitude of 5–10 kJmol^{-1} [32–34].

The intensity increase of the ν_{AH} band by hydrogen bond formation (ΔA) can also be considered as a fundamental characteristic because it is related to the change of the dipole moment of the AH bond during the vibration. The correlation between the square root of the intensity increase $A^{1/2}$ and $-\Delta H$ only holds for very closely related complexes [35] and this is illustrated in Fig. 6 where $-\Delta H$ has been plotted against $\Delta A^{1/2}$ for OH...O=C and OH...S=C systems [26]. This figure shows that for the same $-\Delta H$ value, the square root of the intensity increase is larger for OH...S than for OH...O bonds. This is mainly ascribable to greater charge transfer and dispersion effects, both contributing to an increase of the negative charge at the oxygen atom and positive charge at the hydrogen atom. The perturbation mechanism of Boobyer and Orville–Thomas [36] who consider the intensity peturbation as due to a variation of polarisation of the polarisable lone pair orbital as the AH bond vibrates against it, can also contribute to the increase of the infrared (or Raman) intensity. The contribution of the properties of the ground state and excited state to the experimental infrared intensities are however difficult to separate.

4 Structure of the v_s Band

4.1 Coupling of the v_s Band

In many cases, the v_s band is characterized by several submaxima and the possible origins of this substructure have been thoroughly discussed in Refs. [6] and [7]. All the theories have as their starting point the substantial increase of the mechanical anharmonicity upon hydrogen bond formation.

The vibration term for a given value of the vibrational quantum number v can be written as

$$G_v = \omega_e(v + \tfrac{1}{2}) + X_{11}(v + \tfrac{1}{2})^2 \tag{5}$$

where ω_e is the harmonic frequency and X the anharmonicity constant (expressed in wavenumber units).

The fundamental frequency $v_{01} = G_1 - G_0$, the first overtone is $v_{02} = G_2 - G_0$ and so on

$$v_{01} = \omega_e + 2X_{11} \tag{6}$$

$$v_{02} = \omega_e + 6X_{11} \tag{7}$$

From these, the anharmonicity constant can be deduced

$$X_{11} = \tfrac{1}{2}v_{02} - v_{01} \tag{8}$$

For example, the anharmonicity constant X_{11} for the HF molecule in the gaseous state is equal to $-90 \, cm^{-1}$; in the $(CH_3)_2O...HF$ complex, the experimental frequencies v_{02} ($6485 \, cm^{-1}$) and v_{01} ($3470 \, cm^{-1}$) allow one to calculate a X_{11} value of $-228 \, cm^{-1}$, about a twofold increase from the free molecule [37]. The v_{OH} vibration in free CH_3OH is characterized by an anharmonicity constant of $-90 \, cm^{-1}$ while in the $CH_3OH...N(CH_3)_3$ complex, this value rises to $-210 \, cm^{-1}$ [38], also a twofold increase from the free molecule [38].

Most theories consider the interaction of the fast A—H stretching motion and the slow AH...B bridge stretching motion. There are also other lower frequency bridge vibrations with which v_s can interact.

The combination between the v_{sAH} and $v_{\sigma AH...B}$ stretching motions is often represented in a Stepanov's diagram [30–40]. The $v = 0$ and $v = 1$ potential curves of the v_s vibration are shown in Fig. 7. The energy levels of the combining low frequency vibrations are drawn as vertical straight lines, analogous to Franck–Condon transitions of electronic spectroscopy. This leads to $v_s \pm nv_\sigma$ combinations.

Owing to the anharmonicity of the v_s vibration, the proton will stay on the average, nearer to B in the $v = 1$ state than in the $v = 0$ state and as a

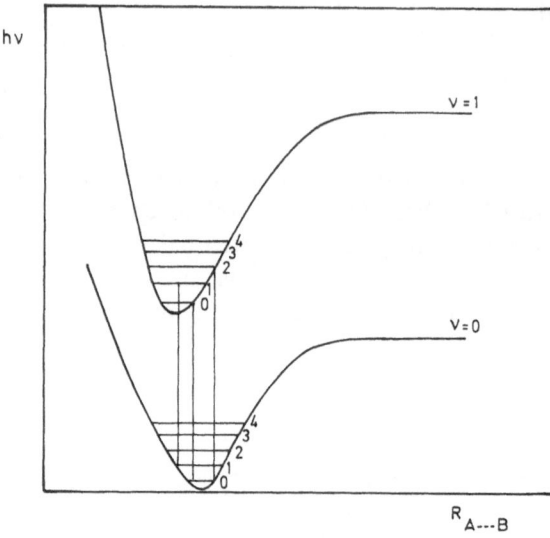

Fig. 7. Stepanov's diagram showing the $v_s \pm n v_\sigma$ combinations. The potential energy is plotted against the $R_{A \cdots B}$ distance in the ground state ($v = 0$) and first excited state ($v = 1$)

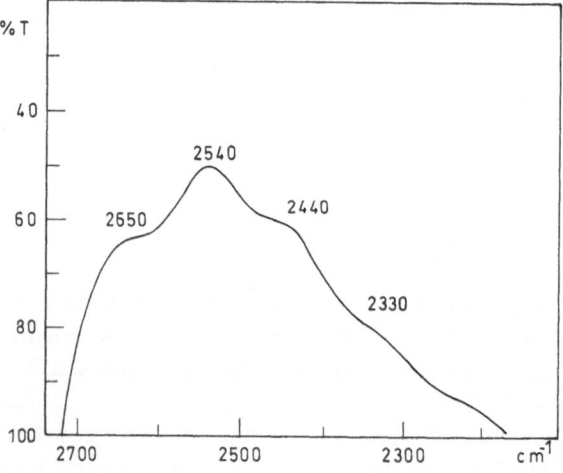

Fig. 8. Infrared spectrum of the complex between methyl n-propylether and HCl in the gas phase. The spectrum is obtained after substraction of the spectra of the components. (Mixture: HCl = 270 nm, ether = 144 nm, 10 cm path length), Reproduced with the permission of the Royal Society of Chemistry of England and of D.J. Millen)

consequence, the hydrogen bond will be stronger in the first vibrational excited state.

Figure 8 reproduces the experimental spectrum of the complex between methyl n-propylether and HCl in the gas phase [41]. The total intensity of the v_s band was found to depend linearly on the product of the partial pressures of the two components so that the possibility of contribution of a 1–2 complex to the band shape was excluded. The spectrum shows a maximum at 2540 cm^{-1} and subbands are 2640, 2440 and 2330 cm^{-1}. These correspond, in order to decreasing wavenumbers to the transitions $v:|0 \leftarrow 1|$ and $v_s:|1 \leftarrow 0|, |0 \leftarrow 0|,$ $|0 \leftarrow 1|$, and $|0 \leftarrow 2|$ (a study of the temperature effect has suggested that the $|0 \leftarrow 0|$ transition is more likely ascribable to the band at 2440 cm^{-1} [42–43]).

The experimental separation of the components is about $110\,\text{cm}^{-1}$, corresponding approximately to the wavenumber of the v_s band, observed in the far infrared region at $115\,\text{cm}^{-1}$ [44]. Millen et al. have studied other systems such as $(CH_3)_2O\ldots HCl$, $(CH_3)_2O\ldots HONO_2$, $(CH_3)_3N\ldots HOCH_3$ also, in the frame of the same theory [45].

It can be concluded that owing to the harmonic coupling the v_s and v_σ are excited by the absorption of one photon and it is therefore interesting to compute the coupling constants between the fast and slow stretching motions.

If we consider a set of quantum numbers v_1, v_2, v_3, the vibrational term G (v_1, v_2, v_3) will depend on the anharmonic constants X_{11}, X_{22} and X_{33} for the three vibrations but also on the anharmonic coupling constats X_{12}, X_{13} and X_{23} between them. This tem can be expressed by adding cross terms to Eq. (5):

$$
\begin{aligned}
G(v_1, v_2, v_3) = {} & \omega_{e1}(v_1 + \tfrac{1}{2}) + \omega_{e2}(v_2 + \tfrac{1}{2}) + \omega_{3e}(v_3 + \tfrac{1}{2}) \\
& + X_{11}(v_1 + \tfrac{1}{2})^2 + X_{22}(v_2 + \tfrac{1}{2})^2 + X_{33}(v_3 + \tfrac{1}{2})^2 \\
& + X_{12}(v_1 + \tfrac{1}{2})(v_2 + \tfrac{1}{2}) + X_{13}(v_1 + \tfrac{1}{2})(v_3 + \tfrac{1}{2}) \\
& + X_{23}(v_2 + \tfrac{1}{2})(v_3 + \tfrac{1}{2}) + \cdots
\end{aligned}
\tag{9}
$$

where ω_{e1}, ω_{e2} and ω_{e3} ar the harmonic frequencies of the three vibrations.

It can easily be shown (8) that the binary combination $v_1 + v_3$ is related to the anharmonicity constants and to the anharmonic coupling constants by the expression

$$
\begin{aligned}
v_1 + v_3 &= \omega_{e1} + \omega_{e3} + 2X_{11} + 2X_{33} + 2X_{13} + \tfrac{1}{2}X_{12} + \tfrac{1}{2}X_{23} \\
&= v_{1(01)} + v_{3(01)} + X_{13}
\end{aligned}
\tag{10}
$$

where $v_{1(01)}, v_{3(01)}$ are the experimental frequencies of the fundamentals of the v_1 and v_3 vibrations.

The coupling constants X_{13} can be computed from these two fundamentals.

The difference tone of v_1 and v_3 is obtained if v_3 is at the $v_3 = 1$ level. The difference between the two terms gives

$$
v_1 - v_3 = \omega_{e1} - \omega_{e3} + 2X_{11} - 2X_{33} + \tfrac{1}{2}X_{12} - \tfrac{1}{2}X_{23} = v_{1(01)} - v_{3(01)}
\tag{11}
$$

Interestingly, the coupling constants cancels out and the frequency of the difference is simply the difference of the frequencies of the two fundamentals.

In the case of other HCl complexes (see Fig. 8), owing to the diffuse character of the $|1 \leftarrow 0|$ band at about $2650\,\text{cm}^{-1}$ the coupling coefficient between the v_s and v_σ vibrations could not be computed.

For the HF complexes, the separation between the subcomponents of the v_s band is greater. For the $(CH_3)_2O\ldots HF$ complex the strong band at $3470\,\text{cm}^{-1}$ is assigned to the v_s fundamental $|0 \leftarrow 0|$ and the side bands at 3710 and 3300 cm^{-1} to the $v_s \pm v_\sigma$ combinations, $|0 \leftarrow 1|$ and $|1 \leftarrow 0|$ respectively [46].

By applying Eqs. (10) and (11) to the v_s and v_σ modes, it can be said that the separation between v_1 and the difference band $(v_s - v_\sigma)$ is equal to $(v_s - v_\sigma)$ but that the separation between v_s and the combination band $(v_s + v_\sigma)$ is equal to $v_s + v_\sigma + X_{s\sigma}$ where $X_{s\sigma}$ is the coupling constant of the v_s and v_σ modes. For the $(CH_3)_2O...HF$ complex, the value of this coupling constant is

$$X_s = (3710 - 3470) - (3470 - 300) = +70 \, cm^{-1}$$

The positive sign reflects the Sheppard effect.

It must be noticed here that other bridge vibrations (in-plane and out-of-plane deformation modes) may combine with v_s.

One of the nicer illustrations of the Stepanov–Sheppard effect is the decrease of $R_{A...B}$ distance in the first excited vibrational level. This distance has been measured by rotational–vibrational spectroscopy for the $CH_3CN...HF$ complex [47]. In the excited state, the $R_{N...F}$ distance is equal to 2.762 Å: this is a 0.034 Å shortening relative to the corresponding ground vibrational state value [48].

The infrared spectrum of the acetic acid dimer in the gas phase shows a central broad OH band at 3027 cm^{-1} and several subbands, the strongest ones being observed at 3130, 3072 and 2958 cm^{-1}. All the observed peaks could be accounted for as combinations of internal modes $(v_{C=O} + \delta_{OH}, v_{C=O} + \delta_s CH_3,$ $v_{C=O} + v_{C-O}$ and so on) [49] not involving the hydrogen bridge vibration observed at 188 and 168 cm^{-1} [50]. In this case the v_s band does not show any progression in the v_σ mode. According to the theory of Marechal, the $v_s - v_\sigma$ coupling is however the central feature in the band shaping mechanism [51]. Models have been proposed to take into account the $v_s - v_\sigma$ anharmonic coupling and the Fermi resonance type interaction between v_s and the internal modes simultaneously. [It must be remembered here that the Fermi resonance originates from an interaction between a fundamental vibration and an overtone (or combination band). When these two vibrational levels of the same symmetry be close together, the presence of cubic or higher order anharmonic terms in the potential energy function results in an interaction between the levels which are displaced. The intensities of the transition involving these levels may also be greatly altered. This effect was first investigated by Fermi in 1931]. The resulting v_s band is broad consists in a progression in the v_s mode but this progression is perturbed at frequencies $2v_i$ (or $v_i + v'_i$), the v'_i values being the internal modes of the proton donor or proton acceptor molecule [7, 52]. In the previously discussed complex between ethers and HCl (or HF), the v_s band does not show any subcomponents that could be assigned to combinations with the internal modes of the ether moiety. The complexes investigated so far in the gas phase are of moderate strength and the breadth of the v_s HCl band is between 300 and 400 cm^{-1}.

In the liquid state, the anharmonic coupling between v_s and v_σ is again a central feature but there is a new factor namely the coupling of v_σ through a fluctuating potential with the surroundings which has the result that the v_s

mode rapidly losses its phase coherence resulting in a broad structureless band. However, if the damping of the v_σ mode is small enough, the progression of the v_σ mode will be observed [53]. When the v_s band is broad and extends to several hundrd wavenumbers, the probability of a Fermi resonance interaction with the internal modes increases. In the infrared spectra of selfassociated phenol derivatives [54] and the complexes involving phenol and pyridine derivatives [55], the v_s band does not show any v_σ progression; this is probbly ascribable to the high damping of this mode. The subbands have been ascribed as arising from overtones or combination bands of the proton donor, whose intensity was increased by Fermi resonance with the v_s band.

The infrared spectra of the complexes between phenol (a), 4-Cl phenol (b) or 3,4-diCl phenol (C) and N-methylimidazole (solvent = carbontetrachloride)

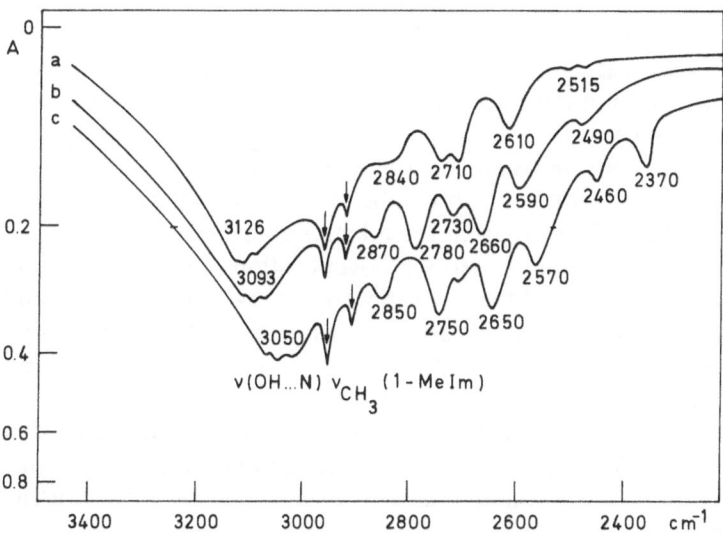

are reproduced in Fig. 9. These spectra show that the v_s (OH...N) values are ordered according the strength of the hydrogen bond, the maximum of the v_s(OH...N) band being observed at $3126\,cm^{-1}$ (phenol), $3093\,cm^{-1}$ (4-Cl

Fig. 9. Infrared spectra in the v_s region ($3400–2300\,cm^{-1}$) of the hydrogen bond complexes between phenol derivatives and N-methylimidazole. Solvent = CCl_4. $T° = 298\,K$

a) phenol; *b*) 4-Cl phenol; *c*) 3,4-diCl phenol (Haulait-Pirson MC, Bogaerts S, Zeegers-Huyskens Th (1977) Bull. Soc. Chim. Belg. 87: 928. Reproduced with the permission of the editor)

Table 4. Spectral data for the complex between 4-Cl phenol and N-methylimidazole. Solvent = CCl$_4$

v_s(OH...N) band		vibrations of 4-Cl phenol (1500–1150 cm^{-1})		
Observed submaxima (cm^{-1})	Possible assignment	Free (cm^{-1})	Bonded (cm^{-1})	Assignment and symmetry[a]
2870	a + c = 2872	1496	1500	v_{cc}(A$_1$)	a
2780	b + c = 2812	1426	1440	v_{cc}(B$_2$)	b
2730	2c = 2744	1322	1372	v_{cc} + δ_{OH}(B$_2$)	c
2660	b + e = 2676	1270	1264	$\sim v_{C\cdots O}$(A$_1$)	d
2590	c + e = 2607	1178	1235	δ_{OH} + v_{cc}(B$_2$)	e
2490	d + e = 2499				

[a] v and δ indicate streching and in-plane deformation vibrations

phenol) and about 3050 cm^{-1} (3,4-diCl phenol). The v_s band is broad and the breadth of the band amounts to about 700 cm^{-1} for the 3,4-diCl phenol complex. Further, the v_s band shows to its low frequency side, between 2900 and 2300 cm^{-1}, several sub bands depending on the nature of the proton donor molecule. This strongly suggests that the internal vibrations of the phenol molecules are involved in the band shaping mechanism. Table 4 lists the observed wavenumbers of the submaxima observed between 2870 and 2490 cm^{-1} for the complex between 4-Cl phenol and N-methyl imidazole. This table also indicates the wavenumbers of some vibrations of 4-Cl phenol free and complexed with the base along with their assignment and symmetry. The para-phenol derivatives are considered as belonging to the C$_{2v}$ symmetry with vibrations of the A$_1$, A$_2$, B$_1$ and B$_2$ species. If we admit that the complex is planar, only the in-plane donor modes combine to interact with v_s. Some possile combinations are indicated in Table 4. There is however a plethora of combinations which could give rise to the observed bands.

The v_i (fundamental) bands can also be observed as transmission windows (Evans holes) which have their origin in the interaction between a broad level of energy E$_1^0$ and a narrow level of energy E$_2^0$. Application of the perturbation redistributes the energy levels and the intensities to yield two new levels E$_1$ and E$_2$. The resulting shifting of the levels leads to a fall in energy-level density. This yields within the broad band a narrow region of increased transmission [57]. Many transmission windows have been observed in hydrogen bonded complexes in the solid state or in solution especially in very strong hydrogen bonds where the v_s band maxima is observed between 1500 and 500 cm^{-1}; this region (the fingerprint region) contains indeed several internal modes of the separated molecules.

The ABC bands observed in very strong and short hydrogen bonds originate from the same mechanism. The minima observed in the infrared (and Raman spectra) have been assigned to Evans windows arising from a strong $v_{OH} \leftrightarrow 2\delta_{OH} \leftrightarrow 2\gamma_{OH}$ interaction [58]. This is illustrated schematically in Fig. 10

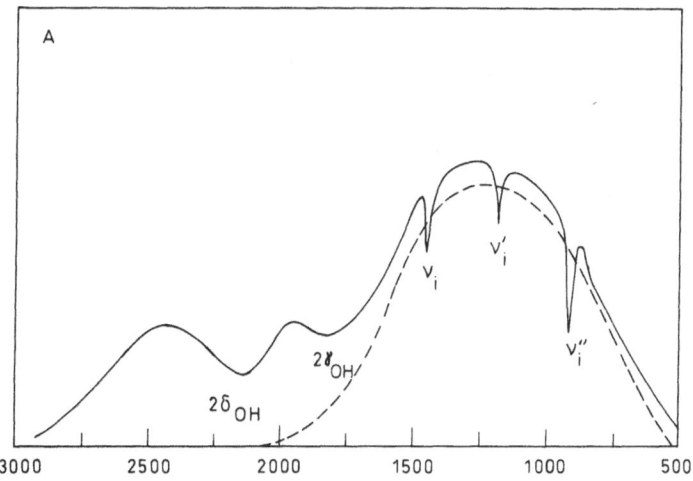

Fig. 10. v_s (OHO) band in a strong hydrogen bond ($R_{O...O} = 2.47 \text{ Å}$)

for a very strong asymmetrical OHO hydrogen bond ($R_{O...O} = 2.47 \text{ Å}$) where the v_{sOHO} absorption extends from about 2500 to 400 cm^{-1}. The maximum of the v_{sOHO} band is observed at about 1400 cm^{-1}; this band is interrupted by narrow transmssion windows between 1500 and 700 cm^{-1}. Further, the two minima observed at 2200 and 1900 cm^{-1} are ascribed to the interaction between the v_{OH} and the $2\delta_{OH}$ and $2\gamma_{OH}$ levels where the fundamental absorptions are observed at about 1150 and 950 cm^{-1}. It is interesting to note that Fig. 10 reproduces the v_{sOHO} for a complex characterized by about the same strength, but without perturbation between the vibrational levels.

Evans windows have also been observed in the vibrational spectra of strong hydrogen bonds where the proton acceptor is a halide anion. Figure 11 reproduces the infrared spectrum between 2800 and 2100 cm^{-1} of triethylammonium chloride in 1,2-dichloroethane. The broad absorption culminating at 2310 cm^{-1} is ascribed to the stretching vibration of the NH$^+$ group bonded to the Cl$^-$ ion. A band with a lower intensity is observed at 2540 cm^{-1} and a minimum at 2510 cm^{-1}. When a phenol derivative is added to the solution, the main band is shifted to higher wavenumbers but is splitted into two main components lying at 2470 and 2580 cm^{-1} (phenol) and 2480, 2580 and 2460 cm^{-1} (4-Cl phenol). The overall shift to higher wavenumbers can be understood by the cooperative effect which weakens

$$N\!-\!H^+\!\cdots\!Cl^-\!\cdots\!H\!-\!O\!-\!\text{(phenyl ring)}$$

the NH$^+$...Cl$^-$ hydrogen bond when a second bond is formed with the Cl$^-$ ion.

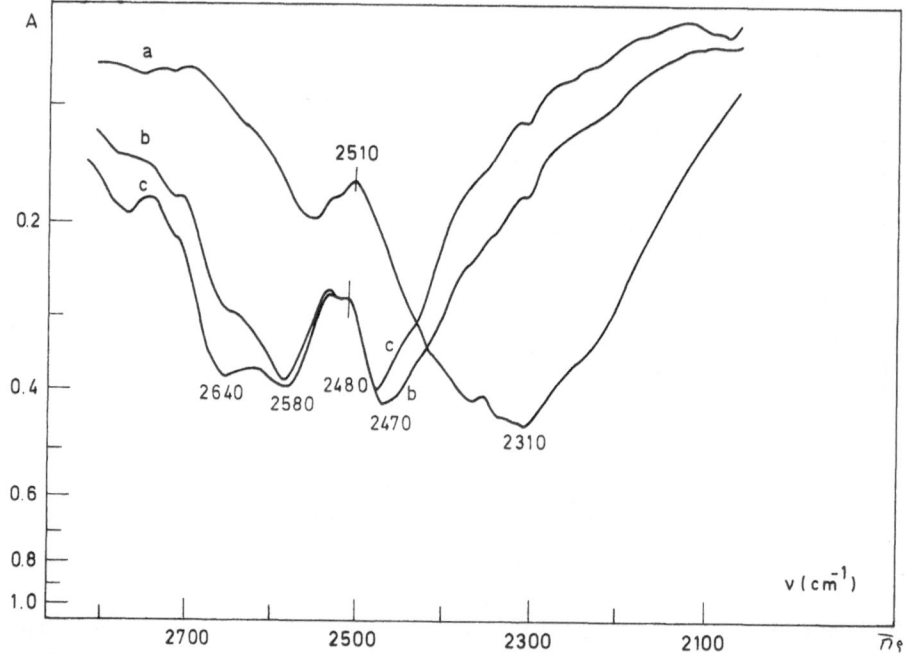

Fig. 11. Infrared spectra (2800–2100 cm^{-1}) of
a) Et$_3$NH$^+$Cl$^-$ in 1,2-dichloroethane (c = 5 × 10^{-2} M)
b) the same solution after addition of phenol (c = 0.22 M)
c) the same solution after addition of 4-Cl phenol (c = 0.22 M)
(Rulinda B, Zeegers-Huyskens Th (1978) Bull Soc Chim Belg 87: 329. Reproduced with the permission of the editor)

Independently on the position of the main v_s(NH$^+$...Cl$^-$) absorption, the minimum is observed at 2510 cm^{-1}. This strongly suggests that this minimum is an Evans hole having its origin in a vibrational mode insensitive to the strength of the NH$^+$...Cl$^-$ bond. This excludes any contribution of the $2\delta_{NH^+}$ or $2\gamma_{NH^+}$ vibrations. The rocking mode of the CH$_3$ group (r_{CH_3}) is observed at 1280 cm^{-1} in the binary and ternary solutions and its first overtone should be observed at wavenumbers slightly lower than 2560 cm^{-1}. Further, the r_{CH_3} is degenerate and its first overtone has the symmetry

$$ExE = A_1 + A_2 + E$$

The A$_1$ component has the same symmetry as the v_s vibration [59].

The formation of Fermi resonance windows is less probable in low-temperature matrices because the v_s levels are usually narrower. The $v_{s(HCl)}$ band of the N,N-dimethylacetamide. HCl complex isolated in an Argon matrix at 10 K is observed at 740 cm^{-1}; this absorption is moderately broad but is however interrupted by a narrow transmission window corresponding to the $v_{C-C} + v_{CNC}$ combination which has the same symmetry (A') as the v_s band [60].

4.2 Tunneling Transition

The splitting of the vibrational levels originating from the tunnel effect is discussed in detail in Chap. V. The splitting of the $v = 0$ and $v = 1$ vibrational levels is schematically shown in Fig. 12 for a symmetrical (a) and slightly asymmetrical (b) double minimum potential function. When the potential function is symmetric, the transitions $|0,0^+\rangle \rightarrow |0,0^-\rangle$ (tunneling transition), $|0,0^+\rangle \rightarrow |1,0^-\rangle$ and $|0,0^-\rangle \rightarrow |1,0^+\rangle$ are allowed. These two last transitions should be observed in the infrared with a separation of a few hundred cm^{-1}. If the double minimum potential function is asymmetric, all four transitions indicated in Fig. 12 are allowed. The difficulties of interpreting experimental infrared spectra in terms of double minimum potential have been discussed by Somorjai and Hornig [61] and as discussed before, several mechanisms can be surimposed on the band splittings. For example, adducts of mono-, di- and trichloroacetic acids with sulfoxides and phosphineoxides have been interpreted initially in terms of $|0,0^+\rangle \rightarrow |1,0^-\rangle$ and $|0,0^-\rangle \rightarrow |1,0^+\rangle$ transitions [62]. In compounds such as KH_2PO_4, the separation of the doublets is too large to be attributed to split vibrational transitions and have been ascribed to a Fermi resonance interaction with the $2\delta_{OH}$ and $2\gamma_{OH}$ vibrations [63].

The $v_{s(BH^+...B)}$ band of many homoconjugated (or heteroconjugated) nitrogen bases shows in many cases two components. The origin of this splitting has been thoroughly investigated by Wood and his coworkers [64–68].

Figure 13 reproduces the infrared spectra in the v_s region of several solutions containing the complex cation $(B_1H^+B_2)$ where the different bases are in order of decreasing basicity: N-methylpiperidine, trimethylamine, N-methylimidazole, 4-methylpyridine, pyridine, thiazole and isooxazole [64]. An increase of the ΔpK_a value (difference of the pK_a values of the two bases) generally parallels a decrease of the hydrogen bond energy and a more asymmetric potential curve for the proton motion. The experimental spectra provide several indications that the doubling of the $v_{s(NH...N)}$ + band is not a direct reflection of proton tunneling. These indications are: a) the widths of the v_s bands. Normally, the $(v_s)_H$ band is found to be broader than the $(v_s)_D$ band; in the present

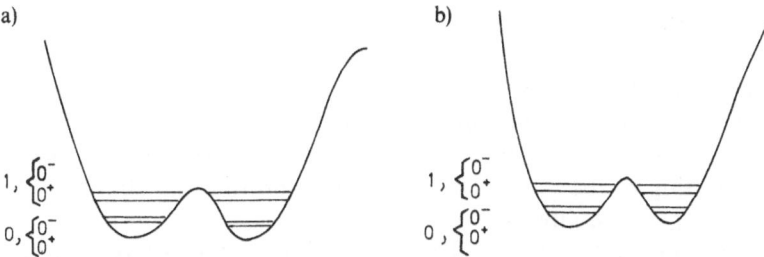

Fig. 12. Splitting of the $v = 0$ and $v = 1$ vibrational levels for a symmetrical **a)** and slightly asymmetrical **b)** double minimum potential

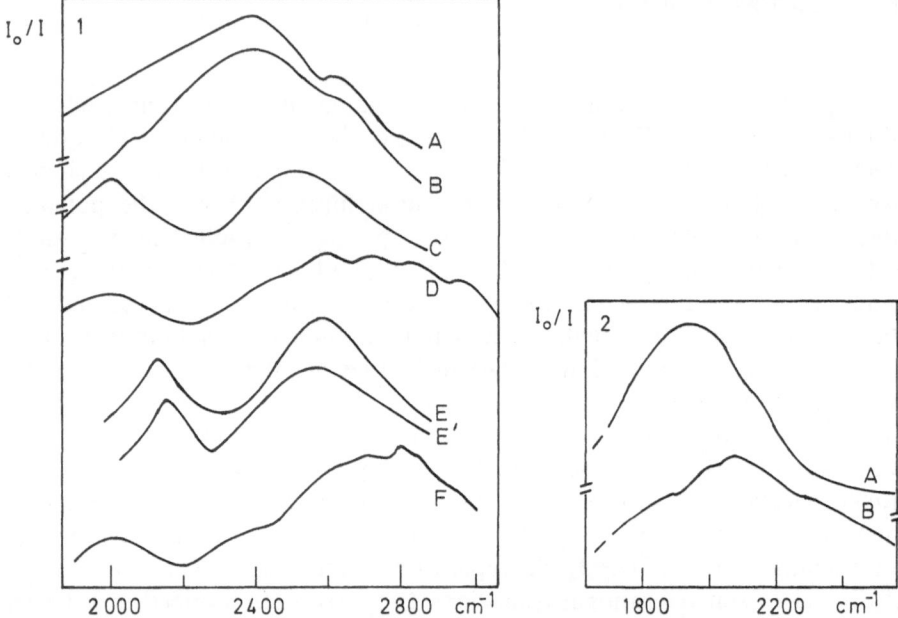

Fig. 13 a. v_s of (*A*) trimethylammonium fluoroborate in NMe imidazole, (*B*) NME piperidinium perchlorate in NMe imidazole, (*C*) NME imidazolinium salts in pyridine, (*D*) NME imidazolinium perchloate in thiazole, (*E*) pyridinium fluoroborate in thiazole, (*E'*) 4-Me pyridinium tetraphenylborate in thiazole, (*F*) thiazolinium perchlorate in isooxazole
b v_s bands of (*A*) NMe imidazolinium (D^+) in pyridine, (*B*) NME imidazolinium (D^+) in thiazole.
(Dean JL, Wood JL (1975) J Mol Struct 26: 215. Reproduced with the permission of Elsevier Science Publ.)

complexes, this is far from the case, the $(v_s)_H$ bands individually being narrower than $(v_s)_D$. b) the intensity ratio of the upper and lower components of v_s decreases with increasing ΔpK_a. As can be seen in Fig. 13, the intensity of the lower component is generally lower when the difference between the frequencies is greater. We can compare as for example spectrum E (NHN$^+$ system with $\Delta pK_a = 2.75$) and spectrum F (NHN system with $\Delta pK_a = 4.55$). These spectroscopic characteristics suggest that in $(v_s)_H$ we are dealing with essentially a single band, the splitting arising essentially from a Fermi interaction (probably with the $2\delta_{NH}$ overtone) [64].

If the doublet arose from a proton tunneling it would be extremely sensitive to the proton potential implying in these systems almost exactly the same potential throughout. This is highly unlikely owing to the strong difference in basicities of the two nitrogen bases forming the complex cations. These examples show that the splitting of the v_s band arising from a proton tunneling is very difficult to prove experimentally.

It has also been shown that the doublet observed in the far-infrared spectra of the acetic acid dimer is not due to a proton tunneling. The structure of the gaseous acetic acid dimer is planar (symmetric group C_{2h}) and six low-frequency intermolecular modes should be observed. The highest frequency band extending from 200–150 cm^{-1} was assigned to the b_u vibration [69]. This band was shown to have a doublet structure with maxima at 188 and 167 cm^{-1} and to disappear in OD dimers. As a result, the doublet structure was ascribed to proton tunneling. A careful reinvestigation has however shown that the doublet still remains in the OD dimer, the frequencies of the maxima being observed at 187 and 163 cm^{-1} [70] and disappearing in carbontetrachloride solution. The spectra have been discussed in terms of the rotational envelope, the most plausible origin being a combination of the effects of Fermi resonance with a broad perpendicular-type structure.

Sobczyk et al. [72] have investigated the infrared spectrum of quinuclidine-3-one hemiperchlorate in the solid state

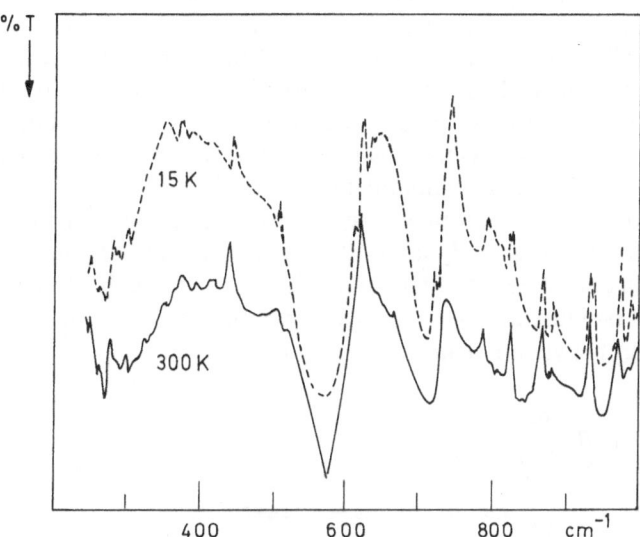

At low tempeature, this cation is centrosymmetric with the proton located midway between the two nitrogen atoms. At room temperature, the bridge becomes longer by about 0.02 Å and the diffraction pattern seems to indicate two maxima for the proton density distribution [71, 72].

Fig. 14. Temperature effect on the low frequency wing of the broad absorption in polycrystalline quinuclidin-3-one hemiperchlorate.
(Grech E, Malarski Z, Ilczyszyn M, Czupinski O, Sobczyk L (1985) J Mol Struct 128: 249. Reproduced with the permission of Elsevier Science Publ. and L. Sobczyk)

The infrared spectrum of this homoconjugated cation in the crystalline state is characterized by a very broad absorption extending from about 2500 to about $200\,cm^{-1}$, interrupted by several transmission windows [73]. No usual temperature effect is observed for the absorption above $1000\,cm^{-1}$. At low temperature (15 K) however a great increase of the intensity for the absorption at about $400\,cm^{-1}$ is observed (Fig. 14). In this case, the decrease of the temperature leads to a symmetrization and shortening of the bridge, and as a consequence to an increase of the intensity of the absorption characterizing the very short symmetric hydrogen bonds. Isotopic substitution causes an elongation of the bridge; deuteration of quinuclidin-3-one hemiperchlorate does not change the position of the centre of gravity of the broad band but strongly decreases the intensity of the absorption at $400\,cm^{-1}$ (Fig. 14). This spectroscopic behaviour suggests that the $400\,cm^{-1}$ absorption may be related to a tunneling transition. However, this conclusion cannot be considered as definitive owing to the extreme breadth of the bands and the other surimposed mechanisms contributing to the band shaping. In this system, the intermolecular stretching mode is observed at much lower frequencies, at about $110\,cm^{-1}$ [73].

5 Application of Infrared Spectroscopy to the Study of Hydrogen Bonds Involving Biological Molecules

The majority of bases involved in important biological processes are characterized by more than one potential acceptor site and in many cases, the vibrational spectroscopy allows one to precise this interaction site. When the vibrational are well localized within a given bond or with other words, when they can be considered as more or less "pure", the perturbations brought about by complex formation may be used to precise the preferential hydrogen bond site(s). For most of the carbonyl derivatives as for example, the band observed between 1800 and $1600\,cm^{-1}$ has a predominant $v_{C=O}$ character and its shift to lower frequency indicates that complex formation occurs at the carbonyl function. Comparison of the thermodynamic data and the shift of the v_{AH} stretching vibration with complexes involving monofunctional systems can also be very useful. This will be illustrated by some examples the first one being the interaction between [(diphenylmethylene)amino]-acetonitrile

DPMAA

and phenol derivative studied in carbontetrachloride solution. The Schiff base has two potential sites available for hydrogen bond formation, the nitrogen atoms of the imine and of the nitrile function. For the complex between phenol and the Schiff base, the formation constants at 298 and 323 K are 6.6 and 3.7 M^{-1}, the enthalpy of complex formation -16.6 kJmol^{-1} and the entropy variation 39 Jmol^{-1}K^{-1} [74]. These values are very similar to those obtained for the complex involving the same proton donor and aliphatc nitriles [4]. The complexes formed between phenol derivatives and imines are characterized by higher formation constants and enthalpies of formation [75]. Further, only one band is observed in the v_{OH} region and its frequency shift of 150 cm^{-1} is very similar to that observed in the complex between phenol and acetonitrile. In Fig. 15, the Badger–Bauer correlation is compared for hydrogen bonds involving hydroxylic proton donors and DPMAA, simple aliphatic nitriles and pyridines. The points for DPMAA and nitriles are situated on the same straight line, the pyridine complexes being characterized by different slope and intercept. The $v_{C\equiv N}$ band observed at 2255 cm^{-1} in the free base shifts to higher wavenumbers in the complexes, while the position of the $v_{C=N}$ band at 1620 cm^{-1} remains practically unchanged. These two vibrations can be considered as sufficiently decoupled from the rest of the molecule. The thermodynamic and spectroscopic data strongly suggest that the preferential hydrogen bonding site is the C$\equiv$N group. This conclusion leads to a failure of the general rule: the energy of the hydrogen bond increases with decreasing s character of the lone pair [76]. The hydrogen bonding site cannot be predicted by energetic factors alone but is also governed by the accessibility of the free lone pair which is in the case of DPMAA higher for the N(sp) than for the N(sp^2) atom [75].

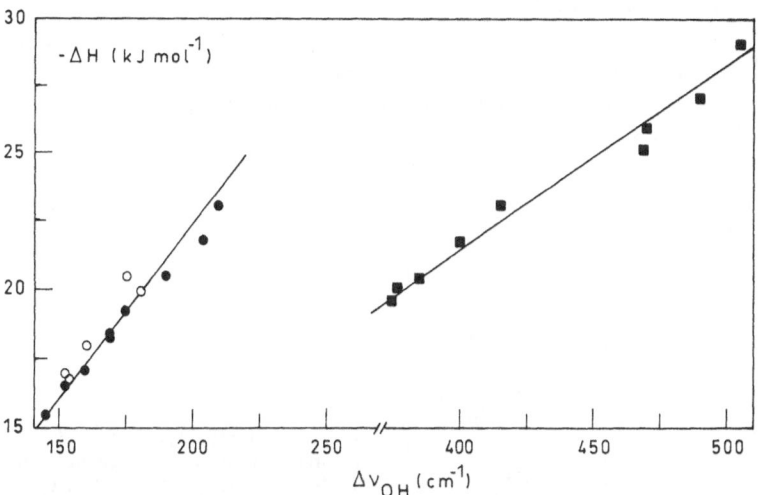

Fig. 15. $-\Delta H$ (kJmol^{-1}) vs Δv_{OH} (cm^{-1}) correlation. Complexes between phenol derivatives and ● DPMAA, O aliphatic nitriles, ■ substituted pyridines. Solvent = carbontetrachloride. The values for DPMAA are from Ref. [75]. The values for nitriles and pyridines are from Ref. [4]

Nicotinamide, methylnicotinate and related derivatives are important biological molecules since they are constituents of the coenzymes dinucleotides which are known to play an important role in a large number of dehydrogenases [77].

These molecules can be considered as pyridine derivatives bearing carbonyl substituents:

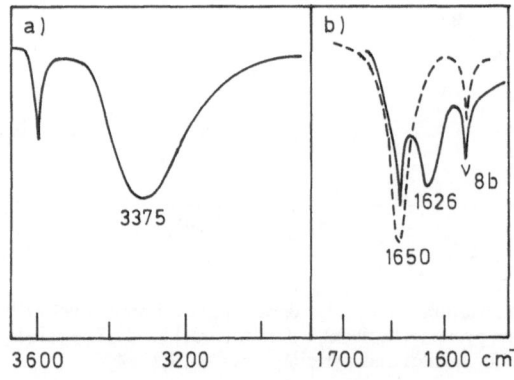

N,N-diethyl-
nicotinamide
(DENA)

methyl-
nicotinate
(MNA)

3-acetyl-
pyridine
(ACP)

These three molecules possess mainly two sites available for hydrogen bond formation; the nitrogen atom of the pyridine ring and the oxygen atom of the carbonyl bond. The wavenumber of the $v_{C=O}$ vibration will decrease by complex formation on the oxygen atom and increase when complexation occurs at the nitrogen atom. As previously discussed, the $v_{OH...}$ stretching vibration is usually less shifted for OH...O than for OH...N hydrogen bonds. Further, when complexation is formed at the nitrogen atom of a pyridine derivative some typical ring vibrations such as the v_1 (990 cm^{-1}), v_{6a} (605 cm^{-1}) and v_{8b} (1580 cm^{-1}) modes are shifted to higher frequencies. The thermodynamic data can also be compared with those available for model systems such as pyridine, aliphatic amides, esters and ketones.

Figures 16 and 17 reproduce the infrared spectra in the v_{OH} and $v_{C=O}$ region of the complexes involving DENA or MNA and a phenol derivative. The spectrum of the DENA complex shows only one band in the v_{OH} region, whose frequency (3378 cm^{-1}) is typical for the OH...O=C interaction; the $v_{C=O}$ band

a)

3375

b)

v_{8b}
1626
1650

3600   3200   1700   1600 cm^{-1}

Fig. 16. Infrared spectrum of the complex between 3Br phenol and DENA in a) the v_{OH} region, b) the $v_{C=O}$ region (——spectrum of the complex, –––spectrum of the free base at the same concentration) (adapted from Ref. [77])

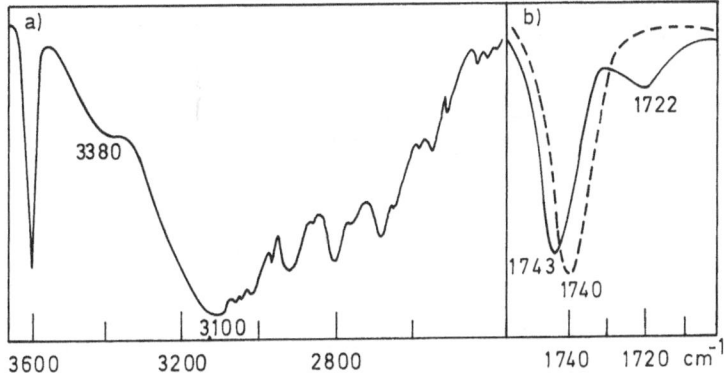

Fig. 17. Infrared spectrum of the complex between 3,5-diCl phenol and MNA in *a*) the v_{OH} region, *b*) the $v_{C=O}$ region (——spectrum of the complex, ――― spectrum of the free base at the same concentration). (Adapted from Refs. [78 and 79])

obseved at $1650\,cm^{-1}$ in free DENA is shifted to $1626\,cm^{-1}$ in the complex and the v_{8b} vibration of the pyridine ring remains unchanged in the complex. Further, the slope and intercept of the Badger–Bauer correlation are similar to these obtained for the complexes between the same proton donors and N,N-dimethylacetamide. The spectroscopic and thermodynamic data strongly suggest that complex formation occurs at the oxygen atom of the carbonyl group [77].

In contrast, the complexes between MNA and phenol derivatives show in the v_{OH} region two absorptions between 3400 and $3300\,cm^{-1}$ and between 3150 and $2950\,cm^{-1}$, assigned to OH...O=C and OH...N (pyridine) hydrogen bonds. Interestingly, the $v_{OH...N}$ band also shows at its low frequency side several submaxima originating from a Fermi resonance with the internal vibrations of the proton donor. In the $v_{C=O}$ region the bands shifted to higher wavenumbers ($\Delta v = +3\,cm^{-1}$) and to lower wavenumbers ($\Delta\tilde{v} = -22\,cm^{-1}$) are assigned to OH...N and OH...O=C complexes respectively. The intensity of the absorptions suggests that more complexes are formed on the pyridine nitrogen than on the oxygen atom. Similar effects are observed for ACP but the concentration of nitrogen complexes seem to be higher than in the case of MNA [78, 79].

The concentration of carbonyl complexes is related to the basicity of the nitrogen atom of the heterocyclic ring and to inductive and mesomeric effects exerted by the $N(R_y)_2$, OCH_3 and CH_3 groups.

It is also noteworthy that the proportion of nitrogen complexes increases with the acidity of the proton donor and that the three molecules are protonated on the pyridine nitrogen. This can be easily deduced from the strong perturbation of the pyridine vibrations such as the v_{8a}, v_{8b}, v_{19b}, v_{19a} and v_{11} vibrations [78].

The proton acceptor ability of the nucleic acid bases has also been studied. These bases have a very low solubility in the usual organic solvents and this

Table 5. Molecules of biological interest, hydrogen bonding site for the interactions with a proton donor and investigated vibrations

Base molecule	Proton donor	Interaction site(s)	Vibration(s)
N,N-diethylnicotinamide	phenol derivatives[a] HCl[b]	C=O N(pyr)	$\nu_{OH}, \nu_{C=O}, \nu_8(pyr)$ $\nu_{NH^+...Cl^-}, \nu_{C=O}, \nu_8, \nu_9(pyr)$
methylnicotinate	phenol derivatives[a] HCl	C=O N(pyr) N(pyr)	$\nu_{OH}, \nu_{C=O}, \nu_{C-O}$ $\nu_{NH^+Cl^-}, \nu_{C=O}$
nicotine	phenol derivatives	N(pyr)	$\nu_{OH}, \nu_{pyridine}$ and $\nu_{pyrrolidine}$
(diphenylmethylene) aminoacetonitrile	phenol derivatives	≡N	$\nu_{OH}, \nu_{C\equiv N}, \nu_{C=N}$
Ethyl-*N*(diphenylmethylene) glycinate	phenol derivatives[a]	C=N, C=O	$\nu_{OH}, \nu_{C=N}, \nu_{C=O}, \nu_{C-O}$
3-methyl-4-pyrimidone	phenol derivatives[a] $(pK_a = 10 - 7.8)$ phenol derivatives[a] $(pK_a = 5 - 3)$ Picric acid[b]	C=O C=O, N(pyr) N(pyr)	$\nu_{OH}, \nu_{C=O}$ $\nu_{OH}, \nu_{C=O}$ $\nu_{NH^+-O^-}, \nu_{C=O}$
N,N-dimethyluracil	phenol derivatives[a]	C_4=O	$\nu_{OH}, \nu_{C=O_2}, \nu_{C=O_4}$

Table 5. *(Contd.)*

Base molecule	Proton donor	Interaction site(s)	Vibration(s)
1,4,4-trimethylctyo-sine	phenol derivatives[a] HBr[b]	$C=O$ N_1	$\nu_{OH}, \nu_{C_2=O}$ $\nu_{NH^+\cdots Br^-}, \nu_{C_2=O}$
N,N,1,9-tetramethyl-guanine	H_2O[a] phenol derivatives[a] HBr[b]	$C=O, N_7$ N_7	$\nu_{OH}, \nu_{C=O}$, ring vibrations $\nu_{C=O}, \nu_{NH^+}, \nu_{NH^+}$ γ_{NH^+}, ring vibra-tions
caffeine	H_2O[a] phenol derivatives[a] H_2O[b]	$C_2=O, C_6=O$ N_9	$\nu_{OH}, \nu_{C_2=O}, \nu_{C_6=O}$ $\nu_{OH}, \nu_{C_2=O}, \nu_{C_6=O}$
9-methylpurine	3,5-dichlorophenol[a]	N_1, N_7	N_1, N_3, N_7 Nitrogen-15 NMR
6,6,9-Trimethyladenine	3,5-dichlorophenol[a]	N_3	N_1, N_3, N_7 Nitrogen-15 NMR

[a] In solution; [b] in the solid state.

difficulty can be overcome by *N*-methylation of the molecules. The complexing ability of 1,3-dimethyluracil, *N,N*,3-trimethylcytosine [80] and *N,N*,1,9-tetramethylguanine [81] towards proton donors has been investigated. The results are summarized in Table 5 and will be discussed more in detail for tetramethylguanine (TMG) which possesses several potential sites available for hydrogen bond formation; the O_6, N_3 and N_7 atoms. The interaction between

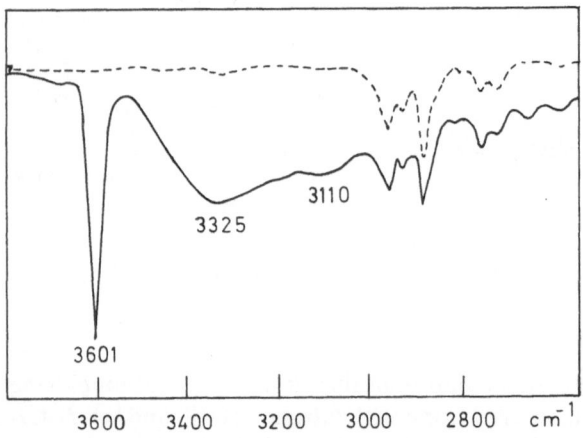

TMG and water or phenol derivatives has been investigated in tetrachloroethylene (v_{OH} region) and 1,2-dichloroethane (v_{C-O} region) by Fourier-transform
infrared spectrometry (FT-IR). Figure 18 reproduces the FT-IR spectrum in the
v_{OH} range of the complex between TMG and 3,5-dichlorophenol. Two bands
are observed at 3325 and 3110 cm^{-1}. These bands are assigned to OH...O=C
and OH...N complexes respectively by comparison with literature data. This
interpretation is strengthened by the analysis of the FT-IR spectrum in the
$v_{C=O}$ region. As shown in Fig. 19, the free v_{C-O} absorption of TMG is observed
at 1691 cm^{-1} (spectrum 1). In the presence of 3,5-dichlorophenol, the intensity
of this band decreases and a new absorption at 1673 cm^{-1} is observed. A
shoulder to the higher frequency band can also be noticed (spectrum 2). The
difference spectrum (3) clearly shows an absorption at 1704 cm^{-1}. The bands
at 1704 and 1672 cm^{-1} are assigned to complexes formed on one of the N
atoms and the C=O function. The PED distribution of the guanine derivative
shows that the C=O vibration is more or less "pure" but that the pyrimidine
and imidazole ring vibrations have a strong "mixed" character [82]. As a
consequence perturbations of the ring modes induced by complex formation
do not allow to make a difference between the OH...N$_3$ and OH...N$_7$ hydrogen

Fig. 18. FT-IR spectrum (3700–2750 cm^{-1}) of the complex between 3,5-diCl phenol and TMG
(solvent = tetrachloroethylene)
– – –spectrum of free TMG
(De Taeye J (1987) PhD Thesis, University of Leuven)

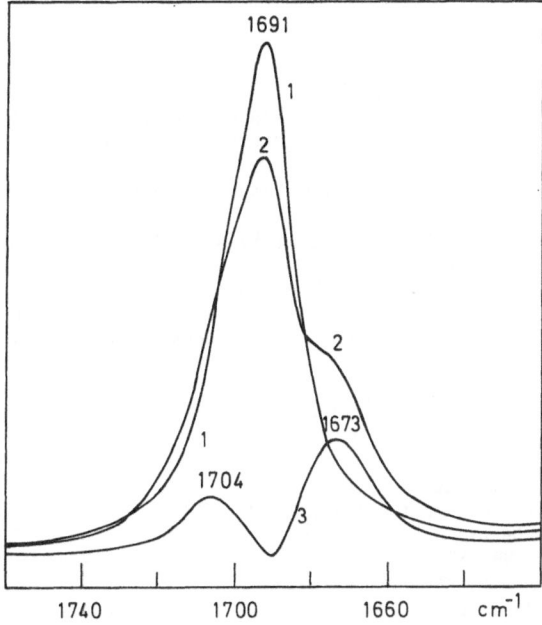

Fig. 19. FT-IR spectrum (1740–1620 cm^{-1}) of
1 TMG
2 TMG + 3,5-dichlorophenol
3 difference spectrum between 2 and 1
(solvent = 1,2 dichloroethane)
(De Taeye J (1987) PhD Thesis, University of Leuven)

bonds. The experimental values of the ionization potentials of the free electron pair of the different N atoms, the charge densities and ab initio calculations on the structure of the guanine-water complex suggests that hydrogen bond formation at N_7 is slightly preferred over N_3 [83].

In adenine dervatives as for example 6,6,9-trimethyladenine

several basic sites are available for hydrogen bond formation mainly the N_1, N_3 and N_7 atoms. Owing to the strong mixed character of the imidazole and pyrimidine vibrations and to the fact that the pyrimidine vibrations are associated with a nuclear motion of the N_1 and N_3 atoms [84], the vibrational spectroscopy does not allow in· this case a determination of the interaction site. This site has been investigated by Nitrogen-15 NMR spectroscopy [85].

6 References

1. Legon AC, Millen DJ (1986) Proc R Soc Lond A404: 1096
2. Pimentel GC, McClellan Al (1960) The hydrogen bond. W.H. Freeman, San Francisco, p 75
3. Vinogradov S, Linnell RH (1971) Hydrogen bonding. Van Nostrand, New York, p 47
4. Joessten MD, Schaad LJ (1974) Hydrogen bonding. Marcel Dekker, New York, p 2
5. Wood JL (1974) In: Yarwood J (ed) Spectroscopy and structure of molecular complexes. Plenum, London, p 303
6. Hadzi D, Bratos S (1976) In: Schuster P, Zundel G, Sandorfy C (eds) The Hydrogen Bond Part II, North Holland, Amsterdam, p 567 and references herein
7. Bratos S, Lascombe J, Novak A (1980) In: Ratajczak H, Orville-Thomas WJ (eds) Molecular Interactions, vol 1, John Wiley & Sons, Chichester Brisbane Toronto, p 301 and references herein
8. Sandorfy C (1984) Topics in Current Chemistry. Springer, Berlin Heidelberg New York, vol 120, p 42
9. Millen DJ (1983) J Mol Struct 100: 351
10. Knözinger E, Schrems O (1987) In: Durig JR (ed) Vibrational Spectra and Structure. A Series of Advances vol 16, Elsevier Amsterdam Oxford New York, p 142
11. Corset J, Lascombe J (1967) J Chem Phys 64: 665
12. Schiöberg D, Luck WAP (1977) Spectr Lett 10: 613
13. Luck WAP, Schiöberg D (1979) Adv Mol Relaxation Interaction Proc 14: 277
14. Kasende O, Zeegers-Huyskens Th (1980) Spectr Lett 13: 493
15. Huyskens PL (1977) J Am Chem Soc 99: 2578
16. Novak A (1974) Structure and Bonding 18: 177 and references herein
17. Olovsson I, Jönsson PG (1976) In: Schuster P, Zundel G, Sandorfy C (eds) The Hydrogen Bond II, North Holland Publ Co, Amsterdam New York Oxford, p 393
18. Ratajczak H (1972) J Phys Chem 76: 3000
19. Szczepaniak K, Tramer A (1967) J Phys Chem 71: 3035
20. Ratajczak H, Orville-Thomas WJ (1973) J Mol Struct 19: 237
21. Ratajczak H, Orville-Thomas WJ (1975) J Mol Struct 26: 387
22. Szczesniak MM, Latajka Z, Scheiner S (1986) J Mol Struct (Theochem) 135: 179
23. Sherry AD (1976) In: Schuster P, Zundel G, Sandorfy C (eds) The Hydrogen Bond III, North Holland Publ Co, Amsterdam, p 1201
24. Rao CNR, Dwivedi PC, Ratajczak H, Orville-Thomas WJ (1975) J Chem Soc Faraday Trans 71: 935 and references herein
25. Zeegers-Huyskens Th (1977) Bull Soc Chim Belg 86: 823
26. Reyntjens-Van Damme D, Zeegers-Huyskens Th (1979) Bull Soc Chim Belg 88: 1003
27. Lippincott ER, Schroeder R (1955) J Chem Phys 23: 1099
28. Snijder WR, Schreiber HR, Spencer JN (1973) Spectrochim. Acta 29A: 1255
29. Allen LC (1975) J Am Chem Soc 97: 6921
30. Sherry AD, Purcell LF (1972) J Am Chem Soc 94: 1972
31. Rao CNR, Dwivedi PC, Gupta A, Randhawa HS, Ratajczak H, Szczesniak M, Romanowska K, Orville-Thomas WJ (1976) J Mol Struct 30: 271
32. Luck WAP (1979) Angew Chem Int Ed Eng 18: 350
33. Luck WAP (1980) Angew Chem Int Ed Eng 19: 28
34. Luck WAP (1981) In: Pullman B (ed) Intermolecular Forces, D. Reidel Publ Co, Dordrecht, p 199
35. Yarwood J (1974) in Yarwood J (ed) Spectroscopy and Structure of Molecular Complexes, Plenum Press London New York, p 174
36. Boobyer JW, Orville-Thomas WJ (1966) Spectrochim Acta 22: 147
37. Beven JW, Martineau B, Sandorfy C (1979) Can J Chem 57: 1341
38. Bernstein HJ, Clague D, Gilbert A, Michel AJ, Westwood A, quoted in reference 8
39. Stepanov BI (1945) Zhur Fiz Khim 19: 507
40. Sheppard N (1959) In: Hadzi D, Thompson T (eds) Hydrogen Bonding, Pergamon Press, London, p 85
41. Bertie JE, Millen DJ (1965) J Chem Soc   : 497
42. Lassègues JC, Huong PV (1972) Chem Phys Lett 17: 441
43. Bertie JE, Falk MV (1973) Can J Chem 51: 1713
44. Belozerskaya LP, Shchepkin DN (1966) Opt Spectr Mol Spectr 11: 146
45. Millen DJ, Zabicky ZC (1965) J Chem Soc   : 3080
46. Arnold J, Millen DJ (1965) J Chem Soc   : 503

47. Kyrö E, Warren R, McMillan K, Evades M, Danzciser D, Shoja-Chaghervand P, Lieb SG, Bevan JW (1983) J Chem Phys 78: 5881
48. Legon AC, Millen DJ, Rogers SR (1980) Proc R Soc London Ser A370: 213
49. Haurie M, Novak A (1965) J Chim Phys 62: 146
50. Jacobsen RJ, Mikawa Y, Brasch Y (1967) Spectrochim. Acta 23A: 2199
51. Marechal Y (1980) In: Ratajczak H, Orville-Thomas WJ (eds) Molecular Interactions vol 1, John Wiley & Sons, Chichester, p 231 and references therein
52. Bratos S (1975) J Chem Phys 63: 3499
53. Robertson GN, Yarwood J (1978) Chem Phys 32: 267
54. Hall A, Wood JL (1967) Spectrochim Acta 23A: 2657
55. Hall A, Wood JL (1967) Spectrochim Acta 23A: 1257
56. Haulait-Pirson MC, Bogaerts S, Zeegers-Huyskens Th (1977) Bull Soc Chim Belg 87: 928
57. Evans JC (1962) Spectrochim Acta 18: 507
58. Claydon MF, Sheppard N (1969) Chem Commun   : 1431
59. Rulinda B; Zeeers-Huyskens Th (1978) Bull Soc Chim Belg 87: 329
60. Mielke Z, Barnes A (1987) J Chem Soc Faraday Trans 2 82: 437
61. Somorjai L, Hornig DF (1962) J Chem Phys 36: 1980
62. Hadzi D, Kabilarov N (1966) J Chem Soc 1: 439
63. Hadzi D (1972) Chemia 26: 7
64. Dean RL, Wood JL (1975) J Mol Struct 26: 215
65. Dean RL, Wood JL (1975) J Mol Struct 26: 197
66. Wood JL (1973) J Mol Struct 17: 307
67. Borah B, Wood JL (1976) Can J Chem 54: 2470
68. Bonsor DH, Borah B, Dean RL, Wood JL (1976) Can J Chem 54: 2458
69. Nakai Y, Hirota K (1959) Bull Chem Soc Japan 32: 769
70. Ginn SGW, Wood JL (1967) Chem Phys 46: 2735
71. Jones DJ, Brack I, Roziere J (1984) J Chem Soc Dalton Trans   : 1795
72. Roziere J, Belin C, Lehman MS (1982) J Chem Soc Commun   : 388
73. Grech E, Malarski Z, Ilczyszyn M; Czupinski O, Sobczyk L (1985) J Mol Struct 128: 245
74. Laureys D, Zeegers-Huyskens Th (1987) J Mol Struct 158: 301
75. Migchels P; PhD Thesis, University of Leuven (in preparation)
76. Van Duijneveldt-Van de Rijdt JG, Van Duijneveldt FB (1971) J Am Chem Soc 93: 5644
77. De Taeye J, Maes G, Zeegers-Huyskens Th (1983) Bull Soc Chim Belg 92: 917
78. Szemik A, Zeegers-Huyskens Th (1984) J Mol Struct 117: 265
79. Kasende O, Vanderheyden E, Zeegers-Huyskens Th (1985) J Heterocyclic Chem 22: 1647
80. Kasende O, Zeegers-Huyskens Th (1984) J Phys Chem 88: 2636
81. De Taeye J, Parmentier J, Zeegers-Huyskens Th (1988) J Phys Chem 92: 4556
82. Tsuboi M, Tokahoshi S, Horada I (1973) In: Duchesne J (ed) Physicochemical properties of nucleic acid vol 2, Academic Press, New York, p 91
83. Port GN, Pullman A (1973) FEBS Lett 31: 70
84. Majoube M (1985) J Raman Spectrosc 16: 98

CHAPTER VII
IR-Overtone Vibration Spectroscopy

W.P. Luck

Some advantages to study intermolecular forces by IR overtone spectra are reported. The overtone intensity is nearly insensitive on interactions contrary to the large intensity increase of fundamental bands. As consequence H-bond equilibria and the structure of H-bonded liquids can be studied easier and in some directions with higher precision. Secondly, the simultaneous excitation of two H-bonded molecules by one quantum and this possibility to study H-bonds are shown. The determination of dipole moment changes by intermolecular interactions is described based on overtone spectra. Last but not least the possibilities to study van der Waals interactions in condensed media are given.

1 Introduction

The theoretical treatment of vibrations with a force proportional to the elongation q gives a fundamental frequency v_0 and in the quantum theory a series of overtones $v\,v_0$ but with the selection rule $\Delta v = +1$ or -1. The classic theory [1] as well as the quantum theory allow overtones with $\Delta v > 1$ if the force F is unsymmetric, for instance, $F = -kq + aq^2$ or $F = dE_{pot}(Morse)/dq$. ($E_{pot}(Morse)$: the so called Morse potential, which could be described in a fairly

Intermolecular Forces
© Springer-Verlag Berlin Heidelberg 1991

good approximation as the interaction of two dipoles perpendicularly oriented to the distance q [2]).

The unsymmetric wave functions derived by the Schrödinger equation with the Morse potential allows deviations from the selection rule $\Delta v = 1$, while the transition moments for $\Delta v > 1$ decrease rapidly with increasing v. As an example Figs. 1 and 2 give the fundamental ($\Delta v = 1$; Fig. 1) and the first overtone ($\Delta v = 2$; Fig. 2) of solutions of CD_3OH in CCl_4 at room temperature. The sharp peak maximum at low wavelength of the overtone spectra has only $1/40 = 0.025$ of the intensity of the fundamental ε_{max}. The broad band of the first overtone spectrum observed at large concentrations, the band of H-bonded OH, is reduced at the maximum intensity by a factor of about $1/770 = 0.0013$ of $\varepsilon_{max}(\Delta v = 1)$. The band area $\int \varepsilon dv$ of the H-bond band of the fundamental vibration v_{01} increases with concentration (with increasing number of H-Bonds) while the area of the first overtone v_{02} remains about constant during H-bond formation [3]. For example the left of Table 1 demonstrates the constancy of $\int \varepsilon dv$ of the methanol overtone v_{02} at supercritical T of about 310°C of different densities, the right hand side shows a very low change of the $\int \varepsilon dv$ of saturated methanol vapour at different T. Even in dilute solutions of CS_2 or CCl_4 at 20°C there are very similar $\int \varepsilon dv$ values.

The higher overtone intensities decrease with their number v. Figure 3 shows, for instance, the overtone and combination bands of liquid water (H-bond bands) ε_{max} of the band maxima of the first (1.45 µm) second (0.98 µm) and third

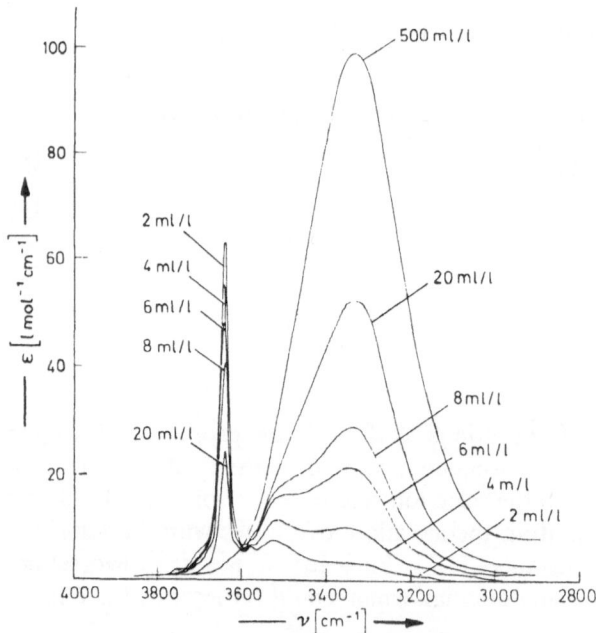

Fig. 1. OH fundamental stretching band of solutions CD_3OH in CCl_4 at 50°C

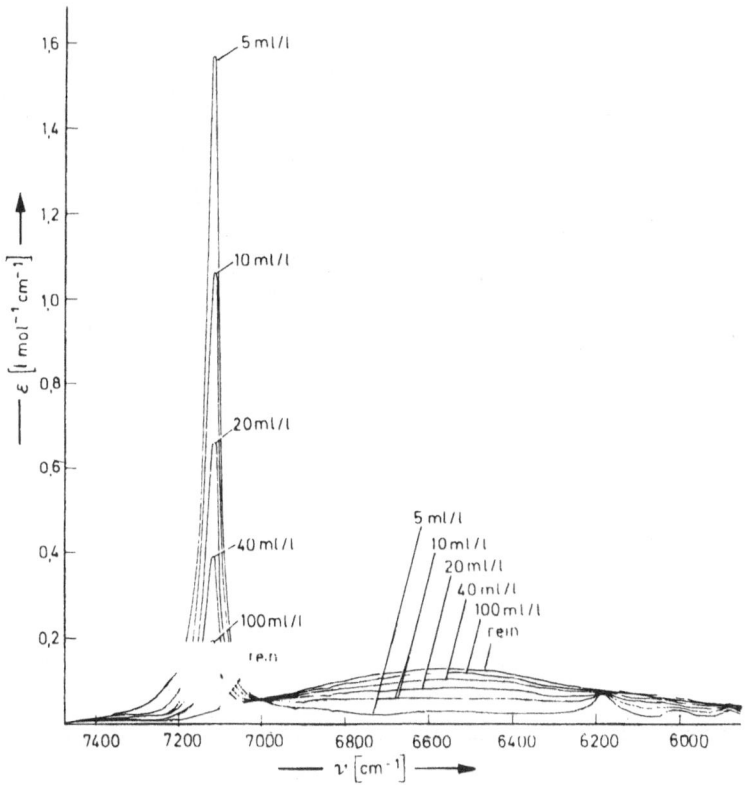

Fig. 2. OH first overtone stretching band of solutions CD_3OH in CCl_4 at 20°C

Table 1. Constancy of $\int \varepsilon dv$ of the first methanol overtone band in relative units [3] of critical densities ρ_K

ρ/ρ_K	T°C	$\int \varepsilon dv$	ρ/ρ_K	T°C	$\int \varepsilon dv$
0.1	314	1040	0.024	174	1145
0.25	307	1097	0.19	201	1150
0.5	310	1066	0.32	220	1100
1	314	1044	1	377	1020
Solutions					
2g/lCCl$_4$	20	1066			
2g/lCS$_2$	20	930			

(0.74 μm) overtones decrease at 10°C $\varepsilon_{max,02}: \varepsilon_{max,03}: \varepsilon_{max,04} = 14: 0.195: 0.011 = 1: 1/71: 1/1265$. This rapid intensity decrease with increasing overtone number is very important for life on earth. Many chemicals, all those which have low dissociation energies—lower than the emission frequencies of solar radiation—are only stable against photoreactions because the low excitation probability for $\Delta v > 1$ and very low for $\Delta v > 2$ of vibration excitation. The highest vibration quantum number is for H_2 where $v_{max} = 14$ and for I_2 v_{max} is about 110, these

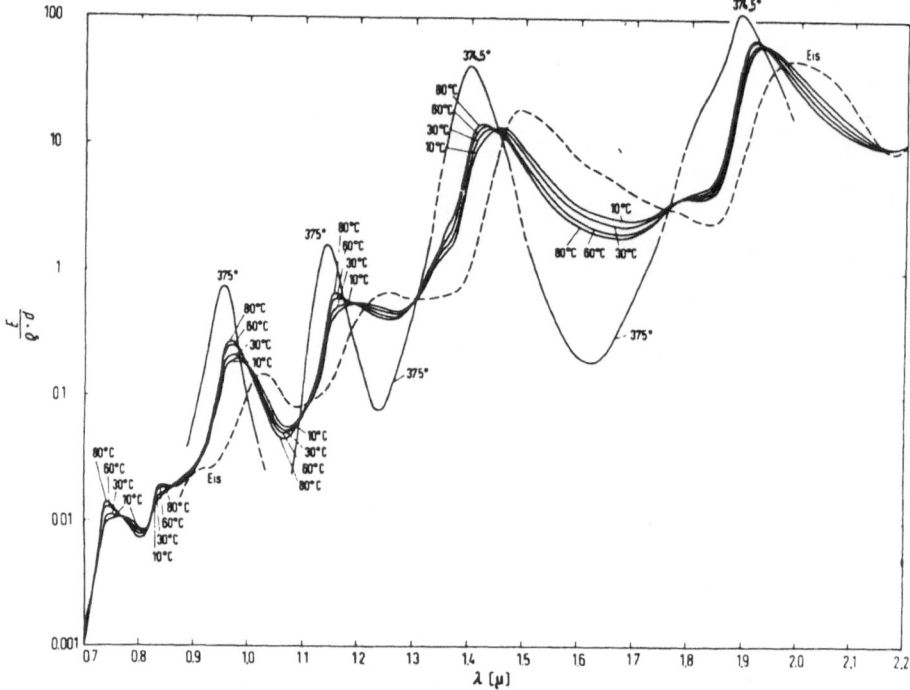

Fig. 3. Overtone and combination bands of liquid water at different temperatures and the spectrum of ice

large ν-values prevent photo-dissociation by vibration excitation. In the blue region of the visible light 1 mol quanta correspond to about 250 kJ/mol. This is a high energy value compared with the dissociation energy of the following molecules (energies in kJ/mol in parantheses): NO_2(33); OH (39); CH_4(85); HCl (92); $(CH_3)O(CH_3)$ (185) and CH_3OH (239) (dissociation in molecules of the elements.)

For the case of IR spectroscopy the low transition probabilities of the overtone spectra have the large advantage that already for v_{02} we need for pure liquids a cell thickness z of about one mm or more, for v_{03} one cm, etc. In contrast, in fundamental IR spectroscopy of pure liquids it is difficult to obtain quantitatively good spectra because of the need of cell lengths of several μm. In addition the photometric accuracy of overtone spectrophotometers is much better than with fundamental instruments.

This difference between fundamental and overtone spectroscopy can also be used to study H-bond equilibria. Fundamental spectra are of an advantage to observe the positions of H-bond bands. Overtone spectra have the advantage to allow quantitative studies of H-bond equilibria constants by determining the content of non H-bonded OH groups precisely (so-called "free" OH_F," "free" meaning free of H-bonds but not always free of van der Waals interactions). Intensity comparisons of fundamental and overtone spectra allow to calculate the dipole moment derivatives and disturbances by intermolecular forces [4, 5]. On the other hand, the combination band (stretching and bending) in the range 1.92 μm (5200 cm^{-1}) differs for water and alcohols by different bending v. This band is therefore useful to study H-bonds of mixtures of water with alcohols (for instance, in biopolymers).

A further advantage of overtone spectra are their larger frequency shifts $\Delta v = v_{gas} - v_{solution}$ by intermolecular interactions. Δv is proportional to the quantum number v. This effect together with the much lower intensity of the H-bonded bands reduces band overlapping effects. Especially in liquids with large band half-width $\Delta v_{1/2}$, the overtone spectroscopy in some cases allows the determination of OH_{Free} or NH_{Free}, if fundamental spectra cannot recognize them. The best procedure for H-bond research is therefore to study both v-ranges together.

2 Determination of Equilibrium Constants

The quotient of the precisely determined absorbances E and the density—corrected concentrations C_0 is the extinction coefficient:

$$\varepsilon_v = E/C_0 z \qquad (1)$$

The change of the density ρ with temperature T is in many cases larger than the precision of the instrument. Differences between equilibrium constants of H-bonds in the literature could be caused by the neglect of the T-dependent densities and by different precisions of instruments and z-values.

By having the equilibria with species of different band maxima as happens in H-bond equilibria, one could determine the monomer concentration C_{Mo} at the maximum of the sharp peak at the lowest wavelength (highest v), (Figs. 1, 2).

For a monomer (non-H-bonded groups) content of $C_{Mo} = \alpha C_0$ and an extinction coefficient ε, it follows that:

$$\varepsilon_v = E/\alpha C_0 z \qquad (2)$$

For extrapolated infinite dilution ($\alpha = 1$) is $\varepsilon = \varepsilon_0$. At finite concentration ε_v of Eq. (1), the appearent extinction coefficient is $\varepsilon_C (C_{Mo}) = \alpha \varepsilon_v = \alpha \varepsilon_0$. We obtain C_{Mo} by:

$$\varepsilon_C(C_{Mo})/\varepsilon_0 = \alpha = C_{Mo}/C_0 \qquad (3)$$

For studies of the reaction enthalpy ΔH by the T-difference of equilibria constants with fundamental bands, extrapolations at C_0 have to be done for every temperature, because the T-dependent molecule distances change the van der Waals and the H-bond interactions and therefore their band maxima [6].

The fundamental OH band intensity increases proportional to the interaction energy and therefore with Δv [7]. Neglecting the change of ε_0 with T could cause some errors in ΔH determinations.

Such effects make it difficult to compare the results obtained by different authors. The simplest H-bond equilibrium occurs in the formation of the cyclic dimer as it happens of lactams or carbon acids. The IR spectra in both cases reveal cyclic dimers [8,9] as they are found for acids in the gaseous state too (Fig. 4). With a low content of CH_2-groups, the amide group $O=C-N-H$ is in its cis configuration [10] and the cyclic dimer with its "6 membered ring" has optimal H-bond angles [11] with the angle between the XH and the lone pair axis equaling zero. H-atoms in H-bonds play a similar role as electrons in chemical bonds; therefore it is useful not to count the H-atoms for the ring

Fig. 4. The cyclic dimer structure of acetic acid in the vapour state (electron scattering result)

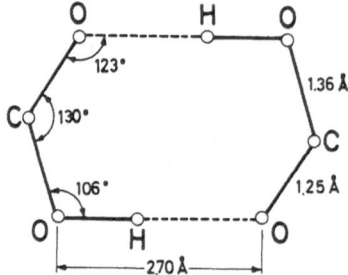

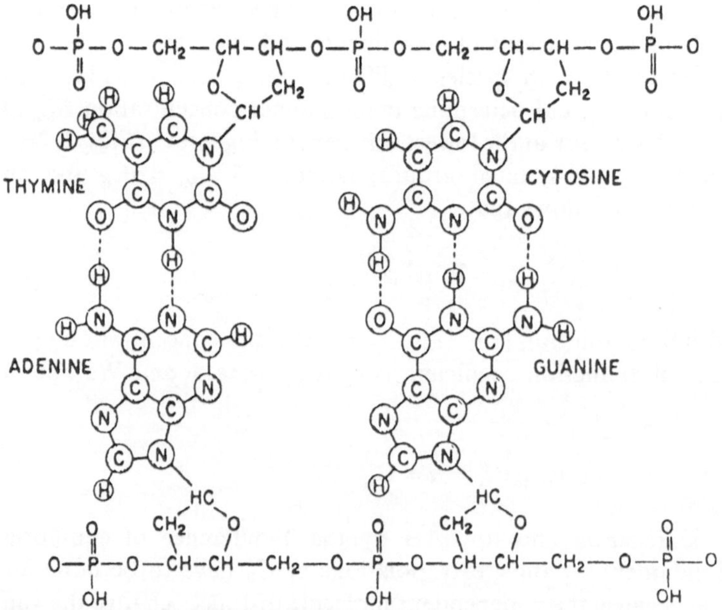

Fig. 5. The H-bond structures with H-bond angles of zero for DNA

size of H-bond aggregates. Six-membered H-bonded rings with their especial stability play a role for cyclic lactam dimers, for carbon acid dimers, for the 6 membered OH ring in ice or in trimers of oxims. Finally such stable 6-membered H-bond also exist in DNA (Fig. 5).

The cyclic dimers, Di, of lactams or carbon acids give [8, 9, 12]:

$$2Mo = Di \tag{4}$$

$$K_{12} = [Di]/[Mo]^2 \tag{5}$$

with: $$C_0 = [Mo] + 2[Di]$$

$$K_{12} = [C_0 - Mo]/2[Mo^2] \tag{6}$$

Equation (6) allows to determine K_{12} with the precise measurable overtone values of ε_C of the monomers (Eq. 3). The ε_0-extrapolation is got by the eq. (6) and (3):

$$K_{12} = [C_0 - \alpha C_0]/2[\alpha C_0]^2 \tag{7}$$

$$K_{12} = (1 - \varepsilon_C/\varepsilon_0)C_0/2(\varepsilon_C/\varepsilon_0)^2 C_0^2 \tag{8}$$

$$2\varepsilon_C^2 C_0^2 K_{12}/\varepsilon_0 = \varepsilon_0 - \varepsilon_c$$

$$\varepsilon_C = \varepsilon_0 - 2\varepsilon_C^2 C_0 K_{12}/\varepsilon_0 \tag{9}$$

In the case of cyclic dimers, plotting $\varepsilon = f(\varepsilon^2 C_0)$ gives a straight line with ε_0 as the axis intercept and $2K_{12}/\varepsilon_0$ as slope (Fig. 6). In overtone spectra ε_0 is nearly T-independent, but is not for fundamental spectra. Figure 7 demonstrates that K_{12} of Eq. (6) is concentration independent in CCl_4-solutions [8, 9] for cyclic lactams —$(CH_2)_n$ CONH- with n = 3 (pyrrolidone), n = 4 (α-piperidone), n = 5 (caprolactam), n = 6 (oenanthlactam and n = 8 (caprylactam) but is C_0 dependent for n = 11 (laurinlactam). For $3 < n < 8$ the amide group is in cis-configuration [10] and the cyclic dimers are energetically preferred so that no higher associates are observed in the whole solubility region in CCl_4. The amide group of laurinlactam is trans [10] and forms linear H-bond chains as Fig. 7 indicates. The equilibrium constants of cyclic lactam dimers are constant in the entire solubility region i.e, with variations of C_0 till a factor of 1000. These dimerization constant are among the best known in physical chemistry. Such simple equilibria of intermolecular forces do not require activity coefficients, which in chemical reactions are typically for overlapping intermolecular interactions, but are not necessary for precisely determining intermolecular equilibria.

The concentration dependence of the spectroscopically determined α-values (Eq. (2)), the content of monomers, is influenced by the type of H-bond

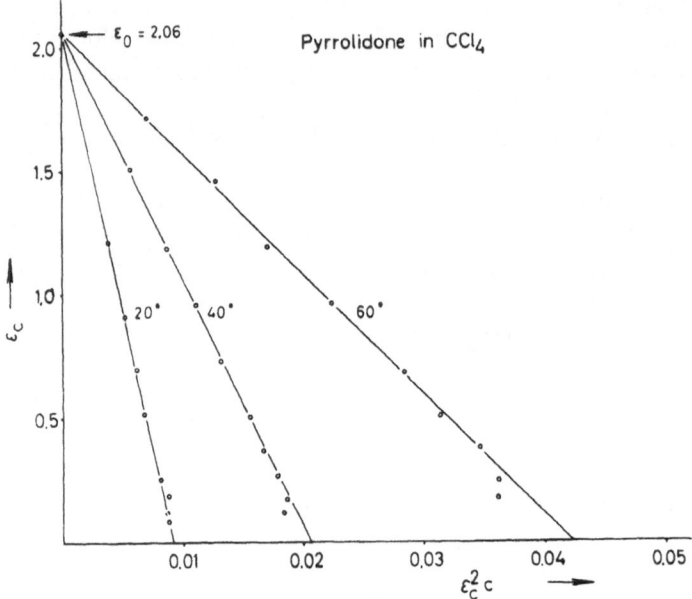

Fig. 6. Example for the determination of the extinction coefficient ε_0 of H-bond dimers. (pyrrolidone in CCl_4—solutions at different T)

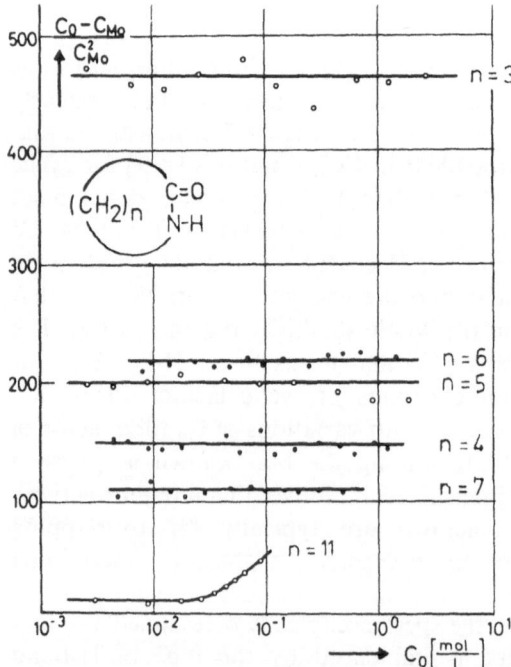

Fig. 7. Constant equilibrium constants of cyclic H-bond dimer formation as a function of the total concentration in CCl_4 solutions. Test for the efficiency of the IR overtone method [8, 9]. (Laurin-lactam forms linear dimers)

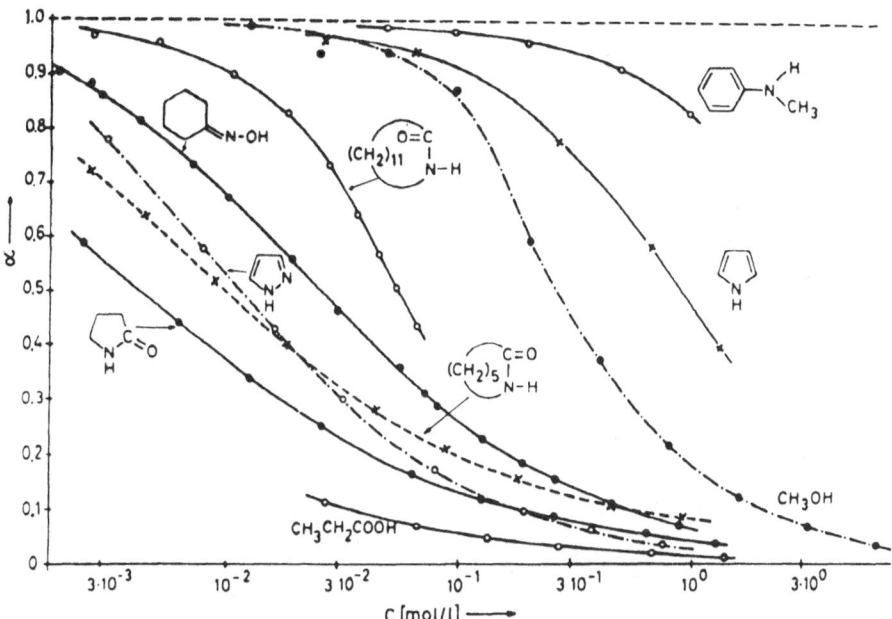

Fig. 8. The fraction α of the total concentration C, non-forming H-bonds (Mo $= \alpha$C) (the so-called "free" groups) in CCl$_4$ solutions at 20°C, determined by IR spectra. Indications of different types of H-bond equilibria [8]

equilibrium. Figure 8 lists data of CCl$_4$-solutions taken with overtone spectra. α-values of pyrrolidone or of caprolactam with cyclic dimers decrease quicker than for laurinlactam with linear associates. The strongest H-bond association is found for propionic acid with a coupled equilibrium of cyclic dimers and some trimers [8, 13, 14]. The α-values of cyclic [15] or linear [16] oximes (CH$_2$)$_n$CH=NOH decrease in Fig. 8 with C$_0$ in a manner between cis and trans lactams. In CCl$_4$ solutions at low C$_0$ these compounds form cyclic dimers and in sufficient concentration also cyclic trimers. This H-bond formation gives a classically coupled equilibrium between dimerization as in Eqs. (4) till (7) and, in addition, the formation of trimers, Tri:

$$3\text{Mo} = \text{Tri} \tag{10}$$

with
$$K_{13} = \text{Tri}/\text{Mo}^3 \tag{11}$$

and
$$C_0 = \text{Mo} + 2\text{Di} + 3\text{Tri} \tag{12}$$

with Eqs. (7), (9) and (10):

$$C_0 = \text{Mo} + 2K_{12}\text{Mo}^2 + 3K_{13}\text{Mo}^3 \tag{13}$$

$$(C_0 - \text{Mo})/\text{Mo}^2 = 2K_{12} + 3K_{13}\text{Mo} \tag{14}$$

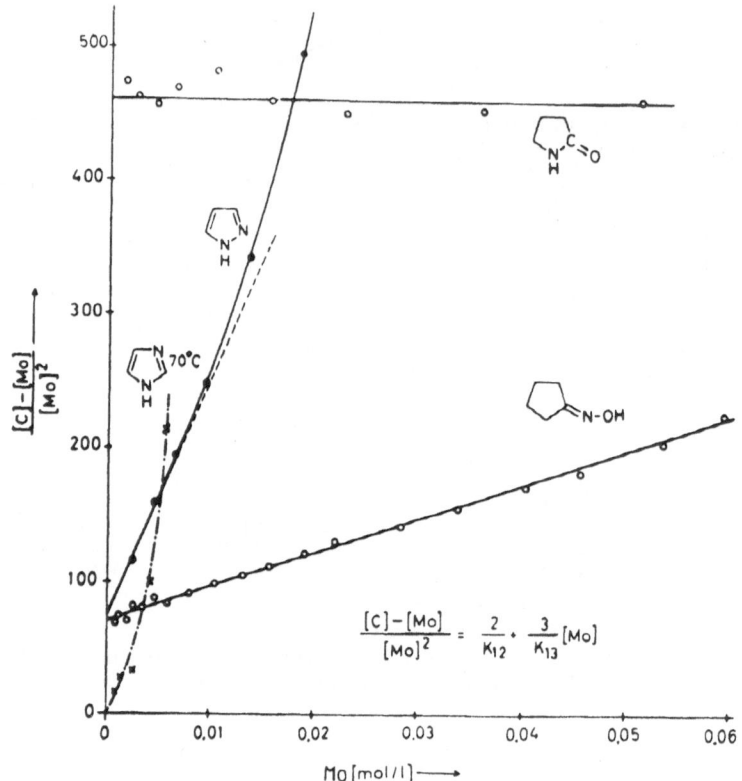

Fig. 9. Plot of Eq. (14) demonstrating that cyclic lactams form only H-bond dimers; oximes form dimers and trimers; pyrazole forms dimers, trimers, and higher aggregates; but imidazole only forms trimers and higher aggregates [8]

In the case of a coupled equilibrium between cyclic dimers and cyclic trimers we have to expect a straight line by plotting the expression of the left side of Eq. (14) in dependence of the spectroscopically determined concentration of monomers, Mo. As Fig. 9 demonstrates this is very well shown by overtone measurements of cyclic pentanonoxime as well as hexanon-, octanon- and dodecanonoximes [15]. The slope in Fig. 9 gives for pentanonoximes $3K_{13}$ and the ordinate axis section gives K_{12}.

Pyrazole data for the $=N—NH$ group in Fig. 9 indicate that at very low C the axis section values K_{12} and at medium C a straight line with a slope for the K_{13} value-indicating that in this region trimers dominate- and finally at higher C we observe higher associates than trimer. In contrast imidazole with the $—N=C—NH$ configuration shows already at low concentration higher association number than 3. In such cases the mean association number n could be determined by:

$$nMo = Ass \qquad (15)$$

$$K_{1n} = Mo^n/Ass = Mo^n/(C_o - Mo) \tag{16}$$

or $$\log K_{1n} + \log(C_0 - Mo) = n \log Mo \tag{17}$$

This double logarithmic plot allows the determination of the mean association number n. Figure 10 gives as example the overtone results of N-ethylacetamide solutions in CCl_4 with n = 2 at low C_0 and n = 4 at larger C_0. Full points indicate that the whole range can be described with a coupled equilibrium of cyclic dimers (with cis-amide configuration) and cyclic tetramers (with trans amide configuration)-Lactams in this plot give a slope of 2, propionic acid of about 2.5, indicating a coupled equilibrium of dimers and trimers. Pyrrole (NH-group), pyrazole and laurinlactam show a slope of 2 at low C_0, which increases rapidly with increasing C_0. Such a case of a linear H-bonded chain is shown in Fig. 11 for methanol in CCl_4 solutions. The slopes start at 60°C with a value of 3 and increase quickly to slopes of 10 and more. At 20°C and 30°C the slope starts with values of 4. Indeed the overtone spectra of methanol in CCl_4 can be described quantitatively in this C_0 region as an equilibrium of monomers and cyclic tetramers [12].

The precisely determined equilibria constants allow calculation of the percentage of different aggregates. Figure 12 shows for N-ethylacetamide solutions the decrease of the monomers and the increase of the tetramers with C_0 and a maximum of the dimers at medium C_0. Similarly for oximes [15] we found such maxima for the percentage of dimers. Both examples illustrates the

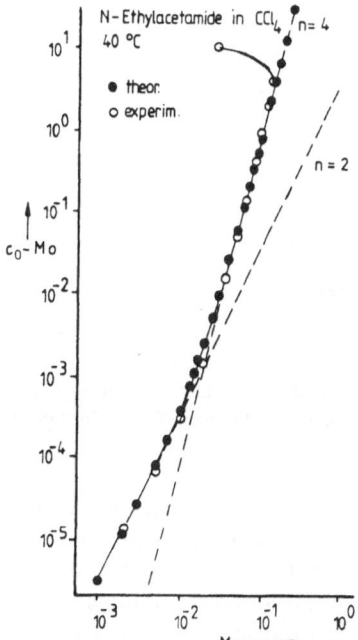

Fig. 10. Determination of the mean association degree n = 2 and 4 of N-ethylacetamide in CCl_4 by Eq. (17) [12]

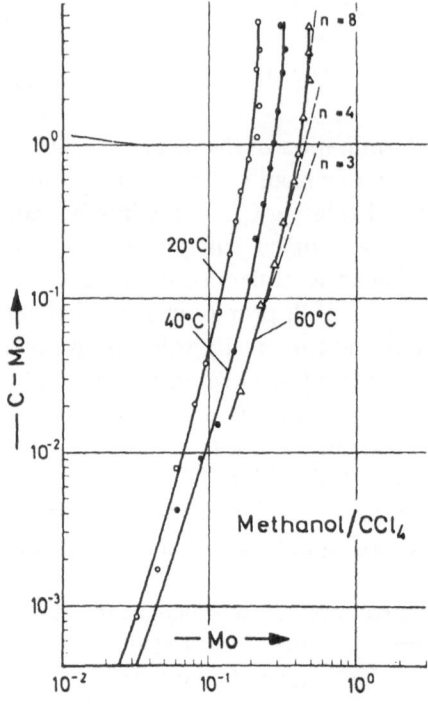

Fig. 11. Determination of the mean association degree n of methanol in CCl_4 at different T. At high concentration, n increases rapidly [12, 15]

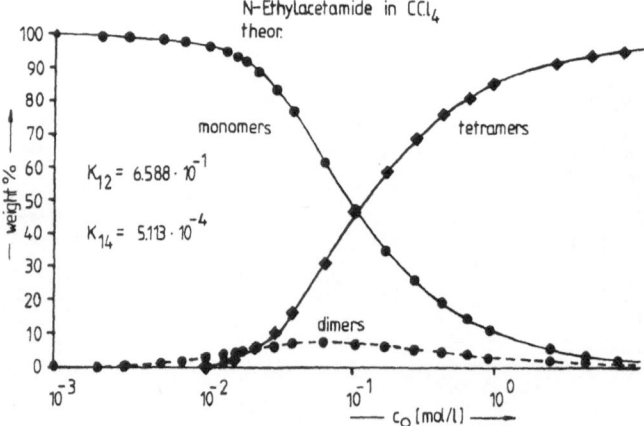

Fig. 12. The relative content of monomers, dimers and tetramers in the coupled H-bond equilibria of *N*-ethylacetamide in CCl_4 at 40°C. The dimer content has its maximum in a small region of the total concentration C_0 [12]

complicated possibilities in the seldomly discussed coupled equilibria. The reason for the higher stability of cyclic oxime trimers is the 6 membered ring —N=O —H...—N=O—H...—N=O—H...with H-bond angles of zero, the dimer —O=N—H...—O=N—H has a unfavoured H-bond angle of about 49°. In this way, precise overtone spectroscopy indicates well the

Table 2. H-bond enthalpies $\Delta H(kJ\,mol^{-1})$; $\Delta H_{12}/2$ and $\Delta H_{13}/3$: ΔH per H-bond determined and estimated H-bond angles β by IR overtone spectra and estimated H-bond angles β

	ΔH_{12}	$\Delta H_{12}/2$	β_{12}	ΔH_{13}	$\Delta H_{13}/3$	β_{13}	n
Imidazole	—	—		(75)	(25)		+
Propionic acid	46	23	0°	62.8	(20.9		
Pyrrolidone	31.8	15.9	0°	—	—		
Caprolactam	29.7	14.9	0°	—	—		
Capryllactam	30.1	15	0°	—	—		
Cyclopentaonoxime	33.5	16.7	49°	55.7	18.6	0°	
Cyclooctanoxime	33.5	16.7	49°	56.5	18.8	0°	
Laurinlactam	(18.8)	(9.4)	0?	(37.3)	(12.4)	0°?	+
Pyrazole	(18.8)	(9.4)	72	(63)	(21)	12°?	+
Pyrrole	(15.5)	(7.7)	90°?				+

+ Higher aggregates with n > 3 exist and the ΔH method is uncertain

preference of the zero H-bond angle as it also is favoured in crystalline hydrates [17]. Table 2 reviews the H-bond enthalpies of these compounds—determined by van't Hoff plots: $\ln K = f(1/T)$.

The oximes indicate that the angle depends on the H-bond energy with a maximum at $\beta = 0$. In this case the increase of ΔH is about 10%.

There is an other interesting method [18, 19] for determining the average association degree $f = \sum nC_n / \sum C_n = C/\int(1/\varepsilon_C)d(\varepsilon_C C)$. This method does not require knowledge of the ε_0 value. Figure 13 shows the results on the reported overtone measurements of oximes [15]. At high total concentration C, the limit value of 2 for the cyclic dimers are also obtained with this method at 20°C.

This precise overtone method for determining H-bond equilibria was established by comparing with the calorimetrically determined heats of mixing ΔH_M of hexanol-hexane or ethanol-hexane mixtures calculated per mole of alcohol. If we plot [21] the ratio of ΔH_M (measured at a mole fraction x_{ROH}) over ΔH_M^0 (extrapolated on infinite dilution) as in Fig. 14 (open symbols) and the α-values

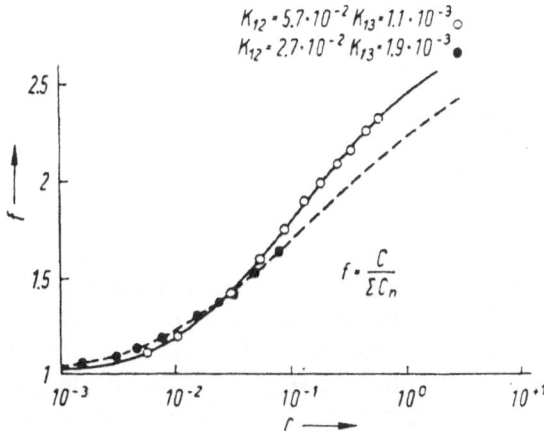

Fig. 13. Determination of the mean association degree f with the integral method without knowledge of the extinction coefficient ε_0 [15] O: cyclopentanone-, ●: cyclododecanone-oxime in CCl_4 at 20°C

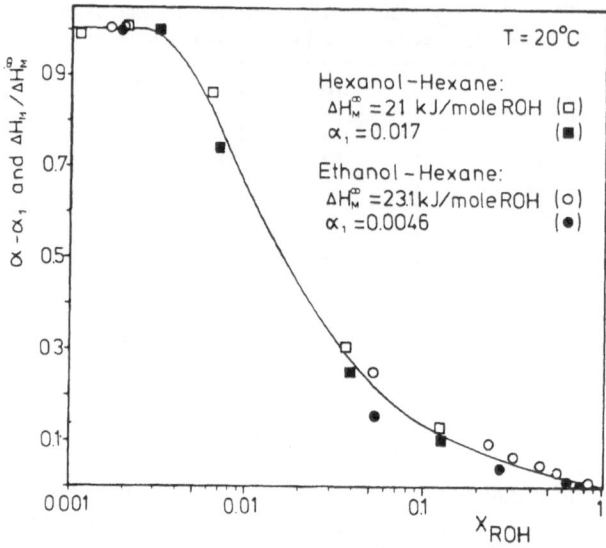

Fig. 14. Comparison between the relative heat of mixing with the overtone *IR-results*. The content of alcoholic monomers as a function of the alcohol mole-fraction in hexane demonstrates the opening of H-bond as the main cause of the heat of mixing and shows the efficiency of the IR-overtone method [21]

(full symbols) of spectroscopically measured monomer content of alcohols (minus the monomer contant α_1 in the bulk alcohol) we get nearly identical curves. This overtone experiment demonstrates that the heat of mixing in these experiments is induced by opening H-bonds during the mixing experiment and that the spectroscopic and calorimetric results are very similar. The reported methods are independent of the type of vibration band and only require slight separation of the XH vibration band of monomers. This method can also be applied when the H-bond bands overlap. It was shown that the α-values determined by different vibration bands are identical but that the overtone band gives the best values.

3 Larger concentrations of H-Bonds

3.1 Liquids

As Fig. 11 already has shown the aggregation number in linear H-bond chains of methanol solutions increases rapidly with concentration. In the case, of overlapping H-bond bands, one can try to estimate the H-bond chain length by apparent ε in the ν region of non H-bonded OH groups. This could not be precisely done by comparison of ε (liquid) with ε in solution because the

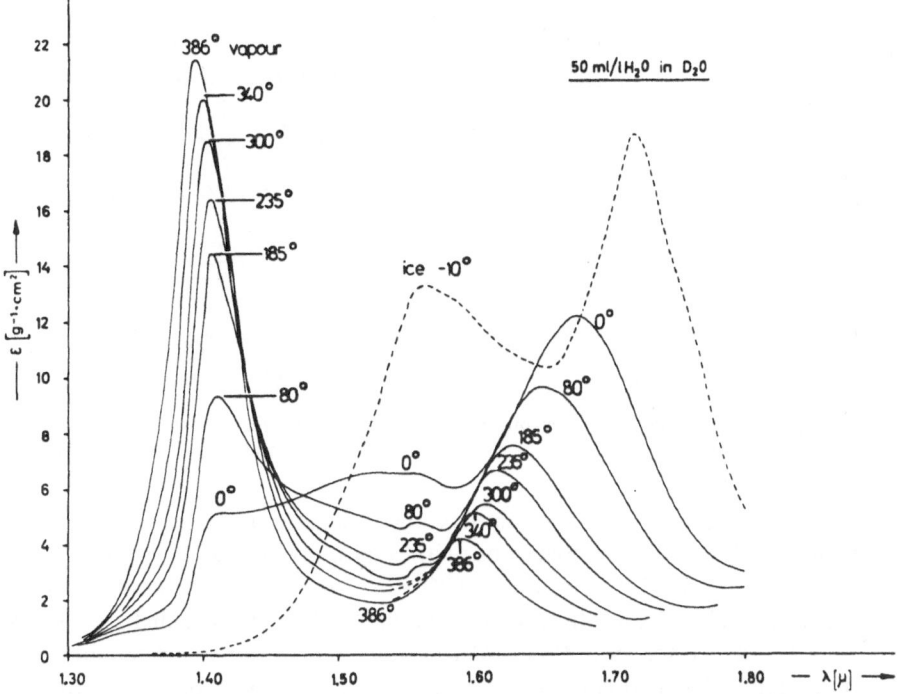

Fig. 15. Overtone OH stretching band of liquid HOD in D_2O as a function of temperature. With the maximum of the sharp band the content of non-bonded OH-groups in liquid water could be estimated, [24]

half-width changes with stronger interactions. In this case, a better approximation is possible by assuming that ε_0 of OH_{free} at high T is similar as in the liquid at low T. This approximation has been done with methanol, ethanol [22] (see Figs. 7–9 in Chap. IX) and water [23, 24] under saturation conditions in a high-pressure cell [23] (see Fig. 15). For such measurements at high T and pressure of some 100 atm, the overtone region has some advantages compared with the fundamental region because the reasonable size of the cell thickness. Above the critical T_C at about 400°C ε of these liquids becomes constant and one could assume this value as ε_0 of OH_F. This approximation method neglects the change of half width and band position with T. With this assumption one can describe the entire T-dependence of these liquid spectra as a decrease of the H-bond content. A band analysis of the HOD spectra of Fig. 15 shows (Fig. 16) that all details can be obtained by the sum: 1. of the ice-like band at $6450\,cm^{-1}$; 2. the band with maximum at $T = 400°C$ at 7100 to $7000\,cm^{-1}$, (this high frequency band has a similar position as OH_{free} in diluted solution (extended confirmed with alcohols [22]) and 3. one needs for the band analysis a third band between the other two with a maximum at about $6850\,cm^{-1}$. The percentage of these three components is given in Fig. 16 [24]. The third medium band corresponds to $\Delta\nu$ of a so-called dimer band observed

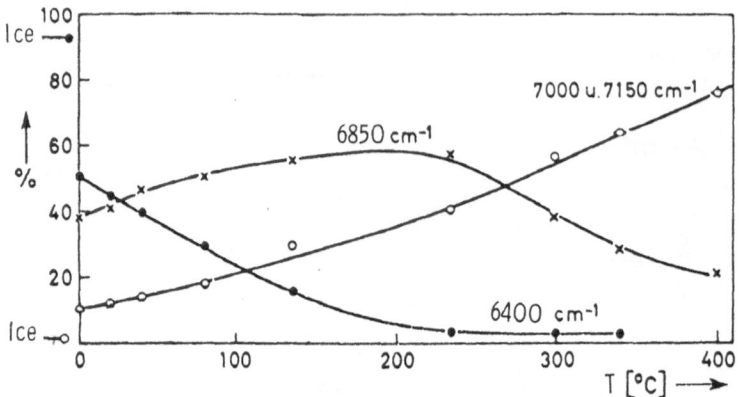

Fig. 16. Band analysis of the HOD overtone band requires at least three bands: 1. free OH at 7000 and 7150 cm^{-1}; 2. medium bonded OH at 6850 cm^{-1}; 3. linear H-bonds at 6400 cm^{-1} [24]

with water or alcohols in rare gas matrix low T spectra (see Chap. VIII). Some arguments such as the comparisons with OH-ether H-bands [25] or with microwave spectra of the water dimer [26] correspond with the assumption of a linear dimer. The O...O distance of this dimer [24] and the distance an ice can lead to the conclusion that in H-bonds polarisations happen in these H-bond chains (so-called cooperatively, see Chap. X) and induce about double Δv values in liquids corresponding to about double ΔH values. Such behaviour can be observed in end groups of H-bond chains. In addition, cyclic dimers and trimers can induce smaller Δv values by unfavoured H-bond angles in matrix spectra [27]. Both types of OH: with reduced cooperativity and with unfavoured angles or distances can result in H-bond chains or networks in liquids. As Fig. 16 demonstrates the content of these unfavoured H-bonds does not change much with T. On the other hand the content of the band with ice-like H-bond energies decreases nearly linearly in liquid water up to 200°C, a temperature up to which the density of liquid water does not change much. Above 200°C this band of optimal H-bonds does not appear and the density decreases more rapidly with T. The half width $\Delta v_{1/2}$ of this medium band and of the band of OH$_F$ increases with T [28] indicating that these bands are representative for average values of a step function of the distribution of H-bond distances, angles and cooperativities. In spite of these partitions the calorimetric properties of liquid water and alcohols were be described nearly quantitatively [29–32] (see Chap. IX) by a two step mechanism:

$$OH_F + \Theta_F = OH_B \tag{18}$$

Θ_F: non-H-bonded (free) lone pair electrons.

OH$_B$: H-bonded OH.

This simplified Eq. (18) seems to be valid as a good approximation at temperature with large H-bond contents. For alcohols at room temperature we estimated H-bond chain lengths of about 100 molecules and in liquid water at 25°C a network of H-bonds of about 400. In these systems small aggregates can not play a big role and the assumption of a reaction of "free" OH groups with aggregates seem to be correct. Equation (18) does not assume free monomers. Free OH should mainly belong to end groups of H-bond aggregates. Therefore, certain arguments against mixture models for liquid water and alcohols seem to be meaningless.

The half-width $\Delta v_{1/2}$ of OH_F of ethanol at 350°C is about twice the $\Delta v_{1/2}$ of diluted ethanol/CCl$_4$ solutions and ε_{max} is at 350°C in liquid ethanol about half the value in CCl$_4$ solutions with about the same $\int \varepsilon dv$ in both cases. If this were to be generalized we could estimate as a rough approximation the content of OH_{free} or NH_{free} in liquids by determining ε and taking as ε_0 in liquids half of the value of ε_0 in diluted CCl$_4$ solutions. This method is demonstrated with the liquid N-ethylacetamide overtone spectrum of Fig. 17 with the result of Fig. 18. At low T of $-50°C$ nearly all NH groups

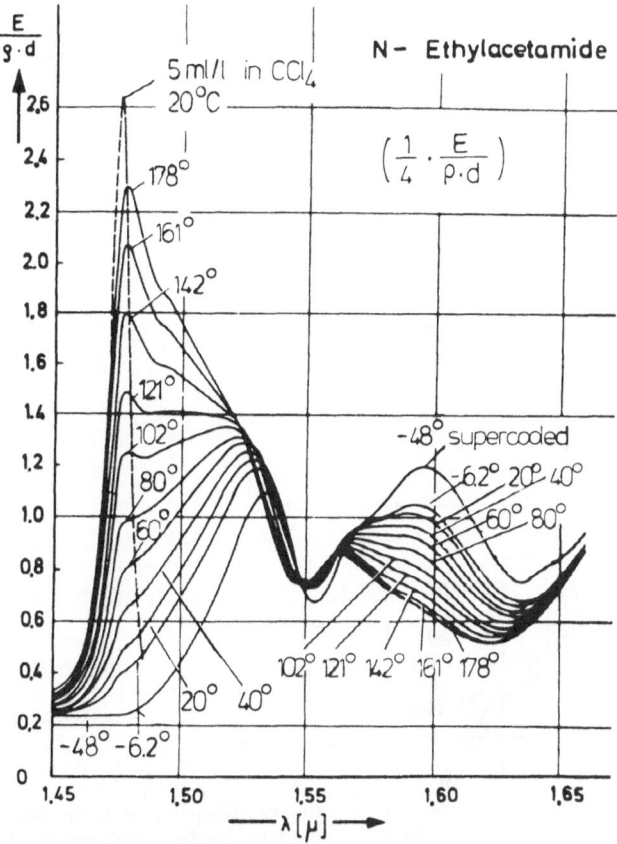

Fig. 17. The first NH-overtone stretching band of liquid *N*-ethylacetamide at different T in comparison with the band of a diluted solution in CCl$_4$; a method to estimate the "free" NH content in liquid *N*-ethylacetamide [64]

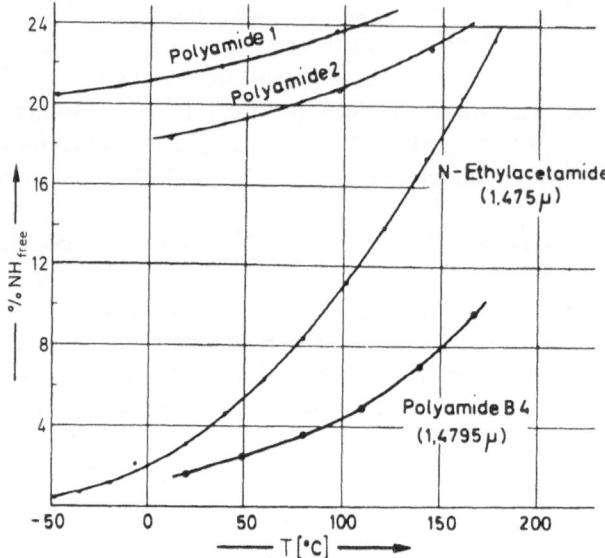

Fig. 18. *Middle*: estimated "free" NH-groups of liquid *N*-ethylacetamide (based on Fig. 17) *Bottom*: estimated "free" NH-groups in solid 6-polyamide B 4 (basd on Fig. 19). *Top*: estimated "free" NH-groups of two polyamide research products

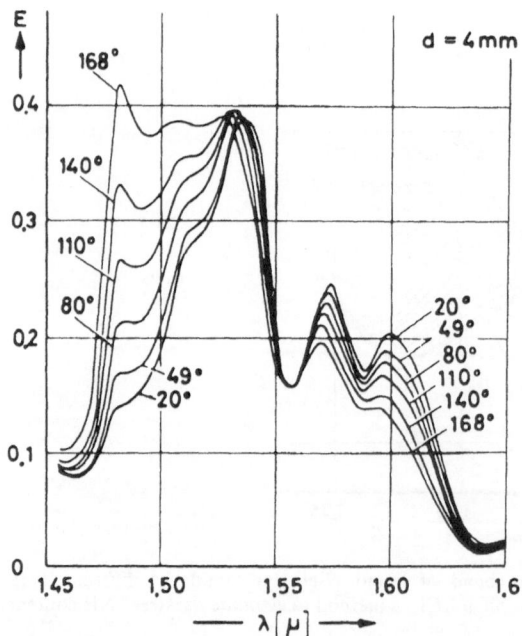

Fig. 19. Temperature dependence of the NH stretching overtone band of solid 6-polyamide. The intensity of the sharp 1.48 μ band is proportional to the content of "free" NH-groups [64]

are H-bonded and open slowly with increasing T. A comparison with the overtone spectrum of solid 6-polyamide demonstrates the possibility to also estimate its H-bond state (Fig. 19). The opening of NH bonds of solid 6-polyamide is similar as in liquid N-ethylacetamide at about 120°C. With the same overtone method it could be also demonstrated that the H-bond content of 6-polyamide changes only slightly during the stretching process. The change of the mechanical properties during stretching seems to be caused by orientation and crystal-like arrangements.

3.2 The Simultaneous Excitation of Neighbouring XH Groups

In overtone experiments with one cell of liquid CD_3OH and a second one of liquid CD_3OD in the comparison/control unit plus two cells of equal thickness with 1:1 mixtures $CD_3OD:CD_3OH$ in the second light beam we obtained [29] the spectrum of Fig. 20. Based on results of other authors [30–32] this can be understood by assumpting a simultaneous excitation of two OH or two OD or one OH plus one OD in two neighboured H-bonded groups by one quantum. The band in Fig. 20 with a maximum at $6766 \, cm^{-1}$ could be a

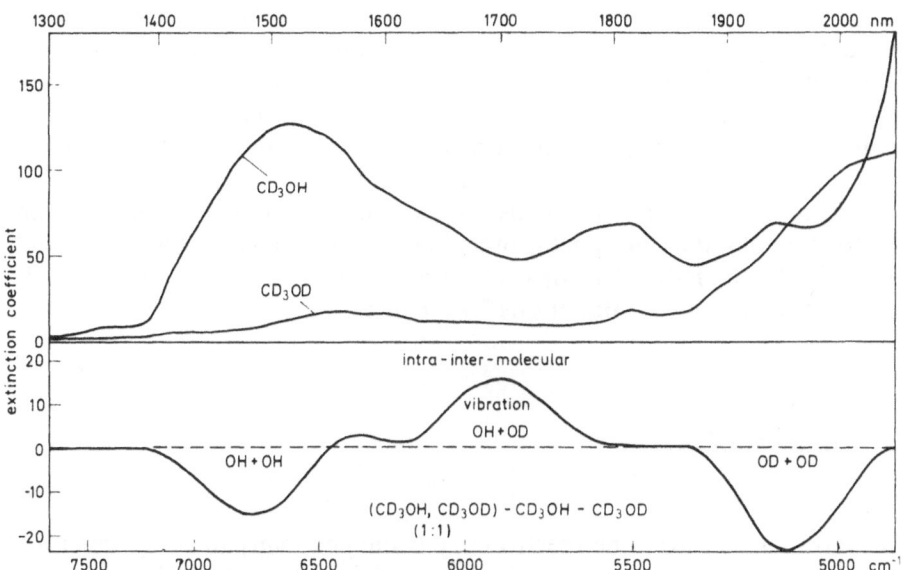

Fig. 20. Above: the spectra of pure liquid CD_3OH and CD_3OD.
Below: The so-called simultaneous bands, obtained by the difference of a mixture OH: OD = 1:1 and two cells of the pure components with half cell thickness [29]
The simultaneous excitation of two neighbouring OH or two OD-groups by one quantum appear as apparent negative bands and the simultaneous excitation of one OH and one OD as positive bands

simultaneous excitation of two OH groups in the cell with pure CD_3OH in the control unit and the band at about $5130\,cm^{-1}$ of two OD groups from the cell with pure CD_3OD. The band maximum at about $5900\,cm^{-1}$ could be correlated with a simultaneous excitation of two neighboured CD_3OH and CD_3OD by one quantum. The small band at about $6330\,cm^{-1}$ may be a simultaneous excitation of a combination $2v_{OD} + v_{def,OH}$. These bands do not have anharmonicities but correspond exactly to the sum of fundamental bands, a fact which establishes the assumption of simultaneous excitations.

The same effect may be the cause of the appearance of a OH + OD combination band of liquid water in Fig. 15 at $1.7\,\mu$ which does not appear in vapour state. In experiments with a mixture of 20 ml H_2O per liter D_2O, such simultaneous OH + OD excitation may also appear. The intensity decrease of this combination band with increasing T could also be established with CD_3OH/CD_3OH [33, 34] mixtures and leads to the conclusion that simultaneous excitation happens in H-bonded neighbours and may depend on the preference of the angle orientation of the OH/OD groups with a parallel component of the OH or OD vectors. The intensity decrease of the simultaneous bands with T-decrease is correlated with a corresponding decrease of the H-bond bands in the fundamental region [34]. This decrease depends on two factors: 1. decrease of H-bonds with T and 2. decrease by $\Delta\overset{.}{v}$ and a corresponding intensity decrease.

The value of ε at constant v of the simultaneous band is a useful measure of the content of H-bonds [29] because it is not so disturbed by band overlapping as the normal spectra with bands of "free" and bonded OH groups. Therefore it could be taken for the estimation of the equilibrium constants of Eq. (18) as a measure of OH_B. It was shown that H-bond energies ΔH for liquid water give similar results of $-15.4 \pm 1\,kJ/mol$ when determined by OH_B with the simultaneous or liquid band [30] or directly only with the OH_F band of liquid water, which gives $\Delta H = -15\,kJ/mol$ [35].

The simultaneous bands are also useful too to study the influence of ions on structure of water [34]. On the other hand, simultaneous OH and CH bands also exist for of H-bond complexes with organic bases [32] which can be used for quantitative studies of OH/base complexes.

3.3 Mixtures of H-bonds

One advantage of overtone spectroscopy is the possibility to study water and alcoholic H-bonds separately using the combination band [v (stretching) + v (bending)] which has its maximum of water in the region around $5200\,cm^{-1}$ ($\lambda = 1.9\,\mu m$) and at about $5000\,cm^{-1}$ ($\lambda = 2\,\mu m$) for alcohols [36–38].

The position of the OH- or NH-stretching bands as well as of this combination band is proportional to the H—bond energy (so called

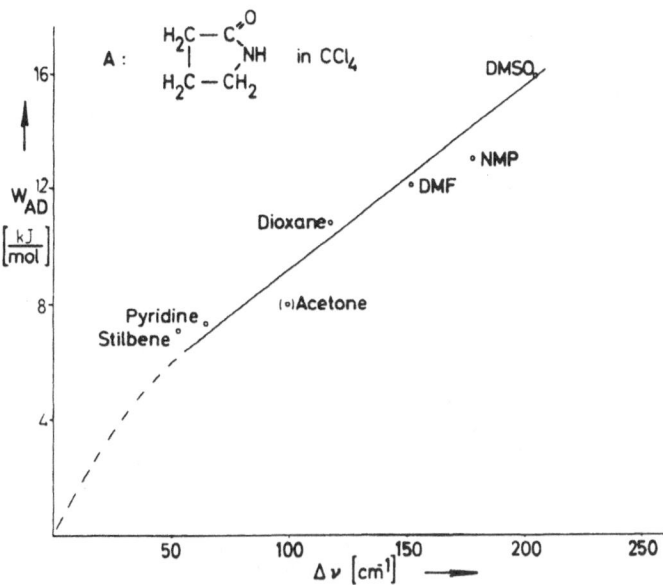

Fig. 21. The H-bond interaction energy W_{AD} between pyrrolidone and different bases in CCl_4 solutions, determined by equilibrium constants obtained by IR-overtone spectroscopy [12]

Badger–Bauer rule). Figure 21 demonstrates this rule with the H-bond energy W_{AD} between pyrrolidone as H-bond donor (D) and different bases as acceptors (A) [8b]. This diagram has been obtained by determing the T-dependence of the equilibrium constant K_{AD} in the coupled equilibrium $A - D$ and the self-dimerisation [8b]. The band maxima are therefore an easy method to estimate the strength of the H-bonds.

To illustrate the efficiency of the combination band of water the center of Fig. 22 shows the maximum of the water band as a function of the relative humidity (rel. h.) of collagen (gelatine or cangoroo tendon) [36, 37]. The bottom of Fig. 22 shows the water uptake in the same samples. Up to about 60% rel. h. the first hydration shell of collagen is formed. During this process the maximum of the water band has a constant v but a lower value compared with water in the bulk. That means, the H-bond strength of the hydrate water of collagen is stronger than in liquid water. At higher rel. h. (> 60%) the water maximum shifts to lower wavelength or higher v. This water corresponds to the multilayer adsorption and becomes more liquid like. If we determine differences of the spectra with different rel. h. (Fig. 23) [38]. One recognizes at low rel. h. a nearly constant band maximum of the hydrate water and at large rel. h. a liquid like water band. — A change of a band of the polymer, as observed during the hydration (Fig. 22, top), may be an indication of a conformer change. —Similar results were also been obtained with bovine nasal cartilage [37].

Another example for this combination band method in the overtone region is the study of the desalination mechanism by celluloseacetate membranes

178   W.P. Luck

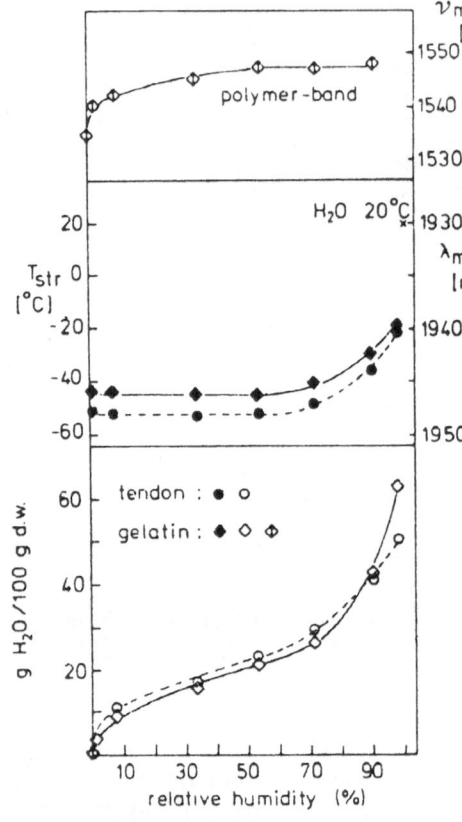

Fig. 22. Bottom: Water uptake of collagen (tendon or gelatin) as a function of relative humidity.

Middle: maximum of the water combination band in both samples (left scale), the structure T_{str}, T of pure water with similar H-bonds.

Top: ν-change of an IR polymer band.

These experiments indicate on stronger H-bond hydrate water up to about 60% rel. h. and multilayer adsorption of liquid water [36, 37]

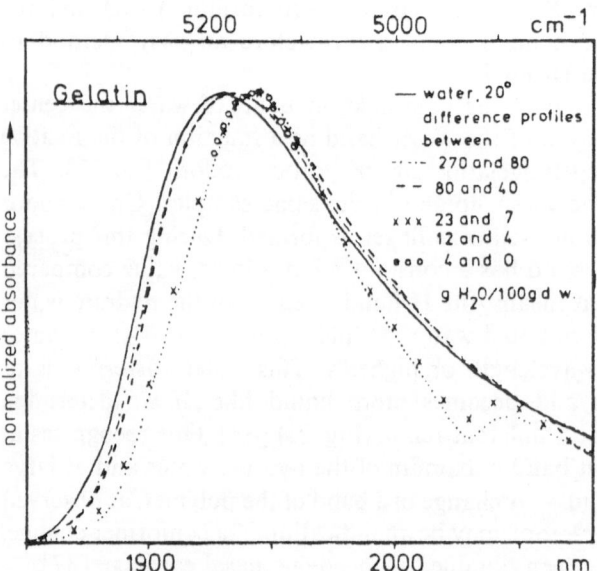

Fig. 23. Differences of the combination band spectra at different rel. h. indicate a slightly different property of the so-called liquid-like water compared with liquid water spectra [38]

[4, 39–41]. In this case we observed contrary to collagen weaker H-bonds in the hydrate layer. The hydrate water band has its maximum at 5220 to 5250 cm^{-1} and the liquid water at 5184 cm^{-1}. By a band analysis at different rel.h. we could estimate the percentage of hydrate and liquid—like water as shown in Fig. 24 [42]. There exists a small amount of a third type of water (rest) with a band maximum at 5200 cm^{-1}. A comparison between two membranes with different acetyl content showed that they differ mainly by the larger uptake of liquid-like water at large rel. h. The membrane T 900 is saturated at about 9 g of water per 100 g of dry membrane.

Different water types also have been found too by NMR methods in polymers [43, 44]. NMR specialist call these two types "bound" and "free" water. The extended IR method demonstrates that the nomenclature hydrate water and liquid like water should be preferred. At first, liquid water is strongly H-bonded, thus the name "free" water is misleading. Secondly bound water seems useful for collagen with stronger H-bonds than liquid water, but in celluloseacetate the hydrate water is less bonded, therefore we prefer the nomenclature "hydrate water" for the first water layer in polymers.

The different types of water may be correlated to the fact that seed can be stored for many years at normal rel. h. about 70% or more without any growth, while they start to grow immediately near 100% rel. h. On the other hand bacteria on food need a high rel. h. to stay alive. The preservation of food therefore mainly requires a reduction of the water activity by adding salt or sugar, etc.

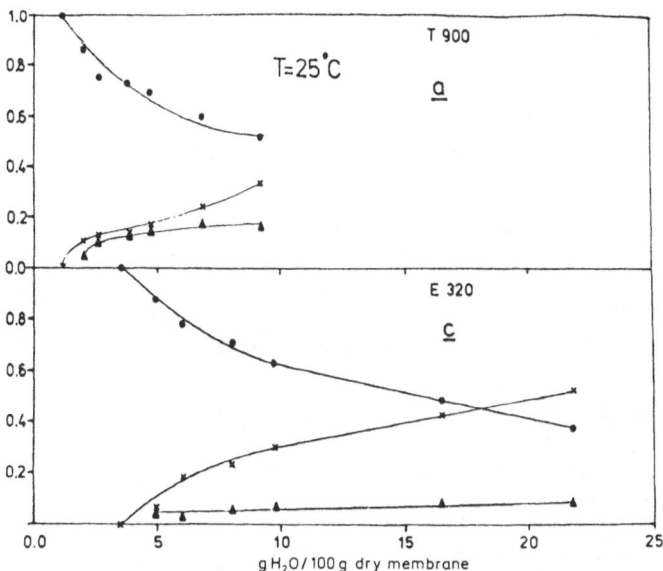

Fig. 24. Content of three different water types in two different celluloseacetate membranes (E320 larger content of hydrophilic groups), determined by the water combination band.
●: hydrate water; ×: liquid-like water; ▲: "rest"-water with properties in between [42]

[45]. This big difference in the properties of water with only a relative small change of the water spectra (Fig. 22 in the middle) could be easier understood, if we add in this Fig. 22 the so-called structure temperature T_{str} (left scale). T_{str} means the temperature at which pure liquid water shows the same v_{max} as the system at 25°C. This heuristic comparison demonstrates that the hydrate water of collagen has such a strength of its averaged H-bonds like supercooled water at about $-45°$ to $-50°C$.

The water H-bonds in desalination membranes [39–42] favour the water flux with lower activation energies. Some H-bonds with alcoholic membrane groups and water as acceptor and larger van der Waals interactions probably by larger coordination numbers of water to the membrane groups compared with liquid water with the averaged coordination number of about 4.4 at room T induce higher adsorption energies of water in such membranes (larger compared with ΔH_{vap} of liquid water [42].

An example of comparing a mixture of two different alcohols by the first overtone is given in Fig. 25 demonstrating the intensity of the free OH overtone in solutions of 0.5 m phenol or 0.5 m methanol and a mixture of 0.25 methanol and 0.25 m phenol in CCl_4 solutions at 20°C. A comparison with the theoretical intensity of the mixture shows a lag of free OH of phenol and an intensity increase of the methanol free OH band in the region 1.48 to 1.411 µm (Fig. 25). In this region there is absorption by OH groups with free OH but H-bonded lone pairs. This Fig. 25 indicates a higher probability to form phenol H-bonds to methanol compared with phenol selfassociates. The steric hindrance of phenol selfassociates may be reduced compared with mixed H-bonds phenol-methanol, or methanol may act as a stronger base than phenol.

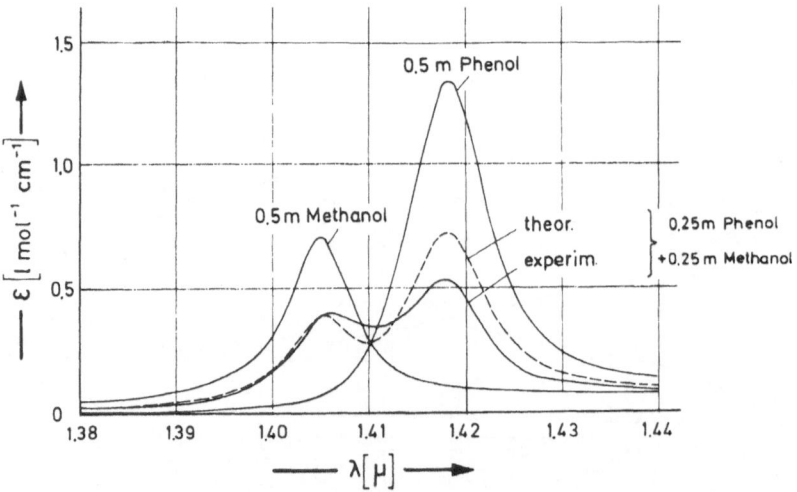

Fig. 25. First OH overtone band of phenol and methanol in CCl_4 and its theoretical mixture 1:1 band. The experimental mixture indicates a higher content of phenol H-bonds to methanol as an acceptor compared with phenol at same total OH concentration as the 1:1 mixture

4 Determinations of Dipole Moment Derivatives by Overtone Intensities

By means of the intensities of the fundamental v_{01} and first overtone v_{02} band it is possible to determine the first and second dipole moment derivatives p_1 and p_2, higher overtones yield those of higher derivatives. If the OH oscillator is treated in approximation as two atomic systems in the dimensionless normal coordinate: $\xi = (r - r_e)/r_e$, with r_e as the equilibrium distance, then one may, in approximation consider [46–49] the dipole moment p in a series of its first and second derivatives $p_1 = dp/d\xi$ and $p_2 = d^2p/d\xi^2$ in the form:

$$p(\xi) = p_0 + p_1\xi + p_2\xi^2 \tag{19}$$

The integrated intensities A_{01} and A_{02} could then be given as approximation by:

$$A_{01} = 2.303 \int \varepsilon_{01}dv = \alpha v_{01}[ap_1 + bp_2]^2 \tag{20}$$

$$A_{02} = 2.303 \int \varepsilon_{02}dv = \alpha v_{02}[-bp_1 + ap_2]^2 \tag{21}$$

with:

$\alpha = (8\pi^3 N_A)/3hc;$	N_A: Avogadro number
$a = \Theta/\sqrt{2} = [2B_e\omega_e]^{1/2}/\sqrt{2};$	B_e: rotation constant
$a' = \Theta^2/\sqrt{2};$	ω_e: harmonic frequency
$b = 5\Theta^2[2x_e/15]^{1/2};$	x_e: anharmonicity constant
$b' = b/5\Theta$	

Unfortunately, A_{01} is proportional to the squares of p_i and therefore for the calculation of the p_i two solutions with different signs, $- -$ or $+ -$, are possible and give an incertainty.

With increasing interactions by H-bonds or by van der Waals type forces the values of p_1 and p_2 also increase [48, 49]. As a consequence, both Eqs. (20) and (21) demonstrate that the intensity of the fundamental band A_{01} increases with interactions as the sum of two increasing terms, while A_{02} is the difference of two similar terms and could be constant as all experiments demonstrate. Solvent interaction can change the intensities. This means that the values for the H-bonds of OH to bases vary if we determine them in the bases as solvent or in ternary systems OH/base/CCl$_4$. Table 3 shows [49] the solvent and temperature dependence of the monomeric OH stretching band of t-butanol (0.012 to 0.02 mol liter^{-1}).

The correspondent v-values at 10°C are [48] in cm^{-1}: vapour state (3643.0; 7114.0); hexane (3622.4; 7064.6); cyclohexane (3620.3; 7064.5); CCl$_4$ (3616.4; 7059.8); trichlorethane (3611.0; 7051.6).

Table 3 reflects the intensity increease of A_{01} by a factor of about 7 with increasing interaction (larger v-shift $\Delta v_{01} = 22$ cm^{-1} but A_{02} increase only by

Table 3. OH stretching band for t-butanol in nonpolar solvents at different T and dipole moments (in Debye) [49]

Solvent	T/°C	$A_{01} \times 10^{-6}$ cm mol^{-1}	$A_{02} \times 10^{-5}$ cm mol^{-1}	p_0	p_1	p_2	p_0	p_1	p_2
					$-\ -$			$+\ +$	
Vapour	35	0.84	1.64	-1.29	0.37	1.66	1.48	0.47	-1.01
Hexane	10	3.02	2.29	-1.46	0.73	2.19	1.81	0.86	-0.95
	60	2.65	2.13	-1.40	0.69	2.09	1.74	0.80	-0.94
Cyclohexane	10	3.29	2.07	-1.38	0.77	2.15	1.73	0.89	-0.84
	60	2.74	1.91	-1.33	0.70	2.03	1.65	0.811	-0.84
CCl$_4$	10	4.24	2.10	-1.36	0.88	2.24	1.77	1.00	-0.77
	60	3.82	2.03	-1.34	0.84	2.18	1.73	0.95	-0.78
Trichlor-	10	5.62	2.05	-1.29	1.03	2.32	1.78	1.14	-0.64
ethane	60	5.20	2.02	-1.29	0.99	2.28	1.76	1.10	-0.66

The correspondent v-values at 10°C are [48] in cm^{-1}: vapour state (3643.0; 7114.0); hexane (3622.4; 7064.6); cyclohexane (3620.3; 7064.5)

25% and $\Delta v_{02} = 62.4$ cm^{-1}). The ratio A_{01}/A_{02} increases from a factor of 5 in vapour with increasing interaction to 27.4 in trichloroethane. In the more probable $+\ +$ solution all p_i values increase with increasing Δv or increasing interaction energy, respectively. Δv as well as A_{0i} and p_i decrease with increasing T or increasing intermolecular distance (weaker interactions). Figure 26 shows estimated dependences of the dipole moment p_0 as a function of the reduced normal coordinate ξ, with the assumption $p_0(\xi = -1) = 0$. These results show the p_0-change by non polar solvents and are correlated with the solvent effects on chemical reactions [50, 51] by solvents. Table 4 shows a similar experiment with methanol/bases H-bonds at 20°C. The spectra have been recorded in ternary systems CH$_3$OH/base/CCl$_4$ and the intensities of the monomeric OH have been extrapolated to infinite alcohol concentrations, the H-bond band intensities has been extrapolated for OH$_{free}$ to zero [48].

The integrated intensity of the v_{01} H-bond band in DMSO is about 9 times larger than for OH$_F$ in CCl$_4$ or about 60 times larger compared with methanol vapour. The overtone band decreases about 50% from CCl$_4$ to DMSO. The ration of the H-bond bands in DMSO $A_{01}/A_{02} = 300$ demonstrates too the large increase of this ratio with increasing interaction.

The corresponding v in cm^{-1} of the systems of Table 4 are [48]: CCl$_4$ 3643; 7117); AN(3540; 6935); DIO(3502; 6850); THF (3480; 6808); DMSO (3372; 6528). The decrease of the integrated intensities of the second overtone v_{03} by methanol H-bonds is given in Fig. 27 [65]. Like v_{02} the change is small compared with the fundamental band, but larger compared with the first overtone band. The influence of the first overtone on the calculation of p_1 is partially reduced taking into account the second overtone. The determination of p_1, p_2 and p_3 is based on the Schroedinger equation (see Ref. [65]). Figure 28 indicates the change of the first dipole moment derivative p_1 with Δv of methanol [52]. The slope $dp_1/d(\Delta v)$ is different between van der Waals interactions and H-bonds, an observation that appears to hold for different spectroscopic properties [52].

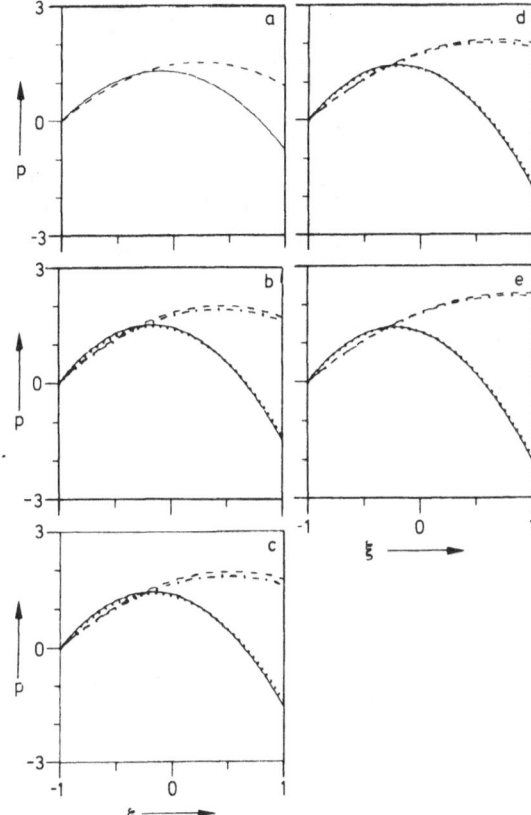

Fig. 26. Dipole moment plots for O—H bond of *t*-butanol in vapour (a), in hexane (b), in cyclohexane (c), in carbon tetrachloride (d), in trichloro-ethane (e). —and...represent—combination at 10° and 60°C, while ——— and —·—·— represent + — combination at 10° and 60°C

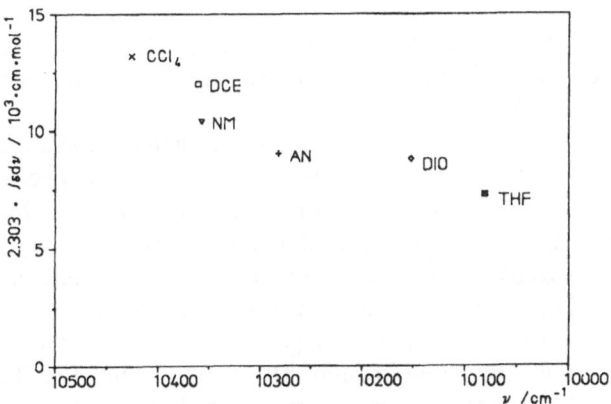

Fig. 27. Decrease of the integrated intensity of the second OH overtone methanol band with H-bond acceptor strength and the corresponding frequency shift

Table 4. Band intensities of OH stretching methanol/base at 20°C and moments (in Debye).

Solvent	$A_{01} \times 10^{-6}$ cm mol^{-1}	$A_{02} \times 10^{-5}$ cm mol^{-1}	p_0	p_1 + +	p_2	p_0	p_1 + −	p_2
CCl$_4$	5.67	3.73	− 1.83	1.00	2.83	2.32	1.16	−1.16
Acetonitrile (AN)	23.33	2.02	− 0.94	2.18	3.12	2.15	2.29	0.14
Dioxane (DIO)	30.9	1.84	− 0.83	2.52	3.35	2.11	2.62	0.51
Tetrahydrofuran (THF)	35.4	1.80	− 0.73	2.71	3.44	2.17	2.82	0.65
Dimethylsulfoxide − d$_6$(DMSO)	50.25	1.70	− 1.10	3.26	4.36	1.78	3.38	1.60

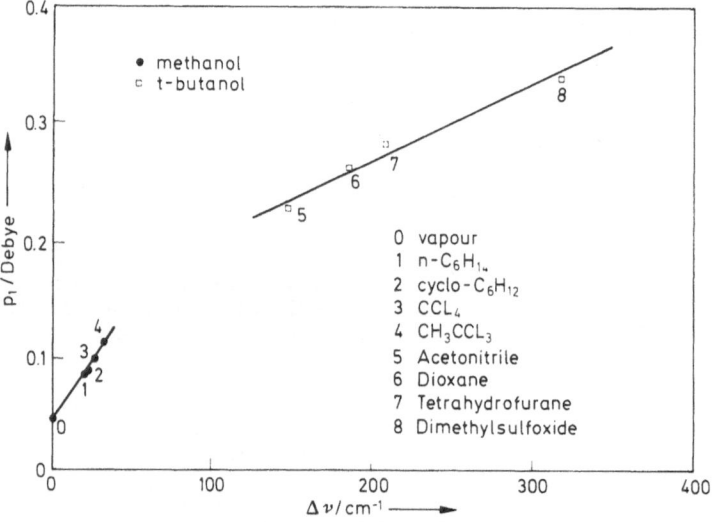

Fig. 28. First dipole moment derivative of methanol in different environments as a function of Δv, [52]. Van der Waals and H-bond interactions induce different slopes in this figure

Figure 29 shows the integrated methanol v_{01} intensities [65]. This figure plots the intensities of bands with different interacting solvents in dependence of the band-maximum position [65].

The broad OH bonded band of overtones give an information in the statistics of ΔH, corresponding to the Badger–Bauer rule the proportionality between ΔH and Δv. The overtone bands have only small intensity changes by interactions, but this is not valid for fundamental bands. They have large intensity changes proportional to Δv. In order to obtain this statistics of different H-bonds energy ΔH by fundamental bands these spectra could be divide by the relation of Fig. 29. Only by this division one can get a view on the statistics of different types of H-bonds by fundamental IR bands. The Fig. 30 demonstrates as example the T-dependence of liquid methanol overtone intensities (at the band maximum from the top to the bottom: 10°C, 20°C, 30°C, 40°C and 50°C). The fundamental bands of the same series in Fig. 31 gives

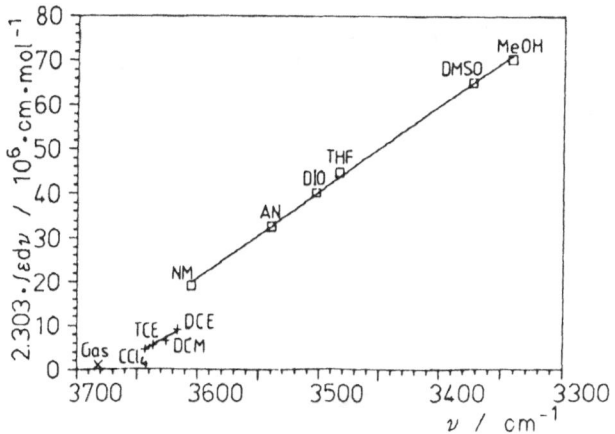

Fig. 29. Integrated OH fundamental stretching band in different environments as a function of the frequency. Different slopes of van der Waals systems and H-bonded systems [52, 65]

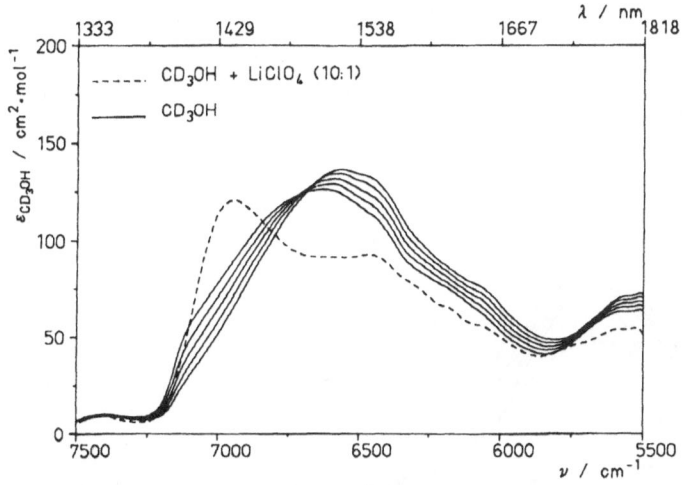

Fig. 30. *Full lines*: Temperature dependence of the first OH overtone band of liquid methanol with a distinct peak in the region of "free OH";
dashed band: Solution spectra methanol: LiClO₄ (10:1) [66]

another T-dependence of the band intensities. Both figures include the methanol spectrum with 10: 1 LiClO₄ additions (dashed). The fundamental band differs from the overtone band for this solution by a different intensity ratio of the maximum at larger ν to the one at smaller ν. Dividing the fundamental intensities of Fig. 31 at every ν with the intensity function of Fig. 29 leads to Fig. 32. These intensity-corrected fundamental spectra are now similar to the overtone spectra. Both now give about similar information on the statistics of interactions (recognizing its ν-values) [66]. The overtone spectra in Fig. 30 are measured with CD_3OH because in the CH_3OH spectrum an intense $(OH + CH)$

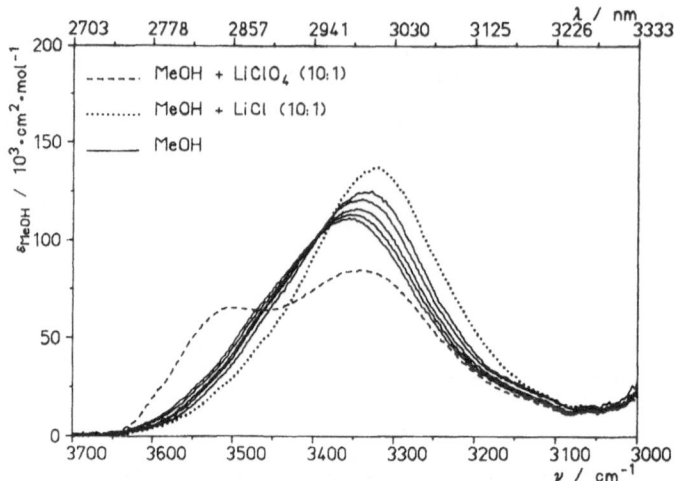

Fig. 31. Fundamental OH band of same systems as in Fig. 30 plus MeOH band of MeOH: LiCl (10:1) (*dotted*)
Spectra differ from Fig. 30 by the large intensity change

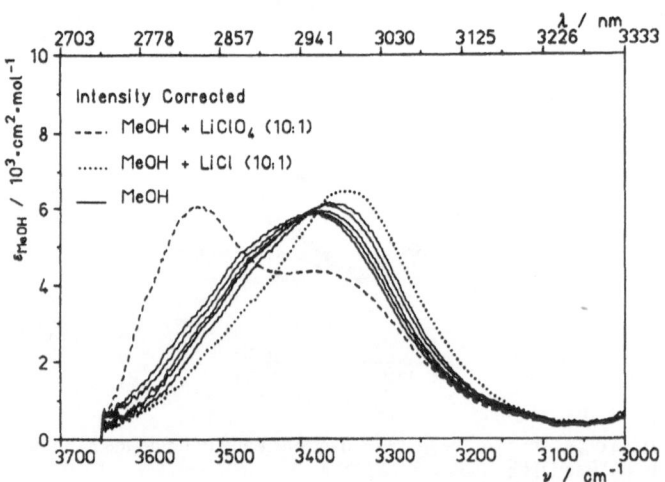

Fig. 32. Fundamental spectra of Fig. 31 corrected with the intensity function of Fig. 29.
These spectra show similar statistics of different H-bonds like the overtone spectra of Fig. 30 [66]

combination band would appear. This comparison can clear some of the discrepancies reported in the literature on IR spectroscopic results of the H-bonded structures of liquids. Especially the link in Fig. 29 between the intensity function of van der Waals and H-bond interactions levels out in Fig. 21 the distinct OH_{free} content in water and alcohols observed with overtone spectra, which plays an important role to understand the structure of water (see Chap. IX).

5 Van der Waals Spectroscopy

Figure 33 shows that intermolecular interactions can be studied with IR-overtone spectra [21, 52]. The heat of vaporisation of ethanol ΔH_{vap} is plotted from infinite diluted solutions in different solvents as function of the first OH overtone frequency. We get two straight lines for the relation $\Delta v = f(\Delta H_{vap})$; one with a larger slope for nonpolar van der Waals interactions and one for

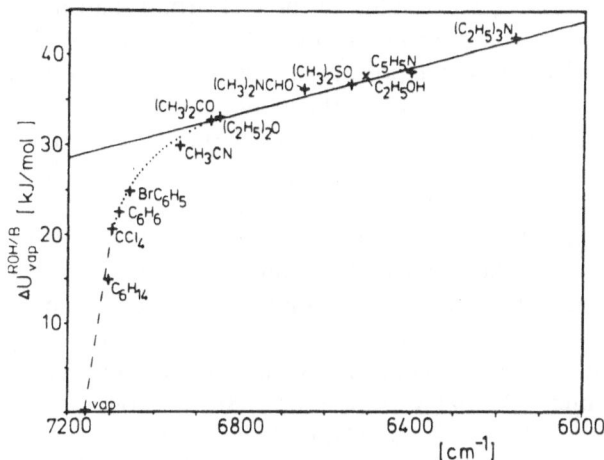

Fig. 33. Heat of vaporization at constant volume ΔU_{vap} of ethanol in different solvents (extrapolated at infinite dilution) as a function of the OH overtone maximum. There exist different proportionalities $\Delta U_{vap} \simeq \Delta v$ for van der Waals and for H-bond interactions [21]

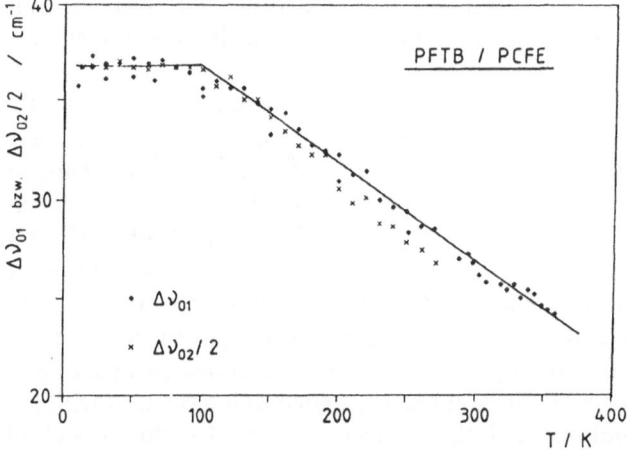

Fig. 34. Temperature shift of the OH fundamental and first overtone maximum of perfluoro-t-butanol (PFTB) in poly-monochloro-trifluoro-ethene (PCFE) solution. (linear for $T > 100$ K and constant at $T < 100$ K) [52, 54]

H-bonds. There are two different Badger–Bauer rules: $\Delta v = a\Delta_{vap} + b$ with different a-values and $b = 0$ for van der Waals interactions, and b not zero for H-bonds. This means that we can establish an IR method to study van der Waals interactions similar to H-bond IR spectroscopy. The shifts by van der Waals interactions are smaller, however. Thus this method requires higher accuracy and extended experiences. The smaller spectroscopic effects caused by van der Waals interactions inclines to use the overtone spectra for such studies because of the higher precision of this method and the double Δv values. Perfluoro-t-butanol (PFTB) has a higher acidity of the OH group than other alcohols by electronic effects of the fluorine groups. The sensitivity of this molecule is much larger by H-bond interactions as well as by van der Waals effects. Figure 34 shows the OH frequency shift of the monomeric PFTB OH in diluted solutions in poly monochloro-trifluoro-ethene (PCFE) with dependence on T [52–54]. The shift of the fundamental band v_{01} is half of the shift of v_{02}. That means $\Delta v_{0i}/v_{0i}$ is constant. Up to about 100 K Δv in Fig. 34 increases linearly with decreasing T (decreasing intermolecular distances). This corresponds to a linear increase of the van der Waals interactions between OH and the weak polar solvent and establishes spectroscopically the linear increase of van der Waals interactions with decreasing T (see Chap. III). At low T the shift becomes nearly T-indepenent. This could mean that the intermolecular zero point energy and the intermolecular vibration levels are so large that its excitation requires fairly high T. At low T, the relation between the interaction and the T-linear intermolecular distances thus needs to be modified. The density of rare gas crystals near $T = 0$ K is also T-independent [55]. This was established by a model calculation based on the dimer van der Waals potentials of rare gases [56]. The OH-PCFE interaction is larger than in rare gases; therefore this T-constant region could be extended further. The linear T-relation could be theoretically estimated [56, 52] by exciting the higher intermolecular vibration levels with increasing T. The agreement between the overtone and the fundamental band in Fig. 34 means that the anharmonicity does not change remarkably by van der Waals interaction.

This result could be established by high pressure experiments up to 10 GPa [57–60]. The decrease of intermolecular distances increases Δv of similar systems almost linearly with p (red shift). The IR van der Waals spectroscopy of monomeric OH groups demonstrates a surprisingly weak interaction between OH and CF-group containing solvents. Δv of PFTB and of 1-H,1-H-heptafluorpropanol (HFB) [42] in perfluoroheptane as solvent shows a maximum at p of about 1 GPa (Fig. 35), Δv decreases with larger p to $\Delta v = 0$, at very high p the frequencies become larger than the HFB-vapour value [58].

The OH band of PFTB and HFB in PCFE as solvent shows two peaks and respectively a shoulder; the first one could be correlated to interactions with the CCl-groups and the second one with the CF-groups. The CF-induced peak of HFB shows a similar behaviour as in perfluoroheptane: at low pressure at first a red shift, then a maximum, a decreasing Δv, $\Delta v = 0$ and at last at very high pressure blue shifts. This behaviour can be understood by the overlapping

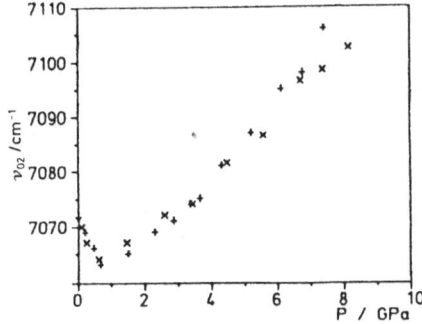

Fig. 35. Pressure-induced shift of the monomeric perfluoro-*t*-butanol OH overtone band in perfluoroheptane solutions [57, 60]

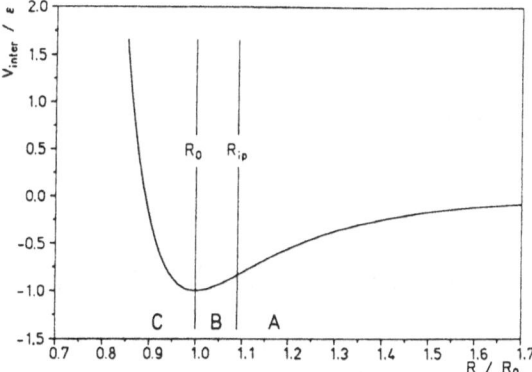

Fig. 36. Scheme of the solvent —OH intermolecular potential curve for understanding Fig. 35. The red shift at low p could mean: the intermolecular distance R is in the region A; at the frequency minimum $R = R_{ip}$, at the turning point with maximum of the intermolecular potential gradient; decreasing v above 1 GPa correlated with R in the region B; observed vapour-v at R_0 and blue shifts in the region C [57, 58]

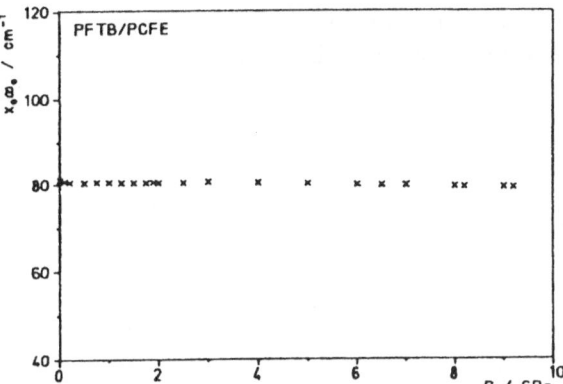

Fig. 37. The pressure independence of the anharmonicity constant of the monomeric PFTB OH band in diluted PCFE solutiions, observed with IR fundamental and first overtone band [60]

between the OH and the van der Waals potentials [42, 58]. The red shift can be correlated with the region A and B in Fig. 36 of the overlapping effective van der Waals potential of the solvent. The Δv-maximum would correspond with the distance R_{ip} of the potential turning point; $\Delta v = 0$ with the potential minimum and the blue-shift region with the repulsion region C in Fig. 36. Also the anharmonicity does not change much at high pressures (Fig. 37). Only the CCl-PFTB interactions indicate a small change of the anharmonicity at high pressure. However, the required band analysis of the overlapping CF and CCl components causes some uncertainty in this case.

Weak OH H-bonds with π-electrons as acceptor react to high pressures only with red shifts [59], but medium strong H-bonds with normal bases show a maximum of Δv as a function of pressure [61]. The method of analysing the pressure effects by the sum of the OH and the interaction potential [52, 58] opens perspectives that IR van der Waals spectroscopy will allow to determine intermolecular potentials in condensed media.

In Chap. III the heat of vaporization ΔH_{vap}, a measure of the intermolecular forces is given for $T < T_B$ (T_B: boiling point) as:

$$\Delta H_{vap} = \Delta H_{vap}(OK) - bT \quad T < T_B \tag{22}$$

or in detail:
$$\Delta H_{vap} = (ZfR/2)[(3T/2) - T] \tag{23}$$

and general:
$$\Delta H_{vap} = [\Delta H_{vap}(OK) - bT][1 - 2x_F] \tag{24}$$

x_F: non linearity of the density (or the hole concentration) (see page 67) Correspondingly we found [54, 62] for $100\,K < T < 400\,K$:

$$\Delta v = \Delta v_0 - bT \tag{25}$$

The v shift at higher temperature has been determined by a pressure cell in the overtone region [63]. A Δv non linear with T has been found near the critical temperature T_C for a series of alcohols in diluted solutions of nonpolar or really polar solvents. Figure 38, for instance, shows the OH overtone v of monomeric 2, 2, 2 trifluoropropanol (TFP) (diluted solution in perfluorobenzene). Up to

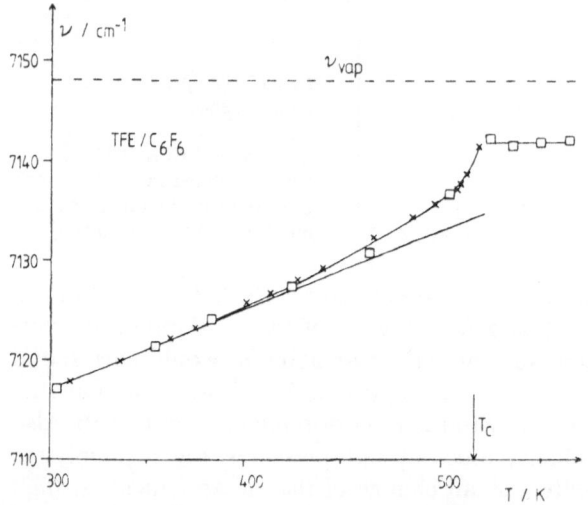

Fig. 38. Maximum of the first overtone OH band of 2,2,2-trifluoropropanol (TFP) diluted solution in perfluorobenzene. $\times$: calculated by Eq. (26)

about 400 K the linear ν shift of Eq. (25) was established. For $400\,K < T < T_C$ the relation:

$$\Delta\nu = [\Delta\nu_0 - b'T][1 - x_F] \qquad (26)$$

corresponds well (crosses x in Fig. 38) to the experimental values (squares). The correction $[1 - x_F]$ instead of $[1 - 2x_F]$ would give in Eq. (24) the total interaction energy in a solution. In contrast the term $[1 - 2x_F]$ in Eq. 24 gives the vaporization in the real gas vapour. These experimentals have been done with an amount of solvent allowing that at T_C the critical density is reached, therefore at $T > T_C$ the shift is constant such as the density. Various experiments on $\Delta\nu$ of alcohols in different solvents in the region of $\Delta\nu - T$ linearity of Eq. 25 gave a relation $b' = d'\Delta\nu_0$ and a similar relation in the reduced T/T_C plot. Therefore:

$$\Delta\nu/\Delta\nu_0 = 1 - d\,T/T_C \qquad (27)$$

Values $0.76 < d < 0.8$ were obtained [54]. Parallel to Eq. (27) we found from Eqs. (22) and (23): $\Delta H_{vap}/\Delta H_{vap}(OK) = 1 - 2T/3T_C$ [28].

This similarity between the two last equations with similar factors of about 0.78 and 0.66 may indicate the relation between van der Waals spectroscopy and the simplified model of normal liquids (see Chap. III). The difference between factor d and 2/3 could be—if it is not an indication of the error limits—caused by the fact that our model assumes the same efficient coordination number fZ for the interaction energy and for the intermolecular specific heat. It could be that f is 1 for the specific heat term.

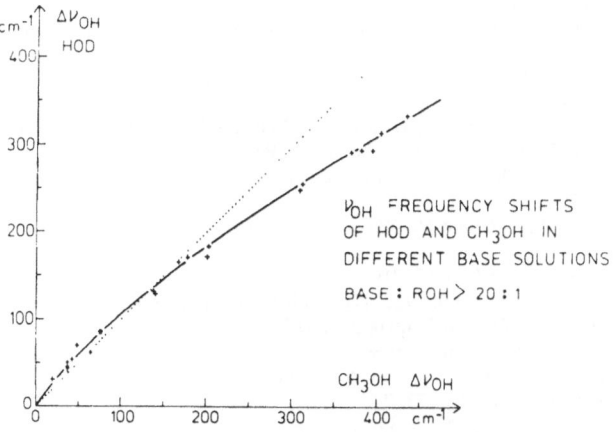

Fig. 39. $\Delta\nu$ of the OH(HOD) in dependence of $\Delta\nu$ (CH_3OH) in diluted solutions in different solvents. With van der Waals solvents ($\Delta\nu < 100\,cm^{-1}$) equal shifts or similar interactions are observed but with H-bond acceptor solvents ($\Delta\nu > 100\,cm^{-1}$) shifts and interactions of CH_3OH compared with HOD are larger

The different spectroscopic behaviour of van der Waals and H-bond interaction (see Figs. 29, 33, 34) may be caused by different potential gradients. A further interesting difference is observed by comparing the solvent shifts of HOD and of methanol. As Fig. 39 demonstrates, the van der Waals interactions ($\Delta v < 100\,\mathrm{cm}^{-1}$) in both cases are very similar, but the H-bond shift by different bases ($\Delta v > 100\,\mathrm{cm}^{-1}$) for methanol is about 25% larger than for HOD.

The good results of this relatively new method give some hope that other interesting results will follow in the near future.

6 References

1. Schaefer C (1929) Einführung in die theoretische Physik, vol I, 3rd edn., de Gruyter, Leipzig, p 153
2. Luck WAP (1982) Die chemische Bindung im Unterricht, Deutsche Physikalische Gesellschaft, Fachausschuß Didaktik der Physik, Vorträge der Frühjahrstagung, Gießen, p 97
3. Luck WAP, Ditter W (1968) Ber Bunsenges Phys Chem 72: 365
4. Singh S, Schiöberg D, Luck WAP (1981) Spectrosc Lett 14: 141
5. Singh S., Fritzsche M, Kümmerle I, Luck WAP, Zheng HY (1985) Spectrosc Lett 18: 283
6. Luck WAP, Zheng HY (1984) J Chem Soc, Faraday Trans 2 80: 1253; Luck WAP, Zheng HY (1984) Z Naturforsch 39A: 888
7. England-Kretzer L, Fritzsche M, Luck WAP (1988) J Mol Struct 175: 277
8. Luck WAP (1965) Naturwissenschaften 52: 25; (1967) 54: 601; Luck WAP (1968) Z Naturforsch 23b: 152
9. Geiseler G, Seidel H (1977) Die Wasserstoffbrückenbindung. Vieweg, Braunschweig
10. Huisgen R, Walz H (1956) Chem Ber 89: 2616
11. Luck W (1976) The angle dependence of hydrogen-bond-interactions. In: Schuster–Zundel–Sandorfy (eds) The hydrogen bond, North Holland, Amsterdam, vol II, Chap. 11, p 527
12. Luck WAP (1973) Infrared studies of hydrogen bonding in pure liquids and solutions. In: Franks F (ed) Water: a comprehensive treatise, vol II. Plenum, New York, p 235
13. Wolf KL, Dunken H, Merkel K (1940) Z Phys Chem 46: 297
14. Reeves LW, Schneider WG (1958) J Chem Phys 34: 314
15. Luck WAP (1961) Ztschr f Elektrochemie 65: 355
16. Geiseler G, Fruwert S (1962) Z Phys Chem 26: 10
17. Falk M, Knop O (1973) In: Franks F (ed) Water: a comprehensive treatise, vol II. Plenum, New York, p 55
18. Kreuzer S (1943) Z Phys Chem 53: 213
19. Hoffmann EG (1943) Z Phys Chem 53: 179
20. Streuer E, Wolf KL (1938) Z Phys Chem 41: 321; Wolf KL, and Harms H (1940) Z Phys Chem B 46: 287; Wolf KL, Klapproth K (1940) Z Phys Chem B 46: 1940
21. Kleeberg H, Kocak O, Luck WAP (1982) J Solution Chem 11: 611
22. Luck WAP (1986) J Mol Liq 32: 41
23. Luck WAP (1965) Ber Bunsenges Phys Chem 69: 626
24. Luck W, Ditter W (1969) Z Naturforsch 24b: 482
25. Luck WAP, Schrems O (1980) Spectrosc Lett 13: 719
26. Dyke TR, Muenter JS (1974) J Phys Chem 60: 2929; Dyke TR (1984) Topics 120: 85
27. Kleeberg H, Luck WAP (1989) Z Phys Chem Leipzig 270: 613
28. Luck WAP (1981) Ber Bunsenges Phys Chem 85: 959
29. Schiöberg D, Buanam-Om C, Luck WAP (1979) Spectrosc. Lett. 12: 83
30. Ketelaar JAA, Hooge FN (1955) J Chem Phys 23: 749
31. Asselin M, Sandorfy C (1970) J Chem Phys 52: 6130
32. Burneau A, Corset J (1972) J Chem Phys 56: 662; ibid (1973) 58: 5188; ibid (1972) 69: 142
33. Fritzsche M (1983) Master Thesis, University of Marburg, AG Luck
34. Fritzsche M (1990) Thesis, University of Marburg, AG Luck
35. Luck WAP (1974) Infrared overtone region, In: Structure of water and aqueous solutions. Sect. III.3, Verlag Chemie/Physik, Weinheim

36. Kleeberg H (1982) thesis, University of Marburg, AG Luck
37. Kleeberg H, Luck WAP (1977) Naturwissenschaften 64: 223
38. Luck WAP, Kleeberg H (1978). In: Metzner H (ed) Photosynthetic oxygen evolution. Structure of water and aqueous solutions, Academic Press, New York, p 1
39. Luck WAP, Schiöberg D, Siemann U (1980) J Chem Soc Faraday Trans II 76: 136
40. Luck WAP (1984) Structure of water and aqueous systems. In: Belfort G (ed) Synthetic membrane processes. Fundamentals and water applications. Academic Press, New York, p 21
41. Luck WAP (1987) Desalination 62: 19
42. Ringriwatananon K (1990) Thesis, University of Marburg, AG Luck
43. Hazlewood CF, Nichols BL, Chamberlain NF (1969) Nature 222: 74
44. Tait MJ, Franks F (1971) Nature 230: 91
45. Rockland LB, Stuwart G (eds) (1981) Water activity—Influences on food quality. Academic Press, New York
46 Chakerian Jr C (1976) J Chem Phys 65: 4228
47. Singh S, Luck WAP (1981) J Mol Struct 74: 49: 65
48. Kriegsmann H, Reklat A, Löffler E, Steiger Th (1990) Z Phys Chem, Leipzig 271: 61
49. Kriegsmann H (1988) Z Phys Chem 269: 1030
50. Reichardt Ch (1988) Solvents and solvents effects in organic chemistry. VCH, Weinheim
51. Gutmann V (1978) The donor-acceptor approach to molecular interactions. Plenum press, New York
52. Luck WAP (1990) J Mol Strucct 218: 281
53. Peil S, Luck WAP (1990) J Mol Struct 224; 175
54. Peil S (1990) Thesis, University of Marburg, AG Luck
55. White GK (1964) Cryogenics. Landolt-Börnstein, vol II/2a, 6th edn, 4: 2, p 186
56. Wess T, Luck WAP (submitted) (1991) Chem Phys Lett
57. Mentel TF (1988) Thesis, University of Marburg, AG Luck
58. Mentel TF, Luck WAP (1990) J Phys Chem 94: 1050
59. Mentel TF, Luck WAP (1990) J Phys Chem 94; 1059
60. Luck WAP, Mentel TF (1990) J Mol Struct 218: 333
61. Luck WAP, Kümmel W, Mentel TF (1990) J Mol Struct, 237; 233.
62. Kümmerle I (1990) Thesis, University of Marburg, AG Luck
63. Luck WAP (1968) Habilitation thesis, University of Heidelberg
64. Luck WAP (1967/68) J Mol Struct 1: 261
65. Englang-Kretzer L (1990) Thesis, Univesity of Marburg, AG Luck
66. England-Kretzer L, Fritzsche M, Luck WAP (1988) J Mol Struct 175: 277

CHAPTER VIII
Intermolecular Interactions at Low Temperature. Matrix Isolation Spectroscopy Applied to Hydrogen-Bonded Complexes and Charge Transfer Complexes

G. Maes

Isolation of molecules, capable of undergoing specific intermolecular interactions with partner molecules, in inert solid matrices at extremely low temperature constitutes a unique experimental test for theoretical models or predictions which are usually based on the assumption of essentially "free" complexes. The technique offers many advantages compared to conventional spectroscopy, especially for H-bonded complexes. Detailed examples of matrix studies in this field are discussed. These include the vibrational correlation diagram for B...HCl complexes, the study of small aggregates of HCl, identification of the interaction site(s) in the polyfunctional bases methyl acetate and uracil, and the IR induced dissociation of the HI complex of dimethylacetamide.

Matrix isolation is less frequently applied to the study of charge transfer complexes. The necessity to use inherently less sensitive Raman spectroscopy for the experimental analysis of this type of intermolecular interaction is one of the main problems. Complexes of the σ^*-acceptors X_2 ($X = Cl$, Br) with n- and π-donors, and of the π^*-acceptors SO_2 and SO_3 with n-donors are described as illustrative examples.

Intermolecular Forces
© Springer-Verlag Berlin Heidelberg 1991

1 Introduction

Chemists or physicists interested in the study of intermolecular interactions between different chemical species usually proceed via one of two different approaches. The choice is either for an experimental investigation of a particular intermolecular complex, or alternatively, theoretical calculations are performed on interaction energy, complex geometry, and derived physicochemical properties. In the very best studies, both methods of investigation are combined.

Theoretical treatments of intermolecular interactions examine the complex in isolation, i.e., free from interactions with other molecules, as outlined in Chaps. II and IV. Most experimental studies, however, are performed in condensed phases—solids, liquids, solutions—at room temperature. The physical nature of the solid and liquid states is mainly determined by many-body ("cooperative") interactions, and 1:1 interactions are certainly not being observed. Solutions in an inert solvent are better, but the results are affected by solute–solvent interactions which may, in some cases, be comparable in strength with the intermolecular interaction under investigation. The only experimental approach allowing to test theoretical predictions is the study of the intermolecular complex in the vapor state at low pressure. As far as vibrational spectroscopy is concerned, such studies pose many practical difficulties—low concentration of the species of interest per unit pathlength in the gas cell, overlapping rotational structure in the spectrum, etc.

A very suitable experimental method for an imitation of the ideal low-pressure vapor state, is matrix isolation. It is a technique for trapping the species of interest in a large excess (100 to 10000 times) of an inert material by condensation at extremely low temperature (4–20 K), so that the diluent forms a rigid solid ("the matrix"). At high matrix-to-solute ratios (denoted as M/S), the solute may be expected to be isolated in the solid matrix cages, where it can be examined at leisure by various spectroscopic methods. Figure 1 depicts schematically the isolation of a solute S in a matrix M.

Noble gases, especially argon, and nitrogen are the most commonly used matrix materials. Although the technique of matrix isolation has been developed originally (1954) to study transient species [1], it has now been used for many other types of study, particularly for the investigation of intermolecular complexes of the H-bond or EDA type. A number of illustrative examples are described in this chapter. A more complete picture of the matrix isolation technique and its application in several fields of chemistry and physics can be obtained from the references given in the general list at the end of this chapter.

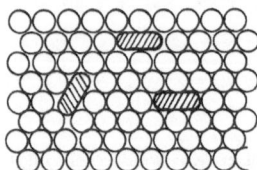

Fig. 1. Schematic representation of a solute S isolated in a matrix M

2 Advantages of Matrix Spectroscopy

From early and more recent studies the following advantages of the combination of matrix isolation and vibrational spectroscopy over conventional infrared or Raman spectroscopy have become apparent:

(i) Most entrapped species, with the exception of very small ones, will be so tightly held as to prevent rotation in the matrix cage, and at the low temperature pertaining (RT is only $170\,\mathrm{J\,mol^{-1}}$ or kT is only $14\,\mathrm{cm^{-1}}$ at 20 K), transitions occur only from the lowest level thermally populated states (Table 1). Matrix spectra are therefore much simpler than those of the corresponding gas phase, and pure vibrational modes appear as sharp bands in absorption (IR) or scattering (Raman) spectra. Even the broad v_σ bands described in Chap. IV are often much narrower, since the anharmonic coupling $v_s \pm v_\sigma$ is restricted to $v_s + v_\sigma$, the fluctuation of $A—H...B$ distances is dampened by the rigid matrix cage, and coupling with neighboring molecules or complexes is strongly reduced at high M/S ratio.

(ii) At room temperature, the formation of an intermolecular complex $A...B$ is governed by the thermodynamical equilibrium:

$$A + B \rightleftarrows A...B$$

for which:

$$\Delta G^0 = -\,RT\ln K = \Delta H - T\Delta S^0$$

K is the formation or equilibrium constant, ΔH the interaction enthalpy, and ΔS^0 the interaction entropy of the complex.

In the temperature range of matrix experiments, the entropy term $T\Delta S^0$ becomes neglectably small. From the physical point of view, this represents the "immobility" of the isolated species A, B, and $A...B$ in the matrix cages. As a consequence, the complex formation "equilibrium" is shifted towards the

Table 1. Population of vibrational and rotational levels at different temperatures[a]

Vibrational levels of SnS			Rotational levels of C_2	
	Percentage Population			
	at 4 K	at 1000 K	T	J with highest population
v	($v = 480\,\mathrm{cm^{-1}}$)		(K)	($B = 1.82\,\mathrm{cm^{-1}}$)
0	100	50	2	0.1
1	10^{-75}	25	4	0.4
2	—	12	20	1–2
3	—	6	77	3
4	—	3	300	7
5	—	2	2500	22
6	—	0.7		

[a] Taken from Ref. [2].

complex species. Any A and B entities isolated in the same matrix host cage will be present as complex species, even in the case of very weak interaction. Matrix isolation spectroscopy is an ideal method to study very weak complexes!

(iii) One expects very little perturbation of the trapped solute species S by the inert matrix cage material M. Furthermore, the probability for isolating two or more A or B species in one and the same matrix cage is particularly low at high M/S ratios, and matrix isolation is a unique method allowing to investigate the pure 1:1 interaction in an inert environment, minimizing both medium ("solvent") and cooperative effects. That is why experimental matrix results constitute the ideal test for theoretical models and predictions.

(iv) By controlled, stepwise increase of the matrix temperature from the initial deposition value (mostly 10 to 15 K in Ar or N_2) to values in the range 25–35 K (a process referred to as "annealing"), it is possible to induce diffusion of the species S through the matrix M and to identify 1:2, 1:3, ..., 1:N complexes progressively. This formation of heterotrimers, -tetramers, ... and -polymers A...$(B)_n$ will of course be accompanied by the simultaneous formation of homomultimers (B_n). Such a controlled formation of complexes of higher stoichiometry is not possible in more conventional vibrational spectroscopy in condensed phases.

There are also some inherent difficulties associated with matrix isolation applied to intermolecular complex spectroscopy:

(i) Frequently the matrix environment causes single frequencies to appear as multiplets, complicating interpretation of spectra. This so-called matrix splitting is usually detected by isolating the species under different experimental conditions (temperature, deposition rate, matrix material).

(ii) It remains extremely difficult to perform reliable intensity measurements in matrices. The main problem is the large uncertainty in the concentration of the species of interest in the matrix as a result of different "sticking coefficients" for solute S and matrix M.

(iii) In the case of a larger polyfunctional partner molecule in the complex being investigated, the great number of different complex species and their corresponding absorptions may render spectral interpretation very difficult.

For strong interactions, the complexes are usually obtained by premixing the electron donor and electron acceptor molecules with the matrix material in the vapor phase and depositing this mixture onto the cold spectroscopic surface (CsI for IR; Cu, Ag, or Pt in Raman) inside a cryostat. Solid complex partners with low vapor pressure are mostly heated in some kind of miniature furnace installed close to the cold window. On heating the compound in the furnace, vapor molecules are captured by the incoming (matrix + other partner) gas stream. Figure 2 illustrates two different types and geometries of mini-furnaces used. In extreme cases of very strong interaction in the gas phase, two separate gas mixtures M/A and M/B have to be spread into the cryostat through two separate inlets in order to avoid reaction (e.g., $NH_3 + HCl$). Very weak complexes can often only be initiated at sufficient concentration after depositing the gas mixture M/A/B and annealing the matrix to allow diffusion.

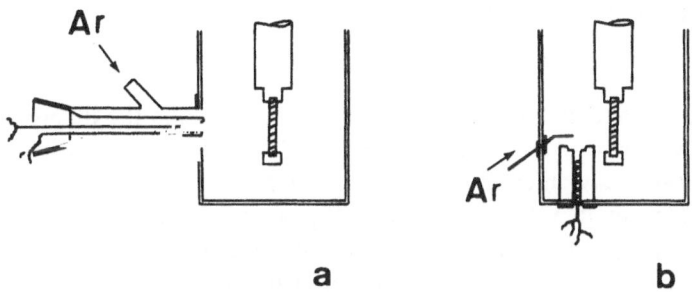

Fig. 2. Incorporation of mini-furnaces in cryostats used for matrix isolation: (a) from Ref. [3], (b) from Ref. [4]; (used with permission)

Some intermolecular complexes are characterized by a very short lifetime at room temperature and are in fact only transition states in reactions between the partners. A typical example is the charge transfer complex between ethylene and chlorine:

At the matrix temperature, the transition state has a lifetime which is sufficiently long to identify the π-complex spectroscopically [5].

It may be clear that the number of intermolecular complexes available for matrix isolation study is considerably larger than at ambient temperature.

3 Hydrogen Bonding in Low-Temperature Matrices

The technique of matrix spectroscopy has found extensive usage in studies of molecular association, especially in the important field of intermolecular H-bonding. These studies are often far more informative than the counterpart investigations in other phases. This kind of experiment has now been performed for a large variety of systems. The aim of the investigations is frequently to obtain one of the following information items:

(i) the *nature and strength* of the H-bonded complex
(ii) the *geometry* of the H-bonded cluster
(iii) the *interaction site* in a polyfunctional molecule
(iv) the behavior of the complex under *irradiation*.

3.1 Nature and Strength of the H-Bond

Matrix isolation of H-bonded complexes has allowed to describe the different steps of the usual acid-base reaction $A-H + B \longrightarrow A^- + {}^+H-B$. These steps are characterized by strongly different properties of the proton donor stretching vibration v_{A-H} (commonly denoted as v_s) which appears as an intense band in the IR absorption spectrum. A widely accepted classification of hydrogen bonds distinguishes three states of interaction or types of H-bonds:

$$\text{type I} \quad A-H\ldots B \quad Cl-H\ldots OH_2$$

$$\text{type II} \quad A\ldots H\ldots B \quad (Cl\ldots H\ldots Cl)^-$$

$$\text{type III} \quad A^-\ldots {}^+H-B \quad Cl^-\ldots {}^+H-N(CH_3)_3$$

Hydrogen bonds of type I correspond to a weak interaction, with interaction energies less than about $50\,kJ\,mol^{-1}$ and relative frequency shifts $\Delta v_s/v_s^0$ up to 20%. Type II is the so-called symmetric H-bond with a more-or-less perfect proton sharing between the two "competing" bases A^- and B. In a very strict sense the term "symmetric H-bond" must be restricted to systems in which A^- and B are the same chemical species (e.g., in HCl_2^-), and the more general term "pseudo-symmetric" should be used for this type of H-bonding characterized by interaction energies of 50 to $80\,kJ\,mol^{-1}$ and extremely large frequency shifts $\Delta v_s/v_s^0$ of the order of 50% to 75%. Type III corresponds to proton transfer, and these complexes are ion pair structures with interaction energies up to $100\,kJ\,mol^{-1}$. Characterization of this type by the parameter $\Delta v_s/v_s^0$ has no sense, since the proton donor vibration v_{A-H} is now replaced by a new internal mode v_{B-H^+}.

Table 2 gathers a large number of IR spectral results obtained for $B\ldots HCl$ complexes in Ar matrices [6].

Introduction of the concept of proton affinity (PA) for the two competing bases A^- and B allows to discuss the nature and strength of H-bonding on a thermodynamical base [7]. In the following reaction scheme the acid-base reaction is subdivided into three subsequent steps, each step being characterized by its own, proper energy balance:

dissociation $\qquad\qquad A-H \longrightarrow A^- + H^+ + PA(A^-)$

association $\qquad\qquad B + H^+ \longrightarrow B^+-H - PA(B)$

ion pair formation $\quad A^- + B-H^+ \longrightarrow A^-\ldots H-{}^+B + \int_{\infty}^{r_e} (-e^2\,dr)/r^2$

or $\qquad\qquad\qquad A-H + B \longrightarrow A^-\ldots H^+-B + \Delta H_{PT}$

with

$$\Delta H_{PT} = PA(A^-) - PA(B) - e^2/r_e$$

Table 2. Matrix frequencies of the HCl stretching mode v_s in hydrogen bonded complexes with different bases [6]

Base	Proton affinity $(kJ \cdot mol^{-1})$	v_s (cm^{-1})	$\Delta v_s/v_s^0$
—	—	2871 (Ar)	—
—	—	2854 (N_2)	—
N_2	495	2863	0.003
CO_2	548	2854	0.006
n-C_3H_8	628	2852	0.007
CHF_3	615	2846	0.009
c-C_5H_{10}	707	2840	0.011
HCl	563	2818	0.018
CO	593	2815	0.020
SO_2	636	2808	0.022
C_6H_6	759	2776	0.033
HI	615	2775	0.033
c-C_3H_6	752	2773	0.034
C_2H_4	680	2753	0.041
CH_2=C=CH_2	779	2737	0.047
n-C_3H_6	751	2726	0.051
CF_3COOH	707	2714	0.055
n-C_3H_4	761	2698	0.060
H_2S	712	2676	0.068
H_2O	697	2664	0.072
CH_3CN	787	2659	0.074
H—CO—H	718	2628	0.085
CH_3—CO—OCH_3	763	2568	0.106
C_2H_5OH	788	2501	0.129
CH_3—CO—OCH_3	828	2450	0.147
CH_3—CO—CH_3	823	2392	0.167
$(CH_3)_2O$	804	2310	0.195
$(CH_3)_2S$	839	2272	0.209
2-F-pyridine	881	2200	0.234
NH_3	854	1371	0.552
CH_3—SO—CH_3	884	1251	0.564
uracil [9]	870	1700	0.408
CH_3—CO—$N(CH_3)_2$	905	740	0.742
pyridine-N-oxide	922	743	0.741
pyridine	924	840	0.707
$(CH_3)N$	942	1486	(v_{N^+-H})
$(CH_3)_3NO$	983	2500	(v_{O^+-H})

representing the total energy balance of the acid-base reaction, in which r_e is the distance between the two charges in the ion pair structure.

As already mentioned before, one of the advantages of studies of H-bonding at very low temperatures is the fact that the entropy contribution to ΔG_{PT} is extremely small at these conditions, and thus it becomes possible to correlate the three types of H-bonds to ΔH_{PT} values:

$$\Delta H_{PT} > 0: \text{ type I is observed}$$

$$\Delta H_{PT} = 0: \text{ type II is observed}$$

$$\Delta H_{PT} < 0: \text{ type III is observed}$$

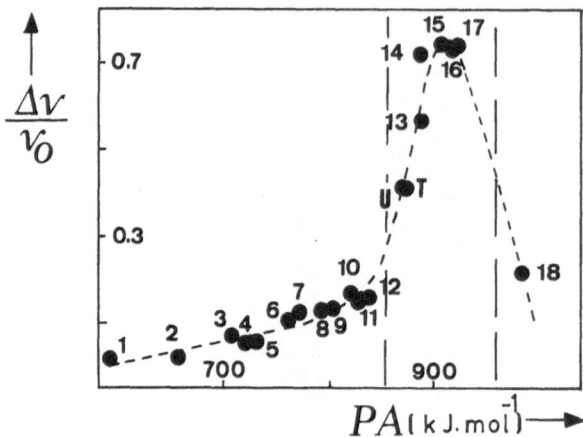

Fig. 3. The vibration correlation diagram for HCl complexed with O-bases in Ar matrices. 1: CO; 2: SO$_2$; 3: H$_2$O; 4: CF$_3$COOH; 5: CCl$_3$CH$_2$OH; 6: CH$_3$COOCH$_3$; 7: CH$_3$OH; 8: C$_2$H$_5$OH; 9: CH$_3$CHOHCH$_3$; 10: (CH$_3$)$_2$CO; 11: CH$_3$COOCH$_3$; 12: (C$_2$H$_5$)$_2$O; 13: (CH$_3$)$_2$SO; 14: dimethylformamide; 15: dimethylacetamide; 16: N-methylpiperidine-2-one; 17: pyridine-N-oxide; 18: trimethylamine-N-oxide; U: uracil; T: thymine

This approach of the prediction of the nature and strength of hydrogen bonding is, however, limited by the large inaccuracy in the knowledge of the value of r_e, and more empirical methods have to be used. Ault et al. [8] have introduced the concept of the *normalized proton affinity difference* Δ which is defined as:

$$\Delta = \frac{PA(B) - PA(A^-)}{PA(B) + PA(A^-)}$$

The idea is that proton transfer in a hydrogen bond should depend systematically upon the difference between the PA values of B an A$^-$. The normalization allows to consider limiting values of $\Delta = \pm 1$; $\Delta = -1$ corresponds to PA(A$^-$) $\gg$ PA(B), or PA(B) $\simeq$ 0, and $\Delta = +1$ corresponds to PA(B) $\gg$ PA(A$^-$) or, hypothetically PA(A$^-$) $\simeq$ 0. In the first limit, v_s approaches v_s^0 of free A—H and in the second limit v_s approaches the frequency of the BH$^+$ ion.

Correlation of the parameter $\Delta v_s / v_s^0$ with Δ yields the vibrational correlation diagram in which the three types of H-bonds are situated in different parts of the curve. For systems containing one and the same proton donor (e.g., HCl), similar diagrams have been established. Figure 3 illustrates such correlation for HCl complexed with oxygen bases [9]. Using this type of vibrational correlation one can easily estimate the nature and strength of the H-bond formed between a particular proton donor and a base not yet investigated in low-temperature matrix environments.

3.2 Geometrical Structure of Small H-Bonded Clusters

Imagine an IR experiment on self-association of a compound HX (e.g., HCl, H$_2$O, NH$_3$) in which, starting from a high M/S ratio and progressively increasing the concentration of the solute S, absorption bands due to solute multimers

$(HX)_m$ of increasing size successively appear in the spectrum. A first problem to solve is the determination of m, the degree of association. One usually compares the growing rates of the different bands to theoretical probability curves $P_m = f(M/S)$, assuming a statistical distribution of monomer, dimer, trimer,... species [10]. This procedure allows to assign the bands to absorbing self-associated species with increasing values $m = 1, 2, 3, ...$. The bands for a particular cluster, e.g., the dimer or the trimer, are then further analysed to obtain information about the geometrical structure of that particular dimer or trimer cluster.

We will briefly describe the results obtained for the dimer $(HCl)_2$ [11, 12]. There are two possible structures for such a dimer: the open-chain structure (sometimes called the "linear" dimer), and the closed, cyclic structure:

open−chain dimer   closed, cyclic dimer

In the open-chain dimer one molecule (a) is the proton donor, and the other (b) is the proton acceptor, whereas in the closed structure both HCl units (a) and (b) behave as proton donor and proton acceptor simultaneously. Figure 4 illustrates the IR spectrum of matrices HCl/Ar at different M/S ratios.

Two very different signals are observed for the dimer $(HCl)_2$, D_1 in the monomer range $(2856\,cm^{-1})$ with low intensity, and D_2 at much lower frequency $(2818\,cm^{-1})$ with an intensity larger than D_1 by a factor of about 8! This result suggests an open-chain structure with a strongly perturbed proton donor molecule (a) (D_2) and a weakly perturbed acceptor unit (b) (D_1). Maillard et al. [11] have demonstrated that the parameters $(\Delta v)_{D2}$ and $(\Delta I)_{D2}$ are both correlated to the rather strong increase of the dipole moment derivative with respect to the H-perturbed molecular (a) stretching coordinate $d\mu/dr_a$. Pressure experiments have revealed a strong variation of the enhancement factor T_μ with the intermolecular distance R (Fig. 5):

$$T_\mu = (d\mu/dr_a)/(d\mu/dr)_{free}$$

The variation of T_μ is explained by the increased charge transfer at smaller intermolecular distance R induced by the higher pressure.

The open-chain dimer geometry has been confirmed from isotopic dilution experiments in matrices. In the spectra of isotopic mixtures HCl/DCl, no new bands are observed between the D_1 and D_2 signals, while D_1 is shifted downwards with $2\,cm^{-1}$ and D_2 with $0.3\,cm^{-1}$. This supports the previous conclusions that, in the homodimer, the two HCl molecules are not equivalent and, as a consequence, only weakly coupled together. This coupling leads to a small frequency shift (about $1\,cm^{-1}$) compared to the values expected from the

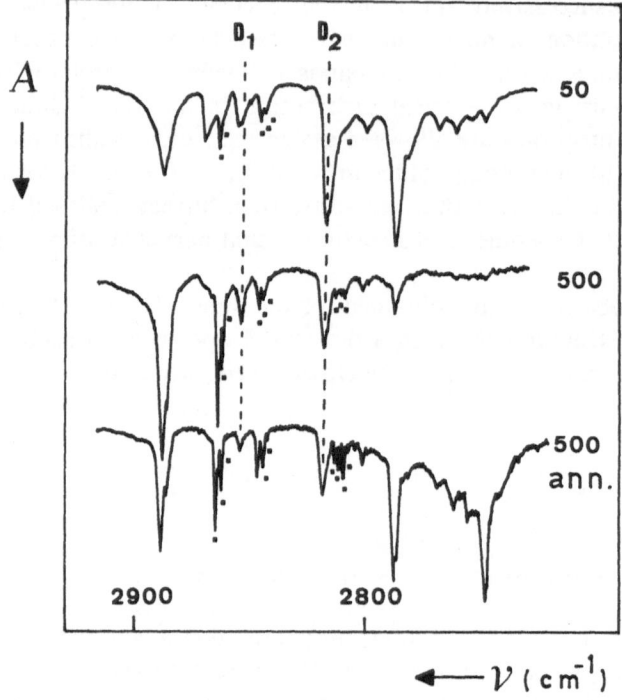

Fig. 4. IR spectra of HCl/Ar matrices at different M/S ratios (D = dimer bands; ∗ = N$_2$ induced bands) [11] (used with permission)

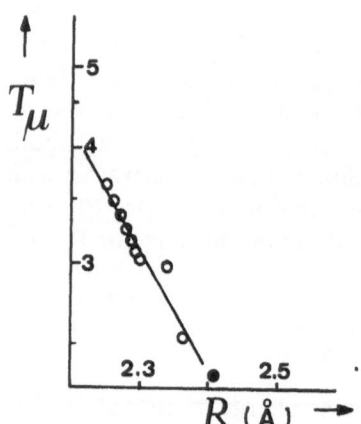

Fig. 5. Variation of the intensity enhancement factor T$_\mu$ with the intermolecular distance R in (HCl)$_2$ [11] (used with permission)

classical expressions:

$$v_a = (F_a/\mu)^{1/2} \quad \text{and} \quad v_b = (F_b/\mu)^{1/2}$$

coupling: $(v'_a)^2 = F_a/\mu + f^2/\mu(F_a - F_b)$

$$(v'_b)^2 = F_b/\mu - f^2/\mu(F_a - F_b)$$

μ is the reduced mass, F_i the stretching force constant of the HCl unit (i), and f the coupling constant between the two oscillators (a) and (b).

In the isotopic mixtures there exist two kinds of mixed dimers, D—Cl... H—Cl and H—Cl...D—Cl with corresponding H—Cl frequencies lying at v_a and v_b, respectively. Thus the decoupling effect is expected to decrease the D_1 frequency and to increase D_2 by the same amount, at least if the F_a and F_b force constants remain unaffected by isotopic substitution. The experimental observation of an unsymmetric decoupling effect ($D_1 - 2\,cm^{-1}$, $D_2 - 0.3\,cm^{-1}$) is explained by the existence of an isotopic effect on intermolecular interaction! From the mixed HCl/DCl dimer experiments the following rule was derived: the force constant F of a hydracid molecule decreases by 0.07% every time that this molecule faces a DX instead of a HX molecule [11]. The mixed dimer frequencies v_{HCl} are then calculated as:

$$(v''_a)^2 = (F_a - \Delta F)/\mu \quad \text{and} \quad (v''_b)^2 = (F_b - \Delta F)/\mu$$

where ΔF represents the force constant variation for H/D substitution. The isotopic H/D frequency shift due to the ΔF effect is of the order of $1\,cm^{-1}$ and nearly equal to the coupling effect. This explains the shifts observed for D_1 ($-1\,cm^{-1}$ for decoupling and $-1\,cm^{-1}$ for ΔF) and for D_2 ($+1\,cm^{-1}$ for decoupling and $-1\,cm^{-1}$ for ΔF).

Theoretical predictions (ab initio) have confirmed the higher stability of the open dimer structure as compared to the cyclic one [13, 14].

	Kollman et al.	*Votova* et al.
	[13]	[14]
	$R = 3.8\,\text{Å}$	$R = 3.81\,\text{Å}$
	$\alpha_1 = 112°$	$\alpha_1 = 97°$
	$\alpha_2 = 11°$	$\alpha_2 < 2°$
	$\Delta E_{lin} = 6.3\,kJ\,mol^{-1}$	$\Delta E = 7.9\,kJ\,mol^{-1}$
	$\Delta E_{cycl} = 3.6\,kJ\,mol^{-1}$	

Table 3. Structure of $(AH)_m$ multimers identified from matrix infrared spectra

Compound/matrix	m	Structure	Ref.
HCl or HBr/Ar or Kr	2	Open chain (C_s)	[11, 12]
	3	Cyclic (C_{3h})	
	4	Cyclic (C_{4h})	
HCN/Ar or N_2	2	Open chain	[15]
	3	Open chain	
H_2O/Ar or N_2	2	Open chain	[16]
	3	Cyclic	
	4	Cyclic (probable)	
NH_3/Ar or N_2	2	Open chain	[17]
	3	Open chain (probable)	
CH_3OH/Ar or N_2	2	Open chain	[18]

Table 3 summarizes the results obtained for several multimers $(AH)_m$ studied by matrix IR spectroscopy.

3.3 The Interaction Site in Polyfunctional Acids or Bases

Many organic polyatomic molecules have several groups or sites available for H-bonding with a proton acceptor or proton donor molecule. A number of these organic compounds play a key role in some of the most important biological processes in living organisms, and the comprehension of their intermolecular interaction channels is frequently crucial to understand the complicated nature of biological processes. The most familiar examples of such compounds are the purine (guanine and adenine) and pyrimidine (cytosine and uracil or thymine) polyfunctional bases, the properties of which are at the origin of the well known structure of DNA and RNA. Matrix spectroscopy is a unique method to study intermolecular interactions of these bases, also because they are highly self-associated and almost insoluble in common spectroscopic solvents.

Consider a rather simple polyfunctional base such as *methyl acetate*. This ester molecule has two possible proton acceptor sites—the carbonyl oxygen and the alkoxy oxygen lone pairs:

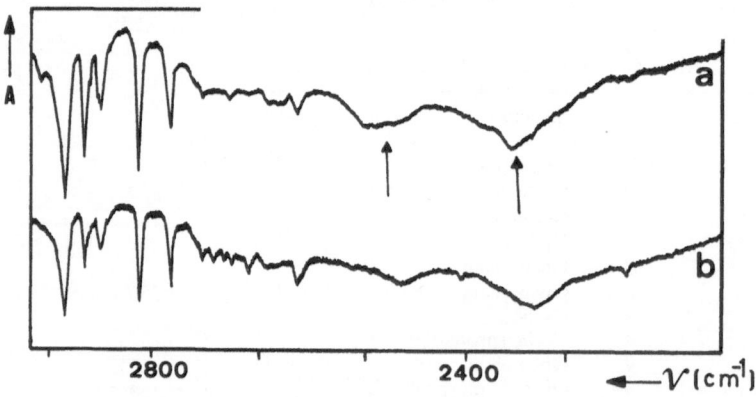

methyl acetate (Z-conformer)

Hydrogen bonding with phenols occurs at the carbonyl oxygen site in solution at ambient temperature. With unsubstituted phenol, $K = 7.9 \, dm^3 \, mol^{-1}$, $\Delta H = -17.1 \, kJ \, mol^{-1}$, $\Delta v_{OH} = -170 \, cm^{-1}$ and $\Delta v_{C=O} = -24 \, cm^{-1}$ at 293 K [19]. The reason why only this complex is identified from the IR spectra in

Fig. 6. IR spectrum of methyl acetate/HCl/Ar (a: 1/2/200; B: annealed)

solution is found from thermodynamics: enthalpy and entropy contributions to ΔG favor the formation of this H-bond over the alternative C—O...H—O one. Latajka has computed an energy difference of 9.2 to 11.3 kJ mol^{-1} for isolated C=O...H—X over C—O...H—X (X = Cl or F) complexes, which illustrates the difference in strength and stability of both hydrogen-bonded complexes [20].

Matrix IR experiments performed on methyl acetate mixed with HCl or H_2O in argon have allowed to identify both types of H-bonded complexes [21]:

In the spectral region of the perturbed HCl vibration, two bands are observed at 2570 and 2460 cm^{-1} (Fig. 6), and both v_s frequencies are correlated with the PA values of separate basic sites in the ester molecule (Fig. 3). The separate PA values have been obtained from literature correlations between PA and O(1s) core electron binding energies [22]. The coexistence of "keto" and "alkoxy" complexes is supported by the appearance of two sets of new $v_{C=O}$, v_{C-O} and v_{C-C} bands, shifted in different directions from the monomer frequencies, e.g., for MeAc/H_2O/Ar:

$$v(\text{C—C}) = 848\,\text{cm}^{-1} \quad 858\,\text{cm}^{-1} \quad 840\,\text{cm}^{-1}$$
$$v(\text{C=O}) = 1760\,\text{cm}^{-1} \quad 1740\,\text{cm}^{-1} \quad 1770\,\text{cm}^{-1}$$
$$v(\text{C—O}) = 1245\,\text{cm}^{-1} \quad 1266\,\text{cm}^{-1} \quad 1228\,\text{cm}^{-1}$$

Complexation at the carbonyl oxygen atom leads to the usual frequency decrease of the carbonyl stretching mode, while formation of a H-bond at the alkoxy site induces a frequency increase of this vibration.

Thus it appears that in solution at room temperature, where equilibrium conditions are reached, the most stable complex is observed. In contrast, a low-temperature matrix is not an equilibrium system and, to a certain extent, the conditions for a thermodynamic control of the intermolecular interaction are not present. If, during rapid freezing of the gas mixture, an ester and an HX species are frozen in the same matrix cage in such a configuration that HX is close to the (weaker) C—O site of the ester, this weak interaction persists unless the rigidity of the cage is reduced by annealing, which enhances the internal energy of the isolated species. That is why annealing a matrix generally results in an increase of the stronger complex at the expense of the weaker one. Figure 6(b) illustrates this for C=O...HCl and C—O...HCl H-bonded species.

We have recently studied H-bonding of *uracils* in low-temperature Ar matrices using FT-IR spectroscopy. For a series of 9 uracils, different H-bonded complexes are identified from the IR spectra of the H_2O-doped uracil/Ar matrices [23], e.g., for 1-CH_3-uracil/H_2O/Ar:

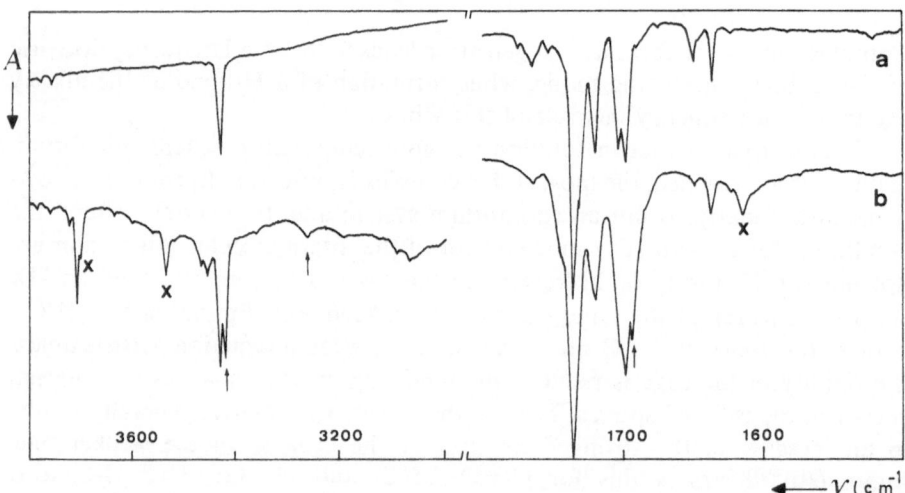

The formation of the $C_4{=}O\ldots H{-}O$ complex is spectroscopically manifested by the appearance of a shifted $\nu(C_4{=}O)$ band ($\Delta\nu = -10\,cm^{-1}$), and by the splitting of the free $\nu(N_3{-}H)$ mode at $3421\,cm^{-1}$ ($\Delta\nu = -8\,cm^{-1}$). On the other hand, the complex $N_3{-}H\ldots O{-}H$ is identified from the new, shifted $\nu(N_3{-}H)$ absorption at $3265\,cm^{-1}$ ($\Delta\nu = -164\,cm^{-1}$) and from a small ($-2\,cm^{-1}$) secondary shift of $\nu(C_4{=}O)$. Figure 7 illustrates these spectral observations.

The interpretation in terms of two different complexes is supported by the appearance of two sets of new H_2O complex absorptions:

$$\nu_3:\ 3705\,cm^{-1}\quad 3711\,cm^{-1}\quad 3726\,cm^{-1}\quad 3720\,cm^{-1}$$
$$\nu_1:\ 3532\,cm^{-1}\quad 3575\,cm^{-1}\quad 3634\,cm^{-1}\quad 3630\,cm^{-1}$$
$$\nu_2:\ 1613\,cm^{-1}\quad 1611\,cm^{-1}\quad 1593\,cm^{-1}\quad 1598\,cm^{-1}$$

(Ref. [16])

Fig. 7. FT-IR spectra of 1-CH_3-uracil/H_2O/Ar matrices at 10 K: ν_{N-H} region and $\nu_{C=O}$ region; a: base/Ar /b: base/H_2O/Ar (H_2O/Ar = 1/200) (↑ indicates shifted N_3H or $C_4{=}O$ mode, X indicates new H_2O absorption)

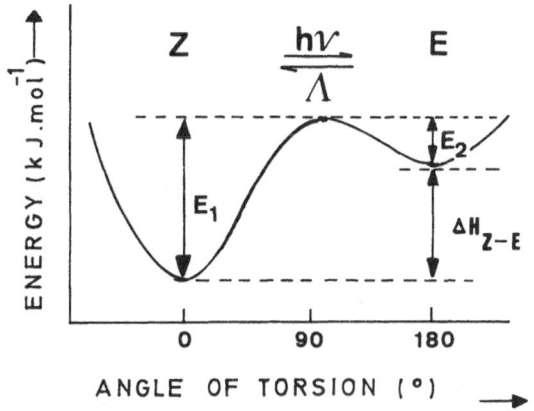

Fig. 8. Energy diagram for methylacetate $Z \longrightarrow E$ conformer interconversion induced by IR irradiation [30] ($E_1 = 56\,kJ\cdot mol^{-1}$; $E_2 \simeq 10\,kJ\cdot mol^{-1}$)

Both sets of frequencies are close to the frequencies found for the proton donor H_2O (a) and proton acceptor H_2O (b) molecule in the H_2O dimer in Ar [16]. The small differences are easily explained by the greater basicity of the uracil $C_4{=}O$ group as compared to $(H_2O)_b$, and the greater acidity of the N_3—H group as compared to $(H_2O)_a$ in the water dimer.

Matrix studies on H-bonding properties of other DNA bases (cytosines) are in current progress in our laboratory.

3.4 IR-Induced Photodissociation of H-Bonded Complexes

A reaction with an activation energy in the range 1 to $25\,kJ\,mol^{-1}$ is generally too fast to be examined experimentally at room temperature. In a cryogenic matrix, however, the extremely low thermal energy available slows down this chemical event, and often the reaction can be suppressed completely under these conditions. In such cases the reaction might be induced by absorption of an IR photon, since mid-infrared photons ($1500-3000\,cm^{-1}$) correspond to energies of a few tens (18–36) of $kJ\,mol^{-1}$. It is well known that this method offers the possibility of exciting selectively and converting quantitatively a stable conformer of a flexible molecule into another, unstable conformer. The former one is usually thermally regenerated in the matrix by annealing (Fig. 8).

After the pioneering work of Pimentel and co-workers on the isomerization of HNO_2 [24], many examples of IR-induced photorotamerizations have been described in the literature, e.g., 1,2-difluoroethane [25], n-butane [26], 1-propanol [27], methyl vinylether [28], methyl formate [29], and methyl acetate [30].

In contrast to the large number of data on unimolecular IR photo-processes in matrices, IR-induced bimolecular processes seem to be quite elusive. The addition-reactions leading to the formation of $Fe(CO)_4$ [31] or CH_2F—CH_2F [32] are well known examples.

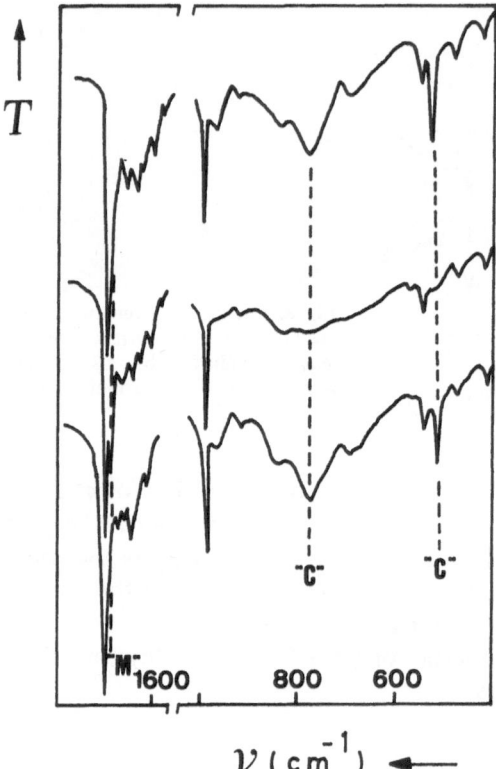

Fig. 9. Effect of IR irradiation on the spectrum of N,N-diCH$_3$-acetamide...HI complex in Ar matrices [34] (C = strong complex, M = metastable complex) (used with permission)

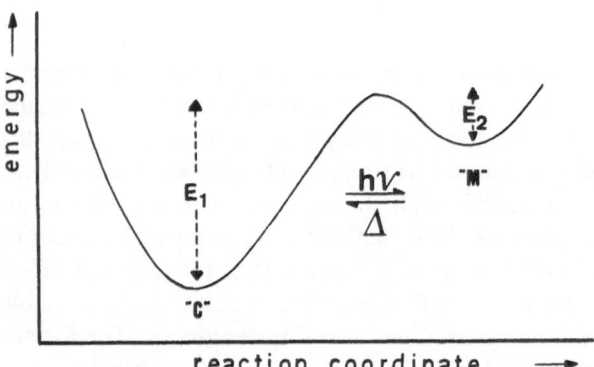

Fig. 10. Energy diagram for the IR-induced photodissociation of the H-bonded complex of N,N-diCH$_3$-acetamide with HI in Ar at 10 K [34] (used with permission)

A third type of IR-induced process occurring in low-temperature matrices is the dissociation and rearrangement of H-bonded molecular complexes. Within the series of hydrogen halide proton donor complexes, this phenomenon seems to be restricted to HI complexes only, and both relatively weak (type I, CH$_3$CN...HI [33]) and strong complexes (type II, N,N-diCH$_3$-acetamide...HI [34]) are sensitive to IR irradiation.

Mielke and Barnes [34] have assigned the broad absorption at $808 \, cm^{-1}$ and the comparatively narrow band at $565 \, cm^{-1}$ observed in the IR spectra of N,N-diCH$_3$-acetamide...HI to the O...H...I antisymmetric stretching and to the bending mode of this strongly H-bonded complex "C" (the shift $\Delta v_s / v_s^0$ is approximately 65%). Figure 9 illustrates that both absorptions disappear on prolonged IR irradiation with a half life of ca. 2 hours in the beam of a conventional dispersive spectro-photometer.

At the same time, a sharp new band grows at $1664 \, cm^{-1}$, close to the $v_{C=O}$ mode of monomer N,N-diCH$_3$-acetamide ($1676 \, cm^{-1}$). Annealing the matrix to a temperature above 30 K regenerates the strong complex absorptions. Evidently, the strong H-bonded complex is dissociated under the influence of the IR beam and another, weaker complex "M" is formed in the same matrix cage. By using different IR interference filters, the authors have demonstrated that irradiation in the $v_{C=O}$ range does not induce the dissociation, and that absorption in the v_{C-H} range is probably responsible for the dissociation process. The nearby lying overtone of the carbonyl stretching mode may provide the route for the energy to be transferred into an intermolecular vibration mode. Thus, the barrier E_1 to dissociation of the strong complex must lie in the range $18-36 \, kJ \, mol^{-1}$ ($1500-3000 \, cm^{-1}$) (Fig. 10). An upper estimate for the barrier E_2 to regeneration of the complex could be derived from the regeneration temperature of 30 K by analogy with the conformational isomerism behavior in matrices [35].

A possible structure for the metastable, weak complex "M" induced by the IR irradiation is a "reverse" or anti-hydrogen-bonded arrangement B...I—H in which the polar base molecule interacts directly with the iodine atom. The unfavorable dipole–dipole interaction would be offset by other intermolecular forces, particularly the rather strong inductive $\bar{\mu}-\alpha$ interaction. Table 4 lists the dipole moments and molecular polarisabilities of the hydrogen halide series. This table shows that the balance between $\bar{\mu}-\bar{\mu}$ and $\bar{\mu}-\alpha$ interaction terms is much more favorable for HI than for the other hydrogen halides. This could account for the observation that only B...HI systems exhibit such photo-dissociation in matrices.

4 Matrix Spectroscopy of Charge Transfer Complexes

Matrix isolation spectroscopy has also found usage in studies of charge transfer (*Electron Donor Acceptor*) molecular interactions. The characteristic *C*harge

Table 4. Dipole moments and polarisabilities within the hydrogen halide series [34]

	HF	HCl	HBr	HI
μ(D)	1.82	1.09	0.83	0.45
α(A$^{\circ 3}$)	2.46	2.60	3.61	5.45

Transfer absorption band of EDA complexes being located in the UV range of the spectrum (see Introduction), experimental observations are mostly made in both the UV and IR spectral ranges.

In many cases these complexes are intermediates on a reaction route for a normally reactive pair of molecules, and it is only at the temperature of the matrix experiments that such charge transfer complexes can be investigated as stable chemical entities.

Raman spectroscopy plays also a great role in this area, since many of the interesting vibrational modes of these molecules, e.g., the halogen stretching modes, are inactive or very weak in the corresponding IR spectrum. The number and types of problems associated with Raman spectroscopy of low-temperature matrices being considerable [36], the amount of matrix data for charge transfer complexes is not very extensive. The main problem arises from the inherently lower sensitivity of Raman as compared to IR spectroscopy. Higher concentrations of electron donor and electron acceptor are needed and part of the advantages of matrix isolation is lost. It is not surprising therefore that the amount of EDA data in matrices is far less extended than the data on H-bonded complexes.

4.1 n-σ* and π-σ* Complexes of the Halogens

In addition to many stable complexes formed between halogen molecules X_2 (X = Cl, Br, I) and n- (e.g., pyridine) or π- (e.g., benzene) electron donors, which have been studied extensively in liquid solutions [37], a number of other X_2 complexes of the n-σ* or π-σ* type cannot be investigated at ambient temperatures because chemical reaction occurs. This is the case for NH_3 with Cl_2:

$$NH_3 + Cl_2 \longrightarrow NH_2Cl + HCl$$

Ribbegård has isolated the complex $H_3N\ldots Cl_2$ in a N_2 matrix at 20 K [38]. In the UV spectrum (Fig. 11) a strong absorption appears, beginning at 2500 Å and extending towards lower wavelengths. Photolysis on the matrix is effective, since this CT band decreases.

In the IR spectrum a doublet band is observed at 460–454 cm^{-1} (Fig. 11). This absorption corresponds to the stretching mode v_{Cl-Cl} which is activated by the charge transfer from the nitrogen lone pair to the halogen antibonding σ_u^* orbital:

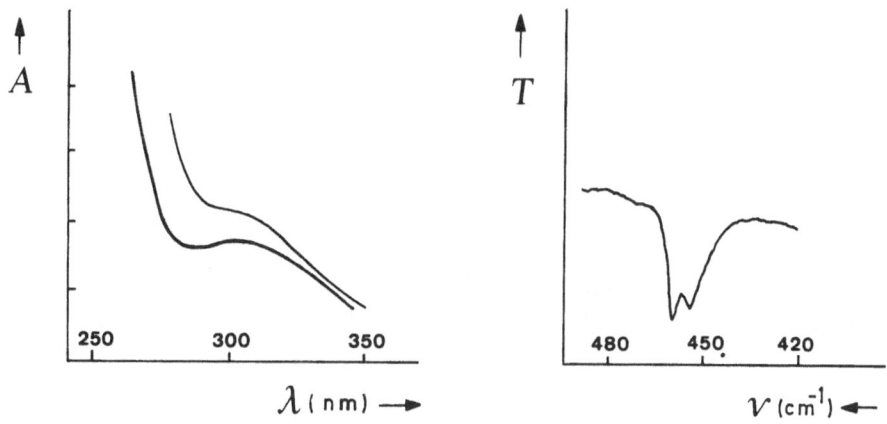

Fig. 11. UV and IR spectrum of $NH_3/Cl_2/N_2$ (M/R = 100) [38]

The band is doubly splitted by the $^{35}Cl—^{35}Cl$ and $^{35}Cl—^{37}Cl$ natural abundance in Cl_2. In the complex with NH_3, the $v_{Cl—Cl}$ frequency is considerably lower than in free Cl_2 (557 cm^{-1} from Raman vapour data [39]). The shift of about 100 cm^{-1} indicates a rather large amount of charge transfer in the complex. The vibration modes of the electron donor molecule NH_3 are only weakly perturbed: Δv_1 (sym. stretch) = +4 cm^{-1}, Δv_2 (sym. bend) = +17 cm^{-1}. The small upwards shift of v_1 excludes a complex structure involving a H-bond between one of the N—H bonds of ammonia and the halogen molecule acting as an electron donor. Table 5 collects data on the $v_{X—X}$ frequencies observed for EDA complexes of Cl_2 and Br_2 with different n- and π-electron donors in matrices.

4.2 n-π* Complexes with SO₂ and SO₃

Several charge transfer complexes of the π-electron acceptors SO_2 and SO_3 have been identified in low-temperature matrix environment. The reaction of SO_3 with H_2O giving rise to the product H_2SO_4 is now recognized as being one of the most important chemical processes in the formation of acid rain. Ammonia reacts with SO_3 to sulfamic acid (NH_2SO_3H) and with SO_2 to thionylimide ($NHSO + H_2O$). The charge transfer intermediates $H_2O...SO_3$, $H_3N...SO_3$, and $H_3N...SO_2$ have all been isolated in matrices.

Sass and Ault have spectroscopically examined the complexes of ammonia with sulfur dioxide and trioxide [45]. The infrared fundamentals of these π-acceptors are weakly shifted by NH_3 in a N_2 matrix (Table 6). In contrast, the deformation modes of NH_3 are strongly sensitive to complex formation, giving rise to shifts up to 350 cm^{-1} (SO_3 complex). These shifts are explained by the strong charge transfer from the N-atom free electron pair to the anti-

Table 5. Halogen stretching frequencies observed for Cl_2 and Br_2 charge transfer complexes in low-temperature matrices (cm^{-1})

Complex	$^{35}Cl_2$	Br_2	Ref.
— (free X_2)	549	318	[39]
$H_2C=O...X_2$	542		[40]
$H_2O...X_2$	535		[41]
$C_6H_6...X_2$	532$^{(°)}$	306$^{(°)}$	[42]
$C_2H_4...X_2$	526	307	[43]
$(CH_3)_2S...X_2$	525		[39]
$H_2S...X_2$	518	285	[39]
$C_2(CH_3)_4...X_2$	464	263	[43]
$H_3N...X_2$	460$^{(°)}$		[38]

Unless otherwise specified ($(°) = N_2$ matrix), results are relative to Ar matrices.

Table 6. Band positions for $H_3N...SO_3$ and $H_3N...SO_2$ complexes in N_2 [45]

Assignment	SO_3complexes free		NH_3	$^{15}NH_3$	SO_2 complexes free	NH_3	$^{15}NH_3$
ν_{as} acceptor	1398		1354	1354	1350	1338	1338
ν_s acceptor	1068		1072	1071	1152	1149	(a)
δ_{as} acceptor	535		594	594	—		
δ_s acceptor	495		552	552	521	530	530
δ_{as} donor	1632	[47]	1565	1561	1632	(a)	(a)
δ_s donor	970	[47]	1317	1312	970	1038	807

(a) Masked or too weak to be observed.

bonding π^*_{S-O} orbital with limited decrease of the force constant k_{S-O} ($9.00 \longrightarrow 9.78$ mdyn Å).

Free SO_3 has D_{3h} symmetry and the symmetrical stretching mode ν_{S-O} is IR inactive. Complexation with ammonia or amines brings about a symmetry decrease in the acceptor molecule and this causes the appearance of this mode in the IR spectrum.

Isotopic substitution H/D or $^{14}N/^{15}N$ for $H_3N...SO_2$ has no effect at all on the frequencies of the SO_2 subunit [45]. This indicates the absence of any coupling between the two subunits in the complex. At first sight this result is in conflict with the large value of the interaction energy (46 kJ mol^{-1}) obtained from ab initio calculations [46]. The explanation lies in the remarkable geometrical structure of the systems $R_3N...SO_2$ (R = H, CH_3). The two subunits are bonded together in an almost perpendicular orientation:

$$-\Delta E\,(\text{kJ mol}^{-1}) = 49$$

$$R\,(\text{Å}) = 2.63$$

$$\beta\,(\text{deg.}) = 85$$

5 Conclusion

We have tried to demonstrate in this short review, which is certainly not comprehensive, that the technique of matrix isolation spectroscopy provides an important—and in many cases even an unique—contribution to the field of experimental research on intermolecular complexes. It is true that the applications for charge transfer complexes are rather limited. On the other hand, in the domain of H-bonding the great value of this method has not yet been exploited fully, and new developments are to be expected in future years.

6 References

General references on the matrix isolation technique

(a) Hallam HE (ed) 1973) Vibrational spectroscopy of trapped species, Wiley, London
(b) Cradock S, Hinchliffe AJ (1975) Matrix isolation, Combridge Univ Press, London
(c) Barnes AJ, Orville-Thomas WJ, Müller A, Gaufrès R (eds) (1981) Matrix isolation spectroscopy, NATO ASI ser. C, vol 76, Reidel Dordrecht
(d) Maes G (1983) Chemie Magazine 9(8):11
(e) Clark RJH, Hester RE (eds) (1989) Spectroscopy of matrix isolation species, Advances in Spectroscopy, vol 17, Wiley, London
(f) Andrews L, Moskovits M (eds) (1989) Chemistry and physics of matrix-isolated species, North-Holland, Amsterdam

References cited

1. Norman I, Porter G (1954) Nature 174:508; Whittle E, Downs DA, Pimentel GC (1954) J Chem Phys 22: 1943
2. Meyer B (1971) In: Low-Temperature Spectroscopy, Elsevier, New York, p 26
3. Barnes AJ, Stuckey MA, Orville-Thomas WJ, Le Gall L, Lauransan J (1979) J Mol Struct 56: 1
4. Graindourze M, Smets J, Zeegers-Huyskens Th, Maes G (1990) J Mol Struct 222: 345
5. Barnes AJ, Cowieson D, Suzuki S (1976) Proc 5th Int Conf Raman Spectrosc, HF Schulz Verlag, Freiburg
6. Barnes AJ (1983) J Mol Struct 100: 259; Barnes AJ Compilation of hydrogen halide complex ν_{HX} stretching wavenumbers, updated 26.8.87 (unpublished results)
7. Perchard JP in General Refs (c) p 551–553
8. Ault BS, Steinback E, Pimentel GC (1975) J Chem Phys 79: 615
9. Graindourze M (1988) PhD Thesis, University of Leuven, Belgium
10. Kreitman MM, Barnett DL (1965) J Chem Phys 43: 364
11. Maillard D, Schriver A, Perchard JP, Girardet C (1979) J Chem Phys 71: 505
12. Maillard D, Schriver A, Foudère F, Obriot J, Girardet C (1984) J Chem Phys 75: 1091
13. Kollman P, Johansson A, Rothenberg S (1974) Chem Phys Lett 24: 199
14. Votova C, Ahlrichs R and Geiger A (1983) J Chem Phys 78: 6841
15. Walsh B, Barnes AJ, Suzuki S, Orville-Thomas WJ (1978) J Mol Spectry 72: 44
16. Ayers GP, Pullin ADE (1976) Spectrochim. Acta A 32: 1629; Engdahl A, Nelander B (1987) J Chem Phys 86: 1819 and 4831
17. Barnes AJ, Orville-Thomas WJ (1978) Proc 6th Int Conf Raman Spectroscopy, Heyden, London, p 257
18. Barnes AJ, Hallam HE (1970) Trans Faraday Soc 66: 1920; Schriver L, Burneau A, Perchard JP (1982) J Chem Phys 77: 2926
19. Vanderheyden L, Zeegers-Huyskens Th (1983) J Mol Liq 25: 1
20. Latajka Z, Ratajczak H, Zeegers-Huyskens Th (1988) J Mol Struct Theochem 164: 201

21. Maes G, Zeegers-Huyskens Th (1983) J Mol Struct 100: 305; Vanderheyden L, Maes G, Zeegers-Huyskens Th (1984) J Mol Struct 114: 165
22. Benoit FM, Harrison AG (1977) J Am Chem Soc 99: 3980; (1978) Org Mass Spectrom 13: 128
23. Graindourze M, Maes G, Smets J, Grootaers T, Zeegers-Huyskens Th (1991) J Mol Struct 243: 37
24. Baldeschwieler JD, Pimentel GC (1960) J Chem Phys 33: 1008; Hall RT, Pimentel GC (1963) J Chem Phys 38: 1889
25. Felder P, Gunthard HsH (1984) Chem Phys 85: 1
26. Räsänen M, Bondybey VE (1984) Chem Phys Lett 111: 515
27. Lotta T, Murto J, Räsänen M, Aspiala A (1984) Chem Phys 86: 105
28. Gunthard HH (1984) J Mol Struct 113: 141; Beech T, Gunde R, Felder P, Gunthard HH (1985) Spectrochim Acta A 41: 319
29. Müller RP, Hollenstein H, Huber JR (1983) J Mol Spectry 100: 95
30. Lenaerts S, Daeyart F, Vanderveken BJ, Maes G (1989) Spectrosc Lett 22: 289
31. Poliakoff M (1987) Spectrochim Acta A 43: 217
32. Hauge RH, Gransden S, Wang J, Margrave JL (1978) Ber Bunsenges Phys Chem 82: 104; (1979) J Am Chem Soc 101: 6950
33. Schriver L, Schriver A, Perchard JP (1986) J Chem Phys 84: 5553
34. Mielke Z, Barnes AJ (1986) J Chem Soc Faraday Trans 2 82: 447; Barnes AJ (1988) Faraday Discuss Chem Soc 86: 86/4
35. Barnes AJ in General Refs (c) p 531–549
36. Ozin GA in General Refs (c) p 373–415
37. Mulliken RS, Person WB (1969) Molecular Complexes, Wiley, New York; Yarwood J (ed) (1973) Spectroscopy and structure of molecular complexes, Plenum Press, London (1973)
38. Ribbegård G (1974) Chem Phys Lett 25: 333
39. Stammreich R, Sala O, Forneris R (1953) Anaes Acad Brasil Cienc 25: 375
40. Agarwal UP, Barnes AJ, Orville-Thomas WJ (1985) Can J Chem 63: 1705
41. Nelander B (1980) J Mol Struct 69: 59
42. Fredin L, Nelander B (1973) J Mol Struct 16: 217
43. Fredin L, Nelander B (1974) Mol Phys 27: 885
44. Frèdin L, Nelander B (1973) J Mol Struct 16: 205
45. Sass CS, Ault BS (1984) J Phys Chem 88: 432; (1986) ibidem 90: 1547
46. Lucchese RR, Haber K, Schaefer HF (1976) J Am Chem Soc 98: 7617; Douglas JE, Kollman PE (1978) J Am Chem Soc 100: 5226
47. Pimentel GC, Bulanin MO, Van Thiel M (1962) J Chem Phys 36: 500

CHAPTER IX
Water—The Most Anomalous Liquid

W.A.P. Luck

A large mushroom of about 100 gram looks solid, but its weight after drying is just a few grams. Water, which was the principal constituent of this beautiful product of nature has been removed. A head of lettuce contains about 97% of water—it is also the principal constituent of flowers. Osmotic pressure and capillary effects cast a spell on this liquid to create all these artistic productions of nature. With the exception of fat the main content of our food is water. It constitutes 76% of lean meat, 81% of fish, 74% of potatoes, about 88% of vegetables, about 80% of fruits etc. [1]. The homo sapiens also consists of 97% water as an embryo and as an adult—depending on age—of 65–70%. What magic role is nature able to play with this special liquid that is water? Water and its solutions is the dominating liquid on our planet. Why does water play this dominant role as the most important component of all biochemical products? Which molecular properties lead to this anomalous liquid which is often neglected in chemistry teaching? We shall try to give an answer in this chapter.

Intermolecular Forces
© Springer-Verlag Berlin Heidelberg 1991

1 Presuppositions

1.1 The Molecule

IR spectral results demonstrate in the vapour state a HOH angle of 105.3° and an OH distance of 0.96 Å. In ice this angle is enlarged nearly to the perfect tetrahedral angle of 109.5° and the OH distance to 0.99 Å. This angle indicates a large heteropolar contribution of the OH bond [2] in agreement with the dipole moment of 1.87 Debye. Quantum chemical calculations show that the axes of the two lone pair electrons (θ) also form a tetrahedral angle. Therefore a tetrahedron can be taken as a model of the water molecule (Fig. 1). We can assume 0.2357 elementary charges with a positive sign at two vertices and equal negative charges at the two others. The distance between the model centre and the positive charges would be 1 Å and for the negative charges 0.8 Å.

This model is useful for computer simulations of liquid water and, if we add a neon-like dispersion potential to the Coulomb field of these four elementary charges, is called the ST2 potential.

Such partial charges could be assumed near the molecular surfaces of all acid protons and of lone pair electrons. The interaction between these two anomalous molecular groups is called a hydrogen bridge or hydrogen bond (H-bond). These H-bonds have two dominant properties:

a. The H-bond interaction energy ΔH_H is larger than most van der Waals energies.
b. H-bond interactions are strongly orientation dependent.

The dispersion interactions are spherically symmetrical, but H-bonds could be attractive if a proton and a lone pair encounter one another or repulsive if two protons or two lone pairs come together. As with the chemical bond we can discuss a bonding and an antibonding interaction between molecules with H-bond abilities. The H-bond interactions could therefore be called "Nebenvalenz" and are specific.

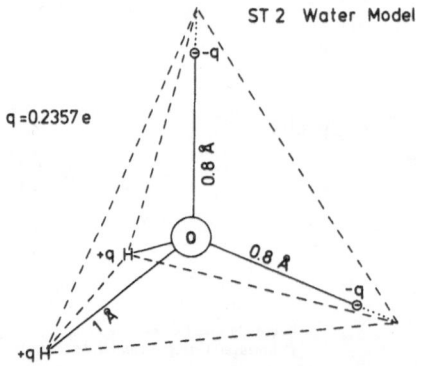

Fig. 1. A point charge (q) model of the water molecule assumed by the so-called ST 2 model

1.2 The H-bond Energy

The size of the dispersion interaction energy is proportional to the number of electrons Z_{el} of a molecule and therefore also to the molecular weight M. The pair potential for dispersion interactions is proportional to the critical temperature T_C. Empirical relations give the ratio $T_B/T_C \approx 2/3$ for the boiling point T_B and the ratio $T_M/T_B \approx 2/3$ for the melting point T_M. Therefore both T_B and T_M are approximate measures for the dispersion forces. If we plot T_B and T_M of molecules of the type XH_2 as a function of Z_{el} (Fig. 2) for water, we find large deviations from the expected proportionality. From the point of view of dispersion interactions we should expect for water with its few electrons: $T_M \approx -90°C$ and $T_B \approx -75°C$. This means that life on earth would be impossible without the anomalous properties of water. A compound of the type XH_2 would need about 100 electrons for dispersion forces to give it a melting point of 0°C and about 200 electrons for a boiling point of 100°C. The anomalous properties of water caused by its H-bonds are one of the important conditions which make evolution possible on our planet. The study of the H-bonds of water therefore seems to be the key for understanding the anomalous behaviour of liquid water. If we compare (Fig. 3) the energy content ΔU or the enthalpy ΔH of water with those of liquid methane (lower curves) in equilibrium with the vapour phase (upper curves) we recognize the large difference between these two molecules of equal electron numbers, and therefore with similar dispersion forces. In Fig. 3, the solid phase at T_M is taken as the zero point of our enthalpy scale. At T_C the enthalpy of water is 36 kJ/mol (or 8.6 kcal/mol) larger than that of methane. The cause of this difference is mainly the H-bond interaction energy which can be estimated by IR spectroscopy as 15.5 kJ/[mol OH] or 31 kJ/[mol H_2O]

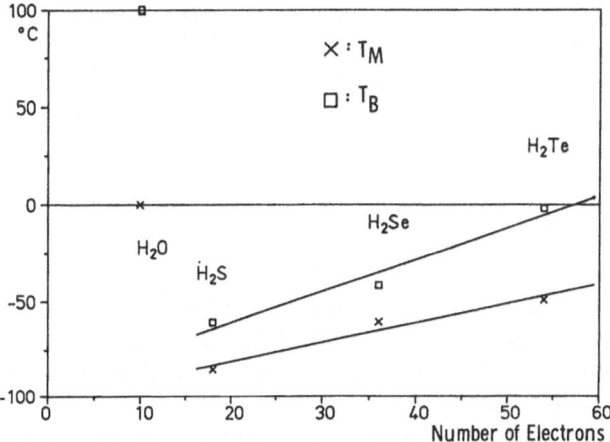

Fig. 2. The melting temperatures T_M and the boiling T_B of XH_2 molecules are proportional to the sum of electrons of the molecules. H_2O deviates severely by its H-bonds

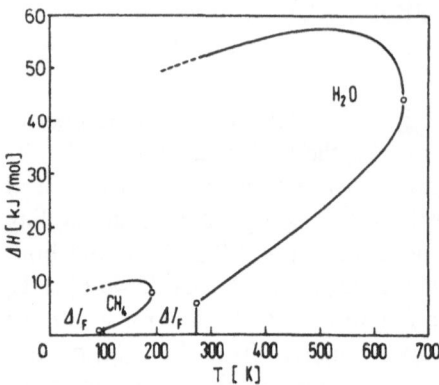

Fig. 3. Comparison of the enthalpy ΔH of liquid (at the bottom) and vapour state (at the top). The larger ΔH-values of H_2O compared with CH_4 with equal electron numbers indicate the influence of H-bonds (1 cal = 4.187 J)

Table 1. A comparison between the heat of sublimation ΔU_S (sum of all intermolecular interactions) the H-bond energy ΔH_H and the van der Waals energies of different liquids in kJ/mol (5)

	$\Delta U_S = \Delta H_S - RT$	ΔH_H	ΔW	$\Delta H_H/\Delta U_S$
Water	48.5	31	17.5	64%
CH_3OH	39.6	16.7	22.9	42%
C_2H_5OH	47.5	16.7	31.0	35%
CH_4	8.4	—	8.4	0%

Table 2. A comparison between the specific heats C_p and heat of evaporation L_V of different liquids. (20°C)

Substance	C_p			L_V		
	$J \cdot mol^{-1}$	$J \cdot g^{-1}$	$T(°C)$	$kJ \cdot mol^{-1}$	$kJ \cdot g^{-1}$	$T(°C)$
H_2O	75.3	4.18	20	44.0	2.45	20
CH_3OH	77.0	2.39	20	39.3	1.20	20
C_2H_5OH	107.6	2.34	20	42.3	1.00	20
$CH_3 \cdot CO \cdot CH_3$	124.3	2.13	20	31.0	0.55	20
$C_2H_5 \cdot O \cdot C_2H_5$	172.0	2.32	20	26.8	0.36	34
$CHCl_3$	112.6	0.92	20	31.4	0.26	20
CH_4	35.5(g)	2.22(g)	20	—	—	—
C_2C_6	71.6	2.39	−100	9.2	0.31	0
C_3H_8	96.7	2.18	−53	15.1	0.35	20

[3, 4] (see Chap. VII). Figure 3 also shows that at 525 K the energy difference between these two molecules is about 50% of that at T_C. If this difference is caused by H-bonds we should expect their contribution in water at 250°C to be only 50% of that at the critical point.

Table 1 compares the H-bond energy ΔH_H of four liquids with their total interaction energy ΔU_S. In water, the H-bond contribution is unusually large, amounting to 2/3 of the total interaction. This is the reason for the anomalous behaviour of this liquid. In addition the small size of the water molecules causes a particularly large energy content per unit volume and the lower molecular weight a large energy content per gram. Table 2 gives the specific heats C_P and the heats of evaporation L_V per mole and per gram at 20°C. Water has the highest C_P and L_V per gram of liquid of all small molecules. Mammals are mobile thermostats; the burning of food increases their temperature and the evaporation of water provides a cooling mechanism. The large L_V value per gram of water gives rise to its high efficiency as a cooling liquid and its high C_P value per gram induces the smallest possible temperature changes by its high heat capacity.

3 The Orientation Dependency of H-bonds

The H-bond energy has its largest value if the axes of the OH groups and of the lone electron pair are antiparallel (Fig. 4a). If we change the H-bond angle to give a cyclic dimer (Fig. 4b), the H-bond energy is reduced to about 40% of the maximum. In the configuration e and f with one proton oriented towards another proton or with one lone pair towards another lone pair, there is a repulsion of similar magnitude to ΔH_H. In configurations c and d between a–e or b–f there is neither H-bond attraction nor H-bond repulsion. In there positions only the normal van der Waals attraction of about $\Delta W = -17 \, \text{kJ/mol}$ HOH (see Table 1) appears. In liquid water we have to take account of a distance distribution in addition to this orientation distribution (Fig. 4a–f). The

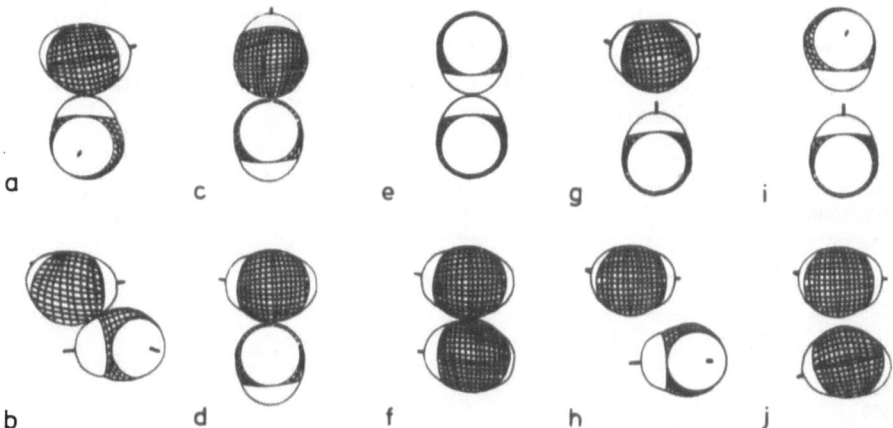

Fig. 4. Orientations of water dimers with different H-bond interactions. **a** Optimal interaction energy: proton axis and lone pair axis antiparallel. **b** Two OH dipoles antiparallel (cyclic dimer) with only 40% energy. **c, d** Positions between **a, e** or **b, f** with no H-bond but only van der Waals interaction. **e** Optimal repulsion: two OH groups antiparallel. **f** Optimal repulsion: two lone pairs antiparallel. **g–j** Orientations like **a, b, e, f** but larger distances; H-bond energies nearly zero

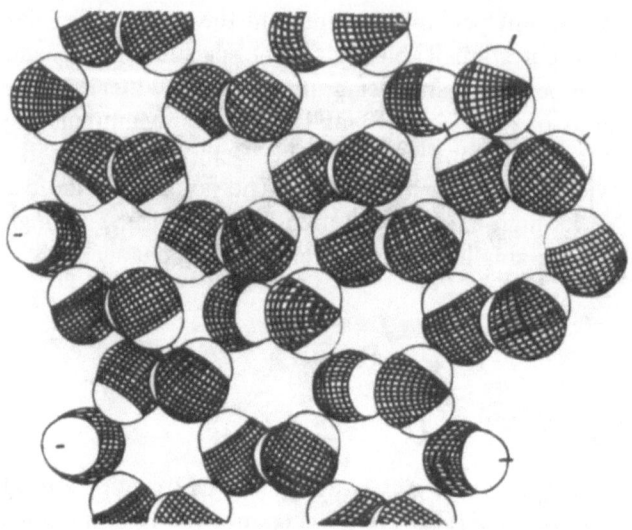

Fig. 5. Model of liquid water as mixture of ordered H-bonded areas with 6 membered rings of optimal H-bonds and areas of non H-bonded OH groups (so-called "free OH")

H-bond energies decrease with increasing distance. At distances corresponding to Fig. 4g–j the H-bond energies decrease to zero.

This angular dependence of the H-bond energy leads to a coordination number Z of 4 instead of the 12 nearest neighbours in van der Waals liquids with spherical molecules. The coordination number 4 leads to the tridymite type of structure of ice I with elementary six-membered rings (Fig. 5). These six-membered rings give rise to the hexagonal symmetry of snowflakes. X-ray scattering results from liquid water give coordination numbers of about 4.4 at room temperature. Table 3 shows that Z increases with increasing temperature in the direction of close packing but decreases to 6 at the temperature of van der Waals liquids like argon.

Table 3. Average   Number   Z   of   Nearest Neighbours

	Ice	Solid argon
Solid	4	12

T/T_c	Liquid water	Liquid argon
0.428	4.37	
0.499	4.39	
0.538	4.42	
0.56		10.5
0.61		7
0.654	4.5	
0.731	4.5	
0.99		6

The tendency to adopt the minimum energy configuration induces only small changes in the coordination number in liquid water near its melting point. To have only four nearest neighbours we should at first sight expect a relation:

$$(\Delta H_H + \Delta W)4/2 > \Delta W12/2 \tag{1}$$

This is not the case: $(15.5 + 8.8)$ kJ/mol $4/2 = 48.5$ kJ/mol is smaller than (8.8×6) kJ/mol $= 52.8$ kJ/mol. In this simple comparison we have neglected the large repulsion energies which would occur in an arrangement with 12 nearest neighbours. This comparison of Eq. (1) demonstrates the important role which repulsions can play in groups with the ability to form H-bonds.

The observed ice-like coordination number 4.4 in water near the melting point is indicative of a small number of opened H-bonds at 0°C. The heat of melting of ice: $\Delta H_M = 6$ kJ/mol can be compared with $\Delta H_M(H_2S) = 2.38$ kJ/mol. On the assumption that the contribution of van der Waals forces on ΔH_M is similar in both compounds, the H-bond contribution of ΔH_M (ice) can be estimated as $(6-2.38)$ kJ/mol $= 3.62$ kJ/mol. A comparison with $2\Delta H_H = 31$ kJ/mol (Table 1) leads us to estimate that about 12% of H-bonds open during melting.

The angular and distance distributions induce a temperature-dependent distribution of H-bond energies. They could be calculated by a computer simulation of liquid water [6]. The simplified water potential of Fig. 1 gives the number of molecules (ordinate in Fig. 6) with a given interaction energy (abscissa in Fig. 6) (full line). Thus Fig. 6 counts all water–water interactions with O...O distances smaller than 3.6 Å at 20°C. There is a maximum at about 19 kJ/mol which corresponds to OH groups in angular configurations that are fairly favourable for H-bond formation. On the other hand, this model calculation demonstrates that about 10% of molecules are in H-bond repulsion orientations as in Figs. 4e and 4f.

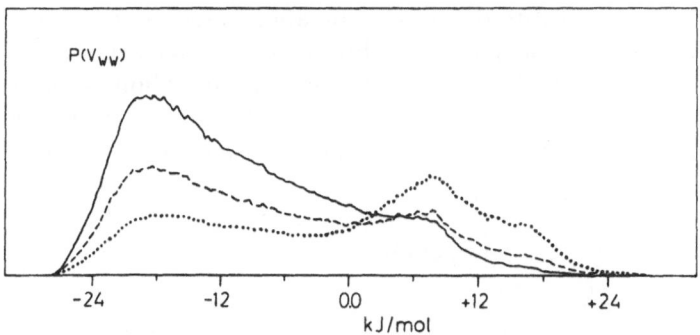

Fig. 6. Pair interaction energy distribution for water–water interaction with O...O distances smaller than 3.6 Å. Full line: pure water; dashed line: hydration shell of ClO_4^- in a 2.2 m $NaClO_4$-solution; dotted line: hydration shell of Na^+ (6)

4 Determination of the H-bond Content

In the preceding sections we have stressed the importance of H-bond content which decreases with increasing temperature. We could try to describe the anomalous properties of liquid water in terms of the temperature-dependent H-bond equilibrium. There are many papers devoted to the study of this equilibrium by measurement of the OH stretching vibration in solution (see for instance [7, 8]). This very efficient method has been able to give equilibrium constants which are really concentration-independent even when the concentration varies by a factor of 1000 [9]. This exactness can be obtained with IR overtone spectroscopy (see chapter VII, page 000). The fundamental IR bands undergo intensity variation by a factor of 30 or more on H-bond formation [10, 11, 12]. This effect causes severe difficulties with quantitative H-bond studies, a fact which is insufficiently recognized in the literature. Overtone spectra are nearly free of this artefact and represent the true statistics on different interactions [13]. The OH overtone or fundamental stretching frequency is shifted to lower values proportional to the H-bond interaction energy (Badger–Bauer rule) [14, 15].

As an example, Fig. 7 demonstrates the first overtone band of methanol solutions in CCl_4. At low concentrations the sharp band is caused by OH groups interacting with the solvent not by H-bond formation but via van der Waals forces. We call such groups "free OH". With increasing temperature this band disappears and a broad band appears at lower frequencies. In such diagrams of the extinction coefficients, the absorption diameters per mole OH are directly proportional to the content of free OH groups OH_F, abbreviated as O_F (orientation defects)[1]. Temperature-dependent spectra of pure liquid alcohols [13] show the appearance of a similar band due to free OH groups at higher temperature (Fig. 8). The same overtone method can be applied to the study of the H-bond content of liquid water [16–17]. Figure 9 shows the temperature dependence of OH_F for alcohols and water [18, 19].

At room temperature the values of OH_F for alcohols are about 2%. Spectroscopic studies show that during the melting of ice about 10% of the H-bonds are opened, in agreement both with the rough estimation based on the heat of melting and with the repulsion configurations obtained by computer simulation (Fig. 6). $OH_F = 0.5$ at about 270°C is in agreement with its estimation in Fig. 3. This means that large H-bonded clusters exist in these liquids at lower temperatures. Therefore we can describe the H-bonds in water and in alcohols roughly as an equilibrium:

$$OH_F + \theta_F \rightleftharpoons OH_B \qquad (2)$$

---

[1] Please differ the symbols: O_F for the content of "free" non bonded OH_F groups and θ_F for the content of "free" non bonded lone pair electrons! For water is valid: $O_F = OH_F = \theta_F$ but for alcohols: $O_F = OH_F \neq \theta_F$.

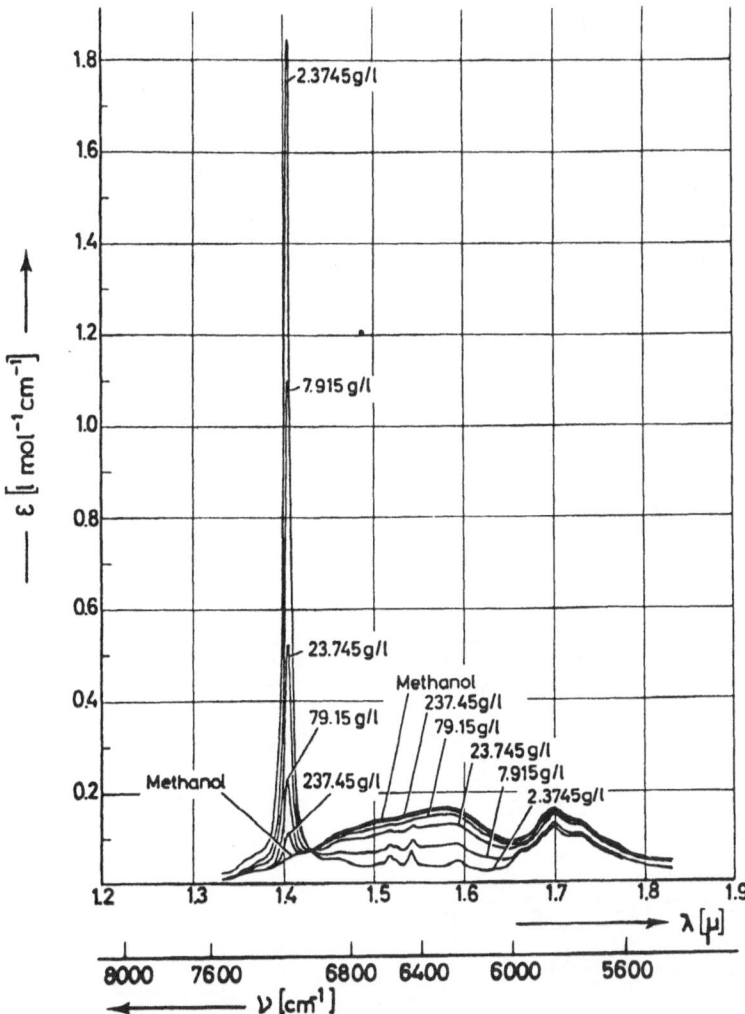

Fig. 7. First OH overtone stretching band of CH_3OH in CCl_4-solutions at different concentrations c. With increasing c the sharp band of non H-bonded OH disappears and a broad band of H-bonded OH appear. (Methanol: abbreviation for liquid methanol room T)

with

θ: lone pair electrons
OH_F: non H-bonded but van der Waals interacting OH groups
OH_B: H-bonded Oh groups.

We define the equilibrium constant:

$$K = \frac{OH_B}{OH_F \theta_F} \qquad (3)$$

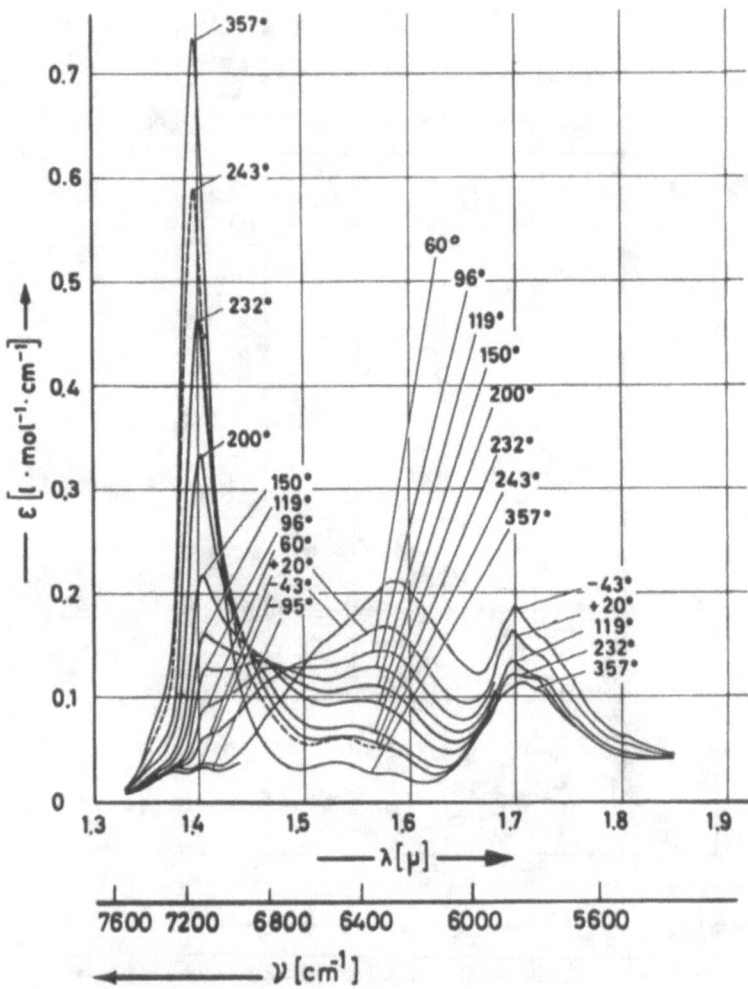

Fig. 8. First OH overtone stretching band of liquid CH_3OH under saturation conditions. At higher T the sharp band at $1.4\,\mu$ appears as indication of the existence of non H-bonded OH groups. (Dotted spectrum at 243°C: vapour spectrum)

For water is: $OH = \theta$ (4)

which gives:

$$K = \frac{OH_B}{OH_F{}^2} \qquad (5)$$

A plot of K as a function of $1/T$ gives a straight line below 200°C and the slope yields $\Delta H = -15.5\,kJ/mol$ for water and $-16.7\,kJ/mol$ for methanol and

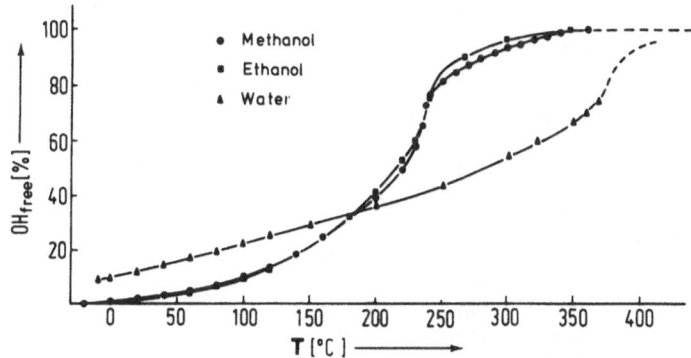

Fig. 9. The spectroscopic determined content of non H-bonded OH groups in liquid H_2O, CH_3OH and C_2H_5OH under saturation conditions (18, 19)

ethanol [13–15]. Equation (4) is not valid for alcohols. The lone pair concentration is twice the OH concentration. This is the reason why below 180°C, OH_F is smaller for alcohols than for water. The higher θ_F content shifts the H-bond equilibrium (2) towards the right.

2 Estimation of the Anomalous Properties of Liquid Water

2.1 Melting and Boiling of Water and Alcohols

The three-dimensional network of H-bonds in water and the branched H-bond chains in alcohols as well as the different concentrations of OH and lone pairs (Table 4) induce different properties in these two types of liquids.

Thirdly, the van der Waals interactions of alcohols are larger than those of water and have a hydrophobic character. The average interactions are therefore smaller in alcohols than in water and induce lower T_M, T_B and T_C (Table 5).

Comparison of the boiling and critical temperatures of alcohols and alkanes show that one OH group increases these temperatures almost as much as 4 to 5 methylene groups. The H-bond of one OH group increases T_C as much as 4 methylene groups. The second OH group of water causes an increase of T_C

Table 4. Concentrations of OH and lone pairs at 20°C in mol/liter

	Water	Methanol	Ethanol
OH	111	24.7	17.1
lone pairs	111	49.4	34.2

Table 5. Melting, boiling and critical temperatures of some liquids and heats of melting and of evaporation at T_B

	T_m [K]	T_B [K]	T_C [K]	$(T_C - T_m)$(K)	ΔH_m [kJ/mol]	ΔU_{vap} [kJ/mol]
CH_3OH	179.2	337.1	512.5	333.3	3.17	34.82
$n\text{-}C_2H_5OH$	155.8	351.6	516.1	360.3	5.02	37.6
$n\text{-}C_3H_7OH$	146.6	370.5	536.8	389.2	5.20	40.4
$n\text{-}C_4H_9OH$	163.6	390.3	561.7	397.1		
CH_4	90.6	109.1	190.7	100.1	0.93	8.1
$n\text{-}C_2H_6$	89.8	184.5	305.3	215.5	2.85	14.1
$n\text{-}C_3H_8$	83.4	231	369.9	286.5	3.52	18.2
$n\text{-}C_4H_{10}$	134.8	272.6	425.4	290.6	4.61	
$n\text{-}C_5H_{12}$	143.4	309.2	470	326.6		
$n\text{-}C_6H_{14}$	178.1	342	507.6	329.5		
$n\text{-}C_7H_{16}$	182.5	371.5	540.1	357.6		
$n\text{-}C_{10}H_{22}$	243.4	447	618.3	374.9		
$n\text{-}C_{11}H_{24}$	247.6	469	642.15	394.5		
$n\text{-}C_{12}H_{26}$	263.5	489.4	660.15			
H_2O	273.15	373	647.3	374.2	6.01	42.25
H_2S	187.5	212.4			2.38	

equal to a further 4.5 CH_2 groups inserted in methanol:

$$T_C(C_{11}H_{24}) < T_C(H_2O) < T_C(C_{12}H_{26})$$

A comparison of the heats of melting shows that ΔH_M is about 1.7 to 2.2 kJ/mol larger for alcohols than compared with the alkane of the same C-number. This difference corresponds to the contribution of about 2–3 CH_2 groups on ΔH_m or of H_2S. The difference between ΔH_{vap} of alcohols $C_nH_{2n+1}OH$ and the corresponding alkanes C_nH_{2n+2} (see Table 5) gives about ΔH_{vap} of the OH groups:

$$\Delta U_{vap}(CH_3OH) - \Delta U_{vap}(CH_4) = (34.82 - 8.1)\,\text{kJ/mol} = 26.7\,\text{kJ/mol}$$

$$\Delta U_{vap}(C_2H_5OH) - \Delta U_{vap}(C_2H_6) = (37.6 - 14.1)\,\text{kJ/mol} = 23.5\,\text{kJ/mol}$$

$$\Delta U_{vap}(C_3H_7OH) - \Delta U_{vap}(C_3H_8) = (40.5 - 18.2)\,\text{kJ/mol} = 22.3\,\text{kJ/mol}$$

These differences consist of about 17 kJ/mol H-bond energy plus about $5.3 - 9.7$ (kJ/mol) van der Waals energy. The presence of the CH_3 group adjacent to the OH of methanol seems to enlarge ΔU_{vap}.

The heat of evaporation of water at constant volume is:

$$\Delta U_{vap}(273.15\,\text{K}) = 42.25\,\text{kJ/mol}; \quad \Delta U_{vap}(373.15\,\text{K}) = 37.2\,\text{kJ/mol}$$

ΔU_{vap} and its contributions could be estimated using the experimentally determined values $(1 - \theta_F) = 90\%$ at 273.15 K and 78% at 373.15 K (values in kJ/mol):

	273.15 K	373.15 K
a) H-bond contribution	27.61 (63%)	24.60 (62%)
b) van der Waals contribution	17.58	17.58
c) contribution of the heat capacity	-1.39	-2.23
$\Delta U_{vap}/kJ\,mol^{-1}$ (estimated)	43.8	39.5

The heat capacity C_v part of the intermolecular degrees of freedom has been estimated on the assumption that all pairs of molecules induce one intermolecular vibrational degree of freedom. The $(1 - \theta_F)$ H-bonded water molecules have a coordination number of four (4/2 pairs) and the free OH have six nearest neighbours, (or 6/2 pairs). $\{(1 - \theta_F)4/2 + \theta_F 6/2\}R = (2 + \theta_F)R$ is useful as an assumption of the intermolecular heat capacity of liquid water. In C_v we must also take into account that the evaporated molecules also have the translation degree of freedom with specific heat 3/2R.

This rough estimation of the heat of vaporisation seems to be efficient and may indicate the validity of our model. We would demonstrate that the anomalous heat of vaporisation of water is based on the H-bond contributions of 63% at 273 K and of 62% at 373 K.

2.2 The Anomalous Density ρ of Water

The density of nearly all liquids and solids decreases with increasing temperature as a consequence of the thermal vibrations around the zero point of the asymmetrical intermolecular potential curves. Three exceptions to this regular behaviour are known: The so called quantum liquid the helium isotope ^4He near 0 K, liquid bismuth and, last but not least, liquid water near its melting point. During the melting of crystals the volume of the liquid increases by about 10% due to a loss of cooperativity. However, during the melting of ice its volume decreases. This volume decrease is related to the loss of about 10% of the H-bonds which increases the coordination number from four. Around the non H-bonded OH groups the coordination number Z will increase towards that of close-packed van der Waals liquids. This increase of Z induces a decrease of volume. This anomalous volume increase during the freezing of ice causes icebergs to have a lower density than water and float in the sea. On the other hand this effect causes water pipes to burst on cold days.

Secondly, the density of liquid water increases between 0 and 4°C and decreases at higher temperature (Fig. 10). This density maximum of water at 4°C is very important for life in fresh water lakes during the winter. During cooling, the water at the surface sinks to the bottom because of the higher density of surface water at the lower temperature. If there were no density maximum, such lakes would freeze from the bottom upwards and fish and other

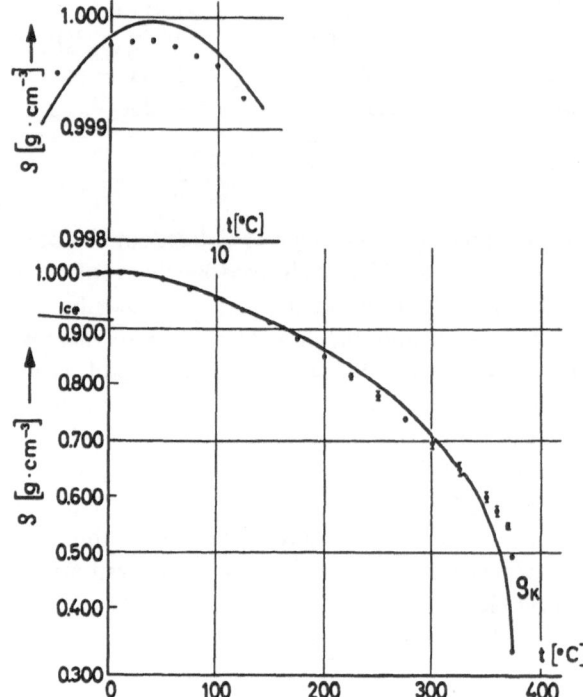

Fig. 10. Experimental density of ice and liquid water (*continuous lines*) and calculated ones (0) with Eq. (8). Above: the ρ-maximum with extended scales (18)

aquatic animals would die. The anomalous density maximum induces a minimum water temperature of 4°C at the bottom. During cooling below 4°C the density is lower than that of the bulk phase case so that the surface freezes first. This ice layer increases only slowly: at first its scattering reduces the cooling rate by radiation, secondly the vapour pressure of ice is lower than that of liquid water, thereby reducing the cooling by evaporation. The density maximum at 4°C causes a low temperature at great depths of even near the equator by convection with the sea near the north and south poles.

The density maximum can be explained both qualitatively and quantitatively by the existence of two contrary processes:

a) The normal density decrease with increasing temperature by the increase of the thermal vibration volume.
b) The anomalous density increase by the opening of H-bonds as the temperature increases (see Fig. 10) and the concomitant increase in the coordination number.

The density maximum at about 4°C can be quantitatively estimated on the basis of the spectroscopically determined "free OH" group content (Fig. 9). As an approximation, we assume an ice-like behaviour for the volume V_B and thermal expansion of H-bonded OH groups:

$$V_B = 1.0350 \ (cm^3/g)\{1 + 1.742 \times 10^{-4} \ (°C^{-1})T(°C)\}. \tag{6}$$

The partial volume V_F of the "free OH" groups and its thermal expansion we have to get the position of the maximum at 4°C to:

$$V_F = 0.64006 \ (cm^3/g)\{1 + 5.7249 \times 10^{-4} \ (°C^{-1})T(°C)\}. \tag{7}$$

We have to assume that Z increases around the "free OH" to such an extent that V_F is about 36% smaller than V_B and that the thermal expansion coefficient is larger by a factor of about 32.9. The weaker van der Waals interaction of "free OH" induces this larger expansion coefficient compared with H-bonded OH. Using Eqs. (6) and (7) we can calculate the density ρ of liquid water to:

$$1/\rho = V = V_B \ (1 - \theta_F) + V_F\theta_F. \tag{8}$$

With the θ_F-values of Fig. 9, we obtain with Eq. (6), (7) and (9) the theoretical ρ values of Fig. 10 in fairly good agreement with the experimental ones.

At about 300°C the expansion coefficients of nearly all liquids are not linear as we have assumed in Eqs. (6) and (7). A density calculation in this temperature region would need non-linear temperature terms in these equations.

Addition of salts causes a concentration-dependent shift of the 4°C ρ maximum due to changes of O_F.

For D_2O $T_M = 3.82°C$ and ρ maximum shifts to 11.18°C. The strength of H-bonds in D_2O is slightly larger, so that it melts at a slightly higher temperature. Near T_M we can assume that heavy water behaves like normal water does at about 4° lower temperature [20]. This can cause problems in the complicated coupled reaction mechanisms in biochemistry. D_2O could therefore be a dangerous chemical for living organisms [21].

2.3 The Specific Heat of Water

The equilibrium between the liquid and saturated vapour phases is a particularly interesting system. We will therefore concentrate our description of liquid water on this equilibrium as we have already done so with Fig. 9.

Figure 11 shows the specific heat C_σ of water under saturation conditions. The ordinate is divided by the gas constant R to give dimensionless units. The plotted vapour values are related to the ideal vapour state. At the melting point the specific heat of liquid water is about twice its value in ice or in the vapour state. We can assume that the number of intermolecular degrees of freedom increases in the liquid state.

To describe the specific heat of condensed phases, we can, as a first approximation, assume a similar C_v^o as for the molecule, but neglecting its translation part $3R/2$. Therefore we can postulate:

$$C_\sigma^l = C_v^o - 3R/2.$$

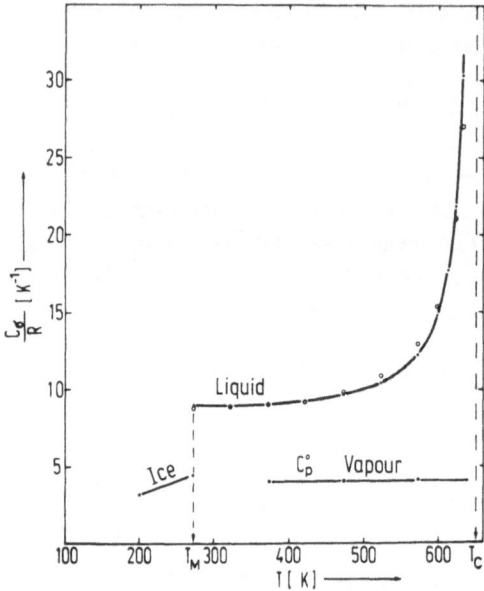

Fig. 11. Upper curve: The specific heat at the saturation line of liquid water over the vapour constant R. Lower curves: below T_M ice and above ideal vapour. Experiments: continuous line and ●; calculated by Eq. (11) (0 in Fig.)

Increasing the temperature excites the intermolecular vibrational levels. This could happen in phase in ordered areas of crystals as phonons. During melting, the Brownian movements of molecules are more or less decoupled. Therefore we can assume that every pair of molecules induces one intermolecular vibrational degree of freedom [22]. We obtain for the second part: $C_\sigma^{II} = RZ/2$, assuming $Z/2$ nearest-neighbour pairs.

In H-bonded parts of liquid water we can assume $Z = 4$ and in non H-bonded OH groups $Z = 6$. Therefore we can assume:

$$C_\sigma^{II} = RZ/2 = ((1 - O_F)4/2 + O_F 6/2)R = \{2 + O_F\}R. \tag{9}$$

A third important contribution has to be taken into account for the specific heat under saturation conditions namely the change of the H-bond equilibrium by a temperature change:

$$C_\sigma^{III} = 2\Delta H_H dO_F/dT. \tag{10}$$

This differential quotient can be calculated from the slope of Fig. 9. The total specific heat of liquid water is therefore:

$$C_\sigma = C_\sigma^I + C_\sigma^{II} + C_\sigma^{III} = C_v^o - 3R/2 + (2 + \theta_F)R + 2\Delta H_H dO_F/dT. \tag{11}$$

At high temperatures near the critical point nonlinearities—in Eyring's model called hole defects—appear in the non H-bonded OH areas [22]. This changes the second part to:

$$C_\sigma^{II} = C_v^o - (3R/2)(1 - x_F) + dx_F/dT \tag{12}$$

with x_F: hole defects.

The specific heats calculated with Eqs. (11) and (12) and Fig. 9 are shown as circles in Fig. 11.

The differential quotient dO_F/dT has a fairly high contribution in Eq. (11) especially near T_C. The turning point at the critical point induces very high values of C_σ at T_C and a maximum. This maximum shifts above T_C and at pressures above p_C due to a shift of the H-bond equilibrium Eq. (2) towards the H-bonded side [23].

If we compare C_σ of water with other liquids this value is not abnormally high because larger molecules have more molecular degrees of freedom and larger C_v^o-values. However, calculation of the specific heat per gram (cf. Table 2) shows this unit to be abnormally large for liquid water.

2.4 The Heat Content at Saturation Conditions

2.4.1 The Heat Content in the Liquid State

Integration of the specific heat (Eq. 11) with respect to temperature gives the internal energy U_{Fl} in the liquid state. The enthalpy H_{Fl} has a similar value. We define the zero point of the U_{Fl} scale as U_{Fl} at T_M:

$$U_{Fl} = \int_{T_M}^{T} C_v^o dT - R(T - T_M)3/2 + (2 + O_F)R(T - T_M) + [(O_F(T) - O_F(T_M))2\Delta H_H]. \tag{13}$$

The different terms in Eq. 13 represent:

1. Heat content of the degrees of freedom of the free water molecules in the ideal vapour.
2. Disappearance of the translational degree of freedom in term 1.
3. Heat content of the intermolecular degrees of freedom (see chap 00).
4. Temperature-dependent energy requires to break H-bonds for equilibrium temperature T. The differences $(T - T_M)$ are due to our definition of the zero point of U_{Fl} at T_M. The fourth term is fairly large and is therefore the main reason of the abnormally large U_{Fl}-values (which we have discussed above in connection with Fig. 3).

The experimental U_{Fl} values for water (x in Fig. 12) are compared with the theoretical values (0) from Eq. 13. The agreement is unexpectedly good. Small deviations near the critical point are due to the neglect of the hole defects x_F

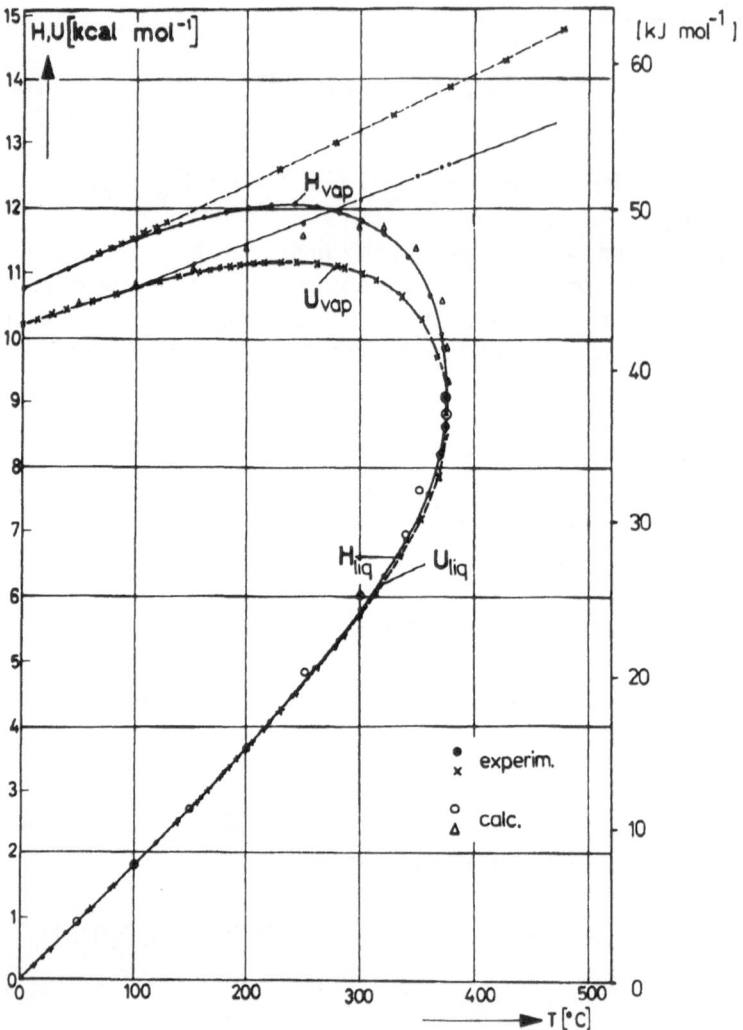

Fig. 12. Internal energy U_{Fl} and enthalpy H_{Fl} of liquid water, U_D and H_D values of saturated vapour. Experimental data: *continuous lines* and x. Calculated by Eq. (13): 0 and Δ (18)

of Eq. 12. This agreement is obtained without adjusting constants. We have simply inserted the spectroscopically measured values of the content O_F of the non H-bonded OH groups in Eq. 13. This agreement demonstrates the efficiency of our estimation method in explaining the properties of liquid water.

2.4.2 The Heat Content of Water Vapour

The heat content U_D of the saturated vapour at T_M differs from that of the liquid state by the heat vaporisation $L_v (T_M)$ at constant volume. We take

$$L_v(T_M) = \Delta U_S - \Delta H_M \tag{14}$$

$\Delta U_S = 48.6 \, \text{kJ mol}^{-1}$: heat of sublimation of ice

$\Delta H_M = 5.99 \, \text{kJ mol}^{-1}$: heat of melting of ice

At temperatures above T_M we have to add the excitation of the thermal energy to Eq. (14) and obtain

$$\Delta U'_D = \Delta U_S - \Delta H_M + \int_{T_M}^{T} C_v^o dT.$$

Equation (15) is shown in Fig. 12 by the straight line at 10.17 kcal/mol. It gives the heat content of ideal water vapour. Above 100°C the experimental points deviate from this straight line (Fig. 12). This difference could be caused by intermolecular energies in real water vapour. Unfortunately, spectroscopic determination of the H-bond content in the vapour state is difficult owing to the rotational fine structure. The difference between the straight line and the experimental U_D or H_D values in Fig. 12 could also be due to van der Waals interaction in the vapour state. This contribution could be estimated by the simplified model of nonpolar liquids reported in chap. 00 (see page 000 or (18)):

$$U_D \text{ (van der Waals)} = -6x_F W + 9RTx_F. \tag{16}$$

The first term in Eq. (16) gives the van der Waals interaction energy of x_F aggregates and the next term its intermolecular heat content. The van der Waals energy of water–water interactions is 8.8 kJ/mol. The theoretical U_D-values obtained by applying the van der Waals correction of Eq. (16) to the straight line discussed above are shown in Fig. 12 as triangles. The differences between these triangles and the experimental data indicate the remaining H-bond energy contribution $\Delta H_{H,D}$ in saturated water vapour (see Table 6).

This rough estimation of the H-bond content of saturated water vapour based on the ΔH_H energy of liquid water is too large, due to the lack of cooperativity in the small H-bond aggregates in the vapour state and therefore gives a little too small values of $(1 - O_F)$. Indeed, at T_C the content of 15.5% is similar but a slightly small compared with the experimental value of about 25% from the overtone spectra of the liquid phase.

We conclude that the anomalous heat contents of water can be explained by the large H-bond energy.

Table 6. Estimated H-bond energy $\Delta H_{H,D}$ and H-bond content $(1 - O_F)_D$ of saturated water vapour

T [°C]	200	300	350	370	T_C
$\Delta H_{H,D}$ [kJ mol^{-1}]	1.26	2.93	4.19	4.6	4.86
vapour pressure [atm]	15.85	87.62	168.01	214.69	221
$(1 - O_F)_D$ [%]	>4	>9.5	>13.5	>15	>15.5

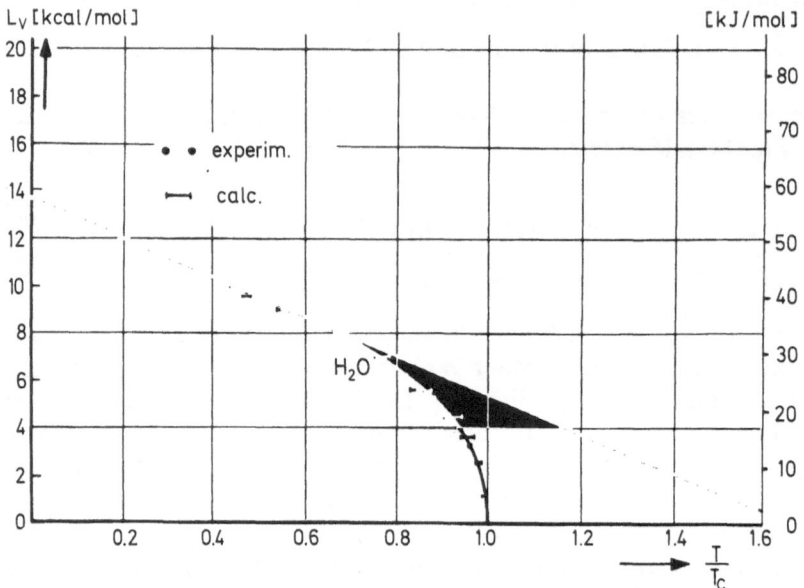

Fig. 13. The heat of vaporisation of water at constant volume. Experiments: 0 and *continuous line*; calculated by Eq. (17): *bars*

2.4.3 The Heat of Vaporization

The heat of vaporization L_v is given as the difference between the heat energy in the vapour state and that in the liquid state:

$$L_v = \Delta U_S(1 - O_F) + 2WO_F - (2 + O_F)RT + RT(1 - x_F)3/2 - \Delta H_{H,D}. \quad (17)$$

The meaning of these different terms is [18]:

1. Energy required to open the H-bonds of the liquid phase.
2. Energy required to solve the van der Waals interactions of the "free" OH in the liquid.
3. Heat content of the intermolecular degrees of freedom of the liquid.
4. Excitation of the translation movement in the vapour minus that lost in vapour clusters.
5. H-bond energy of the remaining of H-bonds in the vapour state near T_C.

Figure 13 demonstrates a fairly good agreement between Eq. (17) and the experiments.

2.4.4 The Vapour Pressure

The knowledge of L_v allows us to describe the temperature dependence of the vapour pressure p_s.

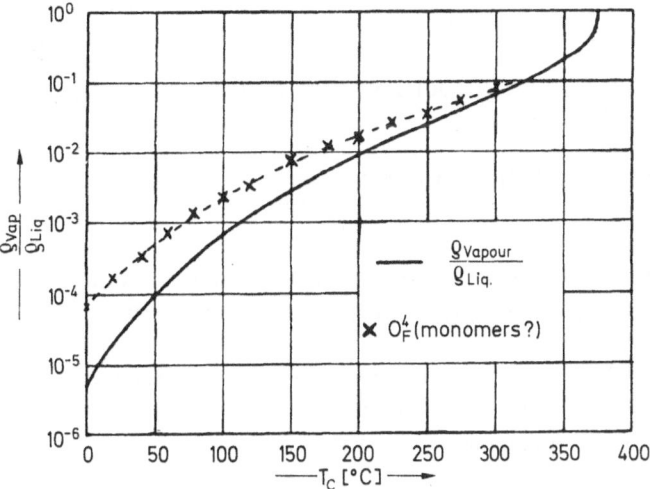

Fig. 14. *Continuous line*: The ratio of vapour and liquid density of water or the ratio of number of molecules in both phases. *Dashed line*: O_F^4 as measure of the monomer content in liquid water

The full line in Fig. 14 gives the ratio of the density in the vapour state to that in the liquid state. This ratio is equal to the ratio of the number of molecules in the two phases. The dotted line shows θ_F^4 from the spectroscopic θ_F values; this quantity determines the formation of molecules with free H-bonds at all four centres and therefore of monomers. The difference between the lines represents the vaporisation process of the monomers, assuming that the monomers in liquid water are in equilibrium with the monomers in the vapour state:

$$N_{vap}/N_{mo} = \rho_{vap}/\rho_{liq}\,O_F^4 = e^{\Delta H_{Mo}/RT} \tag{18}$$

N_{vap}: number of molecules per unit volume in the vapour.
N_{mo}: number of monomers in the liquid state
ΔH_{Mo}: adjusted heat of vaporisation of monomers, this quantity is related to the van der Waals energy ($\Delta W = 7.6\,kJ/mol$) already used.

As shown in Fig. 14, the assumption that the vapour pressure is determined mainly by the equilibrium of monomers in the liquid, is useful. We assume in this approximation that higher aggregates have a larger L_v and therefore possess a negligible p_s. Correspondingly we can assume as an approximation that the solubility of water in nonpolar solvents is mainly determined by the equilibrium between the monomers of liquid water with the monomers in the solvent [24].

Another anomalous property of water is its large critical pressure $p_c = 218.3\,atm$. This is the largest known p_c except for metals. The largest p_c of small organic molecules is $p_c(CH_3OH) = 79.2\,atm$. Table 7 lists those few organic molecules with $p_c > 50\,atm$.

Table 7. Molecules with large critical pressures p_c, values in atm

Organic molecules:
CH_3OH (79.2); C_2H_5OH (63); C_3H_7 (50.3); CH_3OCH_3 (52.5); CH_4 (46.5); C_2H_6 (48.3); C_3H_8 (42.1), C_4H_{10} (37); $C_{11}H_{24}$ (19.9); $C_{12}H_{28}$ (18);

Inorganic molecules:
H_2O (218.3); Br_2 (126); S (116); NH_3 (113.1) and H_2S (90.4)

The van der Waals equation demands: $p_c = 3RT_C/V_C$. Compounds with similar molar volumes at T_C should give a proportionality between p_c and T_C or with the interaction potential.

The molar volume V_C of compounds with H-bonds is relative small: $V_C(NH_3) = 72 \, cm^3/mol$; $V_C(H_2O) = 55.5 \, cm^3/mol$. The anomalous p_c of water is related to its large intermolecular forces and its small V_C.

2.4.5 Surface Tension σ and Surface Energy U_σ

The surface tension σ of water is often described as particularly anomalous. The wetting of textiles by water and the lack of wetting of hydrophobic compounds by water has given rise to the large surfactant and soap industries. One of the biggest surfactant factories in Germany sells washing powders for 10^9 DM per year.

In Lenards wire method to determine σ, the surface tension is often given as the force F to enlarge a layer of liquid by drawing the wire a distance s = 1 cm (σ = F/s). The meaning of σ is more evident if we define it as the isothermal

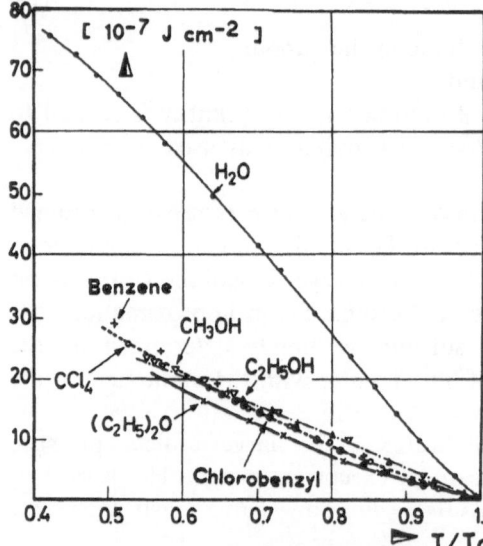

Fig. 15. The surface tension σ of some liquids and its T-dependence

work required to enlarge the liquid surface by unit area:

$$\sigma = Fs/s^2. \tag{19}$$

As Fig. 15, where the reduced temperature T/T_C is used, illustrates this isothermal work (called free enthalpy) σ is really large for water. This large σ-value of water is often described as a consequence of its abnormally high intermolecular forces. This is not completely true. Although scientists are trained to discuss the influence of structure on properties by comparing molar amounts, this is not normally done in the case of σ. Different liquids possess different numbers of molecules per cm^2 and cannot be compared. In addition this number changes with the temperature and the temperature dependence of σ should not be discussed. σ may be of interest for applied problems, but theoretically we have to correct it by the isothermal work σ_M required to bring one mole to the surface:

$$\sigma_M = \sigma V_M^{2/3} N_A^{1/3} \tag{20}$$

with V_M: volume of 1 mol and N_A: Avogadro number.

If we assume 1 mol as a cube (Fig. 16) one edge has the length $V_M^{1/3}$ and one face the area $V_M^{2/3}$. $N_A^{1/3}$ molecules form one edge, therefore the extension of 1 mole at the surface of a liquid is: $V_M^{2/3} N_A^{1/3}$. All σ-values must be recalculated in terms of σ_M for scientific discussions. This has been done with the values of Fig. 15 to give Fig. 17 where water has lost its leading role. As a consequence we have to assume that the leading role of water in Fig. 15 is based on the smallness of water molecules and the large number of water molecules per cm^2. A reduced anomalous value of σ_M (water) still exists in Fig. 17, if we compare it with alcohols or with paraffins. The network of H-bonds in water with two H-bonds per water induces higher σ_M values, but the anomaly here is smaller than for σ.

The isothermal work σ is a free enthalpy in the nomenclature of thermodynamics. During the determination of the isothermal work, exchange of heat energy occurs. To determine the energy U_σ or enthalpy H_σ during this process we have taken into account the Gibbs–Helmholtz equation:

$$H = G - TdG/dT. \tag{21}$$

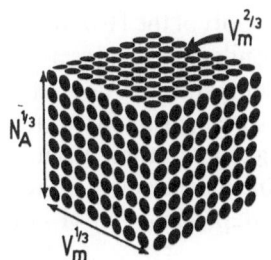

Fig. 16. An ideal arrangement of one mole as cube with $N_A^{1/3}$-layers

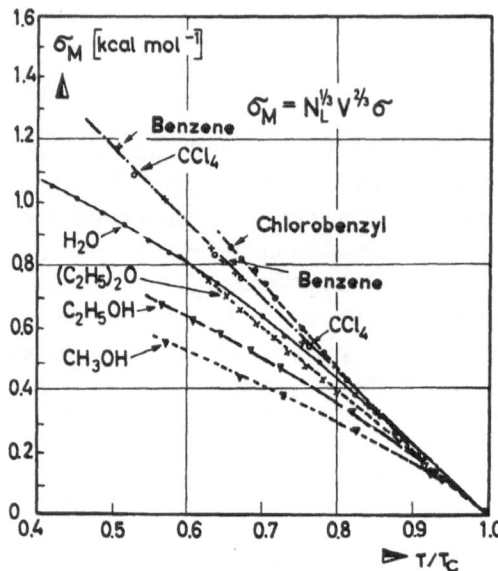

Fig. 17. The molar surface tension σ_M, the isothermic work to bring 1 mole to the surface. (1 kcal/mol = 4.187 kJ/mol)

Similarly valid for $\sigma_M = dG/dF$ (with F: surface area)

$$H_\sigma = \sigma_M - T d\sigma_M/dT. \tag{22}$$

To determine the surface energy H_σ, the energy to bring 1 mol to the surface of a liquid, we have to correct the molar surface tension σ_M by $T d\sigma_M/dT$ given by the slope of the $\sigma_M = f(T)$ diagram. This slope is negative and, up to T/T_C values of about 0.9, linear. As a consequence $H_M > \sigma_M$. During enlargement of the surface the temperature decreases, while during an isothermal process heat from the environment is transported to the surface. In the same way, during shrinking of the surface only a part of the energy can be transformed into work, the rest heats the systems.

Figure 18 gives U_σ for some nonpolar liquids (upper curves) and water and two alcohols (lower curves). For nonpolar liquids U_σ and H_σ are independent of temperature below $0.9\,T_C$. This means that the specific heat at the surface is equal to its value in the bulk. For strong H-bonded liquids σ_M has a maximum at $T \approx 0.9$. A detailed discussion [18, 22] shows that at low temperatures the weakest process occurs with alcohols. The H-bond content does not change but only van der Waals interactions are solved during transport of the H-bonded alcohol chains. In the low temperature region σ_M is independent of temperature as is the case with non-polar liquids. Only at higher temperatures does the H-bond equilibrium change during the transport to the surface. For the water network, this process starts immediately above the melting temperature. A semi-quantitative estimation of the average H-bond content in the surface of water and in the bulk gives an estimation of the temperature-dependence of H_σ of water [18, 25] (triangles in Fig. 18).

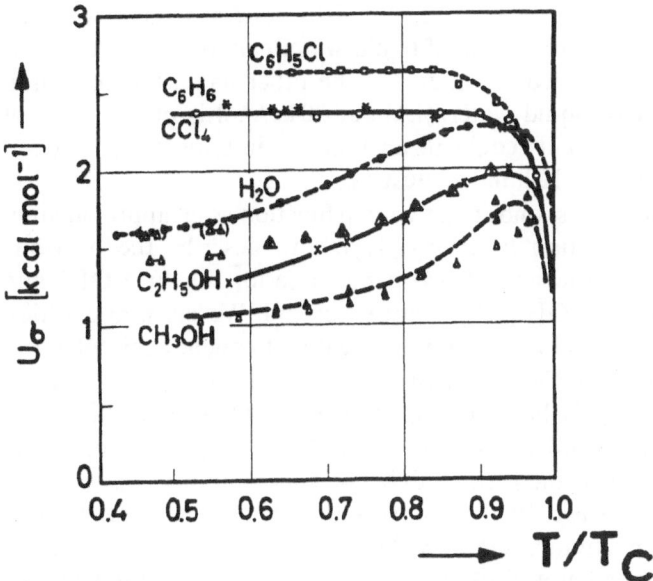

Fig. 18. The molar surface energy H_σ, the energy to bring one mole to the surface. (1 kcal/mol = 4.187 kJ/mol)

The surface energy expressed as the difference between the energy content at the surface and in the bulk is very sensitive to the model used. The approximate description of H_σ demonstrates the usefulness of our postulation of the liquid state of water.

2.4.6 Is Our Simplified Model Correct?

The goal of science is to describe and to predict the behaviour of nature as simply as possible. In the application of scientific method the prediction of unknown properties is more sensitive than the description of known ones. During our research H_σ of alcohols was not known at low temperatures. The prediction of the constancy of H_σ of alcohols at low temperature and the confirmation by our own experiments [18, 26] has increased confidence in our description of H-bonded liquids.

However, we must remember that the agreement between theory and experiment indicates the high quality of the mathematical equations used but not necessarily the correctness of the model. Our description based on the spectroscopically determined H-bond content and the simplified equilibrium of Eq. (2), was unexpectedly good. The argument that the behaviour of a liquid depends on a distribution function of H-bond orientations and of molecular distances is therefore true. As long these two distribution functions cannot be separated and as long as a simplified theory can satisfactorily describe the properties, this simplified description is useful. This statement applies to the calorific properties only. The dynamics of liquid water cannot be tested in this way and need more complicated methods. On the other hand, the activation energy for the viscosity of liquid water of about 20 kJ/mol is of the order of magnitude of one H-bond. This could mean that the intermolecular friction occurs at the surface of large H-bonded clusters.

Our simplified model corresponds to a two step function as an approximation to the complicated distribution functions. Trying to describe the overtone spectra of liquid water by the sum of two bands, one for the "free OH" and one for the linear H-bonded OH [17], is not successful. We need, in addition to the two bands mentioned, at least a third one with a frequency shift of about 40% of the difference between the two main bands. We therefore have to adjust the partition function of the H-bond state by at least a three-step function. The interaction energy of the third medium band is about 40% of the optimum H-bonds. The percentage of these three different contributions is given as a function of temperature in Fig. 19. The contribution of the intermediate state has a flat temperature maximum and may therefore not be very characteristic for the temperature dependence of the water properties. Two or three factors may contribute to an understanding of this third band and may coexist:

1. Angularly unfavourable H-bonds with antiparallel orientation of two OH groups [27, 28].
2. An interruption of the cooperative polarisation of the OH groups by H-bonds. This cooperation has also been studied by IR spectroscopy too [29, 30].
3. The flat maximum of the third state has no great influence on the temperature dependence of the water properties and can therefore be neglected.

The spectroscopically determined H-bond energy could be an average value of the two main H-bond components. A water H-bond in a dimer has a similar IR frequency shift meaning a similar H-bond energy, as a simple water–ether H-bond [31]. However, the cooperativity in a H-bond network of liquid water is so large that the shift and the interaction energy of one OH with the neighbour is as strong as in water-pyridine H-bond. This cooperativity is an additional second order effect described by our simplified water model. Therefore the efficiency of our simplified model is unexpectedly good. One reason could be

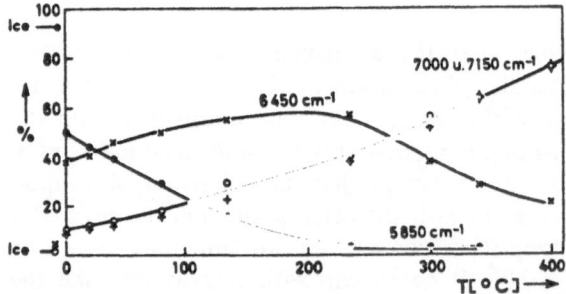

Fig. 19. Band analysis of the OH overtone band with three bands, its percentage contribution to the total absorbance (correct 6450 and 5850 into 6450)

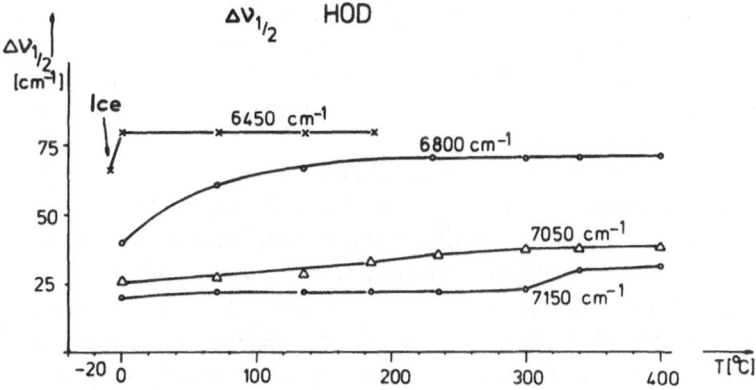

Fig. 20. The HOD half with (curve analysis) of the three bands of Fig. 19. 1. 6450 cm^{-1}: favourable H-bonds; 2. 6800 cm^{-1} unfavourable H-bonds; 3. 7050 and 7150 cm^{-1} "free" OH. The broadening indicates on partition functions around the three states

that the difference in energy between one cooperative optimum H-bond in water and one in an unfavourable orientation or a H-bond with a reduced cooperativity is small, since, during the weakening of an H-bond, the coordination number to nearest neighbours and therefore the van der Waals energy increases in such a way that the energy difference between the two states is small. On the other hand, we have to stress that the H-bond bands discussed possess large half-widths indicating a distribution function. The content O_F of non H-bonded so-called "free OH" groups has been determined at the maximum of the main band at $T > T_C$. The half-widths of the two H-bond bands increase [31] with temperature indicating a broadening of the distribution at higher temperature (Fig. 20). Our two-step or three-step functions represent approximate mean values of two or three idealized states.

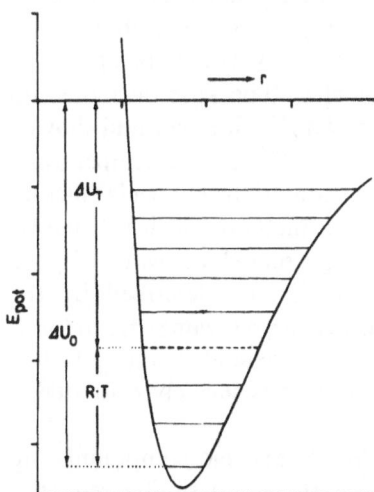

Fig. 21. As useful approximation: the thermal distribution on the intermolecular vibration levels (inducing a distance distribution) could be treated by an averaged vibration level (dashed level)

Trying to find the distribution functions could reduce the courage to attempt a description of water. This point of view seems unnecessary. Our model tries to describe the temperature-dependent statistics of the vibrational levels of the H-bond and the van der Waals potential by an average potential with kinetic energy kT (the dotted line in Fig. 21).

The properties of alcohols can be described in a similar way [18, 25, 23, 32]. Summarizing, we may claim that the structure of liquid water is fairly complicated. There are many parameters involved: the H-bond equilibrium with its distribution of orientations, distances, coordination numbers and lifetimes of the H-bonds. In addition, there are van der Waals interactions between water molecules. This complicated phenomenon can be usefully described by a simplified model—based on Eq. (2)—and the spectroscopically determined H-bond content. This model satisfies the scientific goal of describing the nature as simply as possible. On the other hand a simplified model is better than no model as long as it is able to describe the water properties fairly well quantitatively.

2.5 Aqueous Electrolyte Solutions

Pure water is a fairly complicated substance. Its description has been possible only in approximate terms. This means that the description of solutions is even more difficult. One of the most anomalous properties of water and low molecular weight alcohols is their ability to dissolve electrolytes with fairly high lattice energies. The heat of hydration of salts has to be of a similar order of magnitude as the high lattice energies of some hundred kJ/mol.

One of the first descriptions of this real anomalous effect was given by Walter Kossel [33] who described the complicated water interaction with the average continuous dielectric constant ε which is anomalously large, it is about 80 at room temperature. Kossel stresses that the high lattice energy of ions given by the Coulomb law is reduced in water by a factor of 1/80 because the Coulomb force is given by $1/\varepsilon$. He estimated the interaction of a KBr ion pair and showed that 1/80 of this value is of the order of magnitude of RT at room temperature. The thermal energy could solve the ion pair interaction. In the molecular picture we should discuss the ion-water pair interaction, a method which is done by computer simulations using the four point charge model of water (Fig. 1). The temperature dependence of ε [34, 35] can also be described by the spectroscopically determined O_F-values [36, 37]. The large ε-values are induced by orientation of the OH dipoles at large distances in extended water clusters [34, 35]. At T_C, a region in which the H-bond content is severely restricted, ε falls to 4 [38].

The flocculation of colloids and the solubility of organic compounds are specifically changed in water by salts. In all cases there is a similar sequence of salts, called the Hofmeister or lytropic ion sequence. The IR overtone spectra of water are modified in a similar sequence by different ions [39, 40]. This indicates that the ions change the H-bond content of water. To a rough approximation, the change of water spectra is similar to that caused by a change of the temperature of pure water. This similarity is particularly valid in the frequency range of the band of free water, during the H-bond bands change a little different [41]. In the region of the absorbing "free OH" we can describe the IR overtone spectra by the so called structure temperature T_{str}. The spectra demonstrate that this T_{str} is mainly dependent on the anions. Figure 22 shows such results: the T_{str} of different aqueous salt solutions at concentrations of 1 mol of anion. Large anions like I^-, SCN^- or ClO_4^- induce water spectra similar to those of pure water at higher temperatures. This means that on average the H-bond structure of these solutions is weaker than that of pure water. Such ions are therefore called structure breakers and induce similar changes of NMR spectra as do a rise in temperature of pure water. Small or doubly charged ions could have a T_{str} smaller than the solution temperature, this means on average that such solutions possess a stronger H-bond structure than pure water at lower temperature. Such ions are called structure makers.

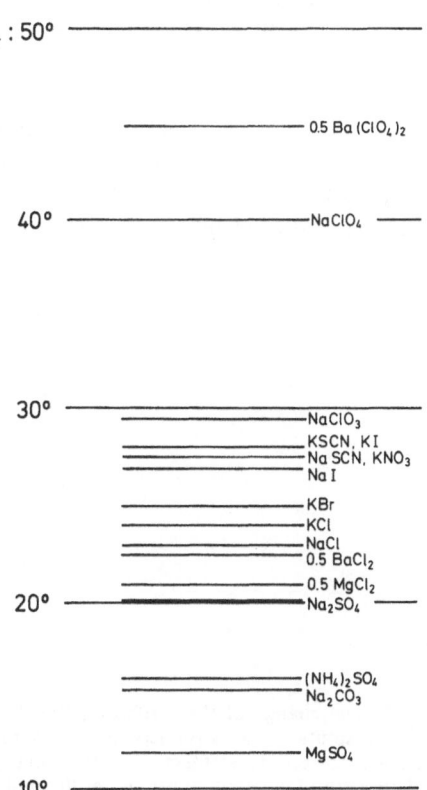

Fig. 22. The so-called spectroscopic determined structure temperature T_{str} of different aqueous salt solutions (anion- concentration: 1 mol/L). Solution T = 20°C. The anion series got by this method corresponds to the lyotropic ion series

This rough description efficiently predicts the behaviour of ternary solutions. For instance, struture making ions can induce a salting-out effect of organic solutes ánd structure breaking ions a salting-in effect [36, 37, 41]. The nomenclature T_{str} seems to be particularly useful in predicting the solubility because the solubility of water often depends on the content of "free OH" groups.

The spectroscopic T_{str} results give an ion sequence too, similar to the Hofmeister ion sequence. This means that empirical lyotropic ion sequences could be understood as series of changing the H-bond structure of water. Aqueous solutions of polyethylene oxide derivatives form immiscible phases above a certain T_K, the turbidity point [39, 41]. T_K depends on the ratio of the number of hydrophilic ethylene oxide groups to that of the hydrophobic groups and on the added salts in the lyotropic ion series. The turbidity temperatures T_K of such solutions are proportional to the spectroscopically determined structure T_{str} [39, 41] indicating that T_K too depends on the change of the water structure by different anions. Figure 23 shows the relation T_K values induced by different anions and the Coulomb field strength of these ions (charge over the ion radious r_{An}). This relation indicates that T_K, T_{str} and the Hofmeister ion sequence depend on the Coulomb field strength of the different ions which disturb the H-bond structure.

The change in water IR spectra due to different anions is very similar to its change due to different organic bases [42]. As these bases can be ordered in a sequence according to their H-bond acceptor strength, the anion sequence gives a similar order of the change of the H-bond equilibrium [43].

Based on the spectroscopy of water we can conclude [44] we can now discuss electrolyte solutions in a new aspect. The lone pair electrons of alkaline molecules seem to be situated near the surface of the molecules and Coulomb interactions of such H-bonds act similarly to the interactions between water and anions. The studies of electrolyte solutions and of H-bonds are very similar from the energy point of view.

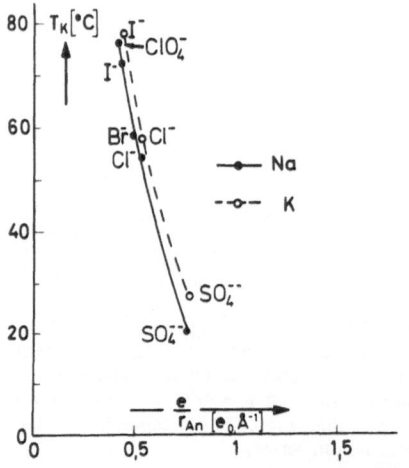

Fig. 23. The change of the turbidity point T_K of aqueous solutions of *p-iso*-octyl-phenol with 9 ethyleneoxide groups as function of the Coulomb field strength of different added anions (0.5 mol)

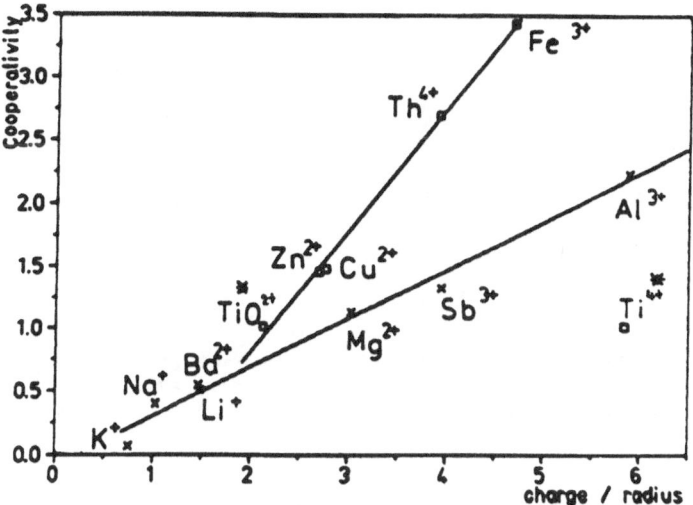

Fig. 24. The cooperativity factor of the strengthening of water H-bonds by different cations as function of the field strength of the cations

The influence of cations on IR spectra is not so well defined as that of anions. One reason is the larger band half-widths. But with a difference spectra technique we can demonstrate a cation-induced band shift of H-bonded OH groups [45, 46]. This effect can be understood as a polarisation of the OH groups in the Coulomb field of the cations. The sequence of cation polarisation action is again parallel to the Hofmeister ion sequence. If we plot the so-called cooperativity factor [45, 30], the factor of the strengthening of the OH frequency shift or the interaction energy, against the Coulomb field strength of different cations (Fig. 24) we obtain one straight line for hard cations and another for soft ones.

A comparison between the K^+ effect and the cooperativity factor for water–water interactions suggests that the hydration water of K^+ polarises the water molecule less than does bulk water, which means that K^+ weakens the water structure. The cooperativity factor of Na^+ is larger than that of water–water. The H-bonds of the Na^+ sphere are stronger than in bulk water. Does this effect play a role in the $K^+ - Na^+$ gradient in biological cells? Is the weakening of the hydration of biopolymers by K^+ necessary to start chemical reactions? On the other hand, the question of the strengthening arises if H-bonds in heavy metal cations are related to their poisonous effects in cells by stronger H-bond structure of the hydrates of biopolymers?.

Modern spectroscopic or computer . simulation studies' of electrolyte solutions confirm the Hofmeister ion sequence that ions not only act as charges but also induce special effects varying from one ion to another. As the computer simulations have demonstrated, the anion–cation hydrations influence each other at medium concentrations. For instance the simulation of a 2.2 mol $NaClO_4$-solution leads to the prediction [6] of an inner sphere complex (anion

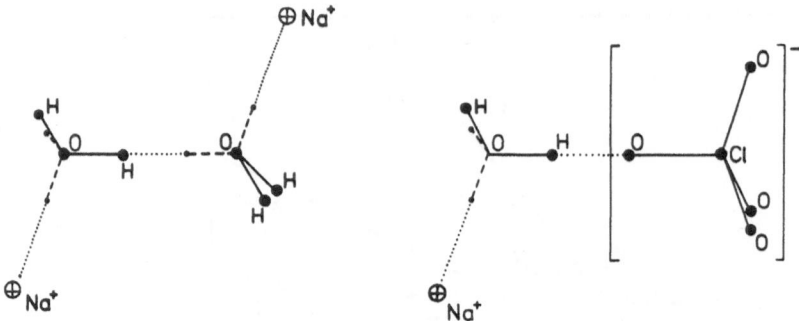

Fig. 25. Average optimal orientation of neighboring ions in the 2.2 m NaClO$_4$ solution (computer simulation [6]). Above, two Na$^+$ connected by two water molecules; beneath, a Na$^+$—ClO$_4^-$ pair connected by one water molecule in the most probeble position

H$_2$O cation) with preferred optimum H-bond angles and distances (Fig. 25). There are indications that chains of complexes anion H$_2$O–cation–H$_2$O anion–H$_2$O cation–H$_2$O etc. may exist separated from bulk water [6].

This model of water and its solutions can be applied successfully to the description of solutions of water in nonpolar solvents [24], the hydration of biopolymers [40, 46], desalination mechanism by membranes [47, 8], surfactants [49], drying processes [50] or on the influence of water on monolayers [51].

I thank the following coworkers for much help in preparing this research of aqueous systems: W. Ditter, F. Bopp, H. Borgholte, C. Buanon-Om, L. England-Kretzer, M. Fritzsche, G. Heinje, T. Kammer, H. Kleeberg, D. Klein, O. Kocak, M. Oberhoffer, K. Ringsriwatananon, D. Schiöberg, O. Schrems, S. Shah, U. Siemann, S. Singh, H.Y. Zheng and A. Zukovskij.

3 References

1. Franks F (1972) Water a comprehensive treatise, Plenum, New York, vol 1, p. 7
2. Luck WAP (1970) Hochschulskripten Nr. 715/715a, Bibliographisches Institut Mannheim–Wien–Zürich, p. 101
3. Luck WAP (1974). In: Structure of water and aqueous solutions. Sect. III.2, Verlag Chemie/Physik, Weinheim
4. Schiöberg D, Buanam-Om C, Luck WAP (1979) Spectrosc. Lett. 12: 83
5. Luck WAP (1987) Pure Appl. Chem. 59: 1215
6. Heinje G, Luck WAP, Heinzinger K (1987) J Phys Chem 91: 331
7. Luck WAP (1965) Naturwissenschaften 52: 25: 49
8. Luck WAP (1973). In: Franks F (ed) Water: a comprehensive treatise. Plenum, New York, vol II, Sect. IV, p. 235
9. Luck WAP (1968) Z. Naturforsch. 23b: 152
10. Luck WAP, Zheng HY (1984) J. Chem. Soc. Faraday Trans. 2, 80: 1253
11. Singh S, Fritzsche M, Kümmerle I, Luck WAP, Zheng HY (1985) Spectrosc Lett 18: 283
12. England-Kretzer L, Fritzsche M, Luck WAP (1988) J Mol Struct 175: 277
13. Luck WAP, Ditter W (1968) Ber Bundenges Phys Chem 72: 365
14. Kleeberg H, Kozak O, Luck WAP (1982) J Solution Chem 11: 611

15. Pimentel GC, McClellan AL (1960) The hydrogen bond, Reinhold Freeman Co, San Francisco
16. Luck WAP (1965) Ber Bunsenges Phys Chem 69: 626
17. Luck WAP, Ditter W (1969) Z Naturforsch. 24b: 482
18. Luck WAP (1980) Angew Chem 92: 29; Angew Chem Int Ed Engl 19: 28
19. Luck WAP (1985) Opt Pur Appl 18: 71
20. Luck WAP (1976) Naturwissenschaften 63: 39
21. Hübner G, Jung K, Winkler E (1970) Die Rolle des Wassers in biologischen Systemen, Vieweg, Braunschweig
22. Luck WAP (1979) Angew Chem 91: 408; Angew Chem Int Ed Engl 18: 350
23. Luck WAP (1986) J Mol Liq 32: 41
24. Luck WAP (1987) Pure Appl Chem 59: 1215
25. Luck WAP (1967) Discuss Faraday Soc 43: 115
26. Luck WAP (1973). In: Chemie, Physikalische Chemie und Anwendungstechnik der grenzflächenaktiven Stoffe. Kongreßband IV. Internationaler Kongreß für grenzflächenaktive Stoffe in Zürich 1972. Carl Hauser, München
27. Luck WAP (1965) Naturwissenschaften 52: 25: 49
28. Luck WAP (1976). In: Schuster–Zundel–Sandorfy (eds) The hydrogen bond. North Holland, vol II, p. 527
29. Kleeberg H, Klein D, Luck WAP (1987) J Phys Chem 91: 3200
30. Kleeberg H, Luck WAP (1989) Z Phys Chemie Leipzig 270: 613
31. Luck WAP (1981) Ber Bunsenges Phys Chem 85: 959
32. Luck WAP (1986) Acta Chim Hung 121: 119
33. Kossel W (1916) Ann Phys 4F 49: 229
34. Hasted JB (1973) Aqueous dielectrics. Chapman and Hall, London
35. Haggis GH, Hasted JB, Buchanan JB (1965) J Chem Phys 20: 3165
36. Luck WAP (1963) Ber Bunsenges Phys Chem 67: 186
37. Luck WAP, Mann B, Neikes T, Schmidt E (1974) Ber Bunsenges Phys Chem 78: 1236
38. Quist AS, Marhall WL (1965) J Phys Chem 69: 3165
39. Luck WAP (1965) Ber Bunsenges Phys Chem 69: 69
40. Luck WAP. In: Rowland StP (ed) Water in polymers. ACS Symposium Series Nr. 127, 1980, p. 43
41. Luck WAP (1964) Fortschr chem Forsch 4: 653
42. Luck WAP, Schiöberg D. (1979) Adv Mol Relaxation Processes 14: 277
43. Kleeberg H, Luck WAP (1983) J Solution Chem 12: 369
44. Luck WAP (1978) Progr Colloid & Polymer Scr 65: 6
45. Kleeberg H, Heinje G, Luck WAP (1986) J Phys Chem 90: 4427
46. Kleeberg H, Luck WAP, Fundamentals Absorption 1984, p. 267; In: Myers AL, Belfort G (eds) Am Inst Chem Engineers, New York, Publ, P-39
47. Luck WAP (1984). In: Belfort (ed) Synthetic membrane processes. Fundamentals and water applications. Academic, New York, p. 21
48. Luck WAP, Schiöberg D, Siemann U (1978) J Chem Soc Faraday Trans II 76: 136
49. Kleeberg H, Luck WAP (1984) In: Surfactants in our world—Today and tomorrow, Proceedings of the world surfactants congress, Munich, Kürle, Gelnhausen, vol IIIC, 1984, p. 247
50. Luck WAP (1960) Angew Chem 72: 57
51. Luck WAP, Shah SS (1978) Progr Colloid & Polymer Sci 65: 53

CHAPTER X
Cooperative Effects Involved in H-Bond Formation

H. Kleeberg

The strength of a H bond can be influenced by other cohesion forces exerted by molecules or ions of the surroundings. Cations C^+ may interact with $O—H———B$ bonds to form $C^+———O—H———B$ complexes. The influence of a given cation on the OH frequency is larger when the $O—H———B$ bond is stronger and a "cooperativity factor" can be deduced from the experimental data. It correlates linear with the ratio charge/radius of the cation. Cooperative effects are also encountered in $X—H———X—H———B$ complexes. New bands appear in the presence of the base B. Correlations are established between the frequency shifts. A model is presented for the estimation of the $X—H$ stretching frequencies in H-bonded chains.

1 Introduction

Hydrogen bonds between proton donors XH and proton acceptors (electron donors) Y (either anions A^- or uncharged molecules B (bases)) and consequently the H-bond equilibrium

$$X—H + Y \rightleftharpoons X—H...Y \tag{1}$$

Intermolecular Forces
© Springer-Verlag Berlin Heidelberg 1991

are influenced by the surrounding medium [1] (solvent effects, for example in the vapor phase, inert gas matrices, solvents), the temperature, and the pressure for a given system XH and Y.

One reason for the considerable interest [2–4] which science takes in H-bonds is the large number of practically important properties (like the temperature and heat of melting and vaporization, the heat capacity) of compounds containing H-bonds (for example, water, alcohols, biomolecules) in comparison to molecularly similar compounds (i.e., similar with respect to the molecular weight and van der Waals interactions as in ethers, aldehydes) which can not form H-bonds.

The strength of the XH...Y interaction will depend on the acidity of XH and the basicity of Y. Comparison of the enthalpies of H-bond formation ΔH_{HB} [i.e., the enthalpy involved in Eq. [1] for different systems in the gas phase (XH: water, alcohols, carboxylic acids and $Y = A^-$: F^-, Cl^-, Br^-, and I^-) by Th. Zeegers-Huyskens [5,6] indicates that a correlation exists between the enthalpy ΔH_{HB} of H-bond formation and the difference in the heterolytic bond dissociation energy (or proton affinity: PA) $\Delta PA = PA(XH) - PA(AH)$ of

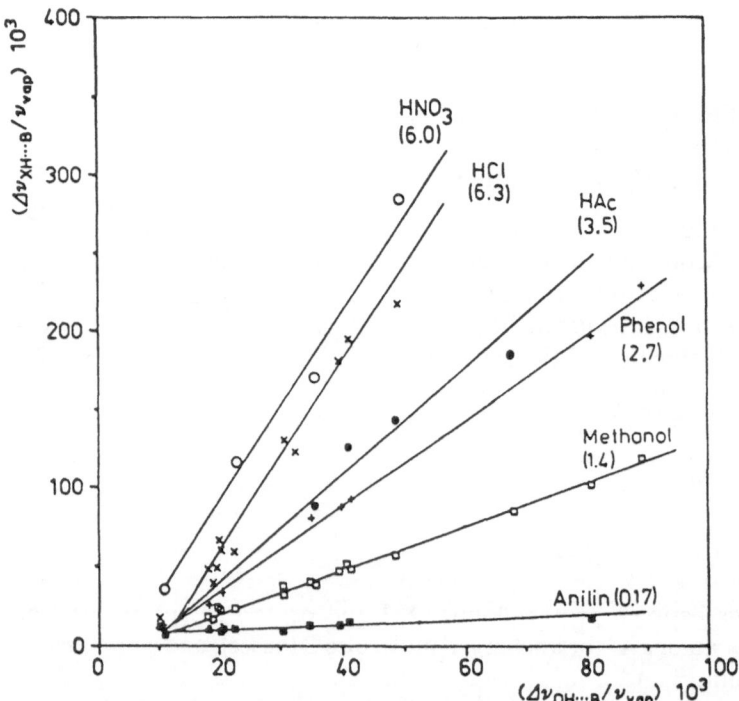

Fig. 1. Correlation between the relative frequency shift $\Delta v_{XH...B}/v_{vap}$ of various proton donors XH with the relative OH frequency shift of HOD $\Delta v_{OH...B}/v_{vap}$ interacting with the same bases B. The slopes of the linear correlations [(acidity factors; see Eq. (4)] are indicated in *brackets* (see also Table 1)

XH and AH like:

$$- \Delta H_{HB} = xe^{-X1\Delta PA} \tag{2}$$

This means the lower the heterolytic bond dissociation energy (i.e., larger acidity) of different XH is, the larger will be $- \Delta H_{HB}$ for a given H-bond acceptor.

Another means to express the H-bond strength in XH...Y complexes, is the comparison of the relative infrared frequency shifts (gas phase XH frequency of XH monomer: $\nu_{XH,vap}$; XH frequency in the XH...B complex: $\Delta\nu_{XH...B}$):

$$\Delta\nu_{XH...B}/\nu_{XH,vap} = (\nu_{XH,vap} - \nu_{XH...B})/\nu_{XH,vap} \tag{3}$$

of different proton donors XH interacting with non-ionic H-bond acceptors B. This relative frequency shift is roughly proportional) to ΔH_{HB}.

In Fig. 1 the relative frequency shifts $\Delta\nu_{XH...B}/\nu_{vap}$ of several proton donors in polar aprotic solvents B are compared with that of DOH...B interactions ($\Delta\nu_{OH...B}/\nu_{vap}$). The linear correlations obtained for each XH exhibit characteristic slopes $m_{W;B}(XH)$ (indicated in brackets in Fig. 1) according to the equation:

$$\Delta\nu_{XH...B}/\nu_{XH,vap} = m_{W;B}(XH)\Delta\nu_{OH...B}/\nu_{OH,vap} + \text{const.} \tag{4}$$

The subscript "W; B" stands for the reference proton donor (W: water, DOH) and the medium (B: H-bond acceptor). Of course the relative frequency shifts of a given proton donor may be taken as a measure of the basicity of proton acceptors.

It has been shown [7] that $m_{W;B}(XH)$ may be used as a measure for the acidity of a proton donor because in $m_{W;B}(XH)$ correlates linearly with the heterolytic XH bond dissociation energy PA(XH) for the cases where the proton donor group is OH (i.e., water, alcohols, acetic acid, nitric acid). For other proton donor groups (namely thiophenol, pyrrole, HCl, HI) slight deviations from this correlation occur.

Another measure of the acidity of XH is its dissociation constant in water. Therefore we compared $m_{W;B}(XH)$ with the pK_a values of the corresponding proton donors in Fig. 2 (points indicated by X; see also Ref. [7]). Obviously a correlation exists, which approaches $m_{W;B}(XH) = 0$ for large pK_a values asymptotically. The slight scatter in this correlation may be due to the cooperative influence of $HO-H...X-H...OH_2....$

We may conclude that $m_{W;B}(XH)$, PA(XH), ΔH_{HB} as well as pK_a (see Table 1 and Ref. [7]) may be used to classify proton donors—especially those containing $-O-H$ groups—with respect to their acidity.

In 1957 Frank and Wen [8] postulated the presence of cooperative effects between H-bonds. This means that an existing H-bond between, for example, an alcohol ROH and an aprotic H-bond acceptor B (i.e., OH...B) would be

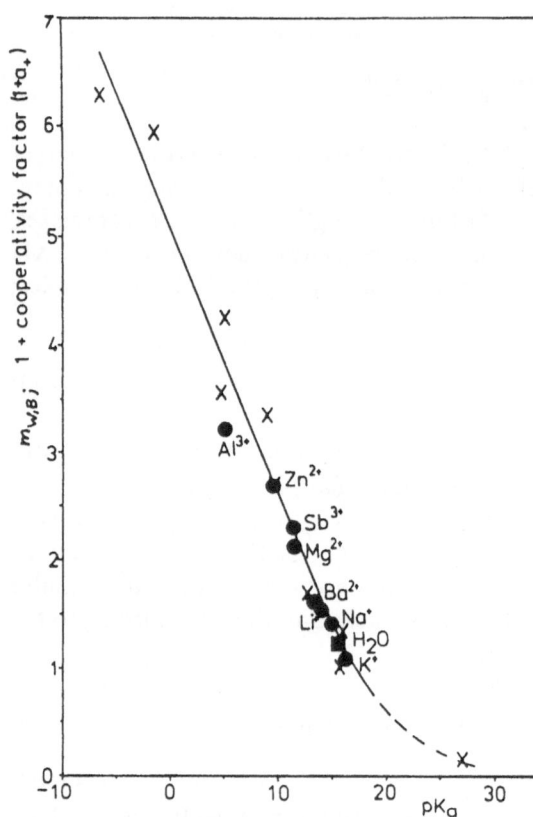

Fig. 2. X: Comparison between the acidity factor $m_{w;B}$ of different proton donors and the pK_a value in water (see also Table 1). •: Comparison between the slopes $(1 + a_+)$ of the linear correlations of Fig. 6 of different cation hydrates and their pK_a values

Table 1. Summary [7] of the heterolytic dissociation energies PA(HX) (in kcal/mol), pK_a values in water, and the acidity factors $m_{w;B}$(HX) [see Fig. 1 and Eq. (4)]

Proton donor (HX)	PA(HX)	pK_a	$m_{w,B}$
Aniline	367	27	0.2
Chloroform	—	—	0.34
Water (DOH)	391	15.7	1.
Methanol	379	16.0	1.4
Pyrrole	361	16.5	1.6
Trifluoroethanol	364	12.7	2.3
Thiophenol	339	—	2.6
Phenol	350	9.86	2.7
Hexafluoroisopropanol	350	9.3	3.4
Acetic acid	349	4.7	3.5
Perfluoro-t-butanol	—	5.3	4.4
HJ	314	−9.48	6.4
HBr	324	−9	6.2
Nitric acid	∼323	−1.34	6.0
HF	371	3.5	—
HCl	334	−7	6.3

strengthened by the formation of a second H-bond between another alcohol ROH and the one already involved in ROH...B (i.e., OH...OH...B). Conversely, the strength of the H-bond between the two alcohol molecules in:

$$O-H...O-H...B$$

with R and R' bonded to the oxygens.

is expected to depend on the H-bond acceptor strength of B. The consequences of cooperative H-bond effects have frequently been discussed on the basis of theoretical calculations [9–14] and molecular dynamics simulations [14, 15] of the hydrate H_2O of cations. Qualitative infrared spectroscopic investigations of the association of OH-groups in solutions [16, 17] and in low-temperature matrices [18], of FH...FH...B complexes in the vapor phase [19] as well as the influence of cation interactions on OH...B complexes [20, 21] shows the important role of cooperative effects. Quantitative experimental investigations on H-bond cooperativity have become available only recently and will be discussed in the following.

2 Cation Cooperativity

Since cations C^+ may interact with $O-H...B$ to form $C^+...OH...B$ complexes we may ask now to how far an extent the acidity of an OH group of water in the first hydration shell of the cation may depend on this cation. Figure 3 shows the infrared spectra (full lines) of water (partially deuterated water HOD is frequently used on behalf of experimental reasons) in some polar solvents B. Due to the formation of OH...B complexes, the frequency of the OH stretching vibration is shifted to lower wave numbers with increasing H-bond acceptor strength of the solvent.

If salts are added to these binary solutions, the cations may interact with DOH to form $C^+...OH...B$ complexes [7, 22, 23]. The IR spectra in the region of the OH streching vibration of these cation complexes may differ considerably from those of the OH...B interaction.

In Fig. 3 the dotted lines represent these spectra for the case of $AlCl_3$ (i.e., $Al^{3+}...O-H...B$). (The two bands observed in the case of the ternary solution containing tetrahydrofurane (see Fig. 3, lower part) are assigned to the OH stretching vibration and the overtone of the HOD deformation vibration, which are coupled by Fermi resonance). In all these solutions the OH frequency shift of cation...OH...B complexes is larger than that due to the corresponding OH...B interaction itself.

Figures 4 and 5 show more examples of this influence on OH...B complexes. In Fig. 4 the full line corresponds to the spectrum of the OH-vibration of HOD in OH...B complexes in tetrahydrofurane [22]. If salt is added to these binary

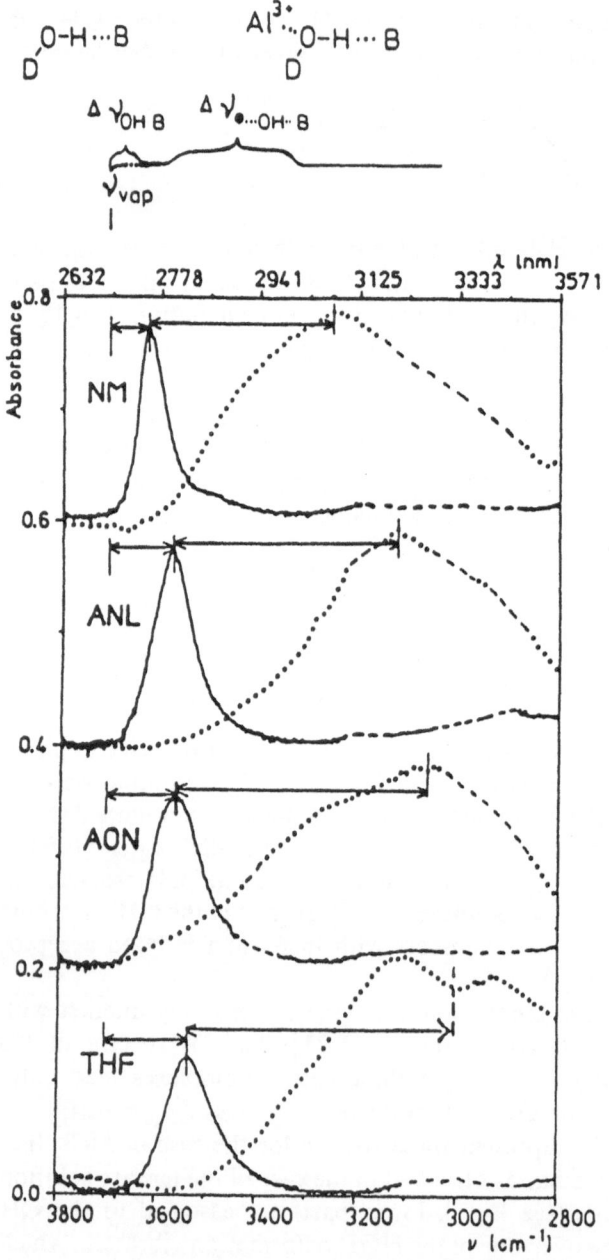

Fig. 3. Cooperative influence of Al^{3+} on OH...B complexes. *Full lines*: spectra of binary mixtures HOD and solvent B (molar ratio water/B $\simeq$ 1/50; NM: nitromethane, ANL: acetonitrile, AON: acetone and THF: d_8-tetrahydrofurane); *dotted lines*: spectra of ternary solutions HOD/AlCl$_3$/B (molar ratio: water/salt = 1/1; salt/B $\simeq$ 1.50) (*dashed lines* indicate regions where solvents show large absorbance)

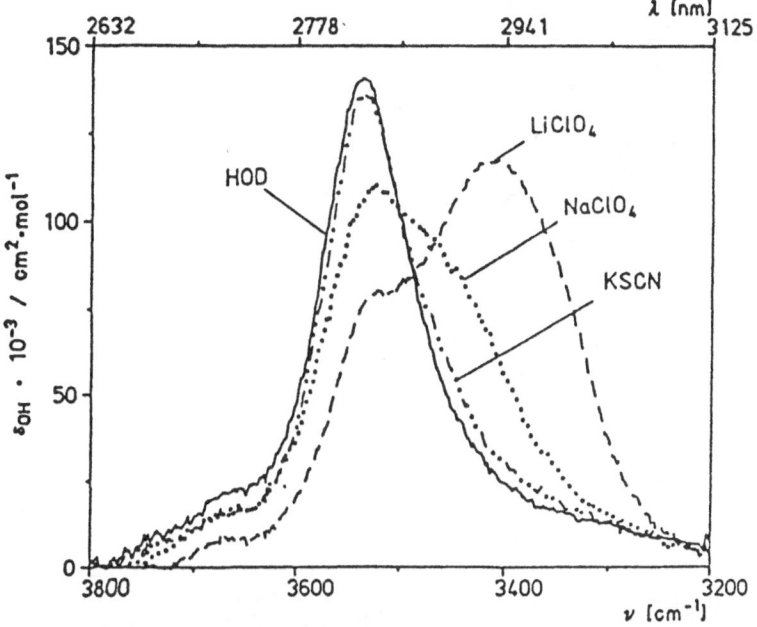

Fig. 4. Infrared spectra of HOD in tetrahydrofurane (THF) before (*solid line*) and after (*broken lines*) addition of salts. The concentrations (mol/1) were: THF: 21.1; water (HOD: 4.3 mol—% H_2O in D_2O): 0.13; $LiClO_4$: 0.21; $NaClO_4$: 0.21 and KSCN: 0.26

solutions, the extinction coefficient ε_{OH} decreases at the maximum of the band corresponding to OH...B complexes and increases at lower wave numbers. This change of the spectra is cation dependent; i.e., the band shift to lower wave numbers is caused by cation...OH_2 interactions.

A direct determination of the exact position of the band maximum of the OH stretching frequency in cation...OH...B complexes from these spectra (see Fig. 4) is complicated by the overlapping of the bands due to the two complexes (namely OH...B and C^+...OH...B) contributing to the spectrum. In order to separate the cation dependent band we subtracted from the experimentally measured spectrum $\varepsilon_{OH}(v)$ of the ternary solutions a fraction $\alpha' < 1$ of the experimental spectrum of the binary solution ($\varepsilon_{OH...B}(v)$; i.e., of the OH...B complexes):

$$\Delta\tilde{\varepsilon}_{OH} = \varepsilon_{OH} - \alpha'\varepsilon_{OH...B} \tag{5}$$

The value of α' was chosen in such a manner that the absorbance at $v_{OH...B}$ is as close to zero as possible but not negative. For Li^+-, Na^+-, and K^+-containing solutions in tetrahydrofurane some results of this procedure is shown in Fig. 5. For a certain range of α' values for each system, a band with nearly constant position of the band maximum $v_{+...OH...B}$ corresponds to cation...OH...B interactions.

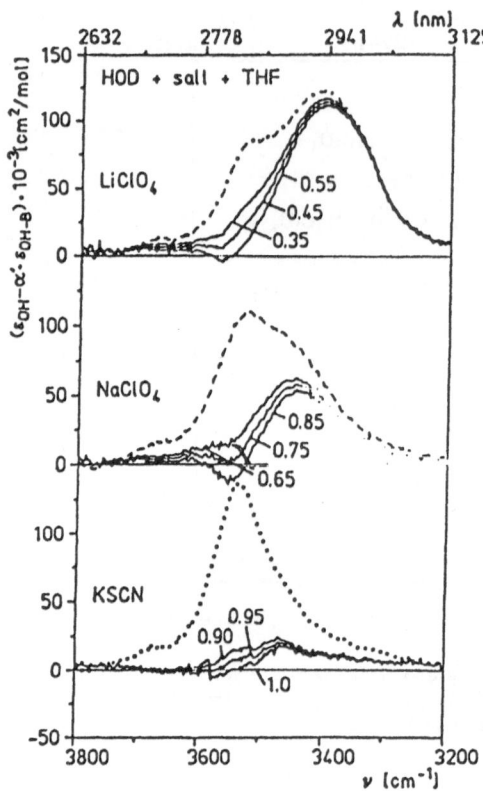

Fig. 5. Determination of the maximum of the cation-influenced OH-band. *Broken lines*: (see Fig. 4) HOD spectra in the ternary solution (water/salt/solvent (THF)); *solid lines*: difference spectra according to Eq. (5) of the OH-spectra in the ternary solution minus fractions α the OH spectrum in THF. The numbers indicate α′ values [see Eq. (5)]

For all cations investigated [7, 22–24], linear correlations:

$$\Delta v_{+...OH...B} = (1 + a_+)\Delta v_{OH...B} + b_+ \tag{6}$$

between the OH frequency shift $\Delta v_{+...OH...B}$ of the $C^+...OH...B$ complexes and the frequency shift $\Delta v_{OH...B}$ of OH...B complexes are found as shown in Fig. 6. This means that the influence of a given cation on the OH frequency is the larger, the stronger the H-bond in OH...B is. On the other hand the influence of cations on a given OH...B complex (i.e., constant $\Delta v_{OH...B}/v_{vap}$) increases with the ratio (cation charge)/radius.[1] These results show quantitatively that the influence of cations on H-bonds in OH...B complexes is not additive but cooperative. Thus, we call a_+ in Eq. (6) the "cooperativity factor".

As shown in Fig. 7, this cooperativity factor a_+ correlates linearly with the ratio charge/radius of the cation as well. This means that the polarization of H_2O by cations and hence the cooperativity increases with the interaction energy between a cation and water. Without the presence of this cooperative infuence, we would expect $a_+ = 0$ (see also Fig. 6). Cooperativity factors a_+,

---

[1] See also Fig. 7 and Ref. [22]

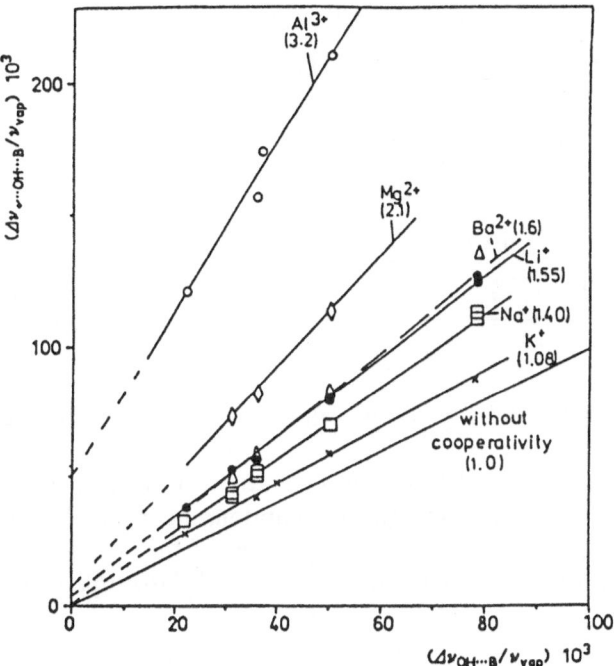

Fig. 6. Correlation between the relative shifts $\Delta v_{+\ldots OH\ldots B}/v_{vap}$ induced by the cations of K-, Na-, Li-, Ba-, Mg-, and Al-salts and the relative shift $[(\Delta v_{OH\ldots B}/v_{vap}$; see Eq. (6)] of OH...B complexes at 25°C. The values of the slopes $(1 + a_+)$ (indicated in *brackets*) and intercepts b_+ of the linear correlations are summarized in Table 2

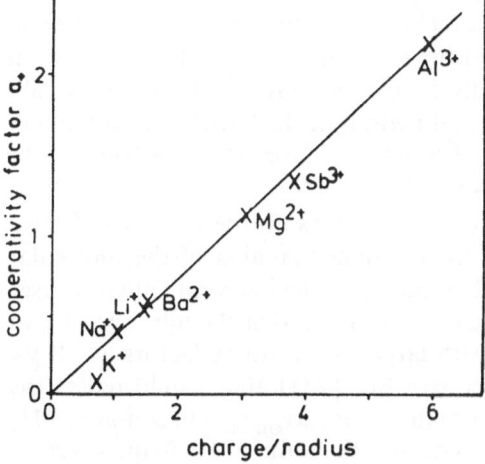

Fig. 7. The cooperativity factor a_+ is compared with the ratio charge/radius of the cations (see Table 2)

Table 2. Summary of experimental properties of cation...OH_2 interactions

Cation	Al^{3+}	Sb^{3+}	Mg^{2+}	Ba^{2+}	Li^+	Na^+	K^+
Cation radii (Å):	0.51	0.76	0.66	1.34	0.68	0.97	1.33
Charge Radius	5.88	3.94	3.03	1.49	1.47	1.03	0.75
Intercept value b_+:	0.051	—	0.007	0.001	0.003	$\simeq 0.001$	—
Cooperativity factor a_+:	2.2	$\simeq 1.3$	1.1	0.61	0.55	0.40	$\simeq 0.08$

intercept values b_+ [see Eq. (6) and Fig. 6], as well as the ratio charge/radius are summarized in Table 2.

This means, with increasing ratio charge/radius of cations the acidity of the hydrate-H_2O will increase. In order to test this assumption, the slopes $1 + a_+$ of Fig. 6 [see Eq. (6)] are compared with the pK_a values of the reaction (in water; the subscript x indicates the typical hydration number and the superscript z the charge of the cation):

$$[C(H_2O)_x]^{z+} \rightleftharpoons [C(H_2O)_{x-1}OH]^{(x-1)+} + H^+ \qquad (7)$$

in Fig. 2 (indicated by: o). This $1 + a_+$ versus pK_a correlation falls on the same line as that of $m_{w;B}(XH)$ versus pK_a.

This result indicates that the increase in the acidity of water molecules due to cation interactions (i.e., $C^+...OH$) is of the same order as the change in the acidity of ROH with varying residue R; this result means that residues R or cations C^+ may polarize the OH group in a similar manner. As a result of this polarization, the acidity of OH increases. For this increase in the acidity of the hydrate water, the slopes of Fig. 6 (i.e., $1 +$ cooperativity factor a_+) are a quantitative and simple measure [7] [compare Eqs. (6) and (4)].

The frequency shift of the OH stretching vibration of hydrate water in aqueous solutions [23, 25, 27] varies in the same manner with the cation as in ternary solutions. In aqueous salt solutions, cation...OH...OH complexes are formed. The H_2O molecules of the second hydration shell (with the subsequent hydration layers present) have a H-bond acceptor strength similar to tetrahydrofurane ($\Delta\nu_{OH...B}/\nu_{OH,vap} \simeq 0.052$ [25]).

Before discussing the cooperative effect of cations on the H-bonds of H_2O in aqueous solutions it seems useful for the understanding of the molecular effects involved to consider the possible meaning of the intercept values b_+ (see Eq. (6), Fig. 6, and Table 2). From Fig. 6 it is obvious that the intercepts b_+ at $\Delta\nu_{OH...B} = 0$ are not zero for cations with large cooperativity factors a_+. If we take these extrapolated values as given (see Ref. [23]), they would reflect the polarization of the water molecules in isolated (since $\Delta\nu_{OH...B} = 0$) cation...OH_2 complexes. Thus $b_+ \neq 0$ seems to indicate that the OH frequencies of cation...OH_2 complexes are shifted in comparison to the vapor phase frequency of the undisturbed H_2O molecule.

Results of theoretical calculations, however, should be compared with the properties of $C^+...OH_2$ complexes in the gas phase. Since we have experimentally verified the validity of Eq. (6) for $0.02 \leqslant \Delta v_{OH/D...B}/v_{OH/D,vap} \leqslant 0.08$ (see Fig. 6), we may assume that a linear extrapolation to $\Delta v_{OH/D...B}/v_{OH/D,vap} = 0$ (i.e. $\Delta v_{OH/D...B} = v_{OH/D,gas}$) reflects the change of the OH frequency of $C^+...OH_2$ complexes in the gas phase. Thus, we may learn about the changes of the properties of $C^+...OH_2$ by comparison of b_+ [see Eq. (6) and Fig. 6] and theoretical results.

Newton and Friedman [26] have calculated the harmonic OH stretching frequencies of $(H_2O)_5$cation...OH_2 complexes (i.e., complexes of cations with a primary hydration shell of six H_2O) as well as of:

$$((H_2O)...)_5\text{cation}...\text{O} \overset{\text{H}...\text{OH}_2}{\underset{\text{H}...\text{OH}_2}{\diagdown}}$$

(i.e., complexes of cations with a primary hydration shell with $6H_2O$, of which one water molecule forms H-bonds to two H_2O of the second hydration shell). As shown in Fig. 8, the frequencies of these calculations exhibit one closely

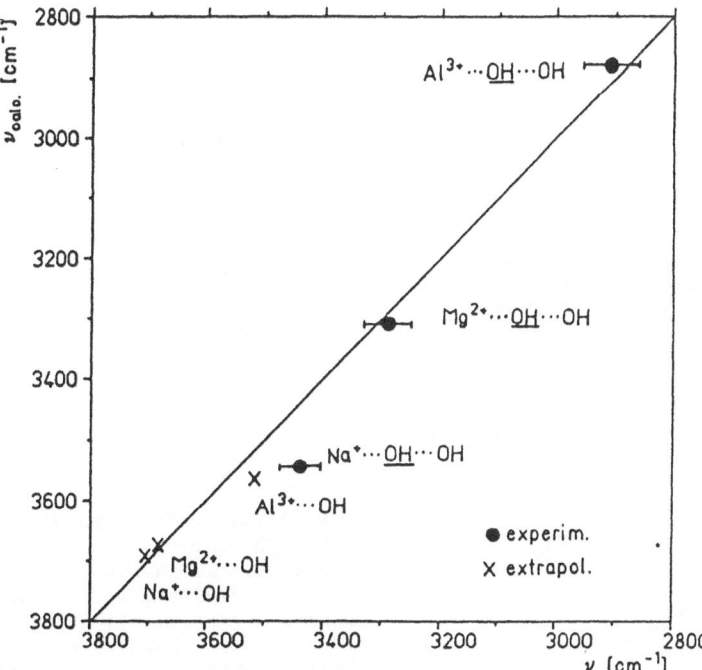

Fig. 8. Comparison between theoretically calculated [26] (v_{calc}) and experimentally determined (v) OH-streching frequencies of cation...OH_2 complexes (see text). X: extrapolated cation shift [see Fïg. 6 and Eq. (6)]; •: experimentally determined in aqueous salt solutions [25, 27]

linear correlation with b_+ values as well as the experimental frequencies of hydrate-water of cations in aqueous solutions. The closeness of the points to the straight line (corresponding to the ideal correlation in Fig. 8) speaks in favor of the accuracy of the calculations and the usefulness of the linear extrapolation of the cation-induced shifts to $\Delta\nu_{OH...B} = 0$ (see Fig. 6).

Thus, we may conclude that the cooperative effect of cations on the H-bonds of water molecules—in ternary solutions or in aqueous salt solutions—is due to the mutual polarization of H_2O by cations as well as the H-bond acceptors. The stronger H_2O is polarized by one of these interactions, the larger will be the effect of the polarization of the second interaction. This "twofold" polarization is called the cooperative effect as discussed originally by Frank and Wen [9]. This may imply that the cation-water interaction energy would depend on the H-bond acceptor interacting with the water molecule, too. This is supported by the dependence of the mutual cooperative influence of both OH groups in complexes like OH...OH...B on the H-bond acceptor B (see

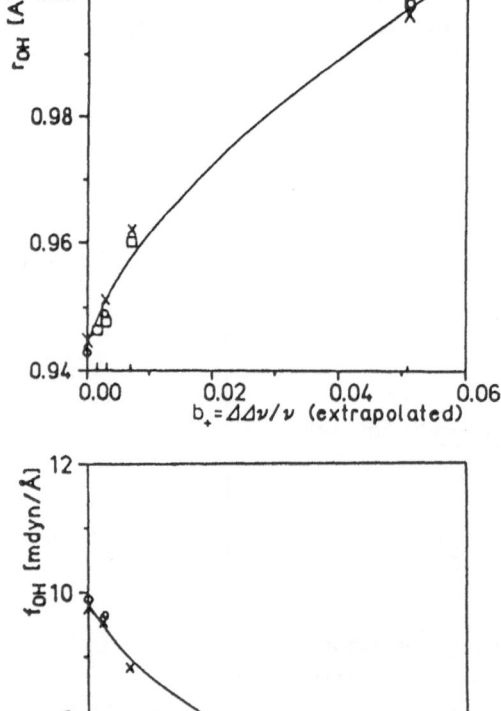

Fig. 9. The theoretically calculated O—H bond length r_{OH} (upper part) and the O—H streching force constant f_{OH} (lower part) of free H_2O and of $C^+...OH_2$ complexes is compared with the extrapolated intercept values (b_+) of Fig. 6. Theoretical data were taken from Hermansson, Olovsson and Lunell [37] ($\times$), Falk, Flakus, and Boyd [38] ($\bigcirc$) as well as Beyer, Karpfen, and Schuster [39] ($\square$)

below). If the ratio charge/radius of cations is sufficiently large, we expect that the polarization of H_2O is so large that dissociation of the O—H bond may occur. This is supported by the increase in the dissociation constant of hydrate water of cations with the cooperativity factor a_+ in aqueous solutions [7] and the formation of hydroxides or (cation-oxygen)$^{(z-2)+}$ ions for highly charged, small cation (like: $(TiO)^{2+}$).

In Figs. 9 and 10 theoretical results on properties of the O—H-bond are compared with b_+. If we assume that b_+ corresponds to the OH frequency of isolated $C^+...OH_2$, we may conclude that the OH force constant f_{OH} decreases (see Fig. 9, lower part) and correspondingly the OH-bond length r_{OH} (see Fig. 9, upper part) increases with b_+.

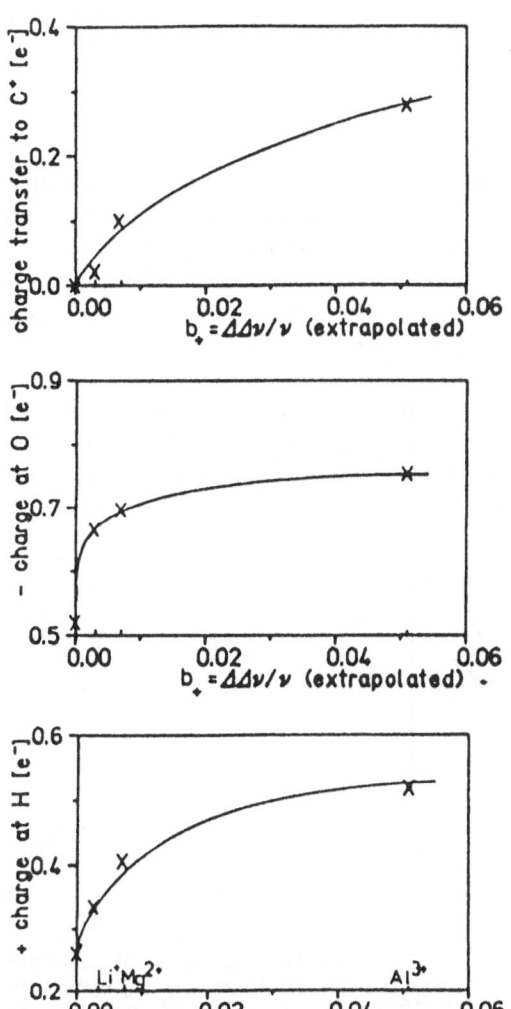

Fig. 10. The theoretically calculated change in the charge distribution of $C^+...OH_2$ complexes (upper part: charge transfer to the cation; middle: increase in negative charge at the oxygen; lower part: increase in positive charge at the hydrogen) is compared with the extrapolated intercept values (b_+) of Fig. 6. Theoretical data were taken from Hermansson, Olovsson, and Lunell [37] ($\times$)

In Fig. 10 the charge distribution of $C^+\ldots OH_2$ according to theoretical calculations [37] is considered. Apart from a slight transfer of electron density from H_2O into the direction of the cation (Fig. 10, upper part), the major changes are reflected by a considerable increase in positive charge density in the region of the hydrogen (Fig. 10, lower part) and in negative charge at the site of the oxygen (Fig. 10, middle) already for small values of b_+.

These theoretical results indicate that cations polarize the electrons of H_2O in such a manner that the O—H bond becomes longer, its force constant decreases and, consequently, the proton is more acidic (i.e., potentially forms stronger H-bonds with a given electron donor) than in isolated H_2O molecules.

As another consequence of the cooperative influence of cations on H_2O interactions, we may expect a cation dependence of the O—O distance R_{oo} between H_2O of the first and the second hydration shell. From the comparison of R_{oo} and shifts of the OD stretching frequencies ν_{OD} of crystalline hydrates, Mikenda [28] found:

$$R_{oo} = (\ln(\Delta\nu_{OD}/B))/A \qquad (8)$$

where A and B are constants for the O—O interaction. Using the OD stretching frequencies assigned to the hydrate water of cations in binary aqueous salt solutions [25,27], we may calculate the O—O distances $R_{oo}(calc)$ which we would expect between O-atoms of H_2O of the first and second hydration shell in solutions with the help of Eq. (8). In Fig. 11 these values $R_{oo}(calc)$ are compared with O—O distances $R_{oo}(exp)$ obtained by neutron or x-ray diffraction studies [29, 30], of aqueous salt solutions.

The full line in Fig. 11 corresponds to full agreement between both sets of data: $R_{oo}(calc) = R_{oo}(exp)$. The closeness of the points to this line proves the usefulness of Eq. (8) for the estimation of O—O distances in aqueous solutions

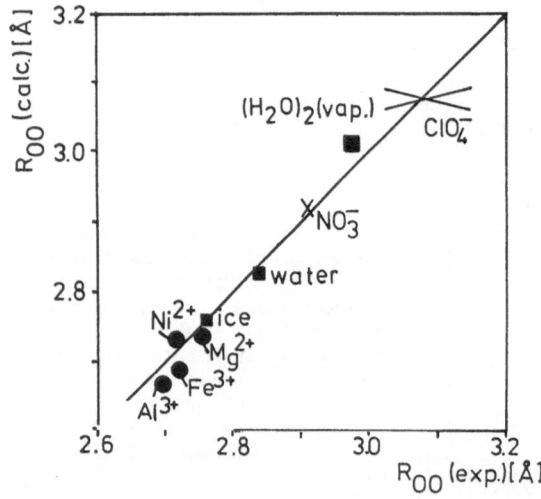

Fig. 11. Comparison of calculated [($R_{oo}(calc)$; see Eq. (8)] and experimental [29–30] ($R_{oo}(exp.)$) O—O distances of aqueous systems

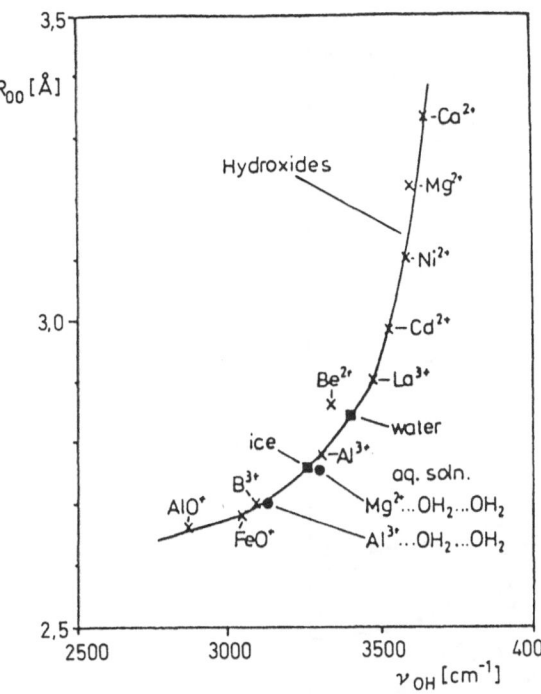

Fig. 12. Different OH...OH interactions (like in crystalline hydroxides, water, ice, or salt-hydrates in solution) may be described with the same $R_{OO}-\nu_{OH}$ correlation, which is very similar to the one found for crystalline hydrates [28] [see also Eq. (8)]

from OD stretching frequencies. The agreement of O—O distances between the first and the second hydration shell of cations supports the idea of a cation-induced strengthening between the hydration shells as discussed above. This conclusion stresses the importance of cooperative effects in aqueous solutions.

A strong cation dependence of the OH-frequency as found in aqueous solution may already be expected from the data of crystalline hydroxides (see Fig. 12). Figure 12 indicates that the hydrate-H_2O of magnesium- and aluminium-ions corresponds to about the same O—O distance R_{oo} versus OH-frequency ν_{OH} correlation as found for the interaction of cations in crystalline hydroxides (i.e., to cation...O—H—...OH-complexes). A common correlation seems to be approximately valid for OH...O interactions generally.

3 Cooperative Effects in XH...XH...B-Complexes

3.1 Proton Donor: OH-Groups

In solutions of 1,4-butanediol in CCl_4 at low concentrations ($C_{diol} \simeq 0.003$ mol/L) a large number of alcohol molecules form intermolecular H-bonds.

Due to the favorable distance of the two OH-groups in 1,4-butanediol the intramolecular H-bonds are closely linear. The spectrum of this diol in CCl_4

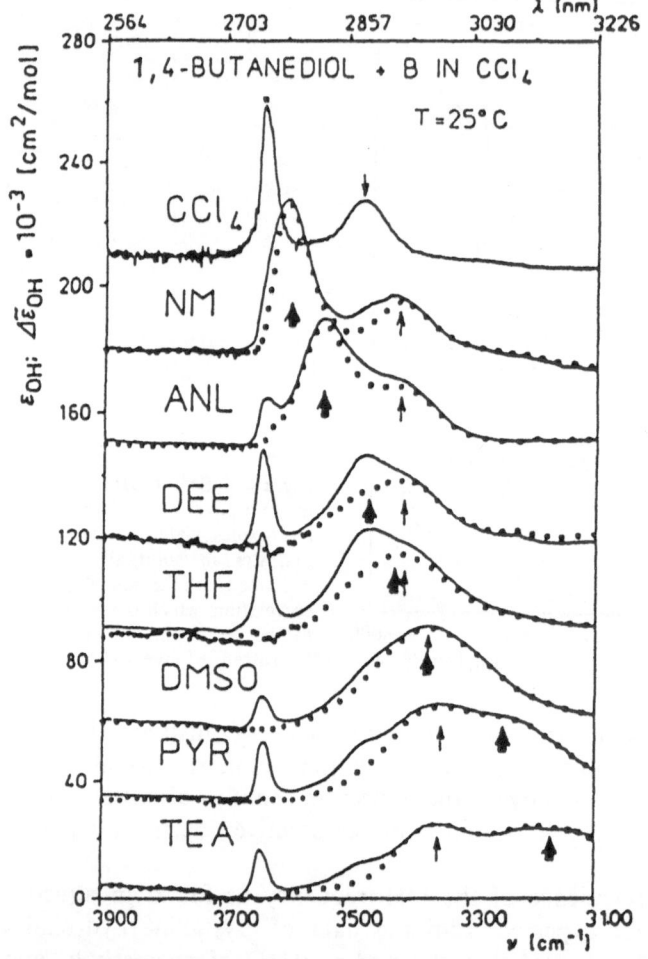

Fig. 13. IR spectra in the region of the OH streching vibration of 1,4-butanediol (0.003 mol diol/L) in CCl_4 without H-bond acceptors B added (at the top) and in the presence of B in CCl_4 (*full lines*, ε_{OH}) at 25°C. *Dotted spectra*: weighted difference spectra [$\Delta\bar{\varepsilon}_{OH}$; see Eq. (5)]. ν_{free}, ν_X and ν_B are indicated by *, ↑ and ↑, respectively. For abbreviations see Fig. 14

(Fig. 13; top) shows two bands which are assigned to free OH and the OH-group involved in the intramolecular H-bond at about 3637 cm^{-1} and 3476 cm^{-1} respectively [23, 24, 31]. OH-groups not involved in H-bonding at all as well as open-end groups (i.e., …OH) will contribute to the absorbance in the region of the free OH on behalf of the equilibrium:

$$
\underset{\substack{\sim 3637 \\ \nu_{free}}}{H-O} \overset{(CH_2)_4}{\diagup} \underset{\substack{3476 \\ \nu_{Ox}}}{O-H} \rightleftharpoons \underset{\substack{\sim 3626 \\ \nu_{open\ end}}}{O-H…OH} \overset{(CH_2)_4}{\diagdown}
$$
$$\nu_{OH}\,(cm^{-1}): \tag{9}$$

If H-bond acceptors B are added to a solution of 1,4-butane-diol in CCl_4, complexes like:

$$(CH_2)_4$$

$$O-H...O-H...B$$

$$\nu_{OH}: \quad\quad \nu_{Ox} \quad\quad \nu_B$$

may be formed. This is evident from the OH spectra of the ternary solutions (see Fig. 13). Two new bands appear in the presence of base B at the expense of the bands present in the binary solutions.

In order to separate the spectra of the complexes containing two H-bonds from that of isolated 1,4-butanediol molecules, we subtracted [see Eq. (5)] a fraction α' of the spectrum of 1,4-butanediol in CCl_4 (see Fig. 13; upper spectrum) from the spectra of the ternary solutions (diol in CCl_4 + B). The condition of the choice of α' was, that the extinction coefficient ε_{OH} in the region of ν_{free} (3637 cm^{-1}) becomes about zero. In these difference spectra ($\Delta\tilde{\varepsilon}_{OH}$; dotted lines in Fig. 13), two separate bands are usually present. Only for H-bond acceptors B of intermediate acceptor strength (namely tetrahydrofurane THF and dimethylsulfoxide DMSO in Fig. 13) one new broad, more-or-less asymmetric and is observed.

Since the OH-frequencies of 1,4-butanediol and 1-butanol in B correlate linearly with a slope of 1.0, we may use the frequencies of complexes of 1-butanol with B ($C_4H_9OH...B$) in CCl_4 for comparison.

In Fig. 14 the frequencies of the diol spectra of Fig. 13 (dotted spectra) are compared with the frequencies ν_{BuOH} of OH...B complexes of 1-butanol in CCl_4 + B.

Two sets of frequencies are obvious in Fig. 14:

—in the first one the frequencies in the diol spectra correlate linearly with ν_{BuOH} with a slope of 1.2. These bands are due to the OH-vibrations of O—H...B in OH...OH...B complexes; these frequencies are abbreviated: ν_B. ν_B depends on the presence of the H-bond between OH...OH; ν_B is shifted by about 20% due to the presence of OH...OH (without cooperativity the slope of ν_B in Fig. 14 would be 1.0). The frequency shift $\Delta\nu_B$ of diol-complexes interacting with different bases B is given by:

$$\Delta\nu_B = (1 + a_B)\Delta\nu_{OH...B} \tag{10}$$

—in the second correlation, the frequencies ν_{OX} in the diol spectra correlate linearly with ν_{BuOH} with a slope of 0.2 and are assigned to the OH vibration of OH...OH...B. The frequencies ν_{OX} depend on the H-bond acceptor strength of B i.e., on the strength of the OH...B interaction (if there were no mutual interactions the slope would be 0.0). The frequency shift $\Delta\nu_{OX}$ in

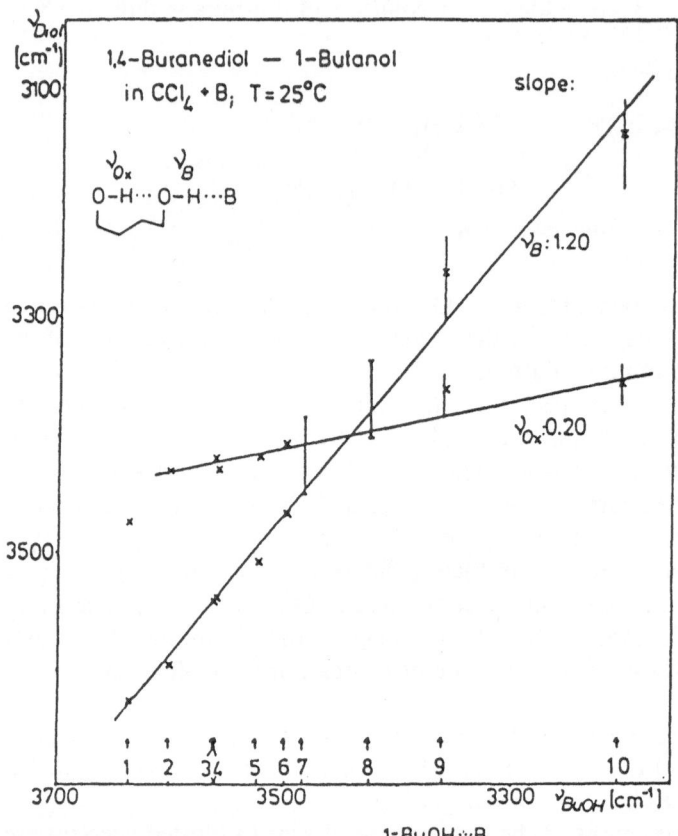

Fig. 14. The OH frequencies of 1,4-butanediol in OH...OH...B complexes as obtained from the weighted difference spectra (see Fig. 13) are compared with OH-frequencies v_{BuOH} of 1-butanol-base complexes. v_{Ox} and v_B refer to the left and right OH group in OH...OH...B complexes, respectively. Abbreviations and solvent numbers: CCl_4 & 1: tetrachloromethane; NM & 2: nitromethane; ANL & 3: cyanomethane; AEE & 4: acetic acid ethyl ester; AON & 5: acetone; DEE & 6: diethylether; THF & 7: tetrahydrofurane; DMSO & 8: dimethylsulfoxide; PYR & 9: pyridine; TEA & 10: triethylamine

Fig. 14 depends on the frequency shift of OH...B complexes according to:

$$\Delta v_{OX} = a_{OX}\Delta v_{OH...B} + \Delta v_{OX}(vap) \tag{11}$$

($\Delta v(vap)$) corresponds to the extrapolated frequency shift for $\Delta v_{OH...B} = 0$). We call a_{OX} and a_B the cooperativity factors of the OH-groups involved in the formation of OH...OH...B complexes.

In the region were both linear correlations (see Fig. 14 and 13) cross, it is difficult to separate both bands experimentally.

Table 3. Summary of the cooperativity factors a_B and a_x, the free X—H stretching frequencies v_{free} as well as the extrapolated frequency shift $\Delta v_x(vap)$ used for the calculations of the shifts involved in the formation of oligomers is also included. The frequencies and frequency shifts are given in cm^{-1}

Proton donor	Solvent	a_B	a_x	v_f	$\Delta v_x(vap)$
H_2O	base, 25°C	0.31	0.17	3707	192
Methanol	base, 25°C	0.33	0.11	3683	209
1,4-Butanediol	base, 25°C	0.22	0.1	3682	248
1,4-Butanediol	CCl_4 + base, 25°C	0.2	0.2	3682	240
Average for OH:		0.25	0.17	—	223
1,4-Butanediol	argon, 5 K	~0.25	~0.25	3682	170
FH	argon, 12 K	0.64	0.34	3962	117
FH	vapor,	0.80	0.35	3962	107
Average for FH:		0.72	0.34	—	112
ClH	argon, 12 K	0.26	0.24	2885	75
BrH	argon, 12 K	0.27	0.26	2559	60

For the sake of clarity it is usefull to study systems where OH...B and OH...OH...B complexes of the same alcohol may be compared. This is possible, for example, by investigating the concentration dependence of the spectra of methanol or water in aprotic solvents B [7, 23, 24, 31]. The results of these experiments are summarized in Table 3.

Comparative experiments at low concentrations of 1,4-butanediol and methanol in aprotic solvents B showed that the H-bond cooperativity is present in these systems to a similar extent as in ternary solutions in CCl_4. [7, 23, 31b] The results are included in Table 3.

3.2 Proton Donor: Hydrogen Halides

In Fig. 15 the results of infrared studies in the vapor phase [19] and in argon matrices at 12 K [32] are summarized for FH...FH...B complexes. Analogous correlations as in the case of OH-donors are observed (compare Figs. 15 and 14) with a_x and $(1 + a_B)$ values according to the equations analogous to Eqs. (10) and (11) for the case of OH proton donors (the meaning of v_F and v_B for FH-complexes corresponds to that of v_{OX} and v_B for OH-complexes, as given above; see Fig. 15) of 0.34 and 1.72, respectively (see Table 3). Thus, the cooperativity of FH interactions is considerably larger than that of OH (see Table 3).

For HCl and HBr the cooperativity may be estimated from investigations of XH...XH...B complexes in argon matrices as well. The corresponding cooperativity factors are included in Table 3.

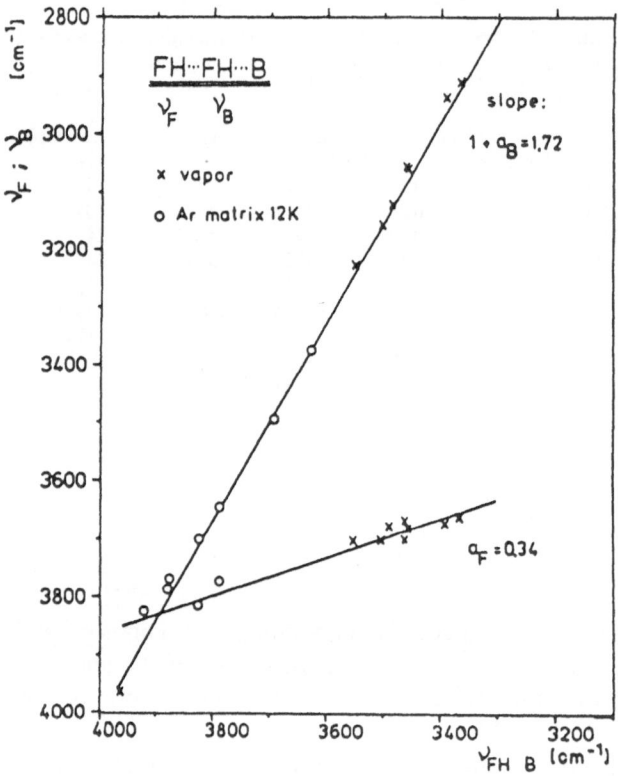

Fig. 15. The FH streching frequencies of FH...FH...B complexes (ν_F and ν_B) are compared with the FH...B frequencies in the vapor phase [19] (×) and is argon matrices [32] (○)

4 Model Considerations

4.1 A Simple Model for the Estimation of X—H Stretching Frequencies of H-Bonded Chains

So far we have discussed the cooperative effect of H-bonded complexes of three interacting molecules or ions as determined experimentally from infrared spectra. Obviously it would be extremely useful for the understanding of condensed matter to have an estimation of the mutual influence of four or more interacting molecules. We will try to extrapolate the experimental results described above for linear H-bonded complexes in the following.

On subsequent additions of XH to another XH or XH-complex we can imagine the formation of linear H-bonded chains:

$$\text{XH} \xrightarrow{+\,\text{XH}} \text{XH}\ldots\text{XH} \xrightarrow{+\,\text{XH}} \text{XH}\ldots\text{XH}\ldots\text{XH} \xrightarrow{+\,\text{XH}} \text{XH}\ldots\text{XH}\ldots\text{XH}\ldots\text{XH}$$

$$v_f \qquad\quad v_{dim}\ \ v_{f,2} \qquad\quad v_{e,3}\ \ v_{2,3}\ \ v_{f,3} \qquad\quad v_{e,4}\ \ v_{3,4}\ \ v_{2,4}\ \ v_{f,4}$$

$$\xrightarrow{+(n-4)\text{XH}} \text{XH}\ldots\text{XH}\ldots\text{X}-\text{H}\ldots\text{XH}\ldots\text{XH}$$

$$v_{e,n}\quad v_{n-1,n}\ \ v_{m,n} \qquad v_{2,n}\quad v_{f,n} \tag{12}$$

The abbreviations for the frequencies of the different XH-groups in linear n-mers are indicated in Eq. (12) below the complexes. The first index refers to the position of XH in the chain and the second to the number of XH-groups in the complex.

We want to estimate now how the frequencies of XH vary with the degree of association, n, and the position of XH in the chain using the information contained in the so-called cooperativity plots (Figs. 14, 15).

As an estimate for the H-bond acceptor strength of the XH monomer with respect to the formation of an H-bond with another XH as the proton donor, the extrapolated frequency shift $\Delta v_x(\text{vap})$ may be used: $\Delta v_x(\text{vap})$ is the frequency difference between v_f (the XH stretching frequency of the XH monomer in the gas phase) and the extrapolated value of v_x for $\Delta v_{\text{XH}\ldots\text{B}} = 0$; i.e., for vanishing influence of the environment (see Figs. 14, 15 and Eq. (11). $\Delta v_x(\text{vap})$ is used for simplicity of calculation; experimentally it is usually close to the shift of the H-bonded XH-group of the dimer XH...XH.

According to Eq. (10) obtained from the cooperativity plots, which may be written for general application in the form:

$$\Delta v_B = (1 + a_B)\Delta v_{\text{XH}\ldots\text{B}} \tag{13}$$

we may calculate now Δv_B taking for $\Delta v_{\text{XH}\ldots\text{B}} = \Delta v_x(\text{vap})$ in order to get the frequency shifts involved in the formation of the XH trimer. This first step gives: $\Delta v_B = \Delta v_{2,3}$ (see above):

$$\Delta v_{2,3} = \Delta v_B(1, \text{step}) = (1 + a_B)\Delta v_x(\text{vap}) \tag{14}$$

Using Eq. (11) in the general form:

$$\Delta v_X = a_X \Delta v_{\text{XH}\ldots\text{B}} + \Delta v_X(\text{vap}) \tag{15}$$

we yield in a similar manner as above (i.e. we take: $\Delta v_{\text{XH}\ldots\text{B}} = \Delta v_x(\text{vap})$):

$$\begin{aligned}
\Delta v_{e,3} &= \Delta v_X(1, \text{step}) = a_X \Delta v_{\text{XH}\ldots\text{B}} + \Delta v_X(\text{vap}) \\
&= a_X \Delta v_X(\text{vap}) + \Delta v_X(\text{vap}) \\
&= (1 + a_X)\Delta v_X(\text{vap})
\end{aligned} \tag{16}$$

For the following steps we take for $\Delta v_{\text{XH}\ldots\text{B}}$: $\Delta v_{e,3}, \Delta v_{e,4}\ldots$, and so on, and thus obtain for the shift of the last but one XH-group of the H-bonded n-mer:

$$\Delta v_{n-1,n} = (1 + a_B)\Delta v_{e,n-1} = (1 + a_B)\Delta v_X(vap) \sum_{i=0}^{n-3} a_X^i \qquad (17)$$

and for the last (X—H-group with lone electron pairs):

$$\Delta v_{e,n} = a_X \Delta v_{e,n-1} + \Delta v_X(vap) = \Delta v_X(vap) \sum_{i=0}^{n-2} a_X^i \qquad (18)$$

The advantage of this calculation—which is based on the consequent application of the experimental results obtained for the XH...XH...B complexes—for the cooperativity in H-bonded chains is its simplicity for the calculation of the last and last but one XH-group in the chain (i.e., $XH_{e,n}$ and $XH_{n-1,n}$).

This means: $v_{2,n} = v_{2,3}$; $v_{3,n} = v_{3,4}$; $v_{4,n} = v_{4,5}$, and so on, or generally: with the exception of the XH-groups $XH_{e,n}$ and $XH_{n-1,n}$, the cooperative effect seems to be independent from the chain length.

For the estimation of the XH-streching frequency $v_{m,n}$ in the middle of a H-bonded chain [see Eq. (12)] we make the (very reasonable) assumption that $\Delta v_{m,n}$ may be estimated from Eq. (17) (corresponding to the last but one XH-group) by taking the cooperativity factor a_B (of the central XH-group of XH...XH...B complexes) as a measure for the cooperative influence on the formation of H-bonded chains. Thus we get for $\Delta v_{m,n}$:

$$\Delta v_{m,n} = (1 + a_B)\Delta v_{m,n-1} = (1 + a_B)\Delta v_x(vap) \sum_{i=0}^{n-3} a_B^i. \qquad (19)$$

According to Eqs. (17–19) we have calculated $\Delta v_{n-1,n}, \Delta v_{e,n}$, and $\Delta v_{m,n}$ for all proton donors for which experimental values for the cooperativity factors have been obtained. In Figs. 16 and 17 the corresponding frequency shifts are plotted for increasing chain ength for the case of OH-groups (with the average values taken: $a_x = 0.17$, $a_B = 0.25$, and $\Delta v_{OH}(vap) = 223$ cm^{-1}) and of HF ($a_x = 0.34$, $a_B = 0.72$, and $\Delta v_{FH}(vap) = 112$ cm^{-1}).

As far as experimental values for linear H-bonded complexes of these proton donors have been found, they are included in Fig. 17 (by: "—"). The experimental frequency shifts fit very well to those calculated; especially the agreement between the data of infinitely long chains and crystalline —OH or HF is excellent.

These results show that the frequency shift which is due to cooperativity is expected to reach a constant value for OH-groups (i.e., beyond the pentamer; see Fig. 16a) and rather gradually in the case of HF (i.e., for $n \simeq 15$; see Fig. 16b) for linear chain association.

One major conclusion which we may draw is that the difference between the frequency of an XH...XH-dimer and an XH-polymer seems to be due to the presence of the mutual cooperative effect involved in H-bond formation.

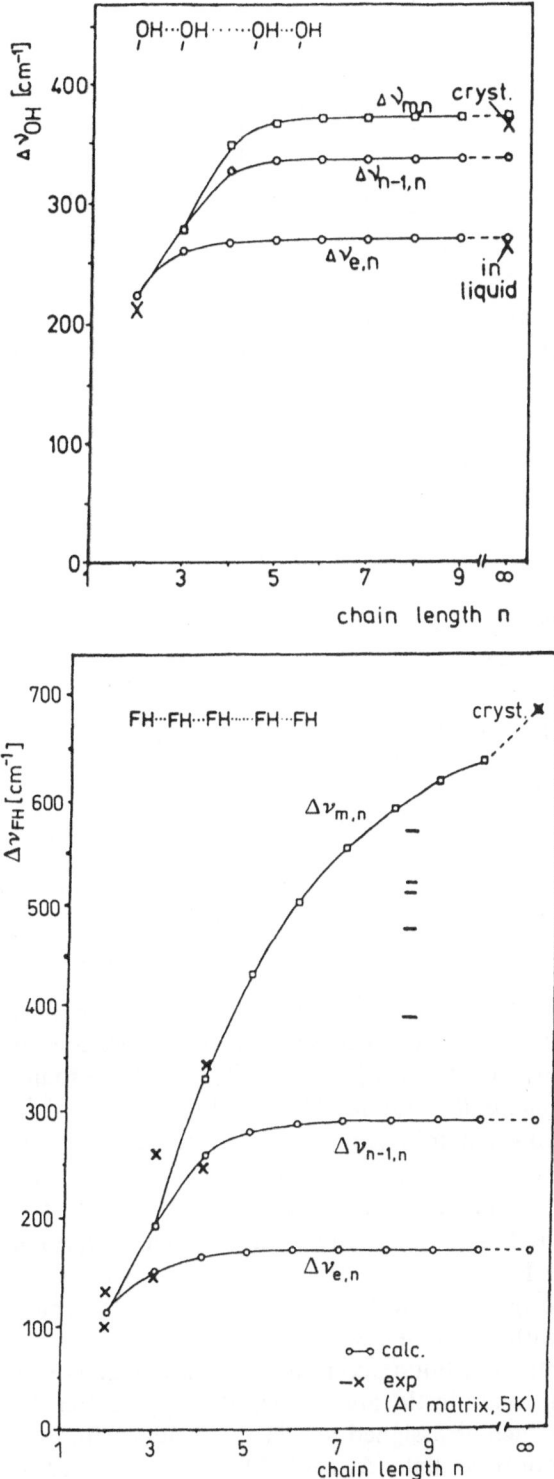

Fig. 16. Dependence of the calculated frequency shifts (see Eqs. (9, 10, and 11)) for the formation of linear H-bonded chains (n-mers) of water (or alcohols). For comparison the OH frequency shifts in the solid and liquid state are included in the figure

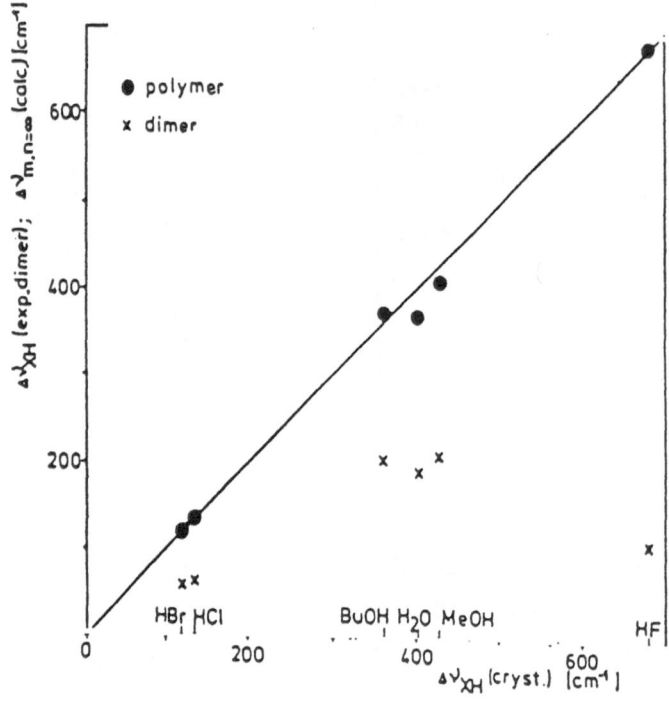

Fig. 17. Dependence of the calculated frequency shifts [see Eq. (9, 10, and 11)] for the formation of linear H-bonded chains (n-mers) of FH. For comparison, the FH frequency shifts in the solid stock and of matrix studies of certain n-mers (x) and other oligomers (-) are included in the figure

4.2 Comparison with Experiment

The frequency shifts calculated according to the above considerations may be compared with experimental frequency shifts (see Fig. 16) of alcanols or water. At a mole fraction of methanol $X_{ROH} = 0.01$ in dichlorethane we found $\Delta v_{dim} \simeq 216\,\text{cm}^{-1}$ and a broad maximum at about $3400\,\text{cm}^{-1}$ ($\Delta v \simeq 280\,\text{cm}^{-1}$). Very probably, n-mers with $3 < n < 5$ dominate in these solutions. For various alcanols and ice, Δv_{OH} in the crystalline state is about $380\,\text{cm}^{-1}$. In solid t-butanol—which crystallizes at room temperature—we found $\Delta v \simeq 360\,\text{cm}^{-1}$ [31b]. These values are in the expected range for long-chain associates according to the considerations above (see Fig. 16).

From temperature difference spectra of water, a band at about $3500\,\text{cm}^{-1}$ is assigned to OH end-groups with free lone electron pairs [23].

By addition of small amounts of H-bond acceptors to liquid methanol or HOD, Symons could show an increase in the absorbance at $\Delta v_{OH} \simeq 280\,\text{cm}^{-1}$ [16] or $240\,\text{cm}^{-1}$, respectively, which he assigned to OH end-groups with lone free electron pairs. The average of these shifts $\Delta v_{OH} \simeq 260\,\text{cm}^{-1}$ is included in

Fig. 16 (indicated by "liquid") and agrees very well with the prediction of the above considerations.

For the self-association of HF, the available experimental frequencies [19, 32] are included in Fig. 17 (these experimental values are indicated by X in Fig. 17). The frequency shifts assigned to linear trimers and tetramers as well as of crystalline HF agree exceptionally well with our predictions.

For bands which could not be assigned clearly for HF oligomers (indicated by "—" in Fig. 17), the above calculations may help to make an assignment possible.

In summary, this comparison shows that the predictions made by the simple model are in excellent agreement with the experimentally determined frequency shifts.

For a clear demonstration of the accuracy of the frequency shifts predicted by Eq. (19) for infinite, linear H-bonded chains, a comparison of the calculated shifts $\Delta v_{m,n=0}$(calc) with experimental shifts for solid proton donors is shown in Fig. 18. The agreement is nearly perfect. There is no correlation, however, between the experimental shifts Δv_{XH}(exp., dimer) of the XH...XH-dimer and the shift Δv_{XH}(cryst) of the polymer; this indicates that it is absolutely necessary to consider the cooperative effects for the understanding of the spectra of protic substances in the condensed phase.

We have summarized a variety of simple properties (like dipole moments, polarizabilities, pK_a-values) of XH groups which may be discussed with respect to the strength of H-bonded associates (see also Table 1). None of these properties seems to correlate with the experimentally determined cooperativity

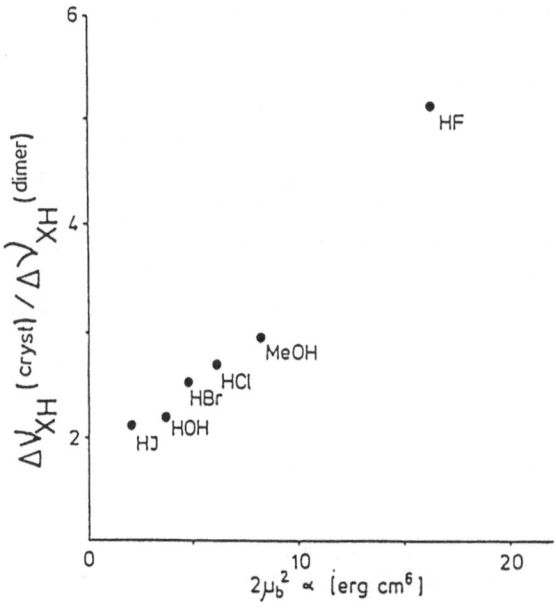

Fig. 18. Comparison of the calculated frequency shift $\Delta v_{m,n=\infty}$(calc.) (*full points*) of linear H-bond associates and of the XH...XH-dimer Δv_{XH}(exp., dimer) (indicated by: x) with the experimental frequency shift Δv_{XH}(cryst.) in solid XH. If cooperative effects were not taken into account, the frequency shift of the dimers should correlate linearly with those in the crystal with a slope of unity

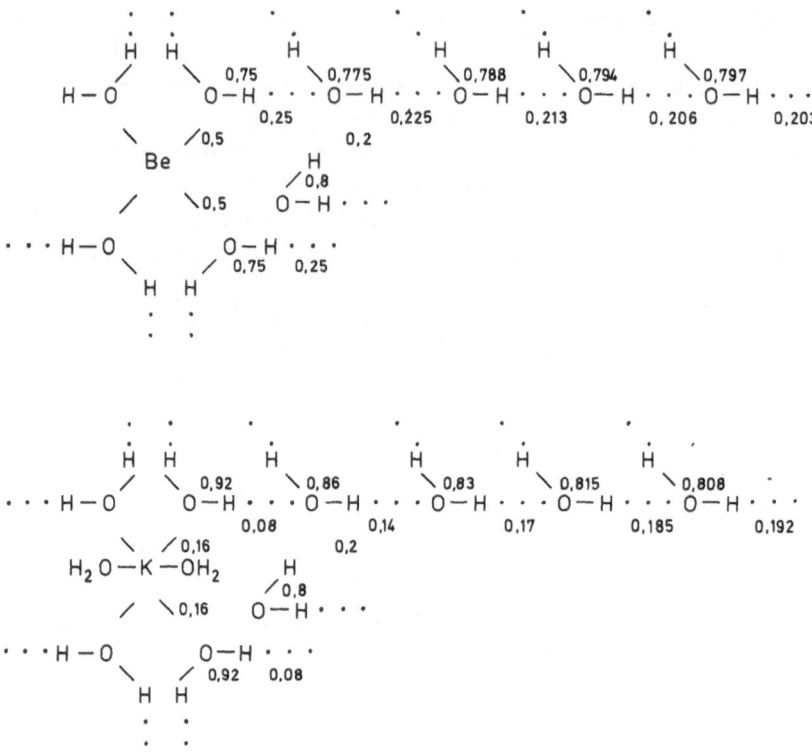

Fig. 19. Comparison of the frequency shift ratio $\Delta\nu_{XH}(\text{cryst})/\Delta\nu_{XH}(\text{dimer})$ and the attraction constant $\Phi_I = 2\mu_b^2\alpha$ (μ_b: bond dipole moment; α: polarizability) of the inductive effect of the proton donors investigated

factors with the exception [40] of the attraction constant of the inductive effect $\Phi_I = 2\mu_b^2\alpha$ (μ_b: bond dipole moment and α: polarizability of XH).

Figure 19 shows that the frequency shift ratio $\Delta\nu_{XH}(\text{cryst.})/\Delta\nu_{XH}(\text{dimer})$ of the polymers and dimers correlates very well with Φ_I. According to the comparatively small bond dipole moment and low polarizability of H_2S or NH_3, for example, we expect that the cooperative effects involved in the formation of higher associates of these compounds will be small but not negligible.

With the result of Fig. 18 we have found a connection between the experimentally determined magnitude of cooperative effects and molecular properties of the proton donors involved. The attraction constant of the inductive effect may help to estimate the size of cooperative effects for systems where experimental determinations are not yet available.

4.3 A Simple Estimation of the Influence of Cations

The fruitful discussion of cooperative effects in the last section may encourage us to find simple tools for the estimation of the cooperative influence of cations on the water structure, since the experimental results (see Chap. IX) indicate that the cooperative effect of cations are in many cases larger than those involved in the self-association of proton donors.

For a rough estimation of the importance of cations on the interactions between water molecules in their hydration shells, we may recall the simple electrostatic valence rule of Pauling [35], which has been used with great success in investigations of the structure of H-bonded crystals [36].

According to these considerations, the valence of a cation is satisfied by the interactions with the counterions or ligands which it is surrounded by. The bond valence v (dimension: valence units v.u.) for a cation interaction is given by: v = cation charge/(coordination number of the cation). The sum of all bond valences around an ion corresponds to its charge.

For beryllium-hydrate and potassium-hydrate, possible structures are schematically shown in Fig. 19a and 19b. For the cation-H_2O interactions of Be^{2+} and K^+ hydration numbers of 4 and 6 seem to be reasonable. The corresponding bond valences for Be^{2+}—H_2O or K^+—H_2O are given by $v_{Be2+} = 2/4 = 0.5$ and $v_{K+} = 1/6 = 0.16$, respectively.

We may now apply the above idea of the bond valence concept of cation-interactions to the oxygen and hydrogen of H_2O. With a valence of 2 v.u. for the oxygen of H_2O after subtraction of the bond valence of the cation-oxygen interaction there will remain $(2-0.5)/2 = 0.75$ and $(2-0.16)/2 = 0.92$ v.u., respectively, for the O—H bond of the hydrate-H_2O of Be^{2+} and K^+. Since the H-atom has a valence of 1, there will remain 0.25 or 0.08 v.u.

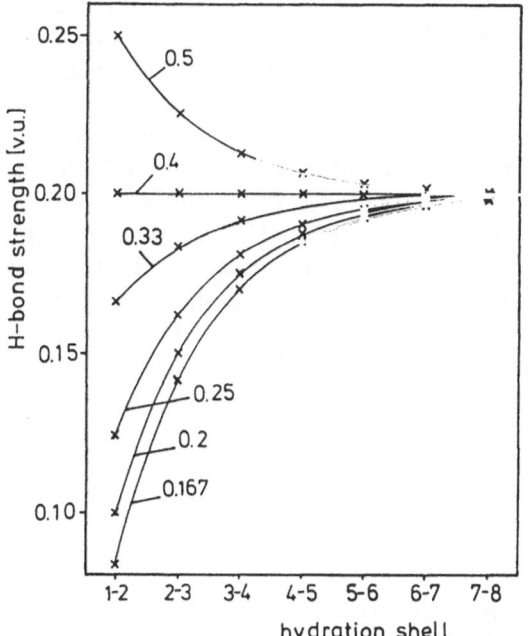

Fig. 20. According to the rigorous application of the electrostatic bond valence rule, the strength of OH...OH interactions in the hydration shells of different cations may vary considerably not only with the cation but also with the distance of the OH-group from the ion

for the H-bond between the first and the second hydration shell of Be^{2+} and K^+, respectively. For H-bonds in ice, a bond valence of 0.2 v.u. has been successfully used [36].

This result shows that some cations may increase and some may decrease the bond valence between not only the first and the second but even between the subsequent hydration shells as indicated in Fig. 20. In Fig. 21 the corresponding H-bond strength (expressed in v.u.) between subsequent hydration shells of cations is given for several ratios cation charge/coordination number. For an application in aqueous solutions, the choice of, for example, 0.15 v.u. for the average H-bond strength between H_2O molecules in water may be preferable over the value determined for ice.

The bond valence concept of Pauling predicts that cations will strengthen the H-bonds between H_2O molecules of the first and second hydration shell according to their ratio charge/coordination number. In comparison with the H-bonds formed between H_2O-polymers, the cation interaction may increase or decrease the strength of the H-bonds. According to empirical findings, cations may act as structure makers or structure breakers, i.e., they enhance or weaken the H-bond structure present in liquid water.

As shown in Fig. 20, the consequent application of the bond valence principle automatically leads to an asymptotic approximation of the H-bond strength in subsequent hydration shells to the value of water polymers. Only beyond the forth or fifth hydration shell the influence of the cation becomes so small that it may be neglected in most cases.

The result of these simple considerations indicates an increasing influence of cation on the H-bond strength of the surrounding H_2O with increasing ratio charge/coordination number. This finding is supported by the experimental findings of the cooperative effects (see Fig. 7; increase with the ratio charge/radius) involved in cation–water interactions. Both results predict the presence of structure makers and structure breakers qualitatively. In addition, they offer an alternative explanation for the far-reaching ionic effects in aqueous solutions, which are important for the understanding not only of solubilities of organic molecules in aqueous salt solutions but also of the salt effect on the structure of biologically active polymers (like enzymes, DNA, glycoproteins).

These results indicate that an understanding of the polarization of proton donors as exhibited by the presence of large cooperative effects in H-bond interactions is not only interesting from the point of view of fundamental sciences but may help to facilitate our description of the very complex systems important for life as well.

5 References

1. Luck WAP, Zheng HY (1984) JCS Faraday Trans 2 80: 1253; (1984) Z. Naturforsch 39a: 888; Kleeberg H, Eisenberg C, Zinn T, J Mol Struc (in press)

2. Eisenberg D, Kauzmann W (1969) The Structure and properties of Water, Oxford Univ Press, London
3. Hadži D, Thompson HE (eds) (1959) Hydrogen Bonding, Pergamon Press, London; Pimentell GC, McClellan (1960); The Hydrogen Bond, Freeman, San Francisco; Schuster P, Zundel G, Sandorfy C (eds) (1976) The Hydrogen Bond, Vol 1–3, North Holland, Amsterdam
4. Luck WAP, Angew (1979) Chem Int Ed Engl 18: 350; (1980) Angew Chem Int Ed Engl 19: 28
5. Zeegers-Huyskens Th (1986) J Mol Struct 135: 93
6. Zeegers-Huyskens Th (1988) J Mol Struct 177: 125
7. Kleeberg H (1988) J Mol Struct 177: 157
8. Frank HS, Wen W-Y (1957) Disc Farad Soc 24: 133
9. Schuster P, Karpfen A, and Beyer A (1980) In: Molecular Interactions (H Ratajczak and WJ Orville-Thomas; eds) p 117; John Wiley & Sons Ltd
10. Schuster P (1981) Angew Chem 93: 532
11. Schuster P (1987) Encyclopedia of Physical Science and Technology; Vol 6; S 520; Academic Press Inc; Beyer A, Karpfen A, Schuster P (1979) Chem Phys Lett 67: 369
11a. Kuring IJ, Sczésniak MM, Scheiner S (1986) J Chem Phys 90: 4253
12. Kollman PA, Allen LC (1970) J Chem Phys 52: 5085; (1970) J Amer Chem Soc 92: 753
13. Clementi E (1976) Lecture Notes in Chemistry, Vol 2, Springer Verlag, Berlin
14. Pálinkás G, Heinzinger K (1986) Chem Phys Lett 126: 251; Spohr E, Pálinkás G, Heinzinger K, Bopp P, Probst MM (1988) J Phys Chem 92: 6754; Radnai T, Pálinkás G, Szász GyI, Heinzinger K (1981) Z. Naturforsch 36a: 1076; Heinzinger K Pálinkás G (1987) In: Interactions of Water in Ionic and Nonionic Hydrates (H Kleeberg, ed) p 1, Springer–Verlag, Heidelberg
15. Tamura Y, Tanaka K, Spohr E, Heinzinger K (1988) Z Naturforsch 43a: 1103; Probst M, Bopp P, Heinzinger K (1984) Chem Phys Lett 106: 317
16. Symons MCR (1981) Acc Chem Res 14: 179; Symons MCR, Thomas VK, Fletcher NJ, Pay NG, (1981) J Chem Soc, Faraday Trans I 77: 1899; Symons MCR, Fletcher NJ, Thompson V (1979) Chem Phys Lett 60: 323
17. Huyskens P (1977) J Amer Chem Soc 99: 2578
18. Luck WAP, Schrems O (1982) Horizons in H-bond Research, Leuven, Aug 22–27
19. Couzi M, le Calvè J, van Huong P, Lascombe J (1970) J Mol Struct 5: 363
20. Baron M-H, de Lozè C (1971) Chim Phys 68: 1293, 1299; (1972) 69: 1084
21. Kuntz JD, Cheng CJJ (1975) J Amer Chem Soc 97: 4852
22. Kleeberg H, Heinje G, Luck WAP (1986) J Phys Chem 90: 4427
23. Kleeberg H (1987) In: Interactions of Water in Ionic and Nonionic Hydrates (H. Kleeberg; ed) p 89, Springer–Verlag, Heidelberg
24. Kleeberg H, Luck WAP (1989) Z Phys Chemie, Leipzig; 270: 613
25. Heinje G, Kammer T, Kleeberg H, Luck WAP (1987) In: Interactions of Water in Ionic and Nonionic Hydrates (H Kleeberg, ed) p 47, Springer–Verlag, Heidelberg
26. Newton M, Friedman HL (1987) J Chem Phys 83: 5210; Friedman HL, Pure & Appl Chem 59: 1063; the author acknowledges the communication of unpublished results of M Newton very much
27. Kristiansson O, Eriksson A, Lindgren J (1984) J Acta Chem Scand, A38: 609, 613
28. Mikenda W (1986) J Mol Struct, 147: 1
29. Caminiti R et al (1980) Z Naturforsch 35a: 1361; Caminiti R, Radnai T (1980) Z Naturforsch 35a: 1368; Caminiti R, Magini M (1979) Chem Phys Lett 61: 40; Caminiti et al (1979) Chem Phys Lett 61: 45; (1979) J Chem Phys 71: 2473; (1976) J Chem Phys 65: 3134
30. Enderby JE, Neilson GW (1979) In: Water a Comprehensive Treatise (F Franks, ed) Vol 6, p 1, Plenum Press, New York; Enderby JE (1981) Physikalische Blätter 37: 107
31a. Kuhn LP (1951) J Amer Chem Soc 74: 2492; (1954) 76: 4323; (1960) 86: 650; Luck WAP (1967) Naturwissen schatten 54: 601
31b. Kleeberg H, Klein D, Luck WAP (1987) J Phys Chem 91: 3200
32. Johnson GL, Andrews L (1982) J A Chem Soc 104: 3043; Andrews L (1984) J Phys Chem 88: 2940
33. Barnes AJ (1983) J Mol Struct 100: 259
34. Barnes AJ, Hallam HE, Scrimshaw GF (1969) Trans Farad Soc 65: 3172
35. Pauling L (1927) J Amer Chem Soc 49: 765; (1929) J Amer Chem Soc 51: 1010
36. Donnay G, Allmann, R (1970) Amer Min 55: 1003; Allmann R (1975) Monatshefte f Chemie 106: 779; Brown ID (1976) Acta Cryst A32: 24; (1978) Chem Soc Rev 7: 359; (1981) In: Structure and Bonding in Crystals (O'Keep M. and Narrotsky A, eds) Academic Press, New York

37. Hermansson K, Olovsson I, Lunell S (1984) Theor Chim A 64: 265
38. Falk M, Flakus HT, Boyd RJ (1986) Spectrochim A, 42: 175
39. Beyer A, Karpfen A, Schuster P (1984) In: Topics Curr Chem; Vol 120, p 1
40. H. Kleeberg, Habilitationsschrift, University of Marburg, Marburg (submitted)

CHAPTER XI

NMR Studies of Elementary Steps of Multiple Proton and Deuteron Transfers in Liquids, Crystals, and Organic Glasses

H.-H. Limbach

Dynamic high-resolution NMR spectroscopy of liquids and solids constitutes a convenient way to study kinetic hydrogen/deuterium isotope and solid-state effects on multiple proton transfer reactions in different environments. In the case of intramolecular double proton transfer reactions, evidence for stepwise reaction pathways is obtained; in each step only one proton jumps, whereas the other remains bound. By contrast, intermolecular double proton transfer reactions behave in a different way. Here, both protons are in flight in the rate-determining reaction step. The origin of the different behavior of both types of reactions is discussed.

It is also found that intermolecular interactions have an influence on the reaction energy surfaces even in cases where kinetic solvent effects are absent. This influence can be monitored when studying the proton dynamics in a timescale of slow molecular motion. It is well known that this timescale is easily reached by IR spectroscopy in the liquid state; it is, however, also reached by NMR spectroscopy when studying fast proton transfers in ordered crystalline or disordered amorphous solids. Recent progress in this field is reported.

1 Introduction

In the past years there has been a special interest for neutral multiple proton transfer reactions including their kinetic hydrogen/deuterium isotope effects

Intermolecular Forces
© Springer-Verlag Berlin Heidelberg 1991

[1, 2]. The simplest double proton transfer reaction of this kind is shown in Eq. (1):

$$AH^* + BH \rightleftharpoons A \overset{H^*}{\underset{H}{\cdots}} B \rightleftharpoons A \overset{H^*}{\underset{H}{\cdots}} B \rightleftharpoons AH + BH^*. \tag{1}$$

These reactions can serve as models for the study of the elementary steps of bond breaking and bond formation in different liquid and solid environments, i.e., of the role of intermolecular interactions on the dynamics of bond breaking and bond formation in condensed matter. Proton exchange reactions according to Eq. (1) are most conveniently studied by dynamic NMR spectroscopy. Using this method, rate constants of multiple inter- and intramolecular double proton transfers have been determined not only in liquid solutions [1–14] but also in the solid state [15–26]. A particular feature of neutral multiple proton transfers is that they can also be induced at cryogenic temperatures by visible light [27, 28]. Recently, also an IR-induced proton transfer has been found [29]. In organic and biochemical systems these processes are related to bifunctional catalysis and biological activity [30, 31]. On the other hand, these reactions are an old topic of theoretical chemistry [32–39] and of the theory of primary kinetic hydrogen/deuterium isotope effects [2–6, 11, 14, 40, 41].

In this paper we present an overview of the problems which are currently studied in our laboratory using liquid and solid-state NMR and IR techniques. Our interest has focused on NH...N proton transfer systems for two reasons: (i) proton transfers between nitrogen atoms are, generally, not as fast as proton transfers between oxygen atoms. They are, therefore, easier followed by NMR spectroscopy; (ii) in contrast to oxygen, nitrogen has a stable isotope ^{15}N with a spin 1/2. Thus, it is possible to obtain information on hydrogen bonding and proton transfer dynamics by studying ^{15}N-labeled compounds in the absence of undesired complications arising from interactions with nuclei having a quadrupole moment.

2 Kinetic HH/HD/DD Isotope Effects Studied by Liquid-State NMR

In order to measure kinetic hydrogen/deuterium isotope effects using conventional kinetic methods, it is necessary to perform "proton inventories" where reaction rates are measured as a function of the deuterium fraction D in the mobile proton sites [30]. In principle, the number of protons m transferred in the rate-limiting step can also be obtained from such isotopic dilution studies. However, in order to extract m from a proton inventory the validity of the "rule of the geometric mean" has, generally, been assumed. This rule states for a double proton transfer that:

$$k^{HD} = (k^{HH}k^{DD})^{1/2} \quad \text{i.e.} \quad k^{HH}/k^{HD} = k^{HD}/k^{DD}. \tag{2}$$

In order to test Eq. (2) experimentally, NMR proton inventory techniques [2–6] have been designed for the direct determination of k^{HH}, k^{HD}, and k^{DD} of inter- and intramolecular reactions. For example, by a combination of ^{1}H- and ^{2}H-NMR measurements it was possible to measure complete sets of kinetic isotope effects for the 1:1 and the 2:1 proton exchange between acetic acid and methanol in tetrahydrofuran (Fig. 1a, b) [3, 4]. Substantial deviations from Eq. (2) were observed for the 1:1 exchange in the sense that replacement of the first H atom by D resulted in a stronger decrease of the rate constants than replacement of the second H atom by D, i.e., $k^{HH}/k^{HD} = 5.1$ and $k^{HD}/k^{DD} = 3$ at 298 K. Even larger deviations, i.e.:

$$k^{HH}/k^{DD} \gg 1 \quad \text{and} \quad k^{HD}/k^{DD} \simeq 1 \text{ to } 2 \tag{3}$$

were found for the tautomerism of meso-tetraphenylporphyrin (TPP) [3] which reacts according to Fig. 2. These deviations from the RGM were originally interpreted in terms of a concerted proton transfer pathway involving tunneling [3, 4, 32].

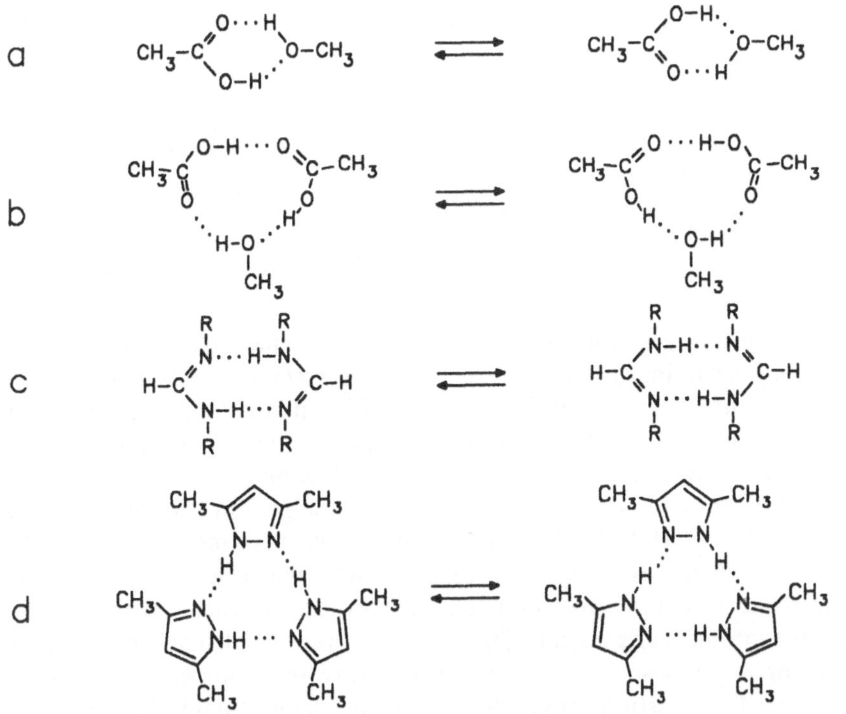

Fig. 1. Intermolecular multiple proton transfer reactions for which kinetic hydrogen/deuterium isotope effects have been studied by dynamic liquid-state NMR spectroscopy. a and b Double and triple proton transfer between acetic acid and methanol according to Refs. [3] and [4]; c double proton transfer in formamidine dimers according to Refs. [5] and [6]; d: triple proton transfer in cyclic trimers of 3,5-dimethylpyrazole according to Refs. [25] and [26]

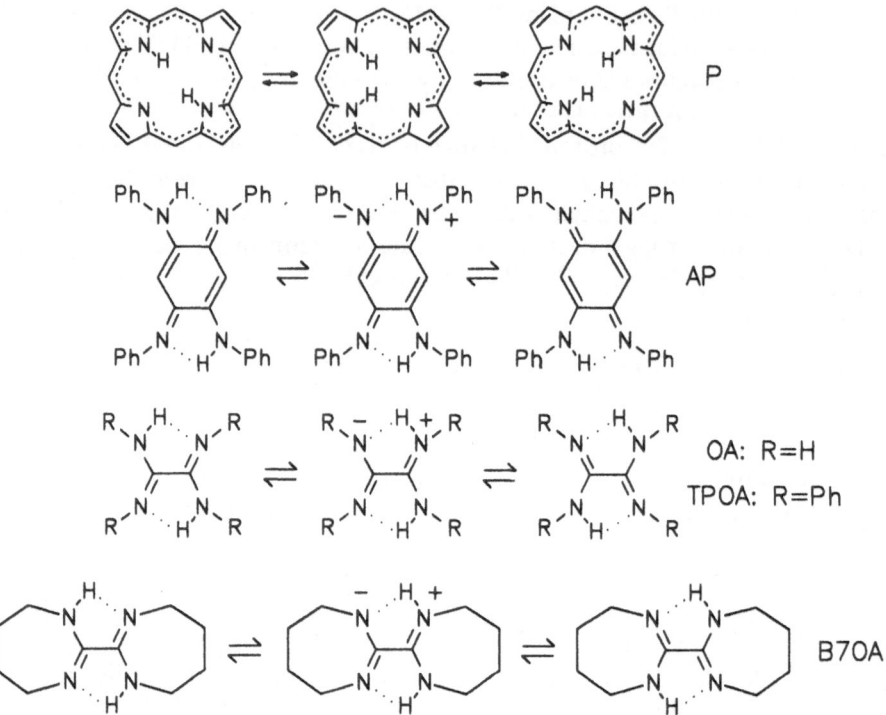

Fig. 2. Intramolecular proton transfer reactions for which kinetic hydrogen/deuterium isotope effects have been studied by dynamic liquid-state NMR spectroscopy. P: porphyrin and derivatives [3]; AP: azophenine [3, 11, 12]; OA: oxalamidine and derivatives [13], B7OA: bicyclic oxalamidine [14]. The dynamics of the tautomerism of P and AP have also be studied by NMR in the solid state (Refs. [15–17] and Ref. [12])

After these initial studies were carried out, it seemed desirable to know whether these deviations are limited to the above systems or whether they are of a general nature. Therefore, the dynamics of different inter- and intramolecular proton and deuteron transfer reactions listed in Figs. 1 and 2 were studied. The results of these studies will be reviewed in the following.

Let us first consider some intramolecular double proton transfer reactions for which full kinetic HH/HD/DD isotope effects have been measured. Actually, let us begin with the example of azophenine (AP), which is subject in liquid solution to a fast intramolecular double proton transfer involving two degenerate tautomers as shown in Fig. 2 [3]. Theoretical calculations [36] gave evidence for a stepwise proton transfer pathway involving a zwitterion as intermediate. The question arose whether this mechanism could be supported experimentally. Therefore, rate constants of this reaction were measured as a function of temperature by applying different methods of dynamic NMR spectroscopy to various isotopically labeled AP species dissolved in different organic solvents [12]. The rate constants did not depend on the dielectric constant of the solvent, which was varied between 2 (toluene) and 25 (benzonitrile). For $C_2D_2Cl_4$ as solvent, the full kinetic HH/HD/DD isotope effects were obtained at different temperatures. The observed kinetic isotope effects of $k^{HH}/k^{HD} = 4.1$ and $k^{HD}/k^{DD} = 1.4$ at 298 K indicate that Eq. (3) is fulfilled as in the case of tetraphenylporphyrin. A theory of kinetic HH/HD/DD isotope effects for degenerate stepwise double proton transfers was presented and showed that these isotope effects were consistent with the stepwise reaction mechanism. A problem with this interpretation was that during this pathway a highly polar zwitterion should be formed, which in turn should give rise to strong kinetic solvent effects on the reaction, by contrast to the experiment.

Attempts were made to resolve this problem by studying the tautomerism of the related oxalamidine (OA) system (Fig. 2). This process was first monitored in a study of the easily soluble derivative TPOA [13]. Unfortunately, TPOA is also subject in the liquid state to isomerism around the CN double bonds and to intermolecular proton exchange [13]. Therefore, the ^{15}N- and ^2H-labeled bicyclic derivative B7OA (Fig. 2) was synthesized for which rate constants including the kinetic HH/HD/DD isotope effects could be measured at 362 K by ^1H-NMR lineshape analysis [14]. Experiments were performed using methyl-cyclohexane-d_{14} (MCY) and acetonitrile-d_3 (AN) as solvents. The kinetic isotope effects were very similar in both solvents. Actually, $k^{HH}_{MCY}/k^{HD}_{MCY} = 2.4$, $k^{HD}_{MCY}/k^{DD}_{MCY} = 1.2$, $k^{HH}_{MCY}/k^{DD}_{MCY} = 3$, $k^{HH}_{AN}/k^{HD}_{AN} = 2.6$, $k^{HD}_{AN}/k^{DD}_{AN} = 1.3$, $k^{HH}_{AN}/k^{DD}_{AN} = 3.5$ at 362 K. Similar values had been obtained for AP at the same temperature [12]. They are again consistent with a stepwise double proton transfer mechanism as shown in Fig. 2. In the case of B7OA this interpretation could, however, be supported by the observation of a substantial kinetic solvent effect of $k^{HH}_{AN}/k^{HH}_{MCY} = 4.5$ at 362 K, indicating a polar transition state as expected. Thus, both the observed kinetic HH/HD/DD isotope and solvent effects point in the same direction. Or, stated in a another way, the observation that Eq. (3) is fulfilled in a degenerate double proton transfer reaction can be taken as a criterion for a stepwise reaction mechanism. This result is important in cases like porphyrins or azophenine where kinetic solvent effects are absent. In these cases the study of the kinetic HH/HD/DD isotope effects represents the only way to establish experimentally a stepwise double proton transfer mechanism.

The absence of kinetic solvent effects on the azophenine tautomerism could be explained in two ways: either the phenyl groups effectively shield the reaction center from the solvent in contrast to B7OA, or the intermediate is an apolar singlet biradical as discussed in Ref. [12].

These results shed also new light on the mechanism of the proton tautomerism in porphyrins, where recent theoretical studies gave evidence for a stepwise proton transfer as shown in Fig. 2 [33, 35, 37]. The previous finding [3] that Eq. (3) is well fulfilled experimentally can now be taken as evidence for the stepwise reaction mechanism. Recent measurements of the low-temperature rate constants of the porphyrin tautomerism as well as of hydro-porphyrins [10] corroborates this interpretation and supports the idea of proton tunneling in this reaction [29].

Let us discuss now the intermolecular double proton transfer between diphenylformamidine (DPFA) molecules in tetrahydrofuran (Fig. 1c). DPFA forms in THF an *s-trans* and an *s-cis* conformer which interconvert slowly on the NMR timescale. According to recent liquid-state ^{1}H-NMR results, only the *s-trans* conformer is able to form cyclic dimers in which a double proton transfer takes place according to Fig. 1c [5]. The energy of activation of this reaction is of the order of $17\,\text{kJ}\,\text{mol}^{-1}$. For the related ^{15}N,^{15}N'-di-*p*-fluorophenyl-formamidine molecule (DFFA) it was possible to measure the full kinetic HH/HD/DD isotope effects at 189.2 K [6]. First, using ^{1}H-NMR spectroscopy, a linear dependence of the inverse proton lifetimes on the deuterium fraction D in the mobile proton site was observed. From this dependence the number of protons transported in the rate-limiting step of the proton exchange was determined to be m = 2, as expected for a double proton transfer in an *s-trans* dimer with a cyclic structure. The full kinetic HH/HD/DD isotope effects on 233:11:1 at 189 K were determined through ^{19}F-NMR experiments on the same samples. This result represents a deviation from the rule of geometric means which is of the same order as in the case of acetic acid/methanol/tetra-hydrofuran [4].

It is clear that the kinetic HH/HD/DD isotope effects obtained for the intermolecular double proton transfer between formamidine molecules and in the system acetic acid/methanol are not consistent with a stepwise double proton transfer because Eq. (3) is not fulfilled. In other words, both transferred protons contribute substantially to the observed kinetic isotope effects by contrast to the intramolecular case. This is in agreement with the observation that the overall kinetic HH/DD isotope effects are larger in intermolecular than in intramolecular double proton transfer reactions. How can one then explain the small deviations from the RGM in the case of intermolecular exchange reactions? In the case of the system acetic acid/methanol/THF these deviations could be interpreted in terms of thermally activated tunneling [3, 4]. The deviations arise because tunneling enhances the reaction rates especially of the light hydrogen isotopes. Further temperature-dependent kinetic studies on the formamidine tautomerism are currently in progress in order to confirm this interpretation also for this system.

A qualitative explanation for the different reaction pathways of intra- and intermolecular double proton transfer systems has been given recently [6]. This explanation is based on the observation that intramolecular proton transfer systems such as porphyrin and azophenine lack the usual flexibility of hydrogen-bonded systems, i.e., the usual low-frequency hydrogen-bond stretching vibration [39] which modulates the hydrogen bond distance. Thus, the molecular frame of heavy atoms in these compounds is relatively rigid and a high energy would be required to reduce the hydrogen bond distance in such systems. This feature is expressed in Fig. 3a by an outer square which schemati-cally represents the molecular frame. It is understandable that it costs too much energy to break the bonds of both protons to their neighboring heavy atoms at the same time, and the proton transfer will be asynchronous. Note that

a

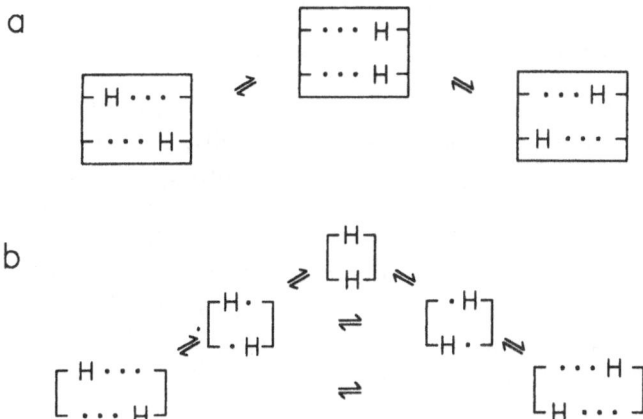

b

Fig. 3. a Stepwise double proton transfer in the case of a fixed molecular frame of heavy atoms; **b** double proton transfer in the case of variable hydrogen bond lengths according to a model proposed in Ref. [4]. Reproduced with permission from Ref. [6]

proton tunneling in this case will always require a minimum energy of activation corresponding to the energy difference between the intermediate and the initial state.

By contrast, the presence of low-frequency hydrogen-bond stretching vibrations in the flexible intermolecular proton transfer systems allows a comparatively easy compression of the hydrogen bond as schematically shown in the model of Fig. 3b. This model has been used for the calculation of the Arrhenius curves of the proton transfer between acetic acid and methanol [3, 4]. In this case the hydrogen bond lengths are variable, i.e., the energy of activation of the proton transfer is pooled into the hydrogen bond stretching vibration which shortens the hydrogen bond length. As a consequence, the barrier for the proton transfer is reduced. At extreme short hydrogen bond lengths the barrier for proton transfer vanishes and, therefore, the difference between a stepwise and a concerted proton transfer mechanism. The imaginary frequency required for a transition state corresponds then to the hydrogen bond stretching rather than to the AH-stretching vibrations. Now, it is well known that the latter are shifted to lower frequencies when the hydrogen bond distance is shortened [42]. Therefore, there will be a considerable loss of zero point energy of both vibrations in the highly compressed transition state. As a consequence, the RGM will be fulfilled at high temperatures. At lower temperatures the transfer may occur by tunneling leading to the above-mentioned deviations from the RGM. Thus, we propose for the intermolecular proton transfer systems a reaction mechanism according to Fig. 3b. Note that one might find intermolecular proton transfer systems with rigid hydrogen bond distances and intramolecular proton transfer systems with flexible hydrogen bonds which could lead to an inverse behavior of kinetic isotope effects.

3 ^{15}N CPMAS NMR Spectroscopy of Proton Transfer Systems in the Crystalline State

One major problem which arises in the theoretical interpretation of kinetic isotope effects of reactions in condensed matter, especially, if tunneling is involved as mentioned above, is the question of how the reaction mechanism is affected by intermolecular interactions in liquid, crystalline, or amorphous glassy environments. In order to obtain information on this problem, one can make use of variable temperature high-resolution solid-state NMR spectroscopy [43]. This method is applicable for spin 1/2 nuclei and exploits line-narrowing techniques such as proton decoupling and magic angle spinning (MAS) in order to remove the effects of dipolar coupling to protons and of the chemical shift anisotropy on the NMR spectra [44]. Cross-polarization (CP) from protons to the nucleus studied enhances the signal-to-noise ratio [44]. Thus, several solid-state hydrogen transfer systems have been studied by natural abundance ^{13}C CPMAS NMR [20, 25, 45, 46]. However, with the exception of hydride transfers in carbonium ions [46], carbon atoms are not directly involved in proton transfers and their NMR lines may not always be sensitive to these processes. Therefore, for NH...N or NH...X proton transfer systems ^{15}N CPMAS NMR is a more suitable method. Unfortunately, as stated above, because of the quadrupole moment of the ^{14}N nucleus, it is necessary at present to enrich the

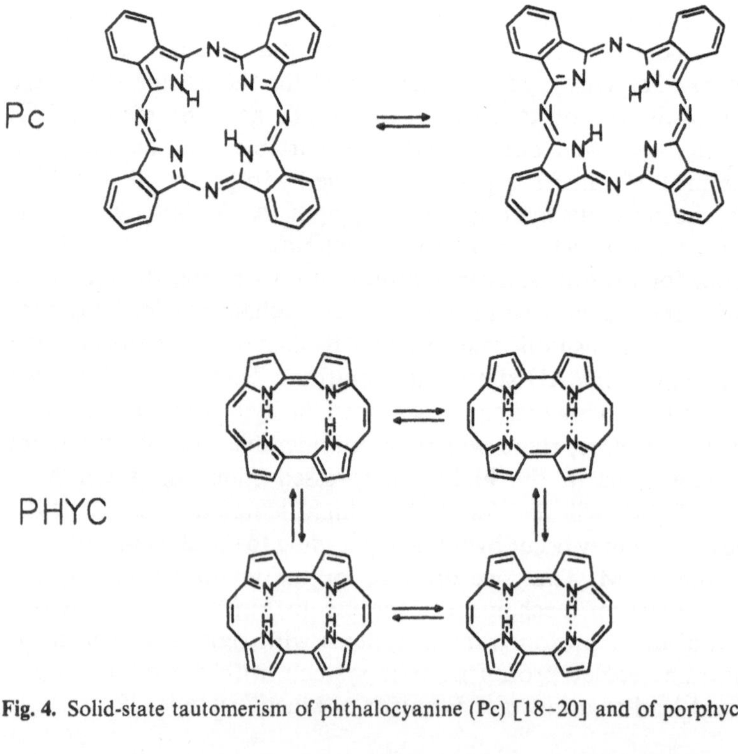

Fig. 4. Solid-state tautomerism of phthalocyanine (Pc) [18–20] and of porphycene (PHYC) [16]

molecules studied with the less abundant spin 1/2 isotope ^{15}N. Using this method, fast proton transfer processes have been monitored in solid porphyrins (Fig. 2) [15–17], phthalocyanine (Fig. 4) [18, 19], porphycene (Fig. 4) [16], azophenine (Fig. 2) [12], tetraazaannulenes (Fig. 5) [21–24] as a function of temperature.

The ^{15}N CPMAS spectra contain a wealth of information about solid-state effects on the proton tautomerism. Let us review only some general results. For this purpose, consider a reaction which is degenerate in the absence of inter-molecular interactions. The reaction profile is then symmetric as indicated schematically in Fig. 6a, where possible intermediates have been omitted for simplicity. However, in the ordered crystalline state this symmetry is, generally, removed by intermolecular interactions and an energy difference ΔE between the two minima of the potential curve for the proton motion arises (Fig. 6b). In the first approximation the potential of the proton motion in a particular molecule does not depend on the tautomeric state of the neighboring molecules. Using ^{15}N CPMAS NMR spectroscopy it was possible to measure ΔE for

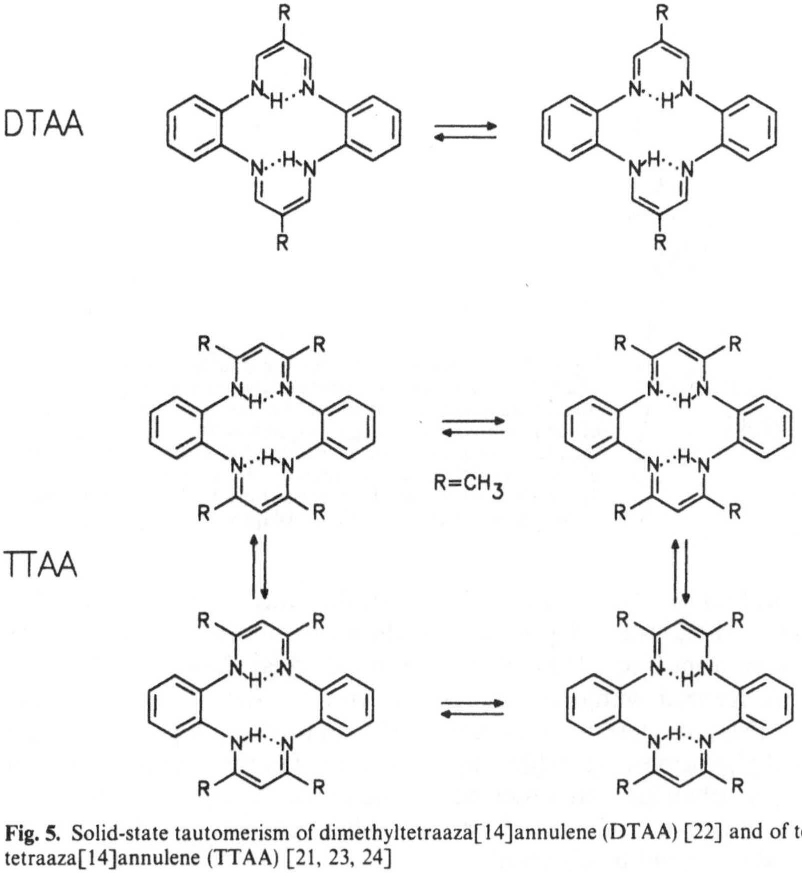

Fig. 5. Solid-state tautomerism of dimethyltetraaza[14]annulene (DTAA) [22] and of tetramethyl-tetraaza[14]annulene (TTAA) [21, 23, 24]

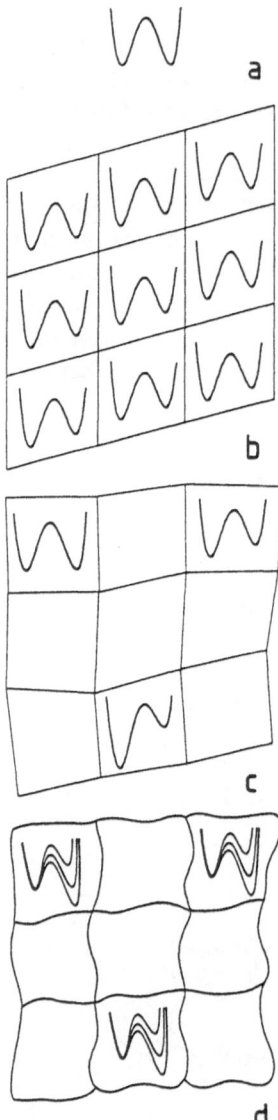

Fig. 6. Perturbation of a symmetric double minimum potential of a bistable molecule by intermolecular interactions. **a** symmetric double minimum potential in the gas phase. **b** Perturbation of the potential in the ordered crystalline state by intermolecular interactions which are the same for all molecules. **c** Perturbation of the potential in the disordered solid state by intermolecular interactions which are the different for all molecules **d** Motional averaged symmetric potentials. (Adapted from Ref. [24])

different porphyrins, azophenine, as well as for the reaction systems shown in Figs. 4 and 5. These values depend on the chemical and the crystal structure; values ranging from 0 to 12 kJ mol^{-1} were found so far. Note that the cases $\Delta E \simeq 0$ were realized within the margin of error for the parent compound porphyrin [16], meso-tetratolylporphyrin [15], and the triple proton transfer in 3,5-dimethylpyrazole (Fig. 1) [25, 26]. In the case of phthalocyanine, different solid-state perturbations were observed depending on the crystal modification [19]. By contrast, no substantial entropy difference between the different tautomeric states could be observed.

From solid-state NMR experiments, not only thermodynamic but also kinetic information on the reaction rates in the solid state can be obtained, as shown for porphyrins, phthalocyanine, and dimethyltetraazaannulene, and pyrazoles [15–26]. It is interesting to note that the rate constants of the tautomerism in solid porphyrin samples coincide with the corresponding solution rate constants [15–17]. Thus, solid-state effects on the thermodynamics of proton tautomerism seem to be more important than kinetic solution solid-state effects provided that the hydrogen bond network is not changed. For strongly hydrogen-bonded proton transfer systems, such as porphycen [16] (Fig. 4) or TTAA [21] (Fig. 5), the rates of proton transfer were found to be too fast in order to be measured by ^{15}N CPMAS NMR. From the observation of temperature-dependent mole fractions of different tautomeric states it was, nevertheless, possible to establish in both cases that the protons move along slightly asymmetric double minimum potentials and not in single minimum potentials. It was also found in both cases that the two NH...N hydrogen bonds of each molecule are perturbed in a different way, which establishes the independent proton motion and the presence of four different tautomeric states shown in Fig. 4 and 5. Note that these NMR results can help to answer the problem of proton localization which arises in the x-ray crystallographic analysis of compounds with labile protons [47–49].

4 ^{15}N CPMAS NMR Spectroscopy of Proton Transfer Systems in the Disordered Solid State

For the crystalline state discussed so far, all molecules of a sample experience the same solid-state perturbation ΔE as expressed by Fig. 6b. This finding is not surprising because of the periodicity of crystals. However, how are the reaction energy surfaces of the proton transfer systems under consideration modulated by an aperiodic amorphous matrix such as an organic glass? Information about this problem is desirable in order to achieve a better understanding of the structure and of rate processes in glasses.

Such information can be obtained by lineshape analysis of suitable high-resolution solid-state spectra. In order to extract thermodynamic and kinetic data from the spectra, first, an appropriate line-shape theory had to be developed [19, 24]. The model on which this theory is based is shown in Fig. 6c. It assumes a continuous distribution of sites characterized by different rate and equilibrium constants of proton tautomerism. The possibility of exchange between the different sites was also taken into account [19, 24]. Actually, bigaussian distributions of the reaction enthalpies and of the enthalpies of activation of proton exchange were employed. Different possibilities, including Marcus theory, of reducing this two-dimensional site distribution function to a one-dimensional distribution were discussed [19].

The theory was then applied to simulate the ^{15}N CPMAS NMR spectra of ^{15}N-labeled amorphous phthalocyanine (Pc) [19]. The spectra show dynamic line-broadening due to proton transfer. Site exchange was found to be absent. The amorphous modification was found to be characterized by a broad distribution of differently perturbed asymmetric double minimum potentials, as expected for a disordered environment. In other words, in a disordered environment not only a distribution of equilibrium constants but also of rate constants were observed [19]. Unfortunately, both distributions are independent from each other and make it difficult to extract a unique set of lineshape parameters by the simulation of the spectra.

Therefore, the effects of a disordered environment on a very fast proton transfer was studied where dynamic line-broadening due to proton transfer is absent. For this purpose, ^{15}N CPMAS NMR measurements were performed on solid solutions of TTAA (Fig. 5) in polystyrene [24]. Since the matrix does not contain nitrogen atoms, the signals of TTAA did not interfere with solvent signals, a problem which arises in ^{13}C studies. We find that the TTAA tautomerism in polystyrene is indeed very fast with respect to the NMR timescale, i.e., much faster than the reorientation of the solvent molecules; therefore, no line-broadening due to slow proton transfer complicates the spectra. However, a distribution of TTAA molecules with different equilibrium constants of tautomerism was found, according to Fig. 6c. Since the latter depend on temperature, this distribution leads to temperature-dependent inhomogeneously broadened lines. Exchange between the different types of TTAA molecules via rotational and translational diffusion is very slow below the glass transition. From the simulation of the NMR spectra, the parameters of a gaussian distribution of free reaction energies of the tautomerism, i.e., the maximum and the width of the distribution, were obtained as a function of temperature [24]. The inhomogeneous line-broadening was confirmed by two-dimensional NMR experiments. Thus, proton transfer dyes such as TTAA are sensitive molecular probes for microscopic order in glasses. In addition, the motional averaging process of the differently perturbed double minimum potential into one effective symmetric double minimum potential within the NMR timescale was observed by performing experiments in the glass transition region [24]. This process indicates that above the glass point the different sites interconvert rapidly within the NMR timesale. This process renders the effective potential of the proton motion apparently symmetric as shown in Fig. 6d.

5 IR Studies of Azophenine in the Liquid and the Crystalline State

The motional averaging of the double minimum potentials described above takes place in a timescale spanned by the correlation time of the molecular motion, which is on the order of nano- to picoseconds for liquids. Since this averaging process is slow with respect to the IR timescale, the model expressed by Fig. 6c should then be valid also for a symmetric bistable molecule in the liquid state, in a timescale characteristic for IR spectroscopy. Consequently, the IR frequencies of vibrations coupled to the reaction coordinate of the proton motion, e.g., NH stretching bands, are expected to be different in the different sites in Fig. 6c, whereas they are similar in the ordered crystalline state modeled in Fig. 6b. Thus, the corresponding IR bands should be inhomogeneously broadened in liquid solution as compared to the crystalline state.

In order to obtain evidence for this model, it was necessary to study a molecule where other IR line-broadening mechanisms are absent and where the proton transfer was established by NMR for the solution and the solid state. It was found that azophenine was an ideal candidate for such studies. Besides the liquid-state NMR measurements described above which established the tautomerism in Fig. 2, solid-state NMR experiments [12] showed that the protons move in crystalline AP in highly asymmetric double minimum potentials according to Fig. 6b, with $\Delta E = 12 \, \text{kJ mol}^{-1}$. IR measurements were performed on azophenine dissolved in liquid CCl_4 and on polycrystalline azophenine [11]. The latter is characterized by surprisingly sharp NH stretching bands without a substructure or line-broadening as observed in other systems with intra-molecular hydrogen bonds [39]. However, a broad NH stretching band was obtained for the liquid solution. Note that this broadening is only observed for the NH/ND stretching vibration; the other vibrations, e.g., the CH stretching vibrations, are not affected or only to a very minor extent. From NMR measurements the rotational correlation time of the NH vector were known to be of the order of nanoseconds; thus, rotational or translation diffusion cannot be responsible for the observed differential liquid–solid IR line-broadening. In the overtone region, the overall line widths of the NH stretching bands are much larger as compared to the fundamental region, and a possible differential broadening of solid-state and liquid-state overtone NH stretching bands could, therefore, not be observed within the margin of error. This overtone broadening is most easily interpreted in terms of short lifetimes of the higher excited NH stretching states. Thus, a plausible explanation for the differential NH stretching line widths in crystalline AP and AP dissolved in CCl_4 is provided by the assumption of a distribution of different solvation sites in which the force constants of the stretching vibration are different, according to the models expressed by Fig. 6b and 6c. In fact, the observation of the differential line-broadening effect provides evidence for a coupling of the NH stretching vibration to the reaction coordinate of tautomerism.

6 Conclusions

It has been shown that neutral double and triple proton transfer reactions as well as their multiple kinetic isotope effects and the effects of intermolecular interactions on the reaction dynamics can conveniently be studied by dynamic liquid and solid-state NMR spectroscopy. Intramolecular and intermolecular proton transfer reaction systems behave in different ways. In the intramolecular cases only one proton contributes significantly to the kinetic isotope effect indicating a stepwise reaction mechanism. Highly polar intermediates lead to kinetic solvent effects if the reaction center is not shielded from the solvent. By contrast, in the intermolecular cases both protons contribute substantially to the kinetic isotope effects, thus increasing their size. The origins of the different behavior is discussed. Tunneling is found to play an essential role in these reactions.

Using novel solid-state NMR techniques, the dynamics of this class of reactions can be studied in a timescale of slow molecular motion, in the crystal-line ordered as well as in the disordered—amorphous and glassy—solid state. It was found that the proton transfer systems studied are subject to static perturbations of the reaction energy profiles of the proton motion, even in cases where kinetic solvent effects are absent. Molecular motions lead to motionally averaged effective reaction surfaces in the liquid state. For one case the findings could be supported by additional liquid/solid-state IR experiments.

Acknowledgements

I would like to thank all my collaborators who have contributed to the work described above: J. Hennig, D. Gerritzen, H. Rumpel, G. Otting, L. Meschede, B. Wehrle, M. Schlabach, G. Scherer, F. Aguilar-Parrilla, J. Braun and Ch. Hoelger. I further thank C.S. Yannoni, IBM, San Jose, C. Djerassi, Stanford, E. Vogel, Köln, H. Zimmermann, Heidelberg for their contribution to this work.

Furthermore, I have to thank the Deutsche Forschungsgemeinschaft, Bonn-Bad Godesberg, the Stiftung Volkswagenwerk, and the Fonds der Chemischen Industrie, Frankfurt, for financial support.

7 References

1. Limbach HH (1983) The use of NMR spectroscopy in the study of hydrogen bonding in solution. In: Gormally J, Wyn-Jones E (eds) Aggregation Processes, Elsevier, Amsterdam, Chap 16 and references cited therein
2. Limbach HH (1990) Dynamic NMR spectroscopy in the presence of kinetic hydrogen/deuterium isotope effects. In: NMR Basic Principles and Progress, Vol 23, Springer, Heidelberg, and references cited therein
3. Limbach HH, Hennig J, Gerritzen D, Rumpel H (1982) Far Disc Chem Soc 74: 229
4. Gerritzen D, Limbach HH (1984) J Am Chem Soc 106: 869
5. Meschede L, Gerritzen D und Limbach HH (1988) Ber Bunsenges Phys Chem 92: 469
6. Limbach HH, Meschede L, und Scherer G (1989) Z Naturforschung, 44a: 459
7. Storm CB, Teklu Y (1974) J Am Chem Soc 94: 1745; Ann NY Acad Sci (1973) 206: 631
8. Hennig J, Limbach HH (1984) J Am Chem Soc 106: 292; Hennig J, Limbach HH (1982) J Magn Reson, 49: 322
9. Schlabach M, Wehrle B, Limbach HH, Bunnenberg E, Knierzinger A, Shu A, Tolf BR, Djerassi C (1986) J Am Chem Soc 108: 3856

10. Schlabach M, Rumpel H und Limbach HH (1989) Ang Chem 101: 84; Int Ed Engl (1989) 28: 76
11. Rumpel H, Zachmann G, Limbach HH (1989) J Phys Chem 93: 1812
12. Rumpel H, Limbach HH (1989) J Am Chem Soc 111: 5429
13. Otting G, Rumpel H, Meschede L, Scherer G, Limbach HH (1986) Ber Bunsenges Phys Chem 90: 1122
14. Scherer G, Limbach HH (1989) J Am Chem Soc 111: 5946
15. Limbach HH, Hennig J, Kendrick RD, Yannoni CS (1984) J Am Chem Soc 106: 4059
16. Wehrle B, Limbach HH, Köcher M, Vogel E (1987) Ang Chem 99: 914; Int Ed Engl (1987) 26: 934
17. Limbach HH, Wehrle B, Schlabach M, Kendrick R, Yannoni CS (1988) J Magn Reson 77: 84
18. Kendrick RD, Friedrich S, Wehrle B, Limbach HH, Yannoni CS (1985) J Magn Reson 65: 159
19. Wehrle B und Limbach HH (1989) Chem Phys 136: 223
20. Meier BH, Storm CB, Earl WL (1986) J Am Chem Soc 108: 6072
21. Limbach HH, Wehrle B, Zimmermann H, Kendrick RD, Yannoni CS (1987) J Am Chem Soc 109: 929
22. Limbach HH, Wehrle B, Zimmermann H, Kendrick RD, Yannoni CS (1987) Angew Chem 99: 241; Angew Chem Int Ed Eng (1987) 26: 247
23. Limbach HH, Zimmermann H und Wehrle B (1987) Ber Bunsenges Phys Chem 91: 941
24. Wehrle B, Zimmermann H, Limbach HH (1988) J Am Chem Soc 11: 7014
25. Baldy A, Elguero J, Faure R, Pierrot M, Vincent EJ (1985) J Am Chem Soc 107: 5290
26. Smith JAS, Wehrle B, Aguilar-Parrilla F, Limbach HH, Foces-Foces MC, Cano FH, Elguero J, Baldy A, Pierrot M, Khurshid MMT, Larcombe-McDouall JB (1989) J Am Chem Soc 111: 7304
27. Völker S, van der Waals JH (1976) Mol Phys 32: 1703
28. Friedrich J, Haarer D (1984) Angew Chem Int Ed Engl 23: 113
29. Butenhoff TJ, Moore CB (1988) J Am Chem Soc 110: 8336; Butenhoff TJ, Chuck RS, Limbach HH, Moore CB (1990) J Phys Chem 94: 7847
30. Gandour RD, Schowen RL, Transition states of biochemical processes, Plenum Press, New York 1978
31. Hermes JD, Cleland WW (1984) J Am Chem Soc 106: 7263
32. Limbach HH, Hennig J (1979) J Chem Phys 71: 3120; Limbach HH, Hennig J, Stulz J (1983) J Chem Phys 78: 5432; Limbach HH (1984) J Chem Phys 80: 5343
33. Sarai A (1981) Chem Phys Lett 83: 50; (1982) J Chem Phys 76: 5554; (1984) ibid 80: 5341
34. Bersuker GI, Polinger VZ (1984) Chem Phys 86: 57
35. Smedarchina Z, Siebrand W, Zerbetto F (1989) Chem Phys 136: 285
36. Dewar MJS, Merz KM (1985) J Mol Struct (Theochem) 124: 183; Holloway KM, Reynolds CH, Merz, KM (1989) J Am Chem Soc 111: 3466
37. Merz KM, Reynolds CH (1988) J Chem Soc Chem Comm 90
38. Bell RP, "The Tunnel Effect in Chemistry", Chapman and Hall, London 1980
39. "The Hydrogen Bond", Schuster P, Zundel G, Sandorfy C, Eds, North Holland Publ Comp, Amsterdam 1976
40. Bigeleisen J (1955) J Chem Phys 23: 2264
41. Albery WJ (1986) J Phys Chem 90: 3773
42. Novak A (1974) Struct Bond 14: 177
43. Lyerla JR, Yannoni CS, Fyfe CA (1982) Acc Chem Res 15: 208; Fyfe CA, Solid State NMR for Chemists, C.F.C. Press, Guelph, Ontario 1983
44. Schaeffer J, Steijskal EO (1976) J Am Chem Soc 98: 1031
45. Szeverenyi NM, Bax A, Maciel GE (1983) J Am Chem Soc 105: 2579
46. Myrrhe PC, Kruger JD, Hammond BL, Lok SM, Yannoni CS, Macho V, Limbach HH, Vieth HM (1984) J Am Chem Soc 106: 6079
47. Robertson JM (1936) J Chem Soc 7719; Karasek FW, Decius JC (1952) J Am Chem Soc 74: 7716; Hoskins BF, Mason SA, White JCB (1969) J Chem Soc Chem Comm 554
48. Webb LE, Fleischer EB (1965) J Chem Phys 43: 3100; Chen BML, Tulinsky A (1972) J Am Chem Soc 94: 4144; Tulinsky A (1973) Ann NY Acad Sci 206: 47; Hamor MJ, Hamor TA, Hoard JL (1964) J Am Chem Soc, 86: 1938; Silvers SJ, Tulinsky A (1967) J Am Chem Soc 89: 3331; Butcher RJ, Jameson GB, Storm CB (1985) J Am Chem Soc 107: 2978
49. Goedken VL, Pluth JJ, Peng SM, Bursten B (1976) J Am Chem Soc 98: 8014

CHAPTER XII
Cluster Research with Spectroscopic Molecular Beam Techniques

K. Rademann

Molecular beams provide a nearly collision-free environment for the spectroscopic investigation of elemental and molecular clusters of different size and composition. Some techniques are reviewed for producing, detecting, and exploring neutral clusters in molecular beams. While scattering experiments are treated by Buck in this volume, we will concentrate on spectroscopic methods specifically developed during the last two decades for the study of clusters. Due to the low intensity of such aggregates in a molecular beam, most of the methods employed are based on laser light sources.

Recent innovations have allowed to use rather conventional techniques like UV/Vis-absorption spectroscopy of clusters and photoelectron spectroscopy. A coincidence variant of the latter method facilitates the study of the energetics and dynamics of stoichiometrically well-characterized, isolated clusters.

1 Introduction

Important experimental information on intermolecular interactions [1] has been obtained by studying spectroscopic properties, transport phenomena, and equilibrium properties of materials in condensed phases. In order to better

Intermolecular Forces
© Springer-Verlag Berlin Heidelberg 1991

understand the interrelationship between macroscopic bulk behavior and microscopic properties (structure, energetics, dynamics, and intramolecular flow of energy) of individual particles, it is necessary to investigate relatively isolated species, for instance in diluted solutions, in rare gas matrices, or in the gas phase at various densities. The highest degree of isolation, however, can be reached only in a molecular beam, where particles move in straight lines through a vacuum chamber. A great variety of experiments on isolated species can be performed and a very broad category of experiments is concerned with elastic, inelastic, or reactive scattering processes (see U. Buck, this volume; and the handbook article by Toennies [9]). Elastic and inelastic scattering data provide directly information on both the attractive as well as repulsive parts of the intermolecular potentials. Partly complementary information can be obtained by spectroscopic beam techniques, which constitute another very broad class of experiments on molecular beam isolated particles. Spectroscopic techniques are particularly useful for investigating intermolecular forces between particles in isolated clusters.

Isolated clusters may be defined as small aggregates consisting of atoms, molecules, ions, or radicals. Depending on composition and size of a cluster, its constituents are held together either by dispersion forces, electrostatic interactions, hydrogen bonding, "conventional" chemical bonds, or metallic interaction. The experimental investigation of molecular beam isolated clusters is of great current interest [2-5] for several compelling reasons. Firstly, and most exciting is the fact, that clusters allow the exploration of intermolecular forces as well as dynamic phenomena as a function of size all the way from atomic or molecular systems to very large aggregates resembling macroscopic matter in condensed phases. Secondly, the level structure will undergo a transition from discrete states in very small, molecular-like clusters to a quasicontinuum of vibrational and electronic states in sufficiently large aggregates. Beyond a certain critical size, new types of elementary excitations like phonons, excitons, polaritons, etc. may be exhibited. Thirdly, the size dependence of geometrical equilibrium structures, isomerisation processes, or phase transitions like melting of clusters can be studied and compared to the behavior of the corresponding bulk material. Fourthly, spectroscopic information on clusters of finite size may provide an ultimate test for the applicability of theoretical methods in cluster science. Finally, cluster research will be most fruitful for the exploration of the chemical dynamics and reactivity of clusters (or of solvated species within a cluster) as a function of size and composition.

We finally remark that beams of charged particles, which may be accelerated up to the speed of light, or beams of particles that do not move with "thermal" velocities, are not customarily designated as molecular beams, but have received great attention in the past. The enormous progress made in the field of ion and cluster ion spectroscopy has been reviewed recently (see, e.g., Castleman and Märk [6] or the multi-author volume edited by J.P. Maier [7]).

2 Isolated Clusters in Molecular Beams

The generic term molecular beam applies to different species like atoms, radicals, molecules, or clusters, which normally travel at "thermal" speeds (a few hundred meters a second or so) under practically collision-free conditions in straight lines through a vacuum chamber.

A molecular beam apparatus usually has at least four components: 1—a source, or reservoir, which is filled with the vapour of the material to be investigated and which is equipped with a small orifice for the expansion of the gas, 2—a system of collimators, which determines the geometrical form of the beam, 3—a detector for the beam particles, and 4—a vacuum system, in which all the other components are arranged, and in which the total pressure is maintained at a very low level, such that the mean free path of the particles is several times the distance between the source and the detector.

The spectroscopic investigation of isolated clusters in molecular beams has some advantages as compared to other experimental systems. Firstly, no interaction with supporting media will complicate the spectra. Secondly, the particles under investigation do not suffer any collisions (this is not to be understood in an absolute sense; the important question is, whether on the time scale of the experiment collisions do occur or not; on the more precise definition of collision-free conditions see Lubman et al. [8]). Thirdly, a broad range of different vibrational and rotational temperatures is amenable to experimental investigation. It is particularly important to note that temperatures in the 1-K region can be reached easily by adiabatic expansion [10].

It has already been mentioned above that the investigation of molecular beam particles may be divided into two major categories of experiments: 1—the study of elastic, inelastic or reactive scattering processes in crossed molecular beams [9] and 2—the exploration of the interaction of particles with electric, magnetic or electromagnetic fields leading to molecular beam spectroscopic experiments [10]. There are, of course, other fascinating applications of molecular beams, like the study of particle-surface interactions or the formation of monolayers and growth of thin films by molecular beam epitaxy [11], to mention but a few. In any case, however, the applicability of the molecular beam technique for the investigation of specific problems depends to a major degree on the sensitivity of the detector and the intensity of the molecular beam.

As for effusive beam sources, which are experimentally as well as theoretically perfectly understood [9], the maximum intensity I_d (in particles per second) at the detector with area A_d (in cm^2) is given by:

$$I_d = 1.118 \times 10^{22} \frac{pA_sA_d}{d^2\sqrt{MT}}, \qquad [\text{molecules/second}] \qquad (1)$$

where M is the molecular weight of the material, T (in K) the temperature in the reservoir, p (in Torr) the reservoir pressure, A_s (in cm^2) the surface of the oven orifice, and d (in cm) the distance between the oven orifice and the detector. The particles in the oven obey a Maxwellian velocity distribution, and the particles in the beam are characterized by exactly the same velocity distribution $f(v)$:

$$f(v) = N4\pi v^2 \exp\left(-\frac{v^2}{\alpha^2}\right), \tag{2}$$

$$\alpha = \sqrt{2kT/m},$$

where m is the particle mass and k the Boltzmann constant; N is a normalization constant.

The total material consumption rate of the effusing gas (total flow through the orifice as measured in particles per second) is given by:

$$I_0 = \frac{p \cdot A_s}{\sqrt{\pi mkT}} \tag{3}$$

or

$$I_0 = 0.0583 \frac{p \cdot A_s}{\sqrt{MT}} \, mol \cdot s^{-1} \tag{4}$$

if p is measured in Torr and A_s in cm^2 (1 Torr = 133.33 Pa).

For the validity of these formulas it is necessary to assume that no collisions between particles occur while they effuse through the orifice of the oven. Deviations will have to be considered if collisions take place (vide infra).

As an interesting example we consider water molecules effusing at 273.15 K and p = 0.01 Torr through an orifice of the area $10^{-2} cm^2$ into a vacuum chamber. The material consumption rate is $8.3 \times 10^{-8} \, mol \, s^{-1}$ and the intensity (see Eq. (1)) at the detector (which may be placed 100 cm away from the orifice and which may have an active area of $10^{-2} cm^2$) is only 1.6×10^{10} molecules per second, but still sufficiently large to be detected.

2.1 Generation of Clusters

In order to produce molecular beams and cluster beams of high intensity, it is necessary to work in the viscous flow regime of gas expansion, which is characterized by the condition that the mean free path λ of the particles inside the reservoir is much smaller than the diameter of the nozzle. According to a rule of thumb, λ is on the order of 10^{-5} cm for a pressure of 1 bar (at room temperature). Adiabatic expansion of a high-density gas (about 1 bar; see Fig. 1) through a small nozzle (diameter less than 1 mm) into a low-pressure ambient

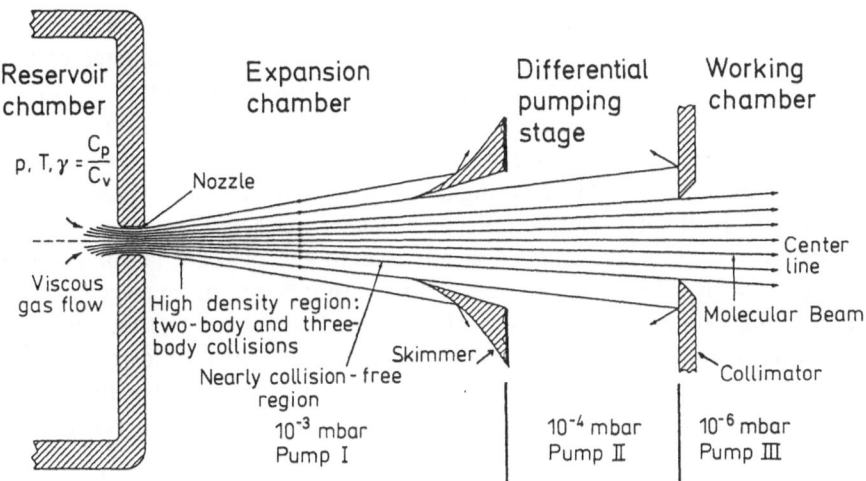

Fig. 1. A highly collimated, intense molecular beam of cold molecules and clusters is extracted out of the high-density region of a supersonic expansion. The viscous gas flow through the nozzle is sustained by a huge density (or equivalently pressure) gradient of many orders of magnitude. Typical pressures p in the reservoir, or "oven" are between 0.1 and 100 bars, but the ambient background pressure in the expansion chamber should be kept below 10^{-3} mbar. Differential pumping is required to further reduce the background pressure such that inside the working chamber a pressure of typically less than 10^{-6} mbar can be maintained during operation of the beam machine.

The beam density, as a function of the distance d from the nozzle, decreases very rapidly. The nearly collision free region is reached already behind 10 or 20 nozzle diameters according to Eqs. (5) and (6). The optimum distance between nozzle and skimmer depends strongly on the total gas flow and on the diameters of both, the nozzle and the skimmer, and is typically between 10 and 100 nozzle diameters. The adiabatic expansion causes a dramatic drop in translational temperature. This cooling effect may be understood immediately by considering Poisson's law for isoentropic expansion of an ideal, monoatomic gas: a pressure ratio of 10^5, for instance, equals a temperature ratio of 100 for $y = 1.67$. All processes, namely pressure drop, translational, rotational, and vibrational cooling, cluster formation and condensation occur in the immediate vicinity of the nozzle, where two- and three-body collisions serve as a microscopic mechanism for energy exchange between particles and all degrees of freedom.

For a thorough discussion of several theoretical aspects and experimental details as well as experimental variants of supersonic molecular beams see Ref. [45]

background (less than 10^{-3} mbar, but see Refs. [14, 15]) results in the formation of a supersonic beam, which is characterized by high Mach numbers (corresponding to narrow velocity distributions or equivalently to low translational temperatures in the mK or K region). Ideally, the centerline intensity per steradian I(0) can be conveniently written in terms of the nozzle flow rate and a peaking factor κ (Beijerinck and Vester [13]):

$$I(0) = (\text{molecules} \cdot \text{s}^{-1}\,\text{sr}^{-1}) = \kappa \dot{N}/\pi \qquad (5)$$

$$\dot{N} = F(\gamma)n \cdot \sqrt{\frac{2\,kT}{m}}\,(\pi D^2/4) \qquad (6)$$

$$F(\gamma) = \left(\frac{\gamma}{\gamma+1}\right)^{1/2}\left(\frac{2}{\gamma+1}\right)^{1/(\gamma-1)} \qquad (7)$$

where n is the particle density (particles per volume) and T the temperature of the gas in the oven, m the mass of the particles and D the diameter of the nozzle orifice; γ is the ratio of the molar heat capacities C_p/C_v. $F(\gamma)$ has values very close to 0.5 [13].

The peaking factor κ, which depends on γ [13], has values larger than 1, but is at best 2.0 for $\gamma = \frac{5}{3}$. The peaking factor κ is 1.0 for an effusive expansion. Therefore, the high intensity of molecular beams extracted out of adiabatic expansions through simple nozzles is not due to the peaking factor (spatial focussing) but rather due to a large nozzle throughput $\dot{N}$, requiring large vacuum pumps for noncondensable gases. It must be pointed out, however, that improvements in nozzle design [17, 18], in particular the development of long conically shaped nozzles [17], have resulted in a gain of beam intensity by about two orders of magnitude. In any case, typical nozzle throughputs can reach 1 or several mol per hour.

The dramatic increase in supersonic beam intensity is accompanied by extensive cooling of rotational and vibrational degrees of freedom as will be discussed below. (Here it may suffice to state that normally $T_{oven} > T_{vib} > T_{rot} > T_{trans}$). However, the most important feature of the adiabatic expansion is the formation of clusters. This process occurs in the high density region of the supersaturated gas in the immediate vicinity of the nozzle, where two-body and three-body collisions provide a microscopic mechanism for cooling and cluster formation. Under conditions of extensive cluster formation, however, cooling of molecules and clusters stops being effective due to the release of heat of condensation. This effect has been studied quantitatively [16] for cooling of aniline with Ar in a mixed, so-called seeded beam, where the particles of interest (aniline) are diluted or seeded in a rare gas.

Several completely different types of cluster sources, which may be operated in a pulsed mode or continuously, have been developed during the last twenty years. Since the pioneering work of Becker and his group in Marburg [19, 20], adiabatic expansion of a rare gas or a mixture of gases through a small orifice into a vacuum chamber is still one of the most convenient ways to produce highly intense, rotationally and vibrationally cooled molecules and neutral clusters. Clusters may also be formed by the gas aggregation technique [21], by sputtering techniques [22], or by laser vaporization methods [23].

Clusters formed by any of these methods always appear as a more-or-less broad distribution of aggregates and normally the mean size of these clusters corresponds roughly to the width of the distribution, although some particularly stable clusters may appear as so-called magic numbers in a mass spectrum of charged clusters. In any case, a great amount of experimental effort is needed to unravel the distribution and to obtain at least semiquantitative information on cluster densities by applying appropriate detection methods.

2.2 Detection of Clusters

Scattering of laser light [24], scattering of high-energy electron beams (diffraction methods, see e.g. Ref. [25]), and scattering of atoms or molecules in crossed beams are diagnostic methods for detecting and studying clusters in molecular beams [17]. However, one of the most useful technique for detecting neutral clusters is mass spectrometry, which requires ionization of molecular beam particles by electron impact ionization, photoionization, Penning ionization laser-induced two-photon or multi-photon ionization. Irrespective of the type of ionization, different mass filters may be coupled to the ionization region of the mass spectrometer. Magnetic instruments, quadrupole mass filters, time-of-flight devices (either a simple Wiley-Mc Laren spectrometer or a more advanced reflectron-type spectrometer) are most common and routinely used in cluster research [45].

Other methods are based on optical detection of clusters, e.g., Laser-induced-fluorescence (LIF) or optical absorption spectroscopy. Bolometric techniques are also often used in detecting molecular beam particles and their energy content [45]. It is also possible in some cases to deposit clusters in a matrix and study the size distribution of these aggregates later on, for example, by electron microscopy.

2.3 Separation of Clusters

It has already been mentioned that cluster generation usually results in a broad distribution of aggregates. Hence, there is a great need to separate neutral cluster beams into beams of singly sized particles. If neutral clusters are charged upon ionization, separation is straightforward by applying mass spectrometric techniques. Successful attempts have been undertaken to neutralize mass-separated clusters by resonant or non-resonant charge transfer methods [26]. Another very promising method, which is based on impulse transfer in crossed molecular beams, has been realized by Buck and coworkers and has been used to separate rare-gas clusters as well hydrogen-bonded clusters [27]. Principally, one could also take advantage of the small velocity slip between clusters of different size. This technique, however, would be applicable only to very small clusters. In order to obtain information on mass-selected aggregates, it is also possible to apply a detection technique, which is truly mass specific if fragmentation processes of ionized clusters can be controlled or neglected. One such technique is resonant two-photon ionization [28–32]. Another technique is based on a coincidence measurement of the cluster ion and the electron, which is released during the course of photoionization [33, 34].

3 Spectroscopic Methods and Results

3.1 Photoionization Mass Spectrometry

This technique is very well established for measuring ion yield curves of atoms or molecules [37], reactive intermediates in effusive beam sources [35], and clusters in supersonic beams [36] as a function of photon energy. A light source in the vacuum ultraviolet spectral range is required. A continuous spectrum may be generated by a discharge in a noble gas, but one of the most suitable sources is an electron storage ring emitting synchrotron radiation [38].

In a series of photoionization experiments on molecular beam-isolated, nonmetallic particles it has been demonstrated that dimers and clusters are excellent precursors for the study of the interaction energies between ions and neutral species. Photoionization of van der Waals molecules is a particularly useful means of determining structures as well as thermochemical quantities. In many cases, the principal information is the (vertical) ionization potential, and in some favorable cases vibrational energies of the ionic final state can be obtained. Furthermore, partial ionization cross sections, binding energies and fragmentation potentials (or appearance potentials) have been studied for many atomic and molecular clusters. From the point of view of chemical dynamics, ionization of dimers and clusters offers a very general route of preparing and investigating reactive complexes as a function of internal energy [36].

As for metal atom clusters, photoionization mass spectrometry is an extremely useful tool to unravel intermolecular interactions as a function of cluster size. Photoionization mass spectra, ionization potentials, and ion yield curves of alkali-metal clusters have been studied extensively for more than twenty years (for recent reviews see Refs. [39] and [40]). A great variety of other metals has been studied (see Refs. [3–5]). One particularly interesting investigation has been the autoionization study of mercury clusters, which has revealed that small clusters (less than 13 atoms per cluster) are only weakly bound by van der Waals forces. Larger aggregates, however, gradually become metallic [41]. This dramatic change of the intermolecular potential greatly influences other microscopic and macroscopic properties of mercury clusters (see below).

3.2 Laser-Induced Fluorescence (LIF)

This method is one of the earliest and one of the most versatile spectroscopic molecular beam technique. It is based on high intensity lasers and on the high efficiency for detecting single fluorescence photons. Several variants exist and beam particles at extremely low concentrations can be detected (for instance, the concentrations of sodium atoms can be as low as one single atom per cm^3). The electronic transition of a jet-cooled particle is probed directly by the fluorescence emitted from the excited state reached by the transition. The exciting laser wavelength is scanned and a so-called excitation spectrum is obtained if the integral fluorescence is collected. These type of spectra closely resemble absorption spectra under the condition that the quantum yield for fluorescence is large. More insight into the dynamics of the excited state and into the energetics of the ground state can be gained by recording energy resolved emission spectra. If the laser is pulsed, it is straightforward to measure fluorescence lifetimes of molecules and clusters in the psec and nsec time domain and to correlate the observed data with intermolecular potentials and intermolecular processes.

LIF and its variants are well-established techniques [10, 42–44] and recent results on large molecules, van der Waals complexes, and hydrogen-bonded clusters have been reviewed by Ito [47].

As far as larger clusters are concerned, it is often necessary to obtain unambiguous information on the nature of the absorbing species. A most direct way to reach this goal is the application of the laser-induced two- or more photon-ionization technique, which in combination with a mass spectrometer can yield spectroscopic, structural, as well as dynamic information on the intermolecular interaction between different particles in a stoichiometrically well-defined cluster.

3.3 Resonantly Enhanced Two-Photon Ionization (R2PI)

The first R2PI-experiments were performed on alkali dimers, but the extension of the technique to larger molecules, to supersonically cooled molecules and to clusters was straightforward [30–32]. The first photon of a laser pulse is resonantly absorbed by an intermediate electronic state and the second photon having the same energy as the first one may depopulate the intermediate state by causing a transition into the ionization continuum, provided the lifetime of the intermediate state is sufficiently large. The ion yield is measured as a function of the laser wavelength and the normalized spectrum thus obtained is called an R2PI-spectrum. In principle, the R2PI technique is useful to obtain dynamic as well as spectroscopic information. One typical example is the determination of the ground state binding energy between an argon atom and styrene molecule (D = 396 cm^{-1}). This experimental result [52] is in good agreement with force field calculations for the benzene·Ar complex [46].

A major advance of the technique, however, was the innovation to use two photons of different color. The two-color R2PI-technique allows for the detailed exploration of the energetics, structure, and dynamics of clusters and large van der Waals-molecules. Ionization potentials of weakly bound aggregates of completely defined composition can be determined with extremely high accuracy and precision, although these species usually exist as part of a distribution of several aggregates. The major advantage of this technique is that it can give in principle direct information on the nature and composition of the absorbing species, which is ionized with a second photon, and finally detected in a mass spectrometer. A representative, recent result is on the determination of the ionization potential of the dimer of benzene [48] in a cluster beam of benzene aggregates. As for the benzene trimer a splitting of absorption lines due to exciton interaction is observed. The severe problem of fragmentation of benzene cluster ions is discussed at length, and a new method, which partly discriminates against these fragmentation effects, is presented [48].

3.4 Absorption Spectroscopy

Researchers have wanted direct measurements of absorption cross sections of jet-cooled molecules and clusters for a long time. Under normal expansion conditions of a supersonic jet, the density of molecules and clusters is extremely low and the optical path length is often too short. As far as large molecules are concerned, progress was made in two laboratories [49, 50]. The use of planar supersonic [49] expansions with pulsed nozzles has facilitated the direct measurement of absorption spectra of large, rotationally and vibrationally cooled molecules like anthracene and its derivatives [49].

Recently, we have advanced a new absorption method for dimers and clusters in a molecular beam [51]. The new absorption technique is based on counting

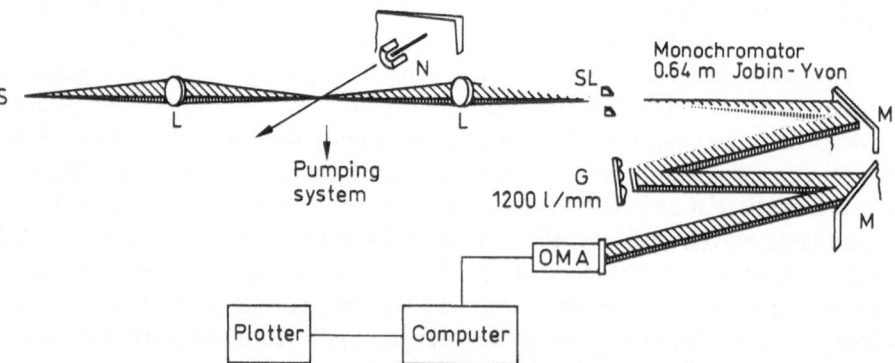

Fig. 2. Schematic diagram of the optical UV/Vis-absorption spectrometer for molecular beam particles formed by supersonic expansion of seeded or non-seeded vapours. The high temperature, pulsed nozzle source is made of stainless steel and can be resistively heated up to 1000 K by thermocoax heaters. The source can be electromagnetically pulsed (10 Hz) with a minimum opening time of about 1 ms. The source is equipped with conically shaped nozzles: minimum throat diameter is 0.3 mm, nozzle length: 1 mm, and opening angle $2\theta = 60°$. S is the light source Deuterium-Tungsten Duplex lamp; L is a quartz lense, and SL the entrance slit of the monochromator, which is equipped with a holographic grating

single photons while the molecular beam is on or off. The detection of photons is performed simultaneously with the aid of an optical multichannel analyzer (OMA) over a certain wavelength range which depends solely on the resolution of the grating employed. Hence, after 20 msec a complete absorption spectrum can be obtained. If the absorption signal is weak, data averaging is performed auto-matically over typically 10000 individual spectra (corresponding to a signal to noise level, which is better than 10^{-4}). In Fig. 2, the set-up of the optical absorption spectrometer is shown schematically. The molecular beam is crossed

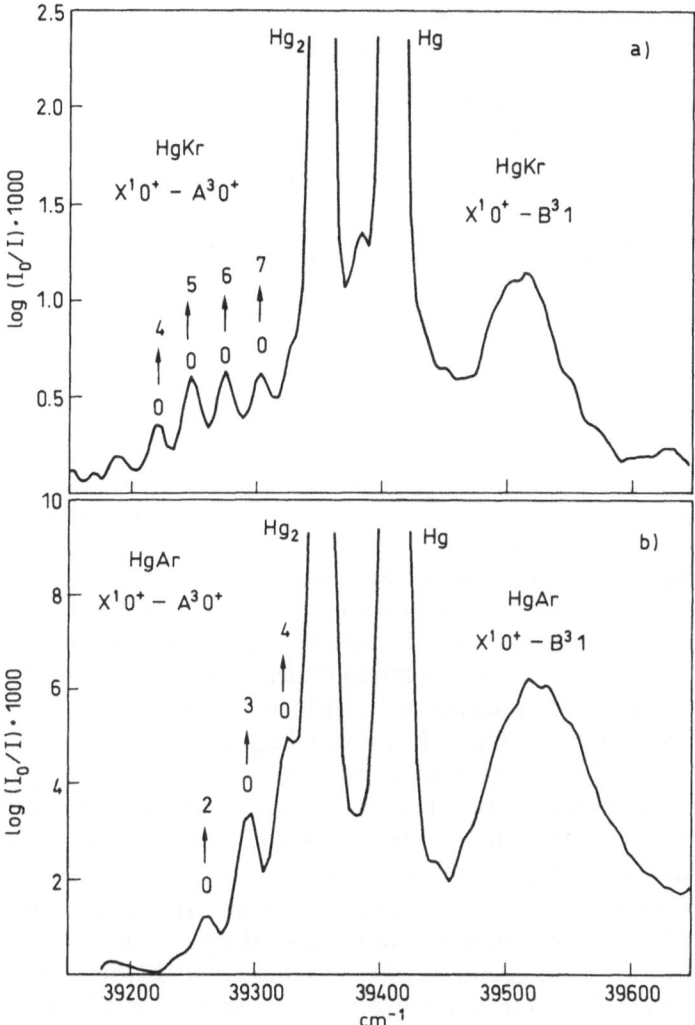

Fig. 3. Figure 3a displays an absorption spectrum of a cluster beam containing mercury atoms, mercury dimers and mixed dimers of mercury and krypton; in Fig. 3b an absorption spectrum of an argon-seeded beam is displayed; for details of assignment see Ref. [51]. The transitions observed correspond to the triplet excitation of the mercury atom at 253.7 nm

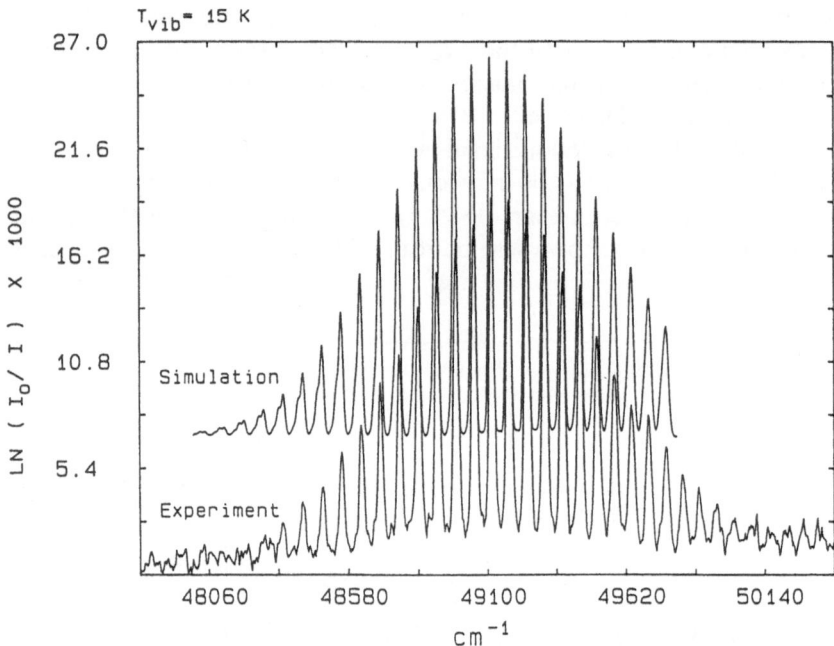

Fig. 4. Singlet absorption spectrum of the mercury dimer in the 204 nm region. The good agreement between computer simulation and experiment has resulted in spectroscopic parameters for the excited singlet state of the mercury dimer [51]. It was also possible to deduce a vibrational temperature for the weakly bound singlet ground state of the dimer

with the focussed, undispersed light emitted by a deuterium (or alternatively tungsten) lamp. The beam diameter (effective path length) is on the order of only 0.3 cm. The transmitted light is then refocussed onto the entrance slit of the monochro-mator and the dispersed light is recorded by the OMA-detector. In Fig. 3 typical absorption spectra for the (weak) triplet excitation of the mercury atom, the mercury dimer and the mixed dimer of mercury and argon is shown. The singlet absorption spectrum of the strongly bound mercury dimer (excimer transition) is displayed in Fig. 4. As for very large mercury aggregates, only a strong absorption band (with maximum at 5.95 eV excitation energy; Fig. 5) is observed. Figure 5 shows the absorption spectrum and the corresponding distribution of mercury clusters in the size range between approximately 20 and 200 atoms. The time-of-flight mass spectrum has been recorded simultaneously during the measurement of the absorption spectrum. The interpretation of the absorption spectrum is straightforward, although we are dealing not with singly sized aggregates, but a rather broad distribution of clusters: While the interaction potential between two mercury atoms leads to a very small binding energy of only 44 meV [51], the bond energies in larger aggregates are significantly higher due to a size-dependent change of the interaction potential, which must be metallic in nature for sufficiently large aggregates. Metallic binding, which is characterized by delocalized valence

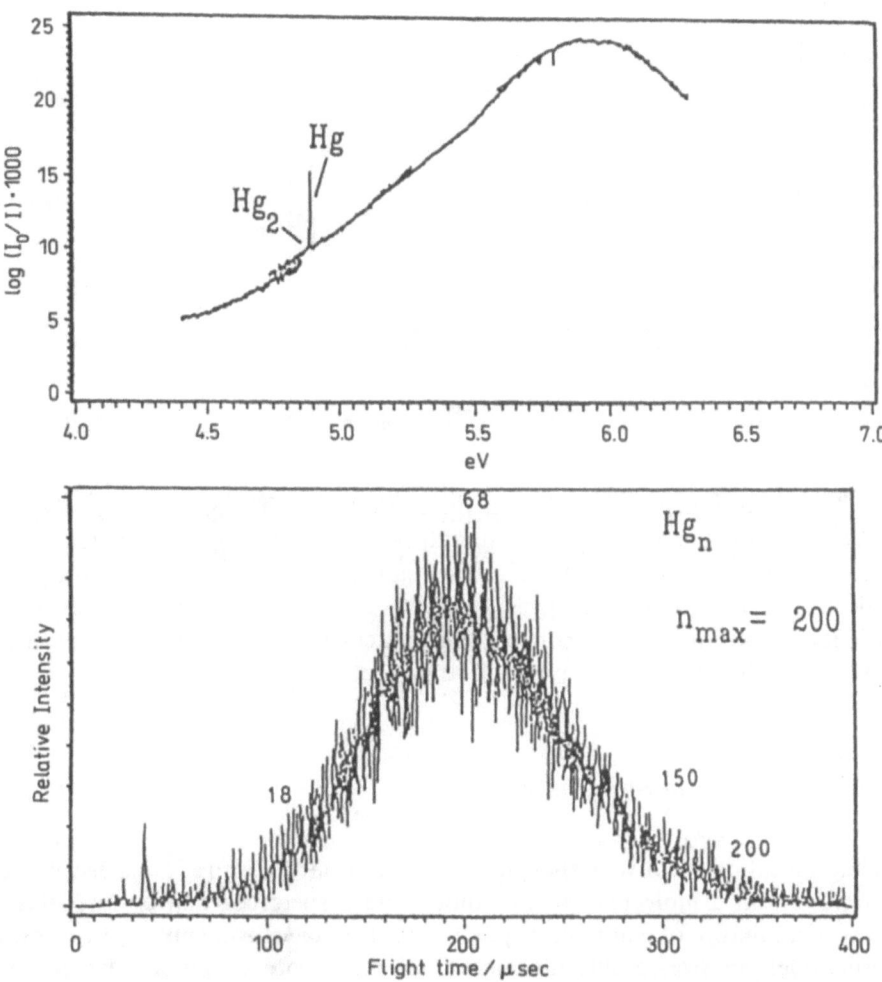

Fig. 5. Simultaneously measured absorption spectrum (upper curve) and time-of-flight mass spectrum of the cluster distribution as obtained by photoionization. The distribution shown is the result of an expansion of Xe and Hg at 620 K and a total reservoir pressure of 2.3 bar

electrons, may give rise to a collective excitation with high oscillator strength. The position of the absorption maximum for metallic clusters depends only on the density of the cluster and is therefore expected to stay constant with increasing cluster size as long as the density of these clusters does not change. The constancy of the peak position has been observed for clusters with more than 60 or 70 atoms. It is remarkable that the absorption spectrum of the distribution shown in Fig. 5 agrees reasonably well with an absorption spectrum calculated for a metallic sphere. This spectrum is displayed in Fig. 6 and it has been obtained by using MIE-theory [53] and the optical constants for metallic, macroscopic, liquid mercury [54].

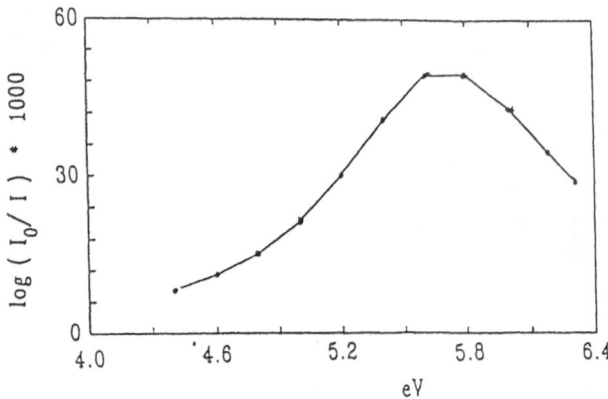

Fig. 6. Calculated absorption spectrum of metallic spheres of mercury according to MIE-theory

In conclusion, we have shown firstly that by applying a new absorption technique it is possible to obtain quantitative information, namely absorption cross sections for low-density molecular beam particles. Secondly, the absorption spectra for large clusters suggest that these aggregates of mercury exhibit a predominantly metallic character. This statement can be quantified by using another technique, namely photoelectron spectroscopy.

3.5 Photoelectron Spectroscopy

The application of conventional photoelectron spectroscopy [37] to a distribution of clusters in a molecular beam would yield a spectrum that corresponds to that of a mixture of compounds possessing itself unknown spectral features. In order to obtain size-specific information, the mass of the cation must be recorded in coincidence with the ejected photoelectron. In a pioneering work on rare-gas dimers and trimers, Poliakoff et al. [33] demonstrated the feasibility of such coincidence studies. The measuring time, however, needed to acquire a single spectrum was on the order of about 100 hours. These long times can be tolerated when working with rare gases, but not when working with metal cluster beams, simply because the material consumption would be, as pointed out in Sect. 2.1, on the order of 100 mol for obtaining a single spectrum. Therefore, we have developed a new technique, which allows us to record simultaneously photoelectron spectra of all clusters present in the molecular beam in total measuring times of less than one hour. The technique has been used to study photoelectron spectra of nonmetallic and metallic mercury clusters, to investigate the photon energy dependence of the partial photoionization cross sections, to determine the bandwidth of the valence electronic structure, and to explore the size dependence of the ionization potentials of mercury clusters with up to 110 atoms. The latter property will be discussed here.

A detailed description of the experimental apparatus employed in the measurements is given elsewhere [34, 55]. Briefly, the photoelectron $\simeq$ photoion-coincidence.cluster beam spectrometer is based on the principle of time-correlated detection of single particles (cations and electrons). It consists of six major components: A differentially pumped three-chamber molecular beam apparatus, a resistively heated stainless steel vapour oven used to generate a seeded or non-seeded cluster beam, a monochromatized vacuum-UV high-repetition nsec-flash-lamp, a time-of-flight mass spectrometer, a magnetic mirror type time-of-flight photoelectron spectrometer, and the coincidence timing and data acquisition electronics. Mass spectra and photoelectron $\simeq$ photoion-coincidence spectra are recorded as follows: The metal to be investigated is co-expanded with a rare gas through a typically 30 µm-diameter conically shaped nozzle into a vacuum chamber kept below 5×10^{-3} mbar. The composition of the doubly skimmed seeded molecular beam is probed 130 mm downstream the nozzle in a vacuum chamber of typically 5×10^{-6} mbar by the monochromatized output of a high repetition rate (25 kHz) vacuum-UV/UV nsec-nitrogen-flashlamp operated as a free-running relaxation oscillator. The cluster cations produced by photoionization are detected in a Wiley–McLaren time-of-flight mass spectrometer [56] equipped with a modified Daly [57] detector. The ejected photoelectrons with a kinetic energy between 0.1 and several eV are energy-analyzed and detected in a newly developed magnetic mirror type 4π acceptance time-of-flight photoelectron spectrometer based on the ideas of Kruit and Read [58] and Cheshnovsky et al. [59]. In our magnetic mirror electron spectrometer the electron trajectories are parallelled by an inhomogeneous, diverging, static magnetic field of $B = 0.5$ kG and are guided by an homogeneous field of 10 G through a 70-cm drift tube towards the scintillation detector. The energy resolution depends on the ratio of the two magnetic fields, the length of the drift tube, the ionization volume, the time duration of the ionizing light pulses, and the energy resolution of the monochromator. The overall resolution achieved in the present set-up is dE/E = 0.15 at 1 eV kinetic energy.

Ionization potentials obtained by threshold spectroscopy as well as photo-electron spectroscopy (PES) at different excitation energies are summarized in Table 1. The data are in very good agreement with ionization potentials by electron impact ionization studies [60, 61].

It is instructive to compare the experimental values of the ionization potentials with the classical electrostatic equation which gives the energy necessary to remove an electron from a uniformly conducting sphere having the same dimension as the cluster. The appropriate expression [62] which connects cluster ionization potentials with the polycrystalline work function by a 1/R term, with clusters having higher ionization potentials for smaller particle size, is:

$$I_p = w + \tfrac{1}{2}(e^2/R), \tag{2}$$

where w represents the work function, R the radius of the equivalent sphere, and e the elementary charge. A spherical geometry is chosen for comparison

Table 1. Ionization potentials of mercury clusters

x	1	2	3	4	5	6	7	8
1	10.4	10.4	10.4	—	—	—	—	—
2	—	9.0	—	—	—	—	—	—
3	—	8.8	—	—	—	—	—	—
4	—	8.6	8.5	—	—	—	—	—
5	—	—	—	8.6	8.4	8.3	—	—
6	—	—	8.4	—	—	—	—	—
7	—	—	—	8.5	8.4	8.2	—	—
8	8.3	—	—	8.4	8.3	8.2	—	—
9	—	—	—	8.2	—	—	—	—
10	—	—	8.1	8.2	8.2	8.1	—	—
11	—	—	—	8.0	8.0	8.0	—	—
12	—	—	—	8.0	—	—	—	—
13	—	—	—	7.9	8.0	8.0	—	—
14	—	—	—	7.8	—	—	—	—
15	—	—	—	7.8	7.9	7.9	—	—
16	—	—	—	7.7	—	—	—	—
17	7.7	—	—	7.7	7.8	7.8	7.6	—
21	—	—	—	—	—	—	—	—
24	7.1	—	—	—	—	—	7.2	—
25	—	—	—	—	—	—	—	—
28	—	—	—	—	—	—	—	—
32	—	—	—	—	—	—	6.6	—
35	6.5	—	—	—	—	—	6.6	—
41	—	—	—	—	—	—	6.4	6.3
44	—	—	—	—	—	—	6.3	—
45	—	—	—	—	—	—	6.3	—
49	6.2	—	—	—	—	—	—	—
51	—	—	—	—	—	—	6.2	6.2
54	—	—	—	6.1	—	—	—	—
55	—	—	—	—	—	—	6.1	—
60	—	—	—	6.0	—	—	6.0	6.0
65	—	—	—	—	—	—	6.0	6.0
68	—	—	—	6.0	—	—	—	—
70	5.9	—	—	—	—	—	—	—
74	—	—	—	5.6	—	—	6.0	6.0
78	—	—	—	—	—	—	5.9	5.9
79	—	—	—	5.7	—	—	—	—
85	—	—	—	5.7	—	—	5.9	5.9
100	—	—	—	5.7	—	—	—	—
109	—	—	—	5.7	—	—	5.8	5.8

x = number of mercury atoms
1 = photoionization mass spectrometry
2 = PES at 10.87 eV excitation photon energy
3 = PES at 10.74 eV excitation photon energy
4 = PES at 10.6 eV excitation photon energy
5 = PES at 10.0 eV excitation photon energy
6 = PES at 9.4 eV excitation photon energy
7 = PES at 8.3 eV excitation photon energy
8 = PES at 7.1 eV excitation photon energy

and the results are shown in Fig. 7. We use the most recent results for the work function, w = 4.49 eV, of Cotti et al. [63]. The size of the mercury atom in the bulk metal is chosen to relate the radius R of an equivalent sphere to the number x of atoms in the cluster, so that R varies as the $\frac{1}{3}$ power of x. It is evident that the classical expression, which is found to be in fair agreement with experimental

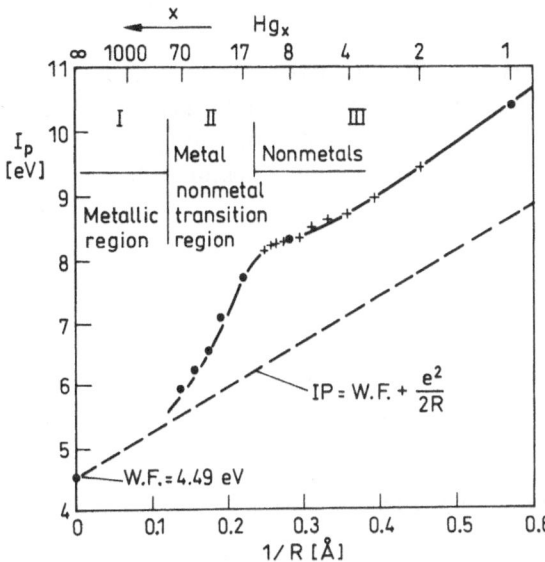

Fig. 7. Comparison of the experimentally determined ionization potential data with the predictions of the classical spherical-droplet model (*straight line*). We plot the ionization potentials vs. 1/R, with R the radius of a sphere with the same volume as an x-atomic metal cluster (assuming atomic volume derived from bulk metallic density)

data for large alkali-metal clusters, [62] is not in accord with the experimental findings for mercury. There is a significant discrepancy of about 2 eV between predicted and experimentally determined ionization potentials of Hg_x clusters for $x \lesssim 13$. However, the most interesting feature of the data is the relatively rapid decrease of the ionization potentials with increasing cluster size, i.e., increasing R, for $x \gtrsim 13$. The deviation between the measured data and the classical spherical droplet prediction decreases continuously. We regard the fact that the I_p of Hg_{109} is approaching the classical straight line in Fig. 7 as an indication that clusters composed of about 109 and more atoms should exhibit an electronic level structure similar to the metallic bulk material.

The results show that for small clusters of mercury atoms one can distinguish between two extreme situations, i.e., the metallic and the van der Waals-type modifications. Mercury atoms have an s^2-atomic configuration that is widely separated in energy from the first unoccupied atomic p-orbital, and so they give rise to small clusters with fully occupied van der Waals-type weakly binding s-bands. The change to the observed bulk metal properties of large Hg_x clusters $(x > 110)$ can only be achieved via overlap of the full s-derived valence band and the empty p-derived conduction band. The experimental results indicate that for Hg_x-clusters, the electronic properties of the bulk metal evolve gradually with increasing particle size in the size range $13 < x < 110$. The strong variation in the electronic structure in course of this transformation from nonmetallic to metallic properties is also evidenced by studies of inner shell 5d-autoionization spectra [41], of optical absorption spectra [64], and of the atomic cohesion energy [61] of mercury clusters as a function of size.

Acknowledgement

It is a pleasure to thank Prof. F. Hensel, Prof. U. Even, Dr. M. Schlauf, Dr. B. Kaiser, and Dr. O. Dimopoulou-Rademann for their valuable help, discussions, and contributions. Financial support by the Deutsche Forschungsgemeinschaft and the German Israeli Foundation (GIF) is gratefully acknowledged.

4 References

1. Varandas AJC (1988) Intermolecular and intramolecular potentials. In Prigogine I, Rice SA (eds) Adv Chem Phys LXXIV, Wiley, New York, p 255 and references therein
2. Jortner J (1984) Ber Bunsenges Phys Chem 88: 188
3. Benedek G, Martin TP, Pacchioni G (eds) (1988) Elemental and Molecular Clusters, Springer Series in Materials Science, Springer, Berlin Heidelberg New York
4. Jena P, Rao BK, Khanna SK (eds) (1987) The Physics and Chemistry of Small Clusters, NATO ASI Series, Plenum, New York
5. Chapon C, Gillet MF, Henry CR (eds) (1989) Small Particles and Inorganic Clusters; Proceedings of the Fourth International Meeting on Inorganic Clusters, Springer-Verlag, Berlin Heidelberg New York
6. Märk TD, Castleman Jr AW (1985): Experimental studies on cluster ions, In: Adv in Atomic and Molecular Physics 20: 65
7. Maier JP (ed) (1989) Ion and cluster ion spectroscopy and structure, Elsevier, Amsterdam
8. Lubman DM, Rettner CT, Zare RN (1982) J Phys Chem 86: 1129
9. Toennies JP (1974) Molecular beam scattering experiments on elastic, inelastic, and reactive collisions. In: Jost W (ed) Physical Chemistry, An Advanced Treatise, Vol VIA, Academic Press, New York, p 227
10. Levy DH (1981) Adv Chem Phys 47: 323
11. Ploog K (1988) In: Scoles G (ed) (1988) Atomic and Molecular Beam Methods, Oxford Univ Press, New York, p 416
12. Ramsey NF (1956) Molecular Beams, Clarendon Press, Oxford
13. Beijerinck HCW, Vester NF (1981) Physica 111C: 327
14. Beijerinck HCW, van Gerwen RJF, Kerstel ERT, Martens JFM, van Vliembergen EJW, Smits MRT, Kaashoek GH (1985) Chem Phys 96: 153
15. Campargue R (1984) J Phys Chem 88: 4466
16. Amirav A, Even U, Jortner J, Birss FW, Ramsay DA (1983) Can J Phys 61: 278
17. Hagena OF (1974) Cluster beams from nozzle sources. In: Wegener PP (ed) Molecular Beams and Low Density Dynamics, Marcel Dekker, New York, p 93
18. Murphy HR, Miller DR (1984) J Phys Chem 88: 4474
19. Becker EW, Bier K, Henkes W (1956), Z Physik 146: 333
20. Becker EW, Klingelhofer R, Lohse P (1962), Z Naturforsch 17A: 432
21. Frank F, Schulze W, Tesche B, Urban J, Winter B (1985) Surf Sci 156: 90, and references therein
22. Begemann W, Dreihöfer S, Meiwes-Broer KH, Lutz HO (1986) Z Phys D3: 183
23. Brucat PJ, Zheng LS, Pettiette CL, Yang SH, Smalley RE (1986) J Chem Phys 84: 3078, and references therein
24. Stein GD, Wegener PP (1967) J Chem Phys 46: 3685
25. Farge J, de Feraudy MF, Raoult B, Torchet G (1981) Surf Sci 106: 95
26. Abshagen, M. Fischer T, Kowalski J, Meyberg M, zu Putlitz G, Stehlin T, Träger F, Well J (1989) J Phys (Paris) 50C2: 169
27. Buck U (1988) J Phys Chem 92: 1023
28. Herrmann A, Hofmann M, Leutwyler S, Schumacher E, Wöste L (1979) Chem Phys Lett 62: 216
29. Hopkins JB, Powers DE, Smalley RE (1981) J Phys Chem 85: 3739
30. Fung KH, Henke WE, Hays TR, Selzle HL, Schlag EW (1981) J Phys Chem 85: 3560
31. Leutwyler S, Even U, Jortner J (1982) Chem Phys Lett 86: 439

32. Rademann K, Brutschy B, Baumgärtel H (1983) Chem Phys 80: 129
33. Poliakoff ED, Dehmer PM, Dehmer JL, Stockbauer R (1982) J Chem Phys 76: 5214
34. Rademann K (1989) Ber Bunsenges Phys Chem 93: 653
35. Rademann K, Jochims HW, Baumgärtel H (1985) J Phys Chem 89: 3459
36. Ng CY (1983) Adv Chem Phys 52: 264
37. Berkowitz J (1979) Photoabsorption, Photoionization, Photoelectron Spectroscopy, Academic Press, New York
38. Baumgärtel H, Jochims HW, Brutschy B (1987) Z Phys Chem 154: 1
39. de Heer WA, Knight WD, Chou MY, Cohen ML (1987) Solid State Physics 40, p 93, Academic Press, New.York
40. Kappes MM (1988) Chem Rev 88: 369
41. Brechignac C, Broyer M, Cahuzac P, Delacretaz G, Labastie P, Wolf JP, Wöste L (1988) Phys Rev Lett 60: 275
42. Smalley RE, Wharton L, Levy DH (1977) Acc Chem Res 10: 139
43. Amirav A, Even U, Jortner J (1980) Chem Phys 51: 31
44. Levy DH (1980) Ann Rev Phys Chem 31: 197
45. Scoles G (ed) (1988) Atomic and Molecular Beam Methods, Oxford Univ Press, New York
46. Jortner J, Even U, Leutwyler S, Berkovitch-Yellin Z (1983) J Chem Phys 78: 309
47. Ito M, Ebata T, Mikami N (1988) Ann Rev Phys Chem 39: 123
48. Schlag EW, Selzle H (1990) J Chem Soc Faraday Trans 86: xxx
49. Amirav A, Sonnenschein M, Jortner J (1986) Chem Phys 102: 305
50. Leopold DG, Vaida V, Granville MF (1984) J Chem Phys 81: 4210
51. Schlauf M, Dimopoulou-Rademann O, Rademann K, Even U, Hensel F (1989) J Chem Phys 90: 4630
52. Dimopoulou-Rademann O, Even U, Amirav A, Jortner J (1988) J Phys Chem 92: 5371
53. Kreibig U, Fragstein CV (1969) Z Phys 224: 307
54. Inagaki T, Williams MW (1981) Phys Rev B23: 5246
55. Rademann K, Kaiser B, Rech T, Even U, Hensel F (1990) Rev Sci Instrum; submitted for publ
56. Wiley WC, McLaren IH (1955) Rev Sci Instrum 26: 1150
57. Daly NR (1960) Rev Sci Instrum 31: 264
58. Kruit P, Read FH (1983) J Phys E16: 313
59. Cheshnovsky O, Yang SH, Pettiette CL, Craycraft MJ, Smalley RE (1987) Rev Sci Instrum 58: 2131
60. Cabaud B, Hoareau A, Melinon P (1980) J Phys D13: 1381
61. Haberland H, Kornmeier H, Langosch H, Oschwald M, Tanner G (1990) J Chem Soc Faraday Trans
62. Eckardt W (1984) Phys Rev B29: 1558
63. Cotti P, Güntherodt HJ, Münz P, Oelhafen P, Wullschläger J (1973) Solid State Comm 12: 635
64. Schlauf M (1989) Doctoral Thesis, University of Marburg

CHAPTER XIII

Molecular Beam Scattering: Method and Results on Intermolecular Potentials

U. Buck

Methods are reviewed for carrying out elastic and rotationally inelastic scattering experiments with molecules. The different kind of detailed and state-selected cross section and its relation to the anisotropic potential energy surface is discussed. Methods are presented for the determination of these potentials from the measured data in combination with ab initio calculations. The scattering of H_2 from H_2, CO, and NH_3 is treated as a typical example for which state-to-state differential cross sections, rotational rainbows in differential energy loss spectra, or both energy loss spectra and state-to-state integral cross sections are used as input information for the determination of the potential.

1 Introduction

One of the most direct ways to obtain information on intermolecular forces is to carry out scattering experiments with the molecules of interest in crossed molecular beams. In these experiments, advantage is taken of the main characteristic properties of beams, namely the nearly collision-free environment and the possibility to introduce perturbations and to measure their effect in a

Intermolecular Forces
© Springer-Verlag Berlin Heidelberg 1991

fully controlled way. By these means single collisions are investigated under well-defined geometrical and energetic conditions. Although this fact was realized very early and first experiments of this kind appeared in the 1930s, the real breakthrough of this technique as a universal tool had to wait until the late 1960s where supersonic jets were used for the production of the beams, ultra-high-vacuum techniques in the detection region, and lasers as a suitable means for preparing and detecting single molecular states [1].

The traditional sources of information on intermolecular forces, the study of equilibrium and transport properties of the matter in the bulk gas or condensed phase have several shortcomings compared with molecular beam data. Condensed-phase studies are found to be affected by the presence of three-body forces which, in turn, require a precise knowledge of the two-body forces for their determination. Gas phase properties suffer from the fact that the interaction potential is buried under a series of averaging processes, so that it is difficult to recover it directly. Of course, spectroscopy provides by far the most precise knowledge of intermolecular forces. It has, however, the short-comings that it gives mainly information on the attractive part of the potential and that the analysis becomes more and more complicated when the number of atoms involved increases. In this respect, scattering experiments can be considered to be a complementary source of information with their sensitivity to both the repulsive and the attractive part of the interaction.

What is needed for the complete determination of the potential is a state-to-state differential cross section which includes elastic and inelastic processes. In this paper we will restrict ourselves to rotationally inelastic processes, since at low collision energies they give by far the largest contributions. Such angular distributions can be directly related to the complete anisotropic, non-central interaction potential as is known from elastic scattering for central potentials [2]. These data are now available for a series of scattering partners [3, 4]. Based on these experimental results and detailed theoretical studies [5], a more-or-less complete picture of the rotational energy transfer has resulted and methods for relating measured cross sections to properties of the anisotropic interaction have been developed. It is the aim of this contribution to report on recent results concerning both detailed experiments on rotational energy transfer and methods for determining the interaction potential from this data. This will be demonstrated by examples measured in our laboratory. For further details we refer, in general, to the book edited by Scoles [6] and special articles on this very topic presented therein [1, 7, 8]. We start with an overview of the methods. After defining the observables, i.e., the different detailed cross sections, the experimental tools are described which are used to measure these cross sections. Then the calculations are discussed from which the information on the inter-molecular potential is derived including the sensitivity to the data. Then results are presented for three typical molecular systems H_2-H_2, H_2-CO, and H_2-NH_3 which illustrate with their increasing complexity the state-of-the-art of crossed molecular beam work and the derived potential energy surfaces.

Two very successful branches of molecular beam work are not treated in this short article. One is concerned with scattering experiments on chemical dynamics, a flourishing field of modern science for which the 1986 Nobel price was granted to D. Herschbach, Y.T. Lee, and J. Polanyi. The other application is the production of clusters [9] and their spectroscopy [10] which when carried out with size-selected species can also contribute detailed information on the determination of potential energy surfaces [11].

2 Methods

2.1 Observables

The quantity which contains most of the information on the intermolecular potential is the state-to-state differential cross section from the initial rotational state j to the final rotational state j':

$$\frac{d\sigma}{d\omega}(j \rightarrow j'; \theta, E). \tag{1}$$

It depends only on the deflection angle θ and the collision energy E. Such a cross section is difficult to measure. Therefore we will include in our discussion also two other quantities for which one of the strict requirements has been relaxed. The first example is the differential energy loss cross section:

$$\frac{d\sigma}{d\omega}(\Delta E; \theta, E). \tag{2}$$

In this case, single rotational states are not resolved and replaced by the energy loss $\Delta E = B_{rot}[j'(j' + 1) - j(j + 1)]$ which is treated as a continuous variable, but the angular dependence is retained. This type of cross section is measured by both velocity and angular analysis of the scattered particles. The second example is the state-to-state integral cross section:

$$\sigma(j \rightarrow j', E) = \int d\omega \frac{d\sigma}{d\omega}(j \rightarrow j'; \theta, E). \tag{3}$$

In this case, the angular dependence is integrated and no longer measured, while the state selectivity is retained. A further quantity of interest is the total differential cross section:

$$\frac{d\sigma}{d\omega}(\theta, E) = \sum_{j,j'} (j \rightarrow j'; \theta, E) \tag{4}$$

which is measured without state selection or velocity analysis and corresponds to the elastic scattering cross section for atomic systems without internal degrees of freedom. Formally it contains a summation over all final rotational states j′ and an average over all possible initial rotational states j.

2.2 Experiments

The general aim of the experiment is to measure state-selected differential cross sections according to Eq. (1). Aside from the usual requirements for high angular resolution in crossed molecular beam experiments, the measurement of this quantity requires, in addition, the preparation of the initial and the detection of the final rotational state. From a larger variety of possible state-selective techniques, essentially two experimental concepts emerged which are used in nearly all experiments.

In the first method, the *energy loss method*, the final state is identified by velocity analysis of the scattered particles. To resolve single rotational states from each other, very high velocity and angular resolution is required for the two intersecting beams, the velocity analyzer, and the detector. In this arrangement the initial state is usually prepared by using a supersonic nozzle beam which in favorable cases generates only molecules in their lowest rotational

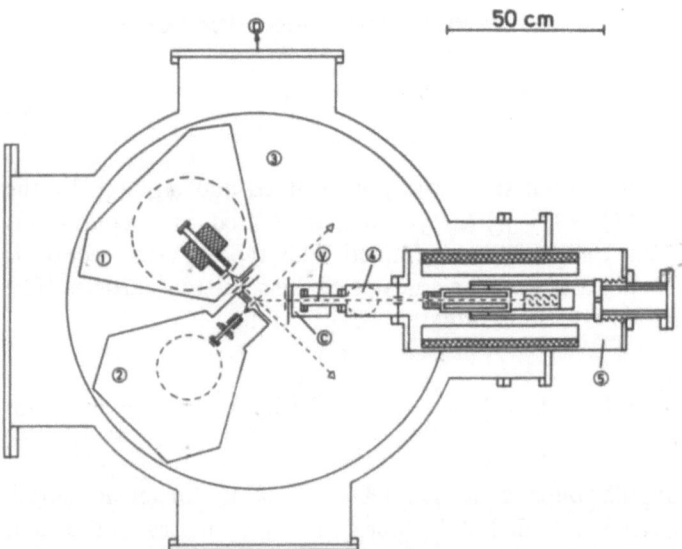

Fig. 1. Schematic view of the crossed molecular beam apparatus for experiments using the energy loss method [12]. (1) Primary beam source chamber, (2) secondary beam source chamber, (3) scattering chamber, (4) detector buffer chamber, (5) detector chamber, (C) pseudorandom chopper, (D) diffusion pump, (V) UHV-valve

state. A typical example of such an apparatus used in our laboratory is shown in Fig. 1 [12, 13]. The two colliding beams which cross at 90° are produced as nozzle beams from two differentially pumped chambers (1) and (2). The angular dependence is measured by rotating the source assembly with respect to the fixed detector unit (4, 5). The scattered particles are detected by a doubly differentially pumped mass spectrometer with electron impact ionization operating at pressures around 10^{-10} mbar. Elastic and inelastic events are separated by time-of-flight analysis of the scattered particles using the pseudo-random chopping technique (C). The flight path varies between 449.5 and 593 mm. In this way the velocity distribution of the scattered particles is measured with a resolution of about 2%, a number which includes effects of the finite ionization volume, the shutter function of the chopper, and the channel width of the electronic storage unit. Collection and processing of the data are executed by a minicomputer. The pumping facilities of the two beam sources allow to use pressures up to 150 bar with 18 μm nozzle diameter. When the two intersecting beams consist of He and Ne atoms or H_2 and D_2 molecules, they have a velocity distribution with a full width at half maximum of less than 5%. In addition, the molecules are in the lowest rotational state. For CO and NH_3 molecules in the secondary beam, the pressure was kept at much lower values around 500 mbar in order to avoid condensation. For NH_3 in the primary beam, a mixture of 8% in He is used to achieve the required resolution of less than 10%.

If the experimental resolution is good enough to separate the time-of-flight (TOF) distributions for different rotational transitions from each other, these intensity peaks are easily converted to state-selected differential cross sections. A typical example is shown in Fig. 2 for the scattering of *ortho*-D_2 by *para*-H_2 at the collision energy of E = 89.1 meV and the laboratory deflection angle $\theta_1 = 60°$ [14]. The two inelastic contributions which correspond to the final rotational states j'(D_2)j'(H_2) = 20 and 02 are small but clearly separated from the elastic peak. They reflect the mutual exitation of D_2 and H_2 from j = 0 to j' = 2 which appears at different ΔE because of the different rotational constants. If the resolution is not high enough for the separation, the TOF-spectra are converted to differential energy loss cross sections according to Eq. (2). The big advantage of this method is the universal applicability to nearly any system, provided that reasonably good supersonic nozzle beams can be produced. The disadvantage is the unspecific detection and the requirement of high quality of the beams.

This problem is avoided in the second method, *the laser method*, in which the state-specific detection method of laser-induced fluorescence (LIF) is utilized. This method only requires a bound–bound electronic transition whose wavelength can be reached with tunable laser radiation. In this way single rotational states can be detected without problems. A modern variant of this method is the resonance-enhanced multiphoton ionization (REMPI) in which the first step is the same as in LIF, but instead of detecting the emitted fluorescence of the excited state, the radiation of one or two other photons is used for the production

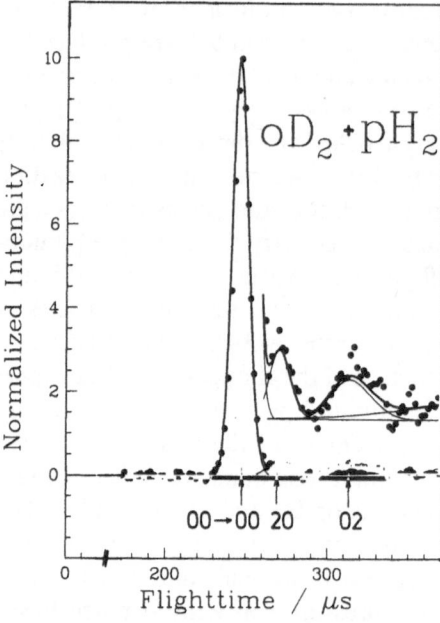

Fig. 2. Measured time-to-flight spectrum for $oD_2 + pH_2$ collisions at $E = 89.1$ meV and $\theta_1 = 60°$ [14]. Rotational transitions are marked by $j(D_2)j(H_2)$. Inelastic transitions are also shown in 5-fold enlargement. *Solid lines* are results of the spectra simulations with cross-section ratios given by the ab initio potential of Ref. [36]

of ions which are then detected. For the initial state preparation, different methods are available. 1—Laser population depletion by optical pumping has been used for measuring state-to-state differential cross sections for Na_2 scattered from rare gases for a large variety of initial and final states between $j = 0$ and $j' = 80$ [15–17]. 2—Deflection in inhomogeneous electric fields has been used for the preparation of molecules with a permanent dipole moment such as LiH $(j = 1)$ [18] and CaCl $(N = 2)$ [19] for which integral cross sections were measured. 3—Finally, also supersonic nozzle beams are employed for the

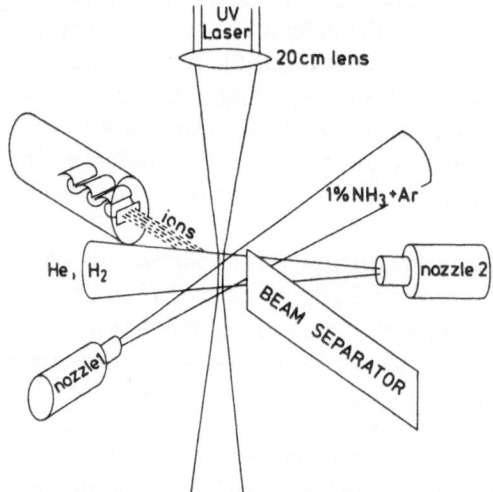

Fig. 3. Schematic view of a molecular beam apparatus for experiments using the laser method [22]

Table 1. State-to-state integral cross sections for $NH_3 + H_2$ collisions at $E = 75\,meV$ in Å^2 starting from $jk\varepsilon = 00 + 1$ and normalized to the sum of the calculated cross section [43]

$j'k'\varepsilon'$			measured		calculated
1	0	+1	12.2		12.2
2	0	+1	5.8		3.5
3	0	+1	1.5		1.6
4	0	+1	0.7		0.3
3	3	+1	2.09		—
3	3	−1	3.34	5.43[a]	6.8
4	3	+1	1.03		—
4	3	−1	1.18	2.21[a]	3.49
5	3	+1	0.19		—
5	3	−1	0.40	0.59[a]	0.65
6	6	+1	0.43		—
6	6	−1	0.25	0.68[a]	0.73

[a] The sum of $\varepsilon' = \pm 1$ final states

initial state selections in measurements of integral cross sections for NO ($^2\Pi$) [20], OH ($^2\Pi$) [21], and NH_3 [22].

An example of such a measurement is the state-to-state integral cross section of $NH_3 + H_2$ [22]. A sketch of the experimental setup is shown in Fig. 3. Two pulsed nozzle beams are crossed at 90°. The rotational cooling of the expansions guarantees that the two molecules are in their rotational ground states before the collision. The final rotational state distribution of the collisionally excited NH_3 molecules is probed by an excimer-pumped dye laser in the ultraviolet spectral range. The laser is focussed into the region where the two beams overlap. The NH_3 molecules are resonantly excited to the $\tilde{B}$ state by two-photon absorption and subsequently ionized by another photon ((2 + 1)REMPI). The formed NH_3^+ ions are extracted from the scattering center to an open particle multiplier mounted inside the chamber. The signal from the particle multiplier in processed through a boxcar integrator and stored in a minicomputer. From the measured signal and the known line strength and Franck–Condon factors of the transition, the final state densities and the integral cross sections are determined. The results are presented in Table 1.

2.3 Determination of Potentials

The great advantage of analyzing differential scattering cross sections is the close connection of the measured angular dependence to the underlying potential energy. Generally speaking, small angle scattering is caused by attractive, and large angle scattering by repulsive forces. In elastic scattering a maximum in the angular distribution occurring near the classical rainbow angle $\theta_r = 2\varepsilon/E$, where ε is the potential well depth and E the collision energy, separates these two regions very clearly [1]. For systems with small well depths, the rainbow is buried under the forward diffraction peak and the cross sections are dominated by diffraction oscillations. The angular positions of these oscillations are a direct measure of the diameter of the repulsive wall R_0:

$$\Delta\theta = \pi/k\Delta l \cong \pi/R_0 \tag{5}$$

to better than 1%. $k = (2\mu E/\hbar)^{1/2}$ is the wave number, μ the reduced mass, and Δl the difference of orbital angular momenta corresponding to $\Delta\theta$. The amplitudes of the oscillations are sensitive to the shape of the potential near the well depth and allow its determination to within 10%.

In order to get a similarly simple picture of the influence of the anisotropy of the potential on the rotationally inelastic cross section, the concept of rotational rainbows has proved to be very useful [5]. For the simple case of the scattering of an atom from a rigid ellipsoid, the effect is related to the maximum in the final angular momentum curve, $j'(l, \gamma)$, as a function of the orientation angle γ of the molecule with respect to the incoming atom for fixed orbital angular momentum l. These relations are demonstrated in Fig. 4. A maximum occurs since at the limiting orientation angles $\gamma = 0$ and $\gamma = \pi/2$ no torque can be transferred to the ellipsoid from the colliding atom. The maximum j'_r is directly related to the anisotropy of the potential ΔR:

$$j'_r = 2k\Delta R \sin(\theta/2) \tag{6}$$

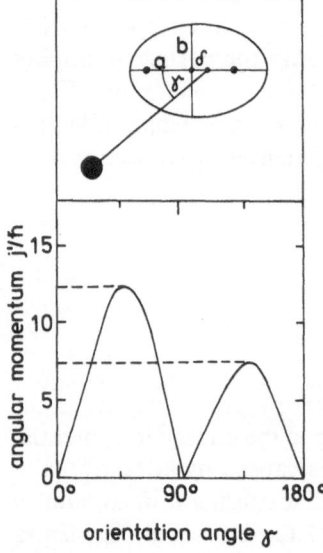

Fig. 4. The final angular momentum j' as function of the orientation angle γ for an off-center rigid shell model to demonstrate the two-peak structure of the rotational rainbow in differential rotationally inelastic cross sections

where k is the wave number and θ the deflection angle. In the present model, ΔR is related to the difference of the long and short semiaxes a and b of the ellipsoid. For heteronuclear systems with a difference δ between the center of mass and the center of symmetry we get $\Delta R = a - b \pm \delta[a(a + b)^{-1}]^{1/2}$ so that two rotational rainbow maxima result since the long semiaxis is different for the two ends of the molecule [23]. Although the expression of Eq. (6) is derived under several approximations, it contains the essential features of the rotational rainbow theory and has proved to be extremely useful, since it relates the measured data to system parameters and the anisotropy of the potential ΔR which can be approximated for realistic repulsive potentials by the difference of the classical turning points for different orientation angles $\gamma = 0°$ and $\gamma = 90°$. Equation (6) also shows that even in state-selective experiments, only the difference of potential terms belonging to two orientations is obtained. For the determination of the absolute scale of the interaction, additional information is needed, for instance from the measurement of the diffraction oscillations. It is noted that in the case of strong attractive anisotropic interactions, the situation gets much more complicated because of a mixture of the normal and the rotational rainbows.

The best way to determine potentials from measured cross sections is to apply so-called inversion procedures which allow the determination without assuming a potential model [2]. For rotationally inelastic scattering, only very limited procedures are available so that, in general, the method of trial-and-error has to be applied. In this procedure, a realistic potential form is assumed and the final interaction is obtained by varying the parameters of the potential function. First, the potential has to be parametrized. A realistic potential surface is conventionally expanded in a series of spherical harmonics for both molecules:

$$V(R) = \sum V_{\lambda_1\mu_2\mu_2}(R)Y_{\lambda_1\mu_1}(\gamma_1, \phi_1)Y_{\lambda_2\mu_2}(\gamma_2, \phi_2). \tag{7}$$

The angles γ, ϕ give the orientation of the molecules in a spherical coordinate system. The number of expansion coefficients $V_{\lambda\mu}(R)$ can be reduced by symmetry considerations, but very often the remaining series converges slowly so that many terms have to be retained in the expansion. In such a case it is advisable not to parametrize each single term $V_{\lambda\mu}(R)$ but to choose a single parametrized form of the potential and to make the parameters P_i angular dependent $V(R) = f(P_i(\gamma_i\phi_i), R)$ [24]. This procedure reduces the number of parameters considerably. Another possibility is to use ab initio calculations of the potential for different orientations of the molecules. Calculations, however, which are precise enough to account for the details of state-selected cross sections are quite rare and are only available for small systems, since they have to be carried out at the configuration interaction (CI) level.

Once a suitable potential function is assumed, the calculation of the cross sections has to be performed by quantum mechanics using either the full "close coupling" method or, at least, the "centrifugal sudden (CS)" approximation in order to account for the detailed features of the cross sections [25]. Both types of calculations are rather time consuming at the computer and not very well suited for extensive fitting procedures. Therefore, sometimes, the more approximate infinite order sudden (IOS) method is used which gives reasonable results if the transferred energy is small compared with the collision energy.

Summarizing, there are three procedures available for comparing measured cross sections with calculations and for determining the underlying intermolecular forces.

1. The measurements are compared with cross-section calculations based on high quality ab initio potentials calculated at the CI level. The comparison is used for final corrections.
2. The suitably parametrized potential is fitted to the measured data. To improve the results also gas-phase data are used as additional constraints for the potential in a multiproperty analysis.
3. For complex systems often both the measurement and the ab initio calculation is difficult to carry out at the most sophisticated level. In such a case, a combined procedure is very helpful. Using reliable self-consistent-field (SCF) Hartree-Fock calculations for the repulsive and perturbation theory for the long-range attractive part of the potential, and adjusting the missing contributions by a fit to the available cross-section data, a very reliable potential results which is, on the one hand, free of arbitrary parametrizations of a complete fit and, on the other hand, also free of problems encountered in full CI calculations.

3 Results

3.1 Overview

There are only a few examples of state-to-state differential cross sections which have been converted to or compared with complete anisotropic potential surfaces. Most of them were obtained for atom–diatom systems. For $Na_2 + Ne$ measured by the laser method, the cross sections were completely reproduced by calculations based on an ab initio potential surface [16]. From the cross sections for the $0 \rightarrow 2$ exitations of D_2 scattered from Ne [13] and Ar [26] and measured by the energy loss method, reliable potential surfaces were derived which are also able to reproduce many other data in a multiproperty analysis [27]. The highest resolution in TOF-experiments was obtained for the systems $He + N_2$ [28] and $He + O_2$ [29] for which $0 \rightarrow 2$ and $1 \rightarrow 3$ transitions were observed, respectively. The agreement with fitted potential surfaces was satisfactory. For the scattering of polyatomic molecules from atoms, results are available for $CH_4 + He$ [30] and Ne [31] as well as for $NH_3 + He$ [32] and $CO_2 + He$ [33]. Aside from $CH_4 + He$ for which the $0 \rightarrow 3$, $1 \rightarrow 3$, and $1 \rightarrow 2$

transitions were observed, all other systems show unresolved energy loss spectra which were used together with information from ab initio calculations in the determination of the potential surfaces using method (3) of Sect. 2.3.

For the interaction of two molecules, not very many system have been treated in the sense of this section. To demonstrate both the possibilities and the shortcomings of the molecular beam scattering method, the three molecular systems H_2—H_2, H_2—CO, and H_2—NH_3 were chosen for a more detailed presentation. Before starting, we would like to comment on the measurement of total differential cross sections [see Eq. (4)] which are available for many systems. A typical example measured in the apparatus discussed in Sect. 2.2 under high resolution conditions is shown in Fig. 5 for $D_2 + NH_3$. For comparison, also the atom–atom system He + Ne is displayed which has nearly the same masses and, therefore, demonstrates the resolution of the apparatus. What is seen are the diffraction oscillations which are severely damped in the case of $D_2 + NH_3$, a consequence of the anisotropy of the potential. To use this information for the determination of the anisotropy without taking into account the inelastic cross sections can, however, be a dangerous task, since this is by no means a unique procedure. On the other hand, these cross sections are an important constraint for the complete potential surface and are quite helpful in the determination of complex surfaces as was shown for $H_2O + H_2O$ [34] and $NH_3 + NH_3$ [35].

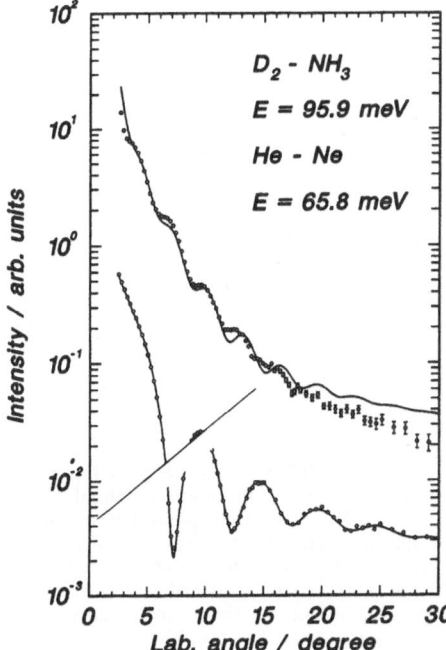

Fig. 5. Total differential cross sections for the scattering of $D_2 + NH_3$ (upper curve) at $E = 95.9$ meV and He + Ne (*lower curve*) at 65.8 meV. Diffraction oscillations are exhibited [43]. *Solid lines* are calculated using experimentally determined potentials. For $D_2 + NH_3$ see Fig. 11

3.2 H$_2$—H$_2$

The molecular hydrogen dimer represents the simplest diatom–diatom system and has attracted a great deal of interest. It is small enough to be tackled in ab initio calculations at the configuration interaction level for a number of orientations. Only very few potential terms of the usual expansion in spherical harmonics [see Eq. (7)] contribute. The results of the best calculation by Meyer and Schaefer are shown in Fig. 6 in space-fixed coordinates [36]. The leading term V_{000} is the isotropic spherically averaged part. The anisotropy is represented by V_{202} which is responsible for the $0 \rightarrow 2$ exitation of one molecule and V_{224} which couples two rotating molecules and which is asymptotically caused by the long-range interaction of the quadrupole moments of the two molecules.

Molecular beam experiments were carried out for this system by measuring differential cross sections for HD + D$_2$ [37, 38] and oD$_2$ + pH$_2$, oH$_2$ [14, 39]. The resolution in the TOF-experiments was sufficient to resolve single rotational transitions $0 \rightarrow 1$ and $0 \rightarrow 2$ for HD and $0 \rightarrow 2$ for D$_2$ and H$_2$ as was already shown in Sect. 2.2. The cross sections for latter system are displayed in Fig. 7 together with exact close-coupling calculations based on the potential of Fig. 6. The figure shows a hierarchy of cross sections for two different initial states, 00 and 01, where the first number refers to D$_2$ and the second one to H$_2$. The inelastic cross sections are more than an order of magnitude smaller than the elastic channels. The most remarkable result is the inelastic cross section for the $0 \rightarrow 2$ transition of D$_2$ in oD$_2$ + oH$_2$ collisions ($01 \rightarrow 21$) which is about a

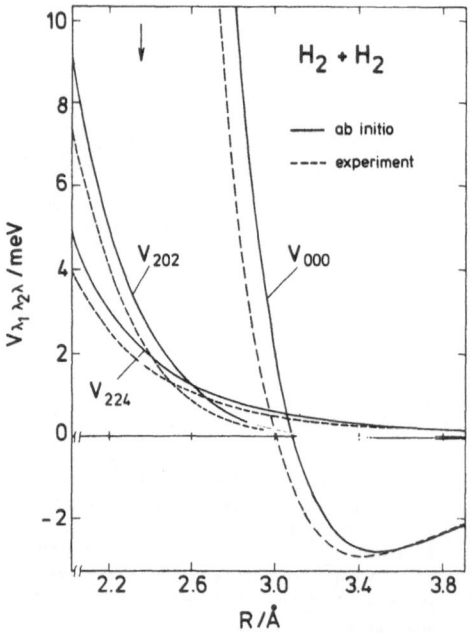

Fig. 6. Potential surfaces for H$_2$ + H$_2$. *Solid lines*: calculated ab initio potential [36]; *dashed lines*: corrected values due to a fit to the molecular beam experiments shown in Fig. 7 [14, 39]. The *arrow* indicates the position of the classical turning point corresponding to the collision energy of the experiment shown in Fig. 7

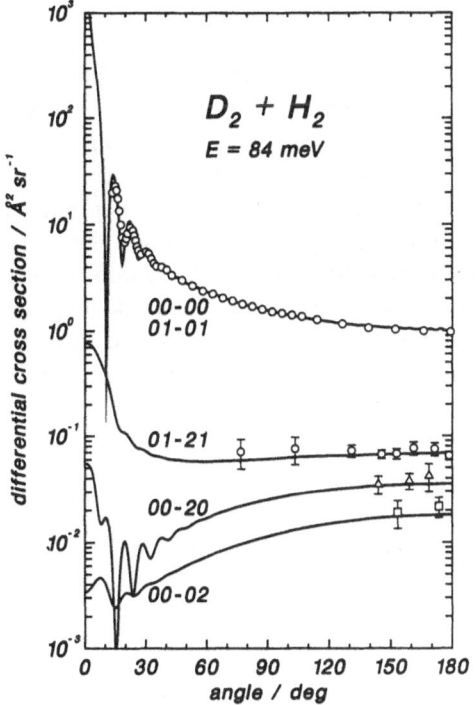

Fig. 7. Measured state-to-state differential cross sections for $D_2 + H_2$ collisions at 84 meV [14]. Transitions are marked by $j(D_2)j(H_2) - j'(D_2)j'(H_2)$. *Solid lines* are calculations with the ab initio potential of [36]. See Fig. 6

factor of two larger than that for the same transition in $oD_2 + pH_2$ collisions $(00 \rightarrow 20)$ in which the scattering partner, H_2, does not rotate before and after the collision. The reason for this behavior is the different coupling for the two processes. The first transition is coupled by V_{202} and V_{224}, while the latter one is coupled only by V_{202}. The smallest measured cross section is that for the $0 \rightarrow 2$ transition of H_2 which requires more energy (43.2 meV) than the same transition of D_2 (22.2 meV), although the same coupling term, V_{202}, is responsiable for both.

The agreement between calculations and measurements is very good. A careful inspection, however, of the diffraction oscillations, not shown in detail in the figure, reveals a small but significant deviation which can be corrected by shifting V_{000} inwards by about 0.08 Å and increasing the well depth by 7%. The resulting corrected potential surface is also shown in the figure. This potential is able to reproduce many of the available data including that of the solid [40]. This example shows the power of the beam method, provided state-to-state data are available.

3.3 H₂—CO

For this system total differential cross sections with well-resolved diffraction oscillations and differential energy loss cross sections measured for $D_2 + CO$ at $E = 87.2$ meV are available [41, 42]. The energy loss cross sections are shown in Fig. 8. Single transitions are not resolved. The dominant feature is the rotational rainbow structure resulting from the CO excitation. As is predicted by Eq. (6), the energy loss increases with increasing scattering angle. The peak at smaller energy transfer correlates with the exitation near the O end (smaller distance from the molecule's center of mass (cm)), while the peak at larger transfer is caused by the exitation near the C end (greater distance from the cm of the molecule). The data have been analyzed in terms of an interaction potential expanded in Legendre polynomials:

$$V(R, \gamma_1, \gamma_2) = \sum_{\lambda_1 \lambda_2} V_{\lambda_1 \lambda_2}(R) P_{\lambda_1}(\cos \gamma_1) P_{\lambda_2}(\cos \gamma_2) \tag{8}$$

The potential is already averaged over the small dependence on the angle $\phi = \phi_2 - \phi_1$. The terms $V_{\lambda_1 \lambda_2}(R)$ are approximated by:

$$V_{\lambda_1 \lambda_2}(R) = V_{\lambda_1 \lambda_2}^{SCF}(R) + V_{\lambda_1 \lambda_2}^{dis}(R) f(R) \tag{9}$$

with $f(R) = \exp[-\alpha(D/R - 1)^2]$. The short range repulsion V^{SCF} is calculated using a large basis set SCF calculation and the long range attraction V^{dis} is obtained from perturbation theory. $f(R)$ is a suitable damping function, the free

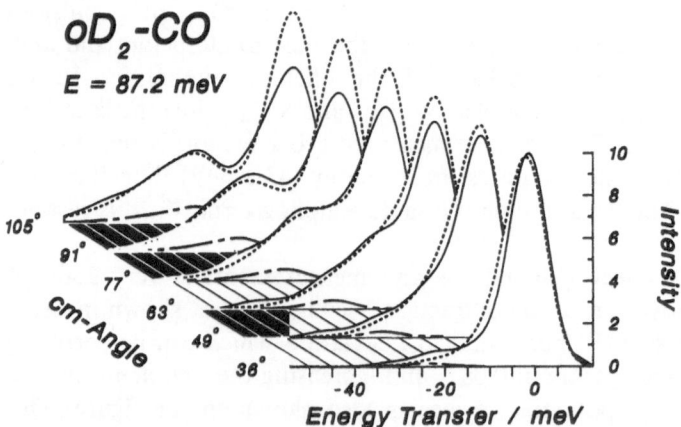

Fig. 8. Measured (*solid*) and calculated (*dashed*) differential energy loss cross sections for $D_2 + CO$ collisions at 87.2 meV [42]. The contribution of the D_2 excitation is shown separately (dashed-dotted). The two-peak rotational rainbow structure is clearly observed at larger angles. See Fig. 3 for an explanation

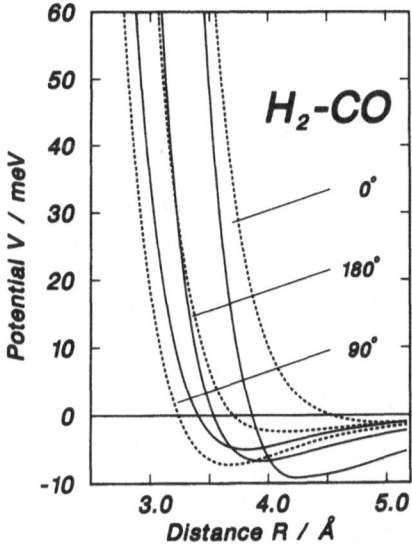

Fig. 9. Potential energy curves for different orientations of the molecules determined from experimental data for $H_2 + CO$ [42]. $\gamma_1 = 0°$ is the approach to the C end, $\gamma_1 = 180°$ to the O end of CO. The different H_2 orientations are given by *solid* ($\gamma_2 = 0°$) and *dashed* ($\gamma_2 = 90°$) lines

parameters of which are fitted simultaneously to the available experimental data. The dynamical calculations are carried out in the CS approximation and the result is shown as a dashed line in Fig. 8. The contribution of the D_2 exitation (dashed-dotted line) is only small compared with the CO exitation. The resulting potential surface is shown in Fig. 9. The repulsive walls are shifted to smaller distances going from the C-approach ($\gamma_1 = 0°$) over the O-approach ($\gamma_1 = 180°$) to the perpendicular approach ($\gamma_1 = 90°$). The differences ΔR which are responsible for the two rotational rainbows result from these curves. The different orientations of the H_2 molecule (solid: $\gamma_2 = 0°$, dashed: $\gamma_2 = 90°$) are only of minor importance and do not change this picture. The parallel configuration of the molecules is the most stable one with a minimum distance of 4.23 Å and a well depth of 9.2 meV. Other potential surfaces for this system could not satisfactorily reproduce the diffraction oscillations, thus indicating a wrong well depth ε and zeropoint R_0, or the rotational rainbows indicating the wrong anisotropy.

3.4 $H_2 — NH_3$

The step from a diatomic collision partner to a polyatomic one causes further complications. In the present case, NH_3 is a symmetric top. Each rotational state is characterized by the rotational quantum number j, the projection on the intermolecular axis, k, and the parity ε. For the present system three different molecular beam measurements are available [43]:

1. Total differential cross sections for $oD_2 + NH_3$ at $E = 95.9\,meV$ and $E = 111.3\,meV$. The results taken at the lower energy were already shown in Fig. 5.

2. Differential energy loss spectra for $ND_3 + D_2$ at $E = 118\,meV$ [41]. In order to increase the resolution and to achieve a lower background signal in the detector, a beam of 10% ND_3 seeded in He is scattering from an oD_2 beam. To avoid possible molecular clusters, those laboratory scattering angles are selected in which only monomers can be scattered. This procedure restricts the measured angular range considerably so that the data cannot be transformed to the center-of-mass and the comparison with calculations has to be carried out in the laboratory system. Typical examples of the TOF spectra obtained in this way are shown in Fig. 10. The measured maximum of the energy transfer is shifted from the elastic transition marked by the two outer arrows. The possible maximum energy transfer is indicated by the arrow in the middle.

3. State-to-state integral cross sections for many rotational transitions at $E = 75\,meV$ starting from $j = 0, k = 0, \varepsilon = +1$ and $j = 1, k = 1, \varepsilon = +1$ [22]. The results for the former are presented in Table 1. It is interesting to note that the cross sections for the simultaneous excitation of j' and k' (33 and 43) are larger than those for pure j' excitation (30 and 40).

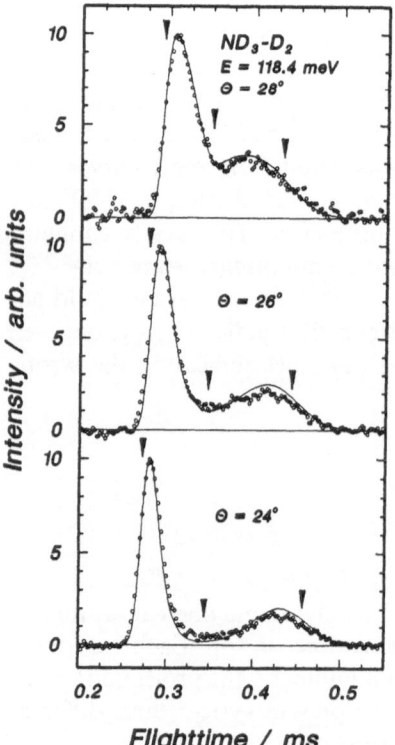

Fig. 10. Measured time-of-flight distributions for $ND_3 + oD_2$ collisions at $E = 118.4\,meV$ and different laboratory angles [43]. The positions of elastic scattering are marked by the two outer arrows. The arrow in the middle indicates maximal energy transfer

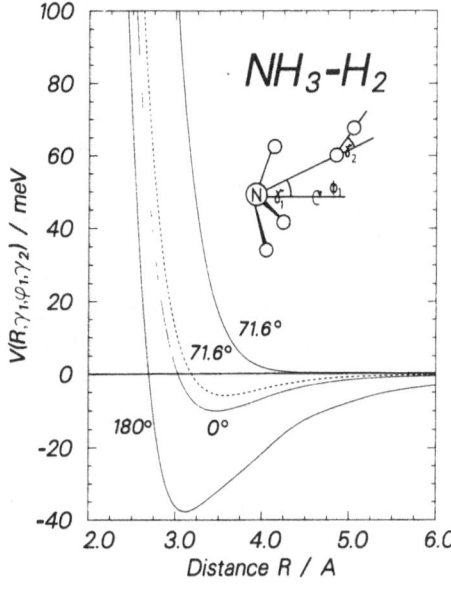

Fig. 11. Potential energy curve for $NH_3 + H_2$ for different orientation angles of the molecules [γ_1 as indicated and $\phi_1 = 0°$ (*solid*) and $\phi_1 = 60°$ (*dashed*) as well as $\gamma_2 = 0°$] determined from experimental data [43]

For the determination of the potential we use the same procedure as is described for $H_2 + CO$ (see Sect. 2.3 method 3). In the potential, ϕ_1 of NH_3 has to be retained. For H_2 we have again averaged over the ϕ_2 orientations:

$$V(R, \gamma_1, \phi_1, \gamma_2) = \sum_{\lambda_1 \mu_1 \lambda_2} V_{\lambda_1 \mu_1 \lambda_2}(R) Y_{\lambda_1 \mu_1}(\gamma_1, \phi_1) P_{\lambda_2}(\cos \gamma_2) \qquad (10)$$

A part of the resulting potential is shown in Fig. 11 for the indicated orientations. The comparison of the calculated data with the measured ones are shown in Figs. 5 and 10 as solid lines and in Table 1. The agreement is very satisfactory if the quite different character and sensitivity of the data is considered. Data set (1) is mainly sensitive to R_0 and ε of the spherically averaged V_{000}. Data set (2) is a measure of the difference ΔR in the repulsive part of different orientations. As can be deduced from Fig. 11, there is indeed a pronounced difference between the approach of the linear H_2 molecules along the H—N bond ($\gamma_1 = 71.6°$, $\phi_1 = 0°$) and the approach between the three H atoms ($\gamma_1 = 0°$, $\phi_1 = 0°$). The other orientations are close to this one. This difference is somewhat reduced in the perpendicular approach of H_2. Nevertheless, the average of this differences causes a peak in the energy loss spectra similar to a rotational rainbow for linear molecules. Data set (3) is essentially sensitive to the attractive part of the anisotropy. The cross sections for the different parities of the same final j'k' values are not well reproduced by the calculation. This is a failure of the CS-approximation. The sum of the two cross sections, however, agrees much better with the predictions. Therefore, the fitted potential is very reliable. By far the deepest configuration is the linear arrangement of H_2 ($\gamma_2 = 0$) towards N of NH_3 ($\gamma_1 = 180°$, $\phi_1 = 0°$) caused by the long-range van der Waals forces.

3.5 Concluding Remarks

Thanks to a combined effort of new experimental and theoretical methods, the cross sections for rotational energy transfer are now well understood. Methods to produce and analyze state-selective data and to compare them with calculations are available and have successfully been applied to derive detailed, anisotropic potential surfaces for molecular systems. The three examples presented here show the power of the method. If data is available on a state-selected level, as is the case for $H_2 + H_2$, it is no problem to determine a reliable potential, in the present example by a slight change of a reliable CI calculation. But even for unresolved energy loss cross sections which display a rotational rainbow pattern, reliable potential surfaces were determined by additional information from large basis set SCF-calculations and damped dispersion forces. This was demonstrated for $H_2 + CO$ and $H_2 + NH_3$. What is of great advantage for deriving potential surfaces from differential cross sections, is the direct relation of certain potential parts to certain cross section features which allows an easy correction in case of a discrepancy. Some details of the derived potential surfaces are compared in Table 2. We have chosen the well depth ε, the zeropoint R_0, and the minimum distance R_m of the spherically averaged potential and the deepest configuration to describe the attractive part. To characterize the repulsive part, the difference ΔR of the potential for two different orientations at $V = 60\,meV$ is used. The potential well depths increase going from H_2 via CO to NH_3, while the minimum distance is largest for $H_2 + CO$. The deepest minimum occurs for $H_2 + CO$ in the linear configuration and for $H_2 + NH_3$ in the $H-H...NH_3$ configuration, a consequence of the large van der Waals forces in this arrangement. Surprisingly, the minimum configuration for $H_2 + H_2$ is the T-shape arrangement, a consequence of the large quadrupole–quadrupole interaction. The repulsive anisotropy ΔR is largest for $H_2 + CO$, for which because of the heteronuclear CO two numbers are available, and smallest for $H_2 + H_2$ which is nearly spherical.

Although H_2 as scattering partner is a special case because of its small anisotropy, the results obtained so far are encouraging enough to serve as guide-

Table 2. Characteristic potential parameters for the molecular systems

Systems	ε/meV	V_{000} R_0/Å	R_m/Å	ε/meV	Deepest term R_m/Å	Configuration	$R(--)-R(\perp)$ at 60 meV/Å
H_2-H_2	2.92	3.00	3.44	4.5	3.33	HH...$\begin{smallmatrix}H\\\\H\end{smallmatrix}$	0.16
H_2-CO	4.37	3.60	4.02	9.2	4.23	HH...CO	0.67, 0.27
H_2-NH_3	6.64	3.24	3.67	37.8	3.11	HH...NH_3	0.35[a]

[a] For this system $--$ is the configuration $H_2...H-N$ and $\perp H_2...NH_3$

lines for further experiments on more complicated molecular systems like two-hydrogen-bonded polyatomic molecules. What is needed for such a procedure is a) the measurement of total and inelastic differential cross sections with at least some of the resolved dynamical feature described above, b) good SCF-calculations, and c) procedures for a quantum calculation of the cross sections for two interacting asymmetric tops.

Acknowledgement

I thank all my coworkers who contributed essentially to the results presented here. I am also grateful to my colleagues Dr. J. Schaefer and Dr. R. Schinke who carried out many of the calculations. Finally, I thank B. Schmidt for carefully reading the manuscript.

4 References

1. Buck U (1988) In: Scoles G (ed) Atomic and molecular beam methods, Oxford, New York, Chaps 18, 20, and 21
2. Buck U (1986) Comp Phys Rep 5: 1
3. Faubel M (1983) Adv At Mol Phys 19: 229
4. Buck U (1986) Comments At Mol Phys 17: 143
5. Schinke R, Bowman JM (1983) In: Bowman JM (ed) Molecular Collision Dynamics, Springer, Berlin, Chap 4
6. Scoles G (ed) (1988) Atomic and molecular beam methods, Oxford, New York
7. Dagdigian P (1988) In: Scoles G (ed) Atomic and molecular beam methods, Oxford, New York, p 569
8. Stolte S (1988) In: Scoles G (ed) Atomic and molecular beam methods, Oxford, New York, p 631
9. Kappes M, Leutwyler S (1988) In: Scoles G (ed) Atomic and molecular beam methods, Oxford, New York, p 380
10. Miller RE (1986) J Phys Chem 90: 3301
11. Buck U (1988) J Phys Chem 92: 1023
12. Buck U, Huisken F, Schleusener J, Schaefer J (1980) J Chem Phys 72: 1512
13. Andres J, Buck U, Huisken F, Schleusener J, Torello F (1980) J Chem Phys 73: 5620
14. Buck U, Huisken F, Kohlhase A, Otten D, Schaefer J (1983) J Chem Phys 78: 4439
15. Bergmann K, Hefter K, Witt J (1980) J Chem Phys 72: 4777
16. Jones PL, Hefter K, Matheus A, Witt J, Bergmann K, Müller W, Meyer W, Schinke R (1982) Phys Rev A 26: 128
17. Serri JA, Becker CH, Elbel MB, Kinsey JL, Moskowitz WP, Pritchard DE (1981) J Chem Phys 74: 5116
18. Dagdigian PJ, Wilcomb BE (1980) J Chem Phys 72: 6462
19. Dagdigian PJ, Bullman SJ (1985) J Chem Phys 82: 1341
20. Andresen P, Joswig H, Pauly H, Schinke R (1982) J Chem Phys 77: 2204
21. Andresen P, Häusler D, Lülf HW (1984) J Chem Phys 81: 571
22. Seelemann T, Andresen P, Schleipen J, Beyer B, terMeulen JJ (1988) Chem Phys 126: 27
23. Bosanac S, Buck U (1981) Chem Phys Lett 81: 315
24. Pack RT (1978) Chem Phys Lett 55: 197
25. Bernstein RB (ed) (1979) Atom-molecular collision theory—A guide for the experimentalist, Plenum, New York
26. Buck U, Meyer H, LeRoy RJ (1984) J Chem Phys 80: 5589
27. LeRoy RJ, Hutson JM (1987) J Chem Phys 86: 837
28. Faubel M, Kohl KH, Toennies JP, Tang KT, Yung YY (1982) Faraday Discuss Chem Soc 73: 205

29. Faubel M, Kohl KH, Toennies JP, Gianturco FA (1983) J Chem Phys 55: 1255
30. Buck U, Kohl KH, Kohlhase A, Faubel M, Staemmler (1985) Mol Phys 55: 1255
31. Buck U, Kohlhase A, Secrest D, Phillips T, Scoles G, Grein F (1985) Mol Phys 55: 1233
32. Meyer H, Buck U, Schinke R, Diercksen GHF (1986) J Chem Phys 84: 4976
33. Beneventi L, Casavecchia P, Vecchiocattivi F, Volpi GG, Buck U, Lauenstein C, Schinke R (1988) J Chem Phys 89: 4671
34. Reimers JR, Watts RO, Klein ML (1982) Chem Phys 62: 95
35. Duquette G, Ellis TH, Watts RO, Klein ML, Scoles G (1978) J Chem Phys 68: 2544
36. Schaefer J, Meyer W, Liu B, unpublished, see Refs [14], [39], and Schaefer J, Köhler WE (1989) Z Phys D 13: 217
37. Buck U, Huisken F, Schleusener J, Schaefer J (1981) J Chem Phys 74: 535
38. Buck U, Huisken F, Maneke G, Schaefer J (1983) J Chem Phys 78: 4430
39. Buck U (1982) Faraday Discuss. Chem Soc 73: 187
40. Norman MJ, Watts RO, Buck U (1984) J Chem Phys 81: 3500
41. Andres J, Buck U, Meyer H, Launay JM (1982) J Chem Phys 76: 1417
42. Schinke R, Meyer H, Buck U, Diercksen GHF (1984) J Chem Phys 80: 5518
43. Andresen P, Buck U, Ebel G, Krohne R, Meyer H, Schinke R, Seelemann T, Diercksen GHF (1990) J Chem Phys 93: 6419

CHAPTER XIV
Molecular Dynamics (MD) Computer Simulations of Hydrogen-Bonded Liquids

P. Bopp

It is the aim of the present lecture to demonstrate how MD simulations are employed to study hydrogen-bonded liquids. It is not attempted to give a comprehensive review of the field, but rather to present an introduction into the method and its possible applications. First, the basic concepts and assumptions of the simulation are briefly recalled. The aspects of the model-building step which are relevant to the hydrogen bonding are then discussed. Selected results, both structural and dynamical, are presented as examples to demonstrate some of the quantities that can be studied in such simulations. The systems treated include pure water and methanol, aqueous ionic solutions and interfacial systems.

1 Introduction

The hydrogen bond is one of the most interesting intermolecular interactions. Due to its strong orientation and to a binding energy which may be of a magnitude comparable to that of intramolecular bonds, it often confers unusual properties to the chemical system where it occurs. In the present lecture we want to discuss how liquids containing hydrogen bonds, most notably water, but also other aqueous and non-aqueous systems, can be studied by computer simulations, and what can possibly be learned from such studies.

Intermolecular Forces
© Springer-Verlag Berlin Heidelberg 1991

Molecular Dynamics (MD) computer simulations are a very powerful tool for the study of liquids in general. Because of the extreme difficulties of an analytical treatment, hydrogen-bonded liquids are specially suited for simulation studies. Furthermore, a large body of experimental data is available for comparison. In this lecture, however, results are presented and discussed only as far as they are relevant to the didactic aim, and the reader is referred to the cited literature for background information about the various systems and for details.

We shall first very briefly review the principles of the Molecular Dynamics (MD) computer simulation method. An excellent monograph has recently appeared [1]. The development of the field has been documented in a collection of reprints [2], and more specific reviews are also available [3]. The model-building step, which has to precede any computer simulation, will be discussed in some detail. We shall then review a few selected results. First, some microscopic properties due to hydrogen bonding in the homogeneous liquids pure water and methanol will be characterized. Then we shall look at the influence of the perturbation resulting from the addition of salt in a finite concentration. As examples, aqueous solutions of the chlorides of monovalent (Li^+, Na^+), divalent (Mg^{2+}, Ca^{2+} and Sr^{2+}), and trivalent (La^{3+}) ions will be mentioned. The vicinity of a surface will also result in a perturbation of the hydrogen bonding in the liquid. This will lead us to considerations about inhomogeneous systems such as water and aqueous solutions near walls, metal surfaces, or membranes.

2 Principles of MD Computer Simulations

It is the aim of a Molecular Dynamics (MD) computer simulation to compute macroscopic properties of a chemical system assuming essentially that the microscopic interaction potentials are known. The classical trajectories of a number of point masses or rigid bodies representing atoms and molecules are first determined numerically. Statistical averages are then computed from these trajectories, leading to the desired macroscopic quantities. One usually wants to study the properties of 'large' systems (on a microscopic scale, say something like 10^{20} particles), but since only a few thousand particles can at best be explicitly treated in the simulation, additional assumptions have to be made. It is the basis of these assumptions that it suffices to consider interparticle interactions only for pairs (or more generally n-tupels) of particles separated from each other by less than a certain correlation length. The method of 'periodic boundary conditions' together with the 'minimum distance convention', which are most often applied in simulations, and other similar approaches, have been developed for this purpose. In most cases the number of particles, the total volume, and the total energy are maintained constant during an MD simulation. This corresponds to the construction of a microcanonic ensemble in statistical mechanics. It is also possible to devise simulation schemes corresponding to the other ensembles of statistical mechanics [2, 4].

The number of numerical integration steps that can be carried out in a simulation is limited. The length of the individual timestep is mostly determined by the steepness and the curvature of the interaction potentials and the masses of the particles, and by numerical considerations concerning the desired accuracy of the trajectories. Even with modern large-scale computers it is difficult to exceed total simulated times of one or two hundred picoseconds, except in selected test cases, for liquids with rather intricate interactions like the ones that we want to study here. It is thus important to recognize that only phenomena occurring at timescales shorter than these values are accessible to our simulations.

The microscopic interatomic or intermolecular interaction potentials are of course the most important 'input' to a computer simulation. They may be taken from ab-initio or other quantum-chemical calculations, from experiments, or from other "ansätze" according to chemical intuition. In some cases intramolecular degrees of freedom are also included in a simulation, and the corresponding potentials have to be provided as well. In order to be used in the numerical calculations, the potentials and their derivatives, the forces, have either to be tabulated with a sufficient accuracy, or have to be expressed mathematically. It is usually assumed that the total intermolecular potential can be expressed as a sum of pair-potentials $V_{ij}(r_{ij})$:

$$V^{tot} = \frac{1}{2} \sum_i \sum_j V_{ij}(r_{ij}) \tag{1}$$

where i and j refer to so-called force-centers and r_{ij} is their separation. A force-center is a site on a molecule, often coinciding with the position of an atom. For instance, an electric charge (usually a fraction of the elementary charge) can be located there to represent an excess electron density. Lennard–Jones potentials are often used to represent an atom or even a group of atoms in a molecule. In intramolecular potentials, and in some instances also in intermolecular ones, three-body contributions, e.g., depending on angles, have to be included in order to obtain satisfactory results from the simulation. The next section will describe some of the potentials, or sets of potentials, used in simulations of hydrogen-bonded liquids. We shall refer to the potentials and the other parameters selected to describe a certain molecule as the 'model' for this molecule.

Several hundred particles have to be included in a computer simulation if a typical extended liquid is to be studied. The potentials describing the intermolecular (and also the intramolecular) interactions must therefore be computationally as simple as possible. The way in which the long-range Coulomb interactions resulting from the point charges are treated (Ewald method, truncation, or other procedures) is also important here. This is because the evaluation of the interaction energies and forces is the limiting factor, in terms of computer time, in such simulations.

3 Modelling of the Hydrogen Bond

Many models, i.e., essentially sets of interaction potentials, have been developed for liquid water and other systems containing water [5]. If the molecule is assumed to be nondeformable (so-called rigid models), moments of inertia have to be assumed as well, while in models where the molecule is allowed internal deformations and motions (flexible models), an intramolecular potential must be supplied. Such flexible models have been developed, for instance, by adding an intramolecular potential to an existing rigid model [6-11].

The most important feature to be reproduced by a model in the present case is, of course, the hydrogen bond. In virtually all models proposed this feature is described by a Coulomb interaction between positive partial charges (usually about one quarter to one half of an elementary charge) located at the positions of the hydrogens and a corresponding countercharge, or counter-charges. In water, these may be located in a way to form a distorted tetrahedron with the hydrogen positions (ST2 model [12]), representing in a way the lone pair orbitals of the molecule, or may be lumped together either at the position of the oxygen (CF [13], SPC [14], and TIPS [15] models) or at a position not coinciding with an atom, on the H—O—H angle bisector (MCY [16], TIP4P models [17]). The dipole moment of the molecule resulting from the partial charges and the geometry is often chosen to be somewhat higher than the observed gas phase dipole moment (1.86 D for H_2O). Of course, in flexible models, the deformation of the molecules due to interactions with their neighbors in the liquid will lead to fluctuations, and in general also to an average change, of the molecular dipole.

All models use empirical functions of Lennard–Jones 12–6 type or other types of simple site–site pair potentials to describe the van-der-Waals part of the intermolecular interaction. The subtle interplay between the steric hind-rances, described essentially by these potentials, and the Coulomb potentials allows to mimic the typical saturability of the hydrogen bond. Goldmann and Backx [18] have shown that such combinations of empirical potentials and charge distributions are a much more effective representation of real water than an equivalent empirical potential augmented by point multipoles at the center.

There is another class of models where the polarisability of the molecule is directly taken into account [19,20]. Simulations with such models require very large computer times, and they have thus not been used in extensive simulations. On the other hand, model fits to ab-initio energies including general three-body contributions have become available recently and first results on water and aqueous solutions have been published [21–23]. They look encourag-ing and further progress is certainly to be expected here. There are also attempts to combine quantum mechanics for certain degrees of freedom in the system with classical mechanics for others, and some of the fundamental papers in this field can be found in Ref. [2]. This approach has not yet been applied to chemical systems as complex as the ones which are of concern here.

Both rigid and flexible models have been developed for other hydrogen bond-forming molecules. The approach is very similar to the examples described above and shall not be discussed here. The results for water and aqueous systems discussed later have mostly been obtained using the following models: The rigid ST2 [12] and TIP4P [17] models, and a flexible model (BJH [6, 11]) developed from the latest version of the CF models (CF3).

Besides the intrinsic virtues and shortcomings of the various models, which are usually known to the investigator, computational considerations also dictate to a certain extent his choice of model. The computer time increases with increasing number of force centers on the molecule (5 in ST2, 4 in TIP4P, and 3 in BJH). In MD simulations rigid models require the solution of three-translational equations and three-rotational ones, thus the necessity of handling them with Euler angles or other equivalent representations. An alternative approach is the introduction of constraint forces to keep the molecular geometry constant. For flexible models the solution of the nine equations (per water molecule) can be carried out in a straightforward manner in Cartesian coordinates. As we shall see, the flexible models yield more information than the rigid ones, but a shorter timestep of integration is necessary because of the fast intramolecular motions. They are thus used in detailed studies where the total simulation time does not have to be extremely long.

4 Results

As examples, we shall now briefly look at results obtained from MD simulations of the following systems:

a. pure water and pure methanol
b. aqueous solutions of NaCl, $MgCl_2$, $CaCl_2$, $SrCl_2$, and $LaCl_3$ at concentrations ranging from infinite dilution to 2.2 molal
c. water and hydrated ions in the vicinity of interfaces, such as a metal surface or a biomembrane.

4.1 Pure Liquids

The results for pure water and methanol will be discussed with two goals in mind:

—to demonstrate the influence of the hydrogen bonding on the structural and dynamic properties of the liquid, and secondly
—to use these results as a reference when the 'perturbed' (by ions, interfaces) ones are studied.

One way to describe the structure of a liquid is to give the atom–atom radial distribution functions (RDFs) for all kinds of atoms present. The RDF $g_{\alpha\beta}(r_{\alpha\beta})$

describes the variation of the local density of particles of type β around a particle of type α relative to the average density expected for homogeneously distributed particles of infinitely small size as a function of the interparticle distance. The integral:

$$n_{\alpha\beta}(r_{\alpha\beta}) = 4\pi\rho_\beta \int_0^{r_{\alpha\beta}} g_{\alpha\beta}(r'_{\alpha\beta}) r'^2_{\alpha\beta} dr'_{\alpha\beta} \tag{2}$$

where ρ_β is the number density of particles β, gives the total number of particles β around a particle α as a function of their distance $r_{\alpha\beta}$. It is sometimes called the running integral, or running integration number, of the RDF. The RDFs often show several oscillations before they approach their limiting value of unity at large distances. The value of $n_{\alpha\beta}$ at the first minimum of $g_{\alpha\beta}$ is a possible rigorous definition of the number of next neighbors of a species, i.e., of the coordination number.

RDFs can in principle be determined experimentally by X-ray, neutron, and electron diffraction. The difficulty resides in the fact that in a system containing N different kinds of atoms, there are $N \cdot (N + 1)/2$ different atom–atom functions $g_{\alpha\beta}$. As many independent experiments are needed to determine them. For pure water the three different diffraction experiments mentioned above have been performed [24, 25]. The most frequently used method, however, is the combination of several neutron diffraction experiments on isotopically substituted samples (method of first and second differences [26]).

Figure 1 shows the typical RDFs $g_{OO}(r_{OO})$, $g_{OH}(r_{OH})$, and $g_{HH}(r_{HH})$ obtained from MD simulations of pure liquid water shightly below room temperature. All models developed for water render these functions reasonably well, we shall thus not go into any detail, but rather highlight the typical features. The results shown here are from simulations with the flexible BJH model at a temperature of 290 K. Due to the low temperature, the structure is quite pronounced in the RDFs of Fig. 1. With increasing temperatures some of the features get washed out [11]. The study of the temperature dependence of the structure is presently an active field of research [27].

g_{OO} The integral over the first peak, i.e., the number of oxygen atoms surrounding a central oxygen, is 4.2. This is much less than the number of 8 to 12 expected for densely packed spheres. The oxygen atoms are kept apart by the hydrogen bonds and the number of neighbors is thus reduced. Four neighbors is also what is expected for a tetrahedral coordination like the one found in ice. At low enough temperatures (like here), a weak second peak is visible in pure water, it disappears with increasing temperature. Generally speaking, there is not much structure beyond 6 or 7 Å, the RDF is equal 1. The pronounced short-range order (the maximum value of g_{OO} is about 3) and the lack of long-range order are typical of liquids.

g_{OH} The first peak (integral = 2) describes the intramolecular O—H distances. If it is better resolved than in the figure here, this peak yields information

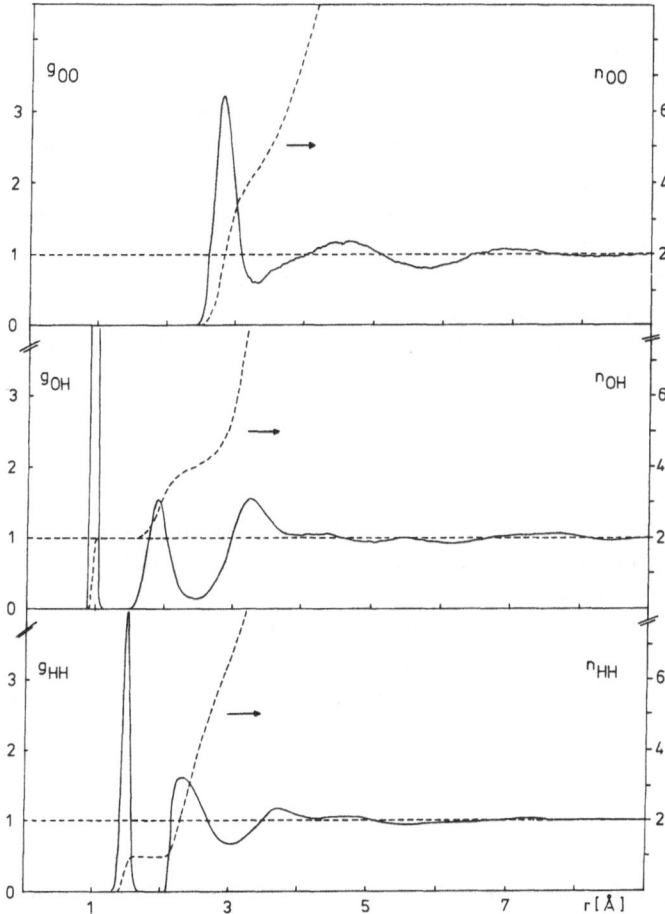

Fig. 1. Radial pair distribution functions g_{OO}, g_{OH}, and g_{HH}, and running integration, for pure water ($T = 290\,K$, $\rho = 1\,g\,cm^{-3}$). Simulation with the BJH model

about the deformation of the molecules due to the interactions with their neighbors. For instance, the average O—H intramolecular distance is increased from $0.9572\,\text{Å}$ in the gas phase to $0.9755\,\text{Å}$ in the liquid. The next peak (integral also about 2) describes the positions of hydrogens of the two neighboring molecules which form (donate) an H-bond to the central molecule. This peak is lower and much wider than the intramolecular peak, a consequence of the distortion (or even lack of) H-bonds in the liquid. Form and shape of this peak are quite sensitive to perturbations of the hydrogen bonds, for instance by solutes in the liquid, as we shall see.

At this point, a short remark is appropriate concerning the term 'hydrogen bond' in connection with the present models. As described above, the hydrogen bond is not contained explicitly 'as such' in these models. The existence of a bond at a certain time during the simulation must be established from geometrical or energetical considerations. What is meant by the statement 'a hydrogen bond exists' is thus that two oxygen atoms and one hydrogen atom are aligned, within certain tolerances, in the way that is expected for a hydrogen bond. As seen in the example discussed here, one always has distributions of distances, angles, and also of interaction energies. Whether certain geometries are considered as 'distorted bonds' or as 'broken bonds' is a matter of definition. It is therefore interesting to compute different properties for 'bonded' and 'unbonded' molecules according to various definitions and to compare them with experimental information in an attempt to reach a consistent definition. As already stated, one may alternatively define the bond as existing if the interaction energy is below a certain limit. It has been shown [28] that the geometric and the energetic definitions are essentially equivalent.

After this digression, let us come back to the discussion of the $O-H$ RDF in Fig. 1. The third peak (at a little over 3 Å) shows that the positions of the second hydrogen atoms belonging to the bond-donating molecules, or of the hydrogen atoms belonging to molecules accepting H-bonds, are also correlated. Information on the relative orientations of molecules can be gained from such considerations. Much more detailed information can of course be determined directly from the simulation. Since the RDFs are also accessible experimentally and used to extract such information, it is very useful to explore in how far such interpretations of the RDFs can be verified (see Ref. [29]). In this context, simulations become a very useful tool to interpret experiments.

g_{HH} Here again we have an intramolecular peak which yields information on the distorted molecular geometries in the liquid. The following, albeit small, oscillations are again in keeping with a mostly tetrahedral coordination of the water molecules.

The tetrahedral coordination of a water molecule by its neighbors has now been mentioned several times, yet the evidence for such a structure that has been presented up to now is only circumstantial, i.e., we have seen that all the RDFs are compatible with such an arrangement. The extent of the tetrahedral order in the system can be explicitly demonstrated by a method proposed by Pálinkás et al. [30]. Let us assume that a water molecule is located in the y–z plane of a Cartesian system with the oxygen on the positive z-axis. From the simulation we determine the positions of its n first neighbors in this molecule fixed system. We then project the positions (say of the oxygen atoms) onto the three planes of the coordinate system and draw contour maps of the resulting densities. Such a map is shown in Fig. 2 for the x–y plane and the first four neighbors from simulations with the ST2 and with the CF3 model.

Since the central water molecule is located in the y–z plane, the density spots seen in Fig. 2 at $x \approx 0$ Å, i.e., on the y-axes, are due to water molecules accepting H-bonds from the central one, while the spots on the x-axes are due to molecules donating bonds to the central one. It is seen that the degree of

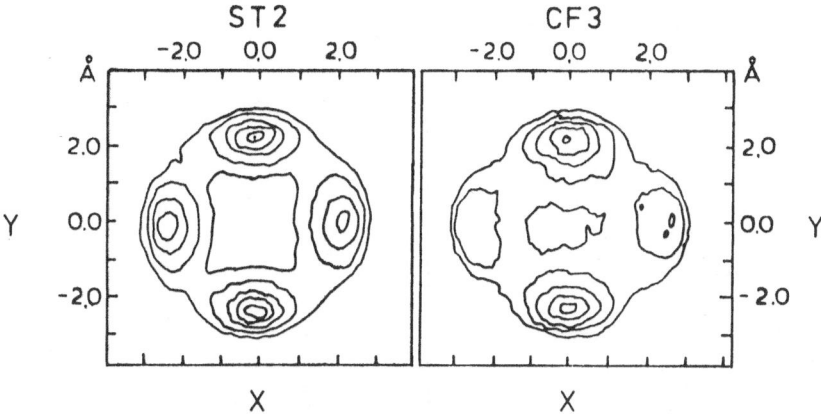

Fig. 2. Two-dimensional representation of the probability to find a next neighbor with respect to a central molecule (located in the y–z plane of a Cartesian system) in pure ST2 and CF3 water

localization of the accepting molecules is about the same in both simulations. The donating molecules, on the other hand, are well localized in the simulation with the ST2 model but quite diffuse with the CF3 model. The projections onto the other planes (not shown here) confirm this finding. This can be understood in terms of the charge distribution of these two models, as explained above. The differences seen in Fig. 2 are an example of the model dependence of simulation results. Presently, the less-pronounced tetrahedral structure of the type found in the CF3-simulation is thought to be more realistic than the one from the ST2 model. This model is therefore sometimes criticized for being 'overtetrahedral'.

We shall now briefly consider the motions of the molecules in the liquid. The function which is used to describe them is the velocity autocorrelation function (vacf) $c_{vv}(t)$:

$$c_{vv}(t) = \frac{1}{NM} \sum_{i=j}^{N} \sum_{j=1}^{M} (\vec{v}_i(t_j)\vec{v}_i(t_j + t)) \tag{3}$$

where $\vec{v}_i(t_j)$ is the velocity of particle i at the time t_j. N is the number of particles over which the average is taken, this can be all particles in the system or only selected particles fulfilling certain conditions. One can, for instance, compare the vacfs computed for the water molecules in the hydration shells of the ions with the ones obtained for water molecules further away from the ions. M in Eq. (3) is the number of 'time origins' t_j over which the vacf is averaged. Instead of using $\vec{v}_i$ in Eq. (3), it is sometimes useful to correlate only certain components of the velocities, or combinations of velocities of several particles belonging to a molecule.

The Fourier-cosine transform of the normalized vacf:

$$\hat{c}_{vv}(\omega) = \int_0^\infty \frac{c_{vv}(t)}{c_{vv}(0)} \cos(\omega t)dt \tag{4}$$

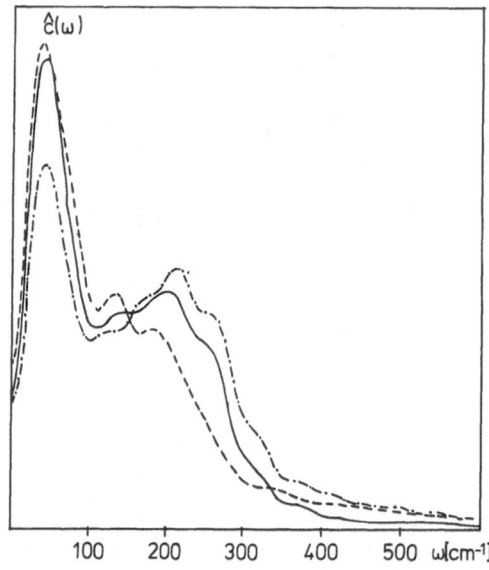

Fig. 3. Spectral densities, in arbitrary units, of the center of mass motions of a water molecule in pure BJH water (*solid line*). *Dashed line*: only motions perpendicular to the plane of the molecule, *dash-dotted line*: only motions parallel to the plane of the molecule

is easier to interpret and more instructive than the vacf itself. This function is directly related to the elastic neutron scattering. Under certain assumptions, it can also be compared with results from infrared and Raman experiments. We shall term $\hat{c}_{vv}(\omega)$ the spectral density of the motions.

Figure 3 (solid line) shows such spectral densities for the center of mass motions of a water molecules in pure water. Two main bands are easily identified, one at about $50\,\mathrm{cm}^{-1}$ and one slightly below $200\,\mathrm{cm}^{-1}$. These two bands have been identified spectroscopically a long time ago [31, 32]. The band at lower frequencies is ascribed to motions of the oxygen atom perpendicular to the O—H...O direction of the bond, i.e., to a bending of the bond, while the band at high frequency is ascribed to the hindered translation of the oxygen mostly in direction of the linear, or only weakly distorted, H-bond [33–36]. Since a molecule takes part in several bonds simultaneously in the liquid, the resulting motions may be quite intricate. Figure 3 shows that they are anisotropic with respect to the plane of the molecule. Motions parallel to the plane of the molecule have more high-frequency components than the ones perpendicular to the plane. This is easily understood from Fig. 2. Motions parallel to the plane occur mostly against the two well-localized acceptor molecules, while only the less well-localized donor molecules oppose the motions perpendicular to the plane. (The differences in structure between CF3 and BJH water are minimal.) Studies like the one sketched here can be extended, both experimentally [36] and in simulations [11, 16, 37, 38], to regions of high temperature and pressure.

The hindered rotations (librations) and the vibrations of flexible molecules in the liquid can be studied by procedures similar to the one discussed for the translations. Details of the requirements with respect to the intramolecular potentials and to the computational methods are given in [6, 39–42]. Two points should nevertheless be highlighted:

1. The determination of vibrational motions of several thousand cm^{-1} from a classical simulation is possible only under certain assumptions. Essentially they are that an effective potential is used which leads to trajectories with the desired vibrational frequencies in simulations of the molecules in the gas phase and in the liquid [6, 39]. Once this adjustment of the potential has been made, the interest is thus not so much in the vibrational frequencies themselves, but in the frequency shifts due to changes in the environment of the molecule in the liquid.
2. The flexible molecules are usually distorted by the interactions with their neighbors in the liquid. The concept of an equilibrium geometry is no longer generally valid. Approximations have thus to be used to determine the rotations and individual vibrational modes [40, 42].

Figure 4 shows the spectral densities obtained by such an approximate method for the librations around the three principal axes of inertia for a water molecule in pure water. The axes are the same as the ones used in Fig. 2. The librations around the x- and z-axes have their maximum around 420 cm^{-1}, while the libration around the y-axis is shifted to higher frequencies. The main reason for this is the low moment of inertia around this axis. We also note that the moments of inertia around the x- and z-axis are not identical. The fact that these two bands are very similar should not be interpreted as being the consequence of an homogeneous environment. The comparison with results from

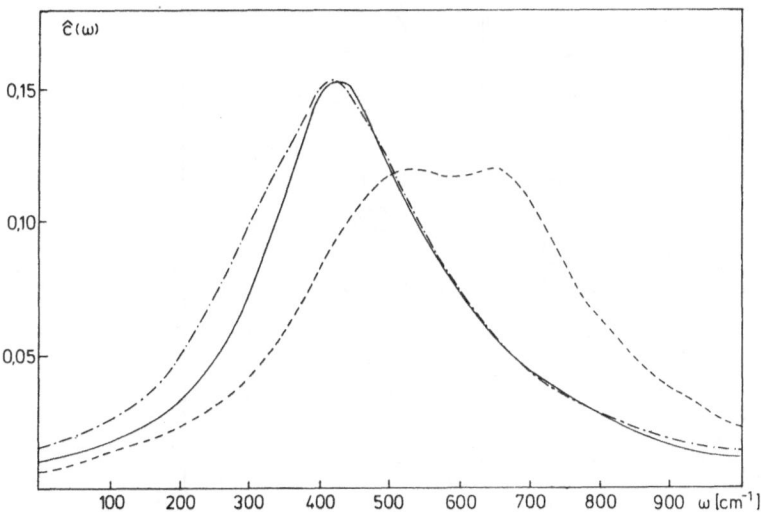

Fig. 4. Spectral densities, in arbitrary units, of the librations around the three principal axes of inertia for pure water. Axes as in Fig. 2. *Solid line*: around the x-axis; *dashed line*: around the y-axis, *dash-dotted line*: around the z-axis

experiments is more difficult in this case, generally frequencies between 475 and 725 cm^{-1} (see e.g. Refs. [33, 34]) are reported for these motions. Simulations with rigid models seem generally to yield higher frequencies, especially for the y-libration (*vide infra*).

As stated above, the interest in the vibrations is more in the frequency shifts due to perturbations of the molecules by different neighbors than in the absolute values of the frequencies. We shall therefore look at these motions in the context of the discussion of ionic solutions below. We note that estimates of the frequency shifts can also be obtained by a perturbation treatment from simulations with rigid models [43].

We shall now turn to methanol as another example of a hydrogen bonded liquid. Similar to water, rigid and flexible models have been developed and several simulation studies have been conducted [44–47]. Figure 5 shows the site–site RDFs for liquid methanol from a simulation with a flexible three-site model [46]. The sites are the oxygen, the hydrogen, and the methyl group, which is treated as one particle in this model. The peak positions and the numbers of next neighbors are consistent with a preference for the formation of hydrogen bonded chain-like structure [46]. Thus, each oxygen has only two other oxygen next neighbors, and the integral over the first intermolecular peak in g$_{OH}$ is about one. Small differences in the peak positions between these RDFs and results from simulations with other models do not affect this finding.

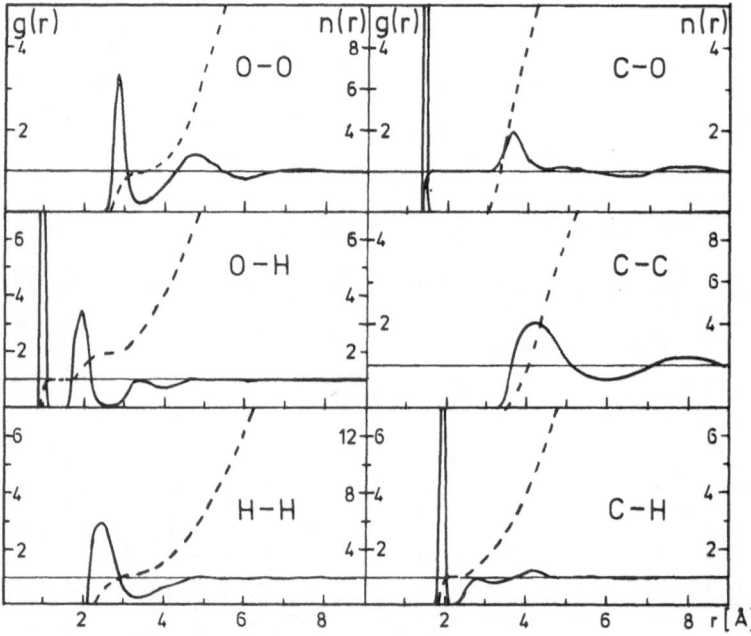

Fig. 5. Site–site radial pair distribution functions and running integration numbers from a simulation of pure methanol at 286 K with a flexible model [46]. The symbol C is used for the methyl group

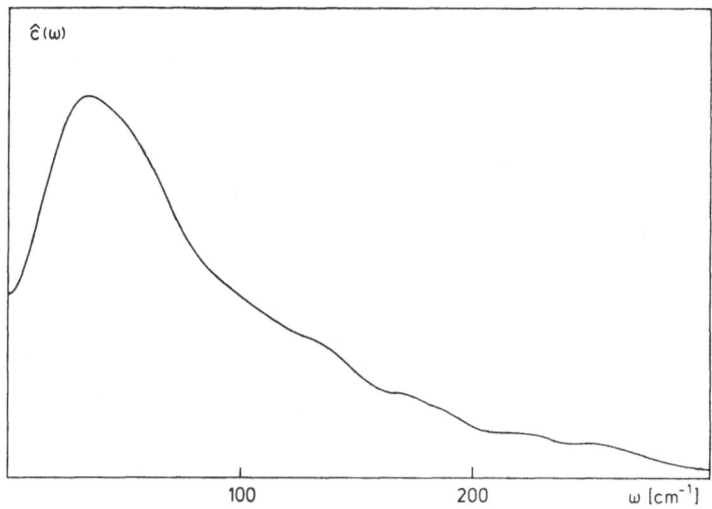

Fig. 6. Spectral density, in arbitrary units, of the center of mass motions of a methanol molecule [49]

Figure 6 shows, analogous to Fig. 3, the spectral density of the center of mass motions of a methanol molecule in liquid methanol. The shape of this spectrum is comparable to the one for the motions perpendicular to the plane of the molecule in water, see Fig. 3. The lack of contributions in the region of higher frequencies, compared to water, is a consequence of the predominantly one-dimensional structure (the hydrogen bonded chain) in methanol compared to the three-dimensional hydrogen bond network in water.

This one-dimensional structure allowed to study the influence of the bonding on the vibrational frequencies of the molecule. Autocorrelation functions and spectral densities, similar to the ones in Eqs. (3) and (4), but referring to the O—H vibrational motion of the methanol molecule, have been computed separately for three subsystems [48, 49]. The subsystems correspond to three bonding situations possible in the liquid. 1—monomers, 2—molecules with free O—H groups (e.g., ends of chains), 3—all other molecules. The hydrogen bond definition used is a geometrical one. We note here that the time a given molecule remains in one of the subensembles (i.e., roughly speaking the lifetime of a bond according to the definition used) is of the order of magnitude of a few pico-seconds, (cf. Ref. [126]) i.e., of the same order of magnitude as the length of the correlation functions. This leads to shapes of the resulting spectral densities, which are shown in Fig. 7. The peaks at high frequencies are the ones of interest here. The peaks at low frequencies are due to the approximations made in the separation of the modes. They are small compared to the main peaks and shall not be considered here.

It is seen that the frequency shifts correlate rather well with the geometrically determined bonding situation. H-bond donating molecules show a peak maximum at about $3330 \, \text{cm}^{-1}$, while unbonded ones show a peak at about

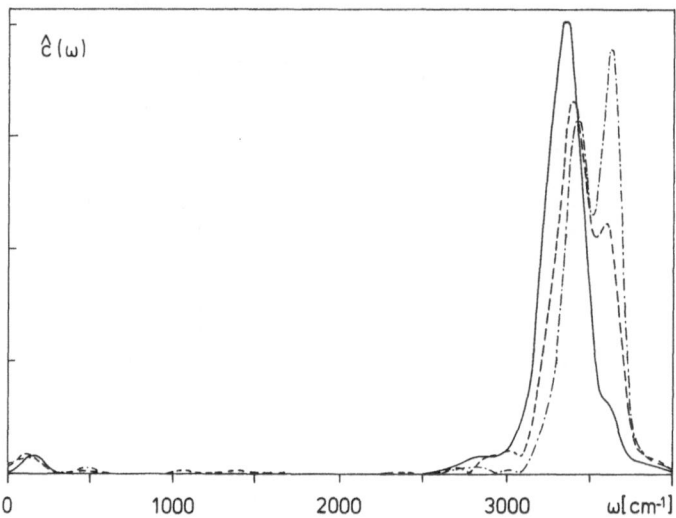

Fig. 7. Spectral densities, in arbitrary units, of the O—H vibrations of the methanol molecule. *Dash-dotted line*: unbonded molecules (monomers), *dashed lines*: molecules with unbonded O—H group, *solid line*: all other molecules

$3610 \, cm^{-1}$. The molecules with unbonded O—H groups occupy an intermediate position. The gas phase O—H vibrational frequency of methanol is $3690 \, cm^{-1}$ with this model. One could thus temptatively conclude that a redshift of about $80 \, cm^{-1}$ can be ascribed to van-der-Waals-type interactions and about $280 \, cm^{-1}$ to the hydrogen bond.

In this section we have reviewed a few results that can be obtained from MD simulations. Many more have not been mentioned, Refs. [50–59] give an overview of some of the aspects neglected here. We have studied how to obtain structural results (e.g., the RDFs, which can also be determined from Monte Carlo (MC) simulations), and dynamical results (e.g., the spectral densities). The same methods will be applied in the following sections to the study of more complex chemical systems.

4.2 Ionic Solutions

The interest in ionic solutions is a very longstanding one in the field of the natural sciences. It is therefore not surprising that computer simulations were immediately applied to this problem [60]. The field has since then been extensively reviewed [3, 61–64]. The often peculiar properties of aqueous ionic solutions are the consequence of the competing influence of the ions and of the hydrogen bond network of the water. One structurally simple ions will be considered here, some more complex ions have been studied by Monte Carlo simulations

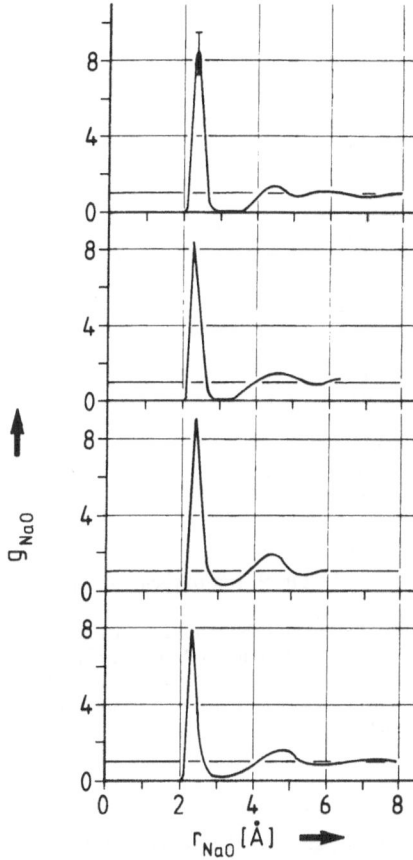

Fig. 8. RDFs g_{Na^+O} from various simulations. From top to bottom: one ion in 215 MCY water molecules [71], one ion in 64 MCY water molecules [72], one ion in 64 TIP4P water molecules [73], 8 Na$^+$ ions and 8 Cl$^-$ ions in 200 CF3 water molecules (2.2 m NaCl solution) [74]

[65]. Solutions of nonpolar solutes have also been studied [66–70], the effects found there are usually smaller and more subtle than in ionic solutions. Again we cannot discuss this field here.

As an example the influence of the Na$^+$ cation on the environing water is shown in Fig. 8. Although the concentration and the temperature are not identical in these simulations, and different ion–water and water–water potentials have been used, there is clear agreement that the structure around the Na$^+$ ion is quite pronounced. The first peak maximum in the Na$^+$—O RDF has a value of about 8, compared with about 3 for the first peak in g_{OO} in Fig. 1. This strongly coordinated layer of water around the ion is the hydration shell. Figure 8 also shows that the influence of the cation extends beyond this first hydration shell. The shape and position of the second peak in g_{Na^+O} are dependent on the ionic concentration and the temperature, and probably also on the model used.

Smaller and/or more highly charged ions develop a much stronger order than the Na$^+$-ion. Figure 9 shows a few examples. The heights of the first peaks are 19.2, 14.0, and 10.6 for Mg^{2+} [75], Ca^{2+} [76], and Sr^{2+} [42], respectively.

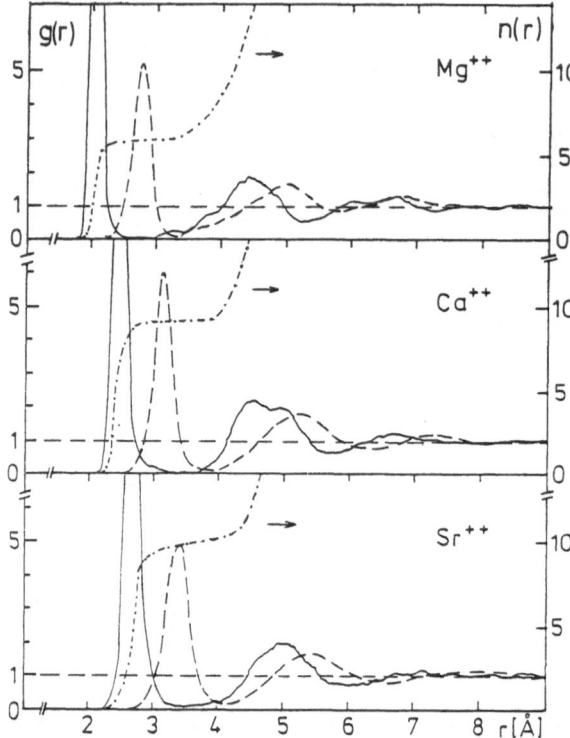

Fig. 9. Ion-oxygen (*solid line*) and ion-hydrogen (*dashed line*) RDFs from simulations of 1.1 m solutions of MgCl$_2$ [75], CaCl$_2$ [76] and SrCl$_2$ [42] with the BJH model. The running integrals refer to the ion−oxygen functions

In all three cases the second peaks in the ion-oxygen RDFs have a height of about 2, indicating the presence of some sort of second hydration shell around these ions. The shift between the ion-oxygen RDFs and the ion-hydrogen RDFs in Fig. 9 shows that the average orientation of the water molecules is also influenced by the ions. In the cases shown in Fig. 9, this order extends well beyond the second hydration shell. The observations made here apply even more to trivalent ions [77, 78]. Ion-water potentials from ab-initio quantum-mechanical calculations and the BJH model for water were used in these simulations.

Anions are comparable to cations with respect to the way in which they structure their environment. Figure 10 shows the g_{Cl-O} and g_{Cl-H} RDFs from simulations of LaCl$_3$ solutions at various concentrations and temperatures [77, 78]. The influence of the salt concentration on the RDF is clearly visible. Generally, the agreement with respect to the hydration around this ion between simulations with different models is at first glance less satisfactory than in the example given above for Na$^+$ [64]. A comparison of simulations of several salt solutions with the same model, on the other hand, reveals that the height, and to a certain extent also position of the first peak in g_{Cl-O} depend on the type of the counterion (cf. e.g. Refs. [42, 74–80]) and on the concentration of the salt (see Fig. 10).

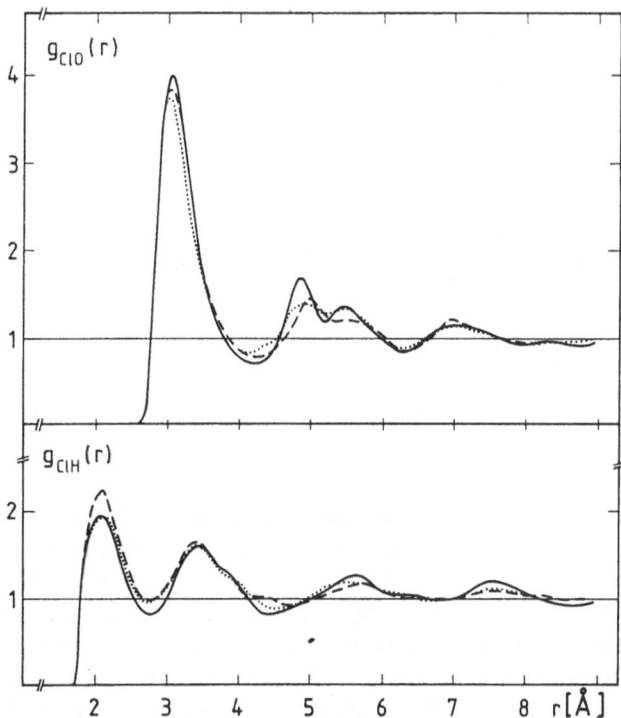

Fig. 10. Cl$^-$-oxygen and Cl$^-$-hydrogen RDF from simulations of LaCl$_3$ solutions [77, 78]. *Solid lines*: 2 m solution at 290 K; *dotted lines*: 2 m solution at 323 K; *dashed lines*: 1 m solution at 329 K

The main features of anionic hydration can nevertheless clearly be seen in Fig. 10. The peak height of the Cl$^-$—O RDF is smaller than for the cations discussed above. In keeping with the size of the ion, it is less than the one found for K$^+$ [81] and comparable to the one found for Cs$^+$ [82]. The shift between the Cl$^-$—H and Cl$^-$—O function is an indication of the orientational order. Here we find that the water molecules in the first hydration shell tend to point one hydrogen toward the ion, i.e., to form an hydrogen bond type configuration with the ion. The distribution of the hydrogen bond angles is rather wide, and its shape and width is concentration dependent [78, 80].

The interplay between the hydrogen bonding and the ion–ion Coulomb interactions leads to another interesting phenomenon. In contrast to the pair potential, the interaction potential averaged over all solvent configurations, say between two Cl$^-$ ions, or between a Na$^+$ and a Cl$^-$, may not be repulsive (attractive) for all ion–ion separations [83, 84]. Oscillations are found which are due to the existence, at least at low concentrations, of complexes where water molecules are favorably bonded to both ions and to the surrounding water, leading thus to a local minimum in the total interaction energy. The quantity used to describe this finding is the potential of mean force (PMF). The same phenomenon has also been found for uncharged solute particles [85].

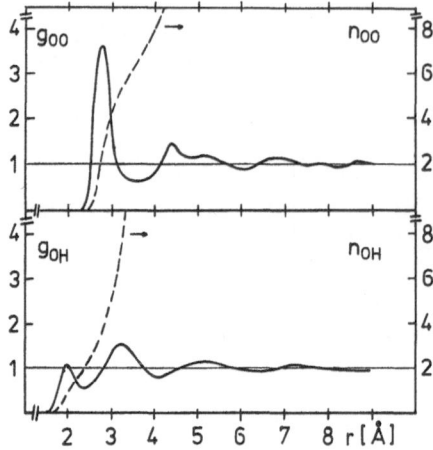

Fig. 11. Oxygen–oxygen and oxygen–hydrogen RDFs from a simulation of a 2 m LaCl₃ solution [77, 78]

The oxygen–oxygen and oxygen–hydrogen RDFs (only the intermolecular part is shown here) in a concentrated ionic solution, here a 2-molal $LaCl_3$ solution, are shown in Fig. 11. The differences between these functions and the ones in Fig. 1 are obvious (the same water model is used). At this concentration roughly 35 and 60 percent of the water molecules are in the first hydration shells of La^{3+} and Cl^-, respectively. Of course, the hydration shells do overlap. Due to the large number of water molecules in the first hydration shell of La^{3+}, the first peak in g_{OO} is shifted to smaller distances and a more pronounced second peak develops. The first peak in g_{OH}, which is to a certain extent a measure of the hydrogen bonding (*vide supra*), is reduced. These changes in the RDF's are accompanied by changes in the average interaction energies between particles. As an example, the average interaction energy of a water molecules with the surrounding water molecules, which is about -5.5 to -6.5 kcal/mole in pure water and solutions at low concentrations, increases to positive values in the 2-molal $LaCl_3$ [78] or in a 13.9-molal LiCl solution [79].

The influence of the ions on the motions of the water molecules is studied in Fig. 12. The spectral densities of the center of mass motions perpendicular and parallel to the plane of the water molecule (analogue to Fig. 3) have been determined separately for the hydration water of Ca^{2+}, Cl^- and nonhydrated water (called here bulk water) [86]. The spectral densities of the bulk water resemble the ones found for pure water (Fig. 3), indicating that the ordering seen for instance in Fig. 9 (middle) is not strong enough to really perturb the motional pattern beyond the first shell in this system. This is also largely true for the hydration water of the Cl^- ion. The only difference is an increase in intensity in the region around 100 cm^{-1} for motions parallel to the plane of the molecule. From a dynamical point of view these water molecules thus are still part of the hydrogen bond network, in keeping with the orientation of the molecules discussed above.

The water molecules in the hydration shell of Ca^{2+} display a different motional pattern. Broad bands appear with maxima at 275 cm^{-1} (parallel to

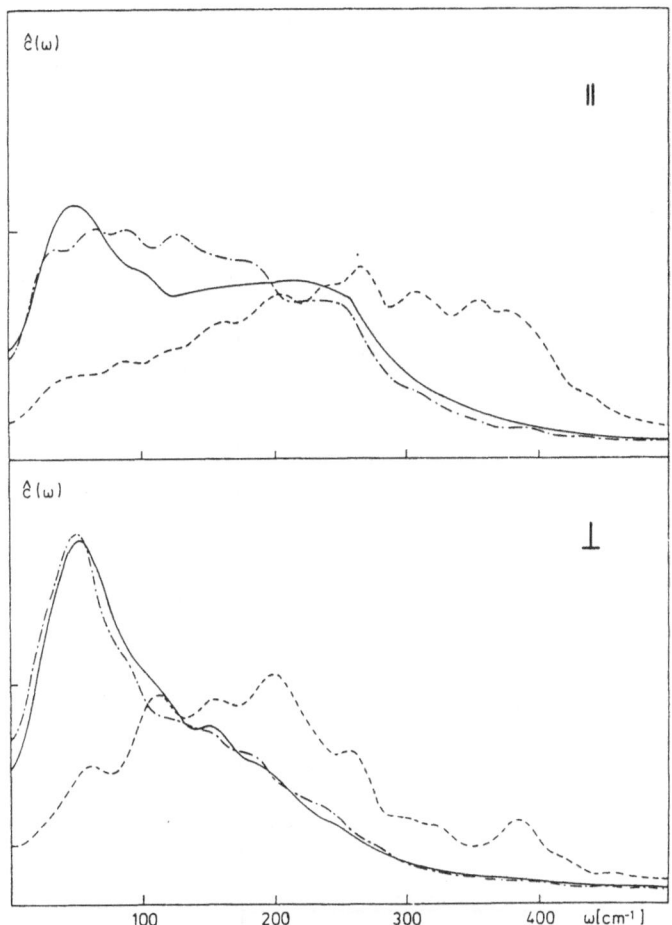

Fig. 12. Spectral densities, in arbitrary units, of the center of mass motions of the water molecules in a 1.1 m $CaCl_2$ solution [86]. *Top*: motions parallel to the plane of the molecule, *bottom*: motions perpendicular to the plane of the molecule. *Solid lines*: bulk water; *dashed lines*: hydration water of Ca^{2+}; *dashed-dotted lines*: hydration water of Cl^-

the plane of the molecule) and $200 \, cm^{-1}$ (perpendicular). Here, the motions are dominated by the ion–water interactions and by the repulsion between water molecules in the hydration shell of the ions. A band position of $290 \, cm^{-1}$ has been reported [87] from Raman spectroscopy for the totally symmetric Ca^{2+}-oxygen vibration. This value agrees reasonably well with the $275 \, cm^{-1}$ found in the simulation for the motions parallel to the plane of the molecule. Superimposing the spectra for the three subsystems in Fig. 12, the net effect of the addition of $CaCl_2$ on the total spectral density of the water molecule motions is a decrease of intensity in the low-frequency region and an increase in the high-frequency region with increasing salt concentration. Similar trends have been found in elastic neutron scattering studies of solutions of aqueous solutions of other salts [88, 89].

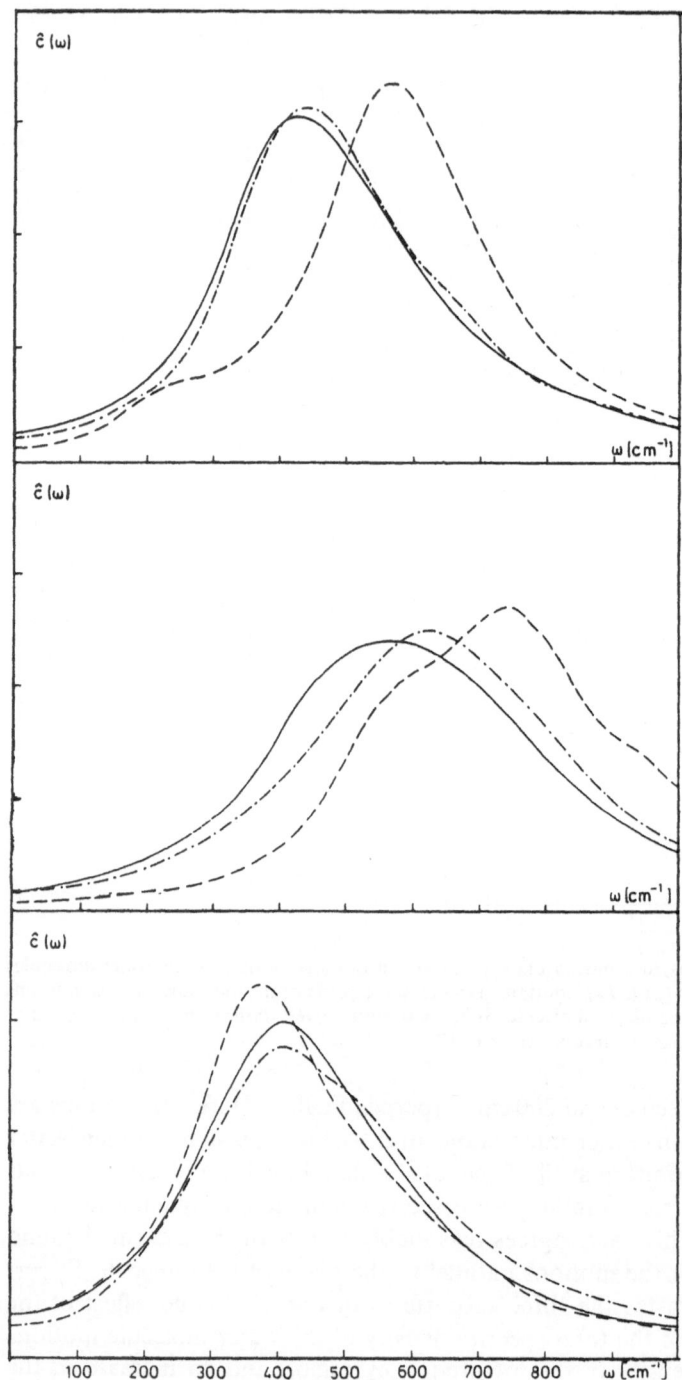

Fig. 13. Spectral densities, in arbitrary units, of the librations around the three principal axes of inertia of the water molecules (see Fig. 4) from the simulation of a 1.1 m $CaCl_2$ solution [86]. *Top*: around the x-axis, *middle*: around the y-axis, *bottom*: around the z-axis. The subsystems are defined as in Fig. 12

We also note in Fig. 12 that the spectral density at $\omega = 0$, which is proportional to the self-diffusion coefficient (Kubo relation), is much lower for the Ca^{2+} hydration water than for the two other subsystems. Trends for the spectral densities and for the self-diffusion similar to the ones discussed here have been found in simulations of salt solutions with other models [90].

The librations of the molecules are also sensitive to the environment and to the bonding situation. Figure 13 shows the spectral densities for the librations around the three principal axes of inertia (see Fig. 4) for the three subsystems used above (Fig. 12). The shapes of the lines may be due to the fact that an exchange of water molecules between the subsystems can take place during the correlation time (cf. the discussion of Fig. 7). The similarity between the bands found for bulk water and the hydration water of Cl^- confirms the interpretation given above for the translational spectra, namely that due to the orientation of the water molecules the Cl^- hydration shell can be accommodated in the H-bond network. The bands for the Ca^{2+} hydration water show marked upshifts for the librations around the x- and y-axes of the molecules, while the band for the z-libration is lightly shifted down. In simulations with the ST2 model, which yields a different preferred orientation for the hydration water of cations [74], all three librations were found shifted up in the hydration shell of Li^+ [90].

The intermolecular vibrations of a water molecule in the liquid, and especially the influence of cations on the vibrational frequencies, have recently been extensively studied experimentally [91–97] and in MD-simulations [98, 40, 42, 43]. Similar to the procedure sketched above, three approximate normal mode frequencies are determined separately for the three subsystems hydration water of the cations, hydration water of the anions and bulk water and compared to the results of infrared spectroscopical measurements [93–97]. Without entering into the details of the discussion it can be stated that surprisingly large redshifts of the O—H vibrations have been established for the hydration water of some divalent cations. As an example, both the symmetric and the antisymmetric O—H stretching vibrations of the hydration water of Ca^{2+} are found to be shifted down by about $300 \, cm^{-1}$ compared to bulk water in the simulation. The H—O—H bending frequency, on the other hand, is not much affected by the different environments of the molecule. Simulations and experiments agree in the trends, but discrepancies remain with respect to specific ions. It remains to be seen whether the inclusion of three body ion–water interaction potentials, which was found necessary from a structural point of view in certain cases [23], will also lead to an improvement here.

Studies similar to the ones discussed here have been undertaken for ionic solutions in other hydrogen bonded solvents like ammonia [99]. It should also be mentioned that besides the MD and MC simulations the method of Brownian Dynamics (BD) [100] has been employed to study ionic solutions. We conclude this section by stressing again that only a few examples could be given here and that many interesting aspects were not covered.

4.3 Interfacial Systems

Solid–liquid [101–106], liquid–gas [107, 108], liquid–liquid [109], and solid–gas [110] interfaces with hydrogen bond-forming molecules on one side have been studied in several instances by computer simulations. This approach has recently been extended to model biosurfaces [111–116]. Interfacial systems are of interest in the context here since they show the effects of a type of perturbation of the hydrogen bonds which is different from what we saw above. Consequently, we shall focus our attention mostly on the side of the interface which contains the H-bonds.

An interesting effect has been shown to occur in the first layer of water near a Pt(100) surface [117, 118]. This simulation was carried out with realistic water–metal potentials and the flexible BJH model for water. The motions of the platinum were also included. Due to the corrugation of the platinum (100) surface, the periodicity of which is close to the nearest-neighbor distance of 2.8 Å in water, the water molecules close to the metal are forced into a square arrangement and develop what might be called a two-dimensional hydrogen bond network. The dynamical consequences of this peculiar bonding scheme have been explored [118] and found to agree with experimental evidence [119]. The translational self-diffusion is strongly reduced in this layer compared to bulk water. In an intermediate layer above this immobilized one, the self-diffusion in the directions parallel to the surface is increased due to a mismatch between the two-dimensional bonding pattern in the vicinity of the metal and the usual water hydrogen bond network.

MD studies of the structure and dynamics of water [111, 113, 115] and ionic solutions [112, 114, 116] in the vicinity of a charged model biomembrane have been undertaken to study the typical phenomena due to the perturbation of the liquid in the interfacial region. A good understanding of the microscopic structure and of the dynamics is an essential prerequisite for the understanding of many elementary biological processes. The approach of an ion toward a membrane and the stripping-off process of its hydration sphere prior to its entry into a transmembrane channel is a case in point. MD studies of the translocation step of an ion through a typical transmembrane channel have shown that a one-dimensional hydrogen bonding pattern can develop under certain circumstances in such a channel [120, 121]. In the vicinity of the membrane, three structuring effects will compete with each other: the structure of the hydrogen bond network, the effects of the (generally charged) membrane and of the channel mouth, and the structure of the hydration shell of the ion. The system pure water/membrane was first studied, later on Na^+ ions were added. The model membrane consisted of COO^- groups representing the headgroups of the membrane molecules, they were free to move in the plane of the membrane and to rotate around their axis of symmetry. The density of the headgroups was varied between 0.042 Å^{-2} and 0.05 Å^{-2}. Since long simulation times are required for these systems, the rigid TIP4P [17] model was used for the water. Effective membrane-water and

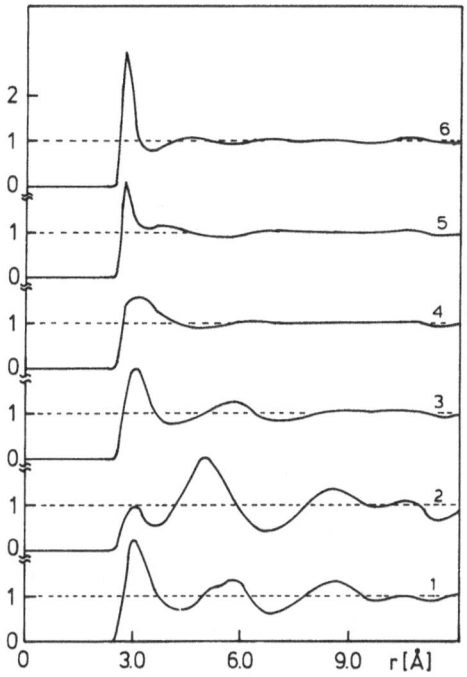

Fig. 14. Oxygen–oxygen distance distribution functions determined for layers of water parallel to a charged model membrane surface [113]. The distances d are counted from the plane of the carbon atoms of the COO$^-$ groups forming the membrane. The first three layers correspond to the peaks in the density oscillations induced in the water by the membrane (see Fig. 16). Layer 1: first layer of water molecules, average density 2.3 g/cm^3, layer 2: average density 1.3 g/cm^3, layer 3: average density 1.7 g/cm^3, layer 4: d = 6.5 – 7.5 Å, average density 1.3 g/cm^3, layer 5: d = 8.5 – 10.5 Å, average density 1.1 g/cm^3, layer 6: more than 12 Å from the membrane, average density 0.9 g/cm^3

membrane-ion potentials parametrized as sums of site–site Lennard–Jones and Coulomb potentials of partial charges [122] were used. Quantum-mechanical calculations, as far as available, confirmed this choice. Ion–water potentials were taken from fits to ab-initio calculation [73].

The strongly corrugated and charged membrane induces density oscillation about 8 Å deep into the water. Three distinct layers of water can be distinguished above the membrane, the average density in these layers is about twice the normal water density (the maximum density is about 5, *vide infra*). Similarly, the water molecules in the interfacial region are also strongly oriented with respect to the membrane. Figure 14 shows the oxygen–oxygen distance distribution functions in the three layers closest to the membrane, in two further layers and at large distances from the membrane (for details, see the caption of Fig. 14). At large distances, the usual oxygen–oxygen function is found (cf. Fig. 1), while a completely different structure, induced by the membrane, is found in the first layers. A detailed study [111, 115] reveals that the molecules in the first water layer reside predominantly between the COO$^-$ groups and the ones in the second layer on top of these groups. The structure found in the first layers (Fig. 14) and the translational and rotational immobilization of the water molecules (and of the membrane headgroups (*vide infra*)) are suggestive of an adsorption phenomenon.

Between the region of membrane-induced structure and the bulk region there is a transition region (layer 4 in Fig. 14) which is almost structureless. Such structureless or structure-broken regions have been postulated to exist

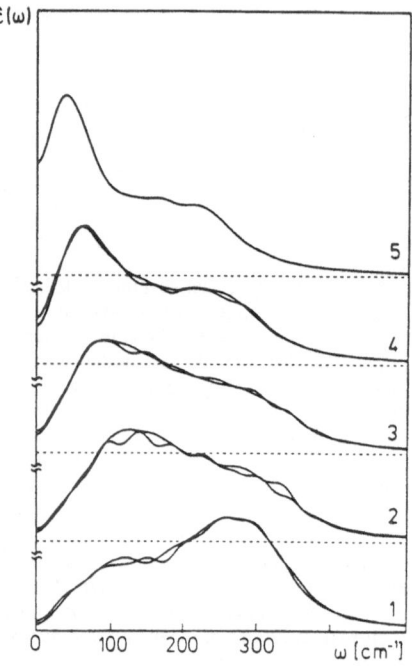

Fig. 15. Spectral densities, in arbitrary units, of the center of mass motions of the water molecules in layers of water parallel to a charged model membrane surface [113, 115]. The layers are defined as follows (cf. Fig. 14): layer 1: $d = 1.00 - 2.75\,\text{Å}$, layer 2: $d = 2.75 - 3.81\,\text{Å}$, layer 3: $d = 3.81 - 5.50\,\text{Å}$, layer 4: $d = 5.50 - 8.00\,\text{Å}$, layer 5: $d > 8.00\,\text{Å}$

around hydrated ions [123, 124]. If the density of the headgroups in the membrane is reduced, this region moves closer to the membrane.

The spectral densities of the center of mass motions of the water molecules (cf. Fig. 3, 6, and 12) in layers parallel to the membrane are shown in Fig. 15. (For statistical reasons the definition of the layers could not be chosen identical to the one in Fig. 14). The transition from a spectral density which is rather similar to the one seen in Fig. 3 (note that a different water model is used, and *vide infra*) in the bulk region (layer 5) to a completely different one in the first layers seems to be continuous. No transition region of the kind of the one found from a structural standpoint (layer 4 in Fig. 14) could be identified. Approaching the membrane, there is a decrease in the self-diffusion constant by more than one order of magnitude. In the first layer the translational diffusional motions are two to three times faster parallel to the membrane than perpendicular to it. It is interesting to note that even at distances of more than 8 Å from the membrane this anisotropy is still visible (about 25% difference).

Figure 16 shows the local density of the water (the position of the oxygen is used), normalized to a density of $1\,\text{g/cm}^3$, as a function of the distance from the membrane surface for pure water, a system with $35\,\text{Na}^+$ ions (vs. $60\,\text{COO}^-$ groups in the membranes), and the completely neutral system with $60\,\text{Na}^+$ ions. The layering of the water in the pure water case discussed above is clearly visible from the density oscillations. When ions are added to the system the strong density oscillations disappear, but the increase in average density by a factor of about 2 within about 3 Å from the membrane remains. Water molecules

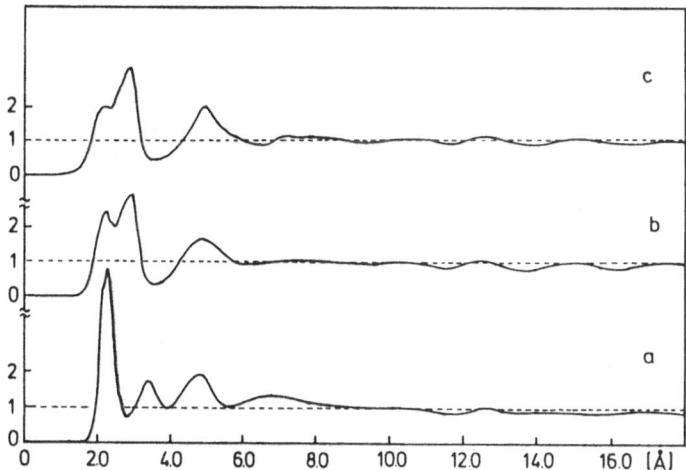

Fig. 16. Density profiles of the water molecule oxygens, normalized to $\rho = 1\,\mathrm{g\,cm^{-3}}$. **a** pure water [113], **b** 35 Na$^+$ ions in system [112], **c** 60 Na$^+$ ions in system [114]

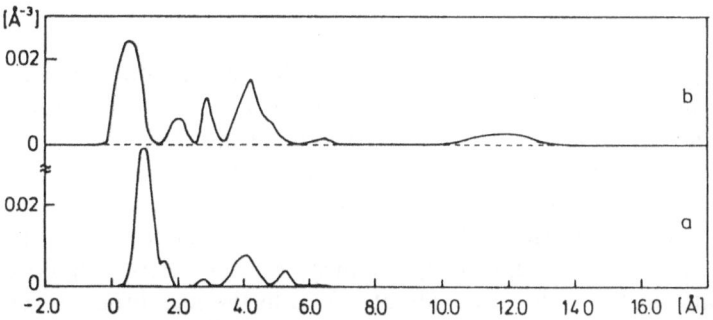

Fig. 17. Density profile of the Na$^+$ ions as a function of the distance from the membrane surface. **a** 35 Na$^+$ ions in system [112]; **b** 60 Na$^+$ ions in system [114]

in the hydration shells of ions have a local density similar to the one found in the first adsorbed layers. Such water molecules attached to ions in the vicinity of the membrane lead to the filling up of the gaps between the layers. Density increases similar to the ones found in the simulations have also been found experimentally [125] in certain systems.

The density profiles of the ions are shown in Fig. 17. The interesting phenomenon here is that the first layer of ions in immediate vicinity of the membrane is saturated when only less than half of the membrane charge is neutralized (24 Na$^+$ ions versus 60 COO$^-$ headgroups in the present simulations, Fig. 17a). This corresponds to a particularly favorable geometric arrangement of these ions with respect to the membrane groups. The subsequent addition of more ions to the system does not change the number of ions in the first layer. The layer broadens and moves a little closer to the membrane. (Some of the

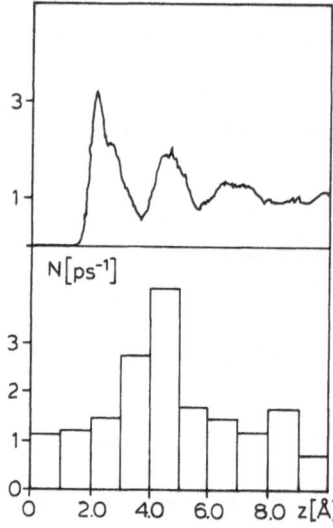

Fig. 18. Density profile of the water (normalized to $1\,g\,cm^{-3}$, cf. Fig. 16) and exchange rate of the hydration water of a sodium ion approaching the membrane as a function of the distance from the membrane [112]

broadening may also be due to the fact that in this simulation [114] the membranes were not kept at fixed positions). Another 30 Na^+ ions are found between the first layer and a distance of 6 Å from the membrane.

The dynamics of the approach of a cation toward the membrane has also been studied [112]. Figure 18 shows the density profile of the water during this specific event (cf. Fig. 16) and the exchange rate of the hydration water of the Na^+ ion as a function of its distance from the membrane. The stripping off of the hydration water at the water layer about 4 Å from the membrane is very obvious.

5 Conclusions

Molecular Dynamics simulations have been able to reproduce many properties found in hydrogen bonded liquids and have thus led to a more detailed and consistent microscopic picture. They have been helpful in the interpretation of experimental results and have opened up new questions. Simulations are increasingly used to study complex systems of pharmaceutical and biological interest. Yet the limitations have to be born in mind. For instance, none of the models available seems to be able to describe satisfactorily and consistently all physicochemical properties of water. The interactions between experiments and simulations have been active and productive in the past. The remaining discrepancies between theoretical and experimental understanding should provide the challenges for future work.

Acknowledgements

It is a pleasure to thank Professors Pierre Huyskens and Thérèse Zeegers-Huyskens for initiating the idea of the ERASMUS lecture cycle on intermolecular forces at the Katholieke Universiteit Leuven. Together with their coworkers, most notably Dr. G. Maes, they made this meeting a most interesting, challenging and pleasant one, both for students and instructors. The European Community is to be commended for sponsoring such international as well as interdisciplinary scientific events.

I thank E. Spohr, Mainz/Irvine, for kindly providing the data for Fig. 1. Many of the simulations reported here were carried out on the CRAY X-MP 24 of the Max-Planck Gesellschaft, the IBM 3090-200E-VF of the Hochschulrechenzentrum der Technischen Hochschule Darmstadt, the IBM 3090-200E of the Institut für angewandte Mathematik (ZAM) der Kernforschungsanlage Jülich, and on the CRAY X-MP 416 of the Höchstleistungsrechenzentrum (HLRZ), also at Jülich. These grants of computer time, and the help and support of the staff at these sites, are gratefully acknowledged.

The author thanks the Deutsche Forschungsgemeinschaft for the award of a Heisenberg fellowship, and financial support by the Fonds der Chemischen Industrie, Frankfurt, is also acknowledged.

6 References

1. Allen MP, Tildesley DJ (1987) Computer simulations of liquids, Clarendon Press, Oxford
2. Ciccotti G, Frenkel D, McDonald IR (eds) (1987) Simulations of liquids and solids, Molecular dynamics and Monte Carlo methods in statistical mechanics, North-Holland, Amsterdam
3. Heinzinger K (1990) In: Catlow CRA et al. (eds) Computer modelling of fluids, polymers and solids, Kluwer Academic Publishers
4. Nosé S (1984) J Chem Phys 81: 511
5. Bopp P (1987) In: Casteiger J (ed) Softwareentwicklung in der Chemie 1, Proceedings des Workshop "Computer in der Chemie", Springer-Verlag, Berlin Heidelberg New York
6. Bopp P, Jancsó G, Heinzinger K (1983) Chem Phys Letters 98: 129
7. Toukan K, Rahman A (1985) Phys Rev B31: 2643
8. Lie G, Clementi E (1986) Phys Rev A33: 2679
9. Demontis P, Suffritti GB, Fois ES, Gamba A (1986) Chem Phys Letters 127: 456
10. Dang LX, Pettitt BM (1987) J Phys Chem 91: 3349
11. Jancsó G, Bopp P, Heinzinger K (1985) Chem Phys 85: 377
12. Rahman A,Stillinger FH (1974) J Chem Phys 60: 545
13. Lemberg HL, Stillinger FH (1975) J Chem Phys 62: 1677
14. Berendsen HJC, Postma JPM, van Gunsteren WF, Hermans J (1981) In: Pullman B (ed) Intermolecular Forces, Reidel, Dordrecht
15. Chanderasekhar J, Jorgensen WL (1982) J Chem Phys 77: 579
16. Matsuoka O, Clementi E, Yoshimine M (1976) J Chem Phys 64: 1351
17. Jorgensen WL, Chanderasekhar J, Madura JD, Impey RW, Klein ML (1983) J Chem Phys 79: 926
18. Goldmann S, Backx P (1986) J Chem Phys 74: 2761
19. Stillinger FH, David CW (1978) J Chem Phys 69: 1473
20. Rullmann JAC (1988) Modelling molecular interactions in proteins and water between quantum mechanics and classical electrostatic, Thesis, Universiteit Groningen

21. Wójcik M, Clementi E (1986) J Chem Phys 84: 5970
22. Corongiu G, Migliore M, Clementi E (1989) J Chem Phys 90: 4629
23. Probst MM, Spohr E, Heinzinger K (1989) Chem Phys Letters, 161: 405
24. Kálmán E, Pálinkás G, Kovács P (1977) Mol Phys 34: 505
25. Pálinkás G, Kálman E, Kovács P (1977) Mol Phys 34: 525
26. Enderby JE (1987) In: Bellissent-Funel M-C, Neilson GW (eds) The physics and chemistry of aqueous ionic solutions, Reidel, Dordrecht
27. Bellissent-Funel M-C, Teixeira J, Bosio L, Dore J, Chieux P (1986) Europhysics Letters 2: 241
28. Blumberg RL, Stanley HE, Geiger A (1984) J Chem Phys 80: 5230
29. Szász Gy I, Dietz W, Heinzinger K, Pálinkás G, Radnai T (1982) Chem Phys Letters 92: 388
30. Pálinkás G, Radnai T, Dietz W, Szász Gy I, Heinzinger K (1982) Z Naturforsch 37a: 1049
31. Segré E (1931), Rend Lincei 13: 929
32. Bolla G (1932), Nuovo Cimento 9: 290
33. Curnutte B, Williams D (1974) In: Luck WAP (ed) Structure of water and aqueous solutions, Verlag Chemie, Weinheim
34. Sceats MG, Rice SA (1980) J Chem Phys 72: 3236, and references therein
35. Walrafen GE, Fischer MR, Hokmabadi MS, Yang WH (1986) J Chem Phys 85: 6970
36. Walrafen GE, Hokmabadi MS, Yang WH, Piermarini G (1988) J Phys Chem 92: 4540
37. Stillinger FH, Rahman A (1974) J Chem Phys 60: 1545
38. Pálinkás G, Bopp P, Jancsó G, Heinzinger K (1984) Z Naturforsch 39a: 179
39. Jancsó G, Bopp P (1983) Z Naturforsch 38a: 206
40. Bopp P (1986) Chem Phys 106: 205
41. Bopp P (1987) In: Kleeberg H (ed) Interactions of water in ionic and nonionic hydrates, Springer-Verlag, Berlin Heidelberg New York
42. Spohr E, Pálinkás G, Heinzinger K, Bopp P, Probst MM (1988) J Phys Chem 92: 6754
43. Heinje G, Luck WAP, Bopp P (1988) Chem. Phys. Letters 152: 358
44. Jorgensen WL (1986) J Phys Chem 90: 1276
45. Haughney M, Ferrario M, McDonald IR (1986) Mol Phys 58: 849
46. Pálinkás G, Hawlicka E, Heinzinger K (1987) J Phys Chem 91: 4334
47. Stouten PFW, Kroon J (1987) J Mol Struct 177: 467
48. Bopp P (1987) Pure & Appl Chem 59: 1071
49. Pálinkás G, Heinzinger K, Bopp P, work in progress
50. Impey RW, Madden PA, McDonald IR (1982) Mol Phys 46: 513
51. Neumann M (1983) Mol Phys 50: 841
52. Berens PH, Mackay DHJ, White GM, Wilson KR (1983) J Chem Phys 79: 2375
53. Kuharsky RA, Rossky PJ (1985) J Chem Phys 82: 5289
54. Kalinichev AG (1986) Intl J Thermophys 7: 887
55. Caillol JM, Levesque D, Weiss JJ (1986) J Chem Phys 85: 6645
56. Levesque D (1987) In: Bellissent-Funel M-C, Neilson GW (eds) The physics and chemistry of aqueous ionic solutions, Reidel, Dordrecht
57. Wójcik M, Clementi E (1986) J Chem Phys 85: 6085
58. Kataoka Y (1987) J Chem Phys 87: 589
59. Tanaka H, Ohmine I (1987) J Chem Phys 87: 6128
60. Heinzinger K, Vogel PC (1974) Z Naturforsch 29a: 1164
61. Heinzinger K (1985) Physica 131B: 196
62. Heinzinger K (1985) Pure & Appl Chem 57: 1031
63. Heinzinger K, Pálinkás G (1987) In: Kleeberg H (ed) Interactions of water in ionic and nonionic hydrates, Springer-Verlag, Berlin Heidlberg New York
64. Bopp P (1987) In: Bellissent-Funel M-C, Neilson GW (eds) The physics and chemistry of aqueous ionic solutions, Reidel, Dordrecht
65. Alagona G, Ghio, C, Kollmann P (1986) J Am Chem Soc 108: 185
66. Geiger A, Raman A, Stillinger FH (1979) J Chem Phys 70: 263
67. Alagona CG, Tani A (1980) J Chem Phys 72: 580
68. Demontis P, Ercoli R, Fois ES, Gamba A, Suffritti GB (1981) Theoret Chim Acta 58: 97
69. Dang LX (1985) Molecular dynamics simulations of nonpolar solutes dissolved in liquid water or trapped in water clathrates, Thesis, University of California, Irvine
70. Dang LX, Bopp P, Wolfsberg M (1989) Z Naturforsch 44a: 485
71. Mezei M, Beveridge D (1981) J Chem Phys 74: 6902
72. Impey RW, Madden PA, McDonald IR (1983) J Chem Phys 87: 5071

73. Bounds DG (1985) Mol Phys 54: 1335
74. Bopp P, Dietz W, Heinzinger K (1979) Z Naturforsch 34a: 1424
75. Dietz W, Riede WO, Heinzinger K (1982) Z Naturforsch 37a: 1038
76. Probst MM, Radnai T, Heinzinger K, Bopp P, Rode BM (1985) J Phys Chem 89: 753
77. Meier W (1989) Molekulardynamische Studien zur Hydratation des La^{3+} Ions, Diplomarbeit, Rheinisch-Westfälische Technische Hochschule, Aachen
78. Meier W, Bopp P, Spohr E, Probst MM, Lin J-1 (1990) J. Phys. Chem. 94: 4672
79. Bopp P, Okada I, Ohtaki H, Heinzinger K (1985) Z Naturforsch 40a: 116
80. Tanaka K, Ogita N, Tamura Y, Okada I, Ohtaki H, Pálinkás G, Spohr E, Heinzinger K (1987) Z Naturforsch 42a: 29
81. Migliore M, Fornili SL, Spohr E, Pálinkás G, Heinzinger K (1986) Z Naturforsch 41a: 826
82. Szász Gy I, Heinzinger K (1983) Z Naturforsch 38a: 214
83. Karim OA, McCammon JA (1986) J Am Chem Soc 108: 1762
84. Dang LX, Pettitt MB (1987) J Chem Phys 86: 6560
85. Tani A (1984) Mol Phys 51: 161
86. Bopp P, unpublished results
87. Kanno H (1987) J Raman Spectry 18: 301
88. Maisano G, Migliardo P, Fontana MP, Bellissent-Funel M-C, Dianoux AJ (1985) J Phys C Solid State Phys 18: 1115
89. Bellissent-Funel M-C, Dianoux AJ, Fontana MP, Maisano G, Migliardo P (1986) In: Neilson GW, Enderby JE (eds) Water and aqueous solutions, Adam Hilger
90. Szász Gy I, Heinzinger K (1983) J Chem Phys 79: 3467
91. Chen S-H, Teixeira J (1986) In: Prigogine I, Rice SA (eds) Advances in Chemical Physics, Vol LXIV, Wiley, New York
92. Luck WAP (1973) In: Luck WAP (ed) Structure of water and aqueous solutions, Verlag Chemie, Weinhein
93. Kleeberg H, Heinje G, Luck WAP (1986) J Phys Chem 90: 4427
94. Kristianson O, Eriksson A, Lindgren J (1984) Acta Chem Scand A38: 609
95. Kristianson O, Eriksson A, Lindgren J (1984) Acta Chem Scand A38: 612
96. Heinje G, Kammer T, Kleeberg H, Luck WAP (1987) In: Kleeberg H (ed) Interactions of water in ionic and nonionic hydrates, Springer-Verlag, Berlin Heidelberg New York
97. Lindgren J, Kristiansson O, Paluszkiewicz C (1987) In: Kleeberg H (ed) Interactions of water in ionic and nonionic hydrates, Springer-Verlag, Berlin Heidelberg New York
98. Probst MM, Bopp P, Heinzinger K, Rode BM (1984) Chem Phys Letters 106: 317
99. Hannongbua SV, Ishida T, Spohr E, Heinzinger K (1988) Z Naturforsch 43a: 572
100. Turq P, Lantelme F, Friedmann HL (1977) J Chem Phys 66: 3039
101. Jönson B (1981) Chem Phys Letters 82: 520
102. Sonnenschein R, Heinzinger K (1983) Chem Phys Letters 102: 55
103. Anastasiou N, Finchham D, Singer K (1983) J Chem Soc Faraday Trans II 79: 1639
104. Spohr E, Heinzinger K (1986) Chem Phys Letters 123: 218
105. Spohr E, Heinzinger K (1988) Electrochim Acta 33: 1211
106. Karim OA, Haymet ADJ (1988) J Chem Phys 89: 6889
107. Matsumoto M, Kataoka Y (1988) J Chem Phys 88: 3233
108. Wilson MA, Pohorille A, Pratt LA (1988) J Chem Phys 88: 3281
109. Linse P (1987) J Chem Phys 86: 4177
110. Spohr E, Heinzinger K (1988) Ber Bunsen Phys Chem 92: 1358
111. Schlenkrich M, Bopp P, Knoblauch M, Skerra A, Brickmann J (1988) In: Jorgensen PL, Varna A (eds) Advances in biotechnology of membrane ion transport, Raven Press
112. Nicklas K (1989) Simulation von wässrigen Ionenlösungen in der Nähe von biologischen Membranen, Diplomarbeit, Fachbereich Physikalische Chemie, Technische Hochschule Darmstadt
113. Schlenkrich M (1989) Struktur und Dynamik von Wasser in der Nähe von biologischen Grenzschichten, Diplomarbeit, Fachbereich Physik, Technische Hochschule Darmstadt
114. Böcker J (1989) Molekulardynamische Simulation von Wasserlamellen, Diplomarbeit, Fachbereich Physikalische Chemie, Technische Hochschule Darmstadt
115. Schlenkrich M, Nicklas K, Brickmann J, Bopp P (1990), Ber Bunsen Phys Chem 94: 133
116. Schlenkrich M, Nicklas K, Böcker J, Bopp P, Brickmann J (1990) In: Rivail J-L (ed) Modeling of Molecular Structures and Properties, Studies in Physical and Theoretical Chemistry Vol. 71 Elsevier Amsterdam

117. Spohr E (1989) J Phys Chem 93: 6171
118. Spohr E (1990) Chem Phys, 141: 87
119. Ibach H, Lehwald S (1980) Surf Sci 91: 187
120. Skerra A, Brickmann J (1987) Biophys J 51: 969
121. Skerra A, Brickmann J (1987) Biophys J 51: 977
122. Jorgensen WL, Gao J (1986) J Chem Phys 90: 2174
123. Bernal JD, Fowler RH (1933) J Chem Phys 1: 515
124. O'M Bockris J, Reddy AKN (1970) Modern Electrochemistry, Plenum Press, New York
125. Joosten JGH (1988) In: Ivanov IB (ed) Thin liquid films, Marcel Dekker, New York
126. Matsumoto M, Gobbins KE (1990) J Chem Phys 93: 1981

CHAPTER XV
The Energy of Intermolecular Interactions in Solution

G. Somsen

The word solution is used to describe a liquid or solid phase containing more than one substance, when the amount of one of the substances is much larger than the amount of other components. The substance present in excess, which may itself be a mixture, is called a solvent. Generally it is treated differently from the other substances which are called solutes.

From a molecular point of view the description of liquid solutions involves two kind of intermolecular interactions.

The first is that between the solute molecules or ions and the solvent, generally called solvation or, when the solvent is water, hydration. Since in most cases solutions are considered at constant pressure the corresponding energetic quantity is the enthalpy of solvation or hydration.

The second type of interaction is that between solute molecules and/or solute ions mutually. It is called solute–solute interaction. In electrolyte solutions the term interionic interaction is used. The energetic quantities involved at constant pressure are the enthalpies of interaction and the various enthalpic interaction coefficients.

1 Enthalpies of Solvation and Related Quantities

The change in enthalpy of the process

$$A \text{ (perfect gas)} \longrightarrow A \text{ (infinitely diluted solution)}$$

Intermolecular Forces
© Springer-Verlag Berlin Heidelberg 1991

is called the (standard) enthalpy of solvation, $\Delta_s H^0$. Seldom is it possible to measure $\Delta_s H^0$ directly by calorimetry. In most cases the enthalpy of solvation is determined with the aid of cycles like

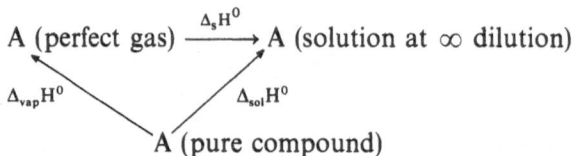

where $\Delta_{vap} H^0$ is the enthalpy of vaporization or sublimation of pure A in the liquid or solid state and $\Delta_{sol} H^0$ is the (standard) enthalpy of solution of pure substance A. Both quantities can be determined experimentally and the enthalpy of solvation is obtained by:

$$\Delta_s H^0 = \Delta_{sol} H^0 - \Delta_{vap} H^0 \tag{1}$$

If the solute is a non-associated electrolyte MX the process of solvation of cations M^+ and anions X^- is reflected by

$$v_+ M^{z+}(pg) + v_- X^{z-}(pg) \longrightarrow v_+ M^{z+}(\infty \text{ dilution}) + v_- X^{z-}(\infty \text{ dilution})$$

with $v_+ z_+ + v_- z_- = 0$. Here, two processes of solvation are involved, that of the cation and that of the anion. Due to the condition of electroneutrality they can not be separated experimentally. For a "simple" electrolyte MX the total enthalpy of solvation can be determined from the cycle

$$M^+(pg) + X^-(pg) \xrightarrow{\Delta_s H} M^+(\text{solution}) + X^-(\text{solution})$$
$$\underset{\Delta_{lat} H^0}{\searrow} \qquad \underset{\Delta_{sol} H}{\swarrow}$$
$$MX \text{ (crystal)}$$

in which $\Delta_{lat} H^0$ is the (known) lattice enthalpy of crystalline MX, while $\Delta_{sol} H$ can be determined experimentally. With the aid of experimentally determined enthalpies of dilution or by extrapolation of values of $\Delta_{sol} H$ at different concentrations to zero concentration, the standard enthalpy of solution, $\Delta_{sol} H^0$, can be obtained. The standard enthalpy of solvation $\Delta_s H^0$ follows from

$$\Delta_s H^0 = \Delta_{sol} H^0 - \Delta_{lat} H^0 \tag{2}$$

Sometimes the enthalpy of solvation of a compound cannot be determined along these lines because it is impossible to realize the gas phase of the compound. Examples are the non-electrolyte glucose and the electrolyte tetrabutylammonium bromide. In such cases it is convenient to replace the gaseous state by a solvent which shows no peculiar specific interaction with the solute, a so-called reference solvent. For ions or polar non-electrolytes dissolved in water or other protic solvents often reference solvents have been used which are aprotic, have a low acidity or basicity and a dielectric constant high enough to ensure complete dissociation of 1-1 electrolytes. Examples are propylene-carbonate, dimethylformamide en dimethylsulphoxide. The enthalpy change in the process of transferring the solute from the reference solvent to the solvent under investigation is called the enthalpy of transfer, $\Delta_{tr}H$.

1.1 Non-Electrolytes

1.1.1 Enthalpies of Solvation

The solvation of non-charged particles in solution is due to Van der Waals interactions. Also, dependent on the combination "solute-solvent", specific effects like hydrogen bond formation and charge transfer may play a part. As most of these interactions are of a short-range type, the different parts of a solute molecule are more or less independently solvated. This is demonstrated by the observation that the enthalpies of solvation of homologous series of compounds consist of additive molecular group contributions. The limited amount of known enthalpies of solvation in non-aqueous solvents show group additivity very clearly. Much more data are available for water as the solvent. In Table 1 the enthalpies of hydration $\Delta_s H^0$ are presented of several series of organic compounds.

Except for the first members of the different homologous series, the values of $\Delta_s H^0$ can be represented as a function of the number of carbon atoms n_C by the following equations:

$$\text{alkanes} \qquad \{-14.1 - (2.81 \pm 0.15)n_C\} \, \text{kJ mol}^{-1}$$
$$\text{alkylamines} \quad \{-47.2 - (3.03 \pm 0.13)n_C\} \, \text{kJ mol}^{-1}$$
$$\text{alkanols} \qquad \{-47.0 - (3.55 \pm 0.09)n_C\} \, \text{kJ mol}^{-1}$$

for $n_C \geqq 2$.

The polar groups in these compounds affect the local solvent structure around a dissolved molecule. Consequently the additivity is approximative and the CH_2 increments are not identical. Assuming a constant contribution of the methyl group in the series of alkanes (equal to half the enthalpy of hydration of ethane) and using the CH_2 increments from the equations, a calculation of the contributions of the other functional groups leads to the following values:

$$CH_3 \quad -9.87 \, \text{kJ mol}^{-1}; \quad NH_2 \quad -40.3 \, \text{kJ mol}^{-1}; \quad OH \quad -40.7 \, \text{kJ mol}^{-1}.$$

These results show that the enthalpies of transfer from the gaseous to the

Table 1. Standard enthalpies of hydration of several series of organic compounds at 25°C (values in kJ mol^{-1})

alkane	$\Delta_s H^0$	amine	$\Delta_s H^0$	alkanol	$\Delta_s H^0$
methane	−13.8	methylamine	−45.3	methanol	−44.7
ethane	−19.7	ethylamine	−53.7	ehtanol	−52.4
n-propane	−22.5	n-propylamine	−56.0	n-propanol	−57.6
n-butane	−26.0	n-butylamine	−59.0	n-butanol	−61.6
n-pentane	−27.2	n-pentylamine	−62.1	n-pentanol	−64.7
n-hexane	−31.4	n-hexylamine	−65.8	n-hexanol	−68.2
n-heptane	−33.9			n-heptanol	−72.1
n-octane	−39.7			n-octanol	−74.1

aqueous phase are exothermic for hydrophilic groups (OH, NH$_2$) as well as hydrophobic groups (CH$_3$, CH$_2$). The interaction of OH and NH$_2$ with water is strong in comparison with hydrophobic groups.

1.1.2 Enthalpies of Transfer

The change in enthalpy for the process of transferring a solute from one solvent to another contains group contributions also. However, due to the smaller magnitude of enthalpies of transfer in comparison to enthalpies of solvation, the first reflect more clearly the influences of specific effects. Enthalpies of transfer can be easily calculated as the differences of the enthalpies of solution in the corresponding solvents.

Table 2 presents the enthalpies of solution of a series of alkanols in the solvents propylenecarbonate (PC), dimethylsulphoxide (DMSO), dimethylformamide (DMF) and water. It shows that all enthalpies of transfer $\Delta_{tr} H^0$ from water to a non-aqueous solvent are positive. Also their magnitudes are larger than those of $\Delta_{tr} H^0$ for two non-aqueous solvents.

Table 2. Standard enthalpies of solution $\Delta_{sol} H^0$ of alkanols in several solvents at 25°C (values in kJ mol^{-1})

alkanol	PC	DMSO	DMF	H$_2$O
methanol	+6.3	−1.4	−0.6	− 7.3
ethanol	+8.4	+1.2	+1.3	−10.2
n-propanol	+9.5	+2.6	+2.2	−10.1
i-propanol	+9.9	+3.6	+3.0	−13.1
n-butanol	+10.6	+4.1	+3.0	−9.2
i-butanol	—	—	+3.2	−9.2
s-butanol	—	—	+3.6	−13.0
t-butanol	+10.5	+5.0	+3.5	−17.3
n-pentanol	+11.6	+5.4	+3.8	−7.7
t-pentanol	+10.3	+4.9	+3.1	−18.3

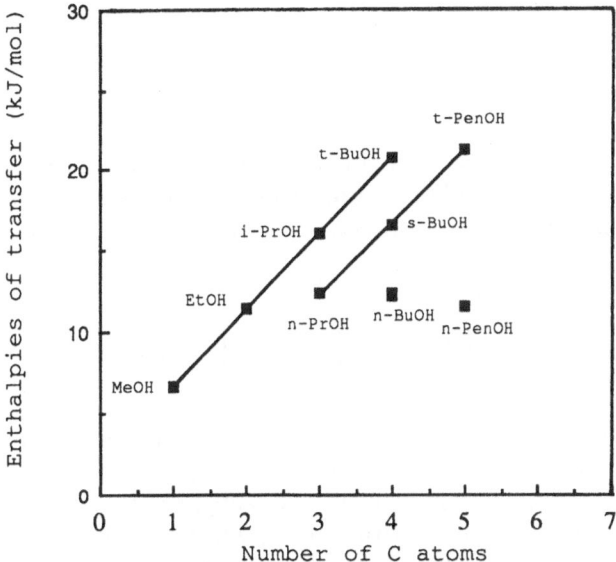

Fig. 15.1. Enthalpies of transfer from water to DMF of several alkanols

In addition, striking differences can be observed between the isomeric alkanols. For transfers between non-aqueous solvents the values of $\Delta_{tr}H^0$ of isomeric alkanols are almost equal. However, for transfers from water to a non-aqueous solvent the secondary and tertiary alkanols show much more endothermic values of $\Delta_{tr}H^0$. This is visualized in Fig. 1 for water and DMF. As this figure shows, $\Delta_{tr}H^0(H_2O \longrightarrow DMF)$ of n-alkanols levels off for $n_C \geqq 2$. Also the value for i-butanol is practically equal to that of n-butanol. But the values of $\Delta_{tr}H^0$ of the non-primary alkanols show regular trends in the series methanol $\longrightarrow$ ethanol $\longrightarrow$ i-propanol $\longrightarrow$ t-butanol and the series n-propanol $\longrightarrow$ s-butanol $\longrightarrow$ t-pentanol. Substituting a α-H atom by a methyl group gives a change in $\Delta_{tr}H^0(H_2O \longrightarrow DMF)$ of approximately $+4.6\,kJ\,mol^{-1}$.

The leveling-off of $\Delta_{tr}H^0$ with increasing n_C for the n-alkanols may be due to compensating effects. On the one hand the contribution to $\Delta_{tr}H^0$ $(H_2O \longrightarrow DMF)$ on transfer of CH_2 en OH groups becomes more negative with increasing n_C (compare the values for n-BuOH and n-PenOH), on the other hand the structural contribution from the peculiar (hydrophobic) hydration of the alkanol molecules becomes more positive with increasing n_C. This last contribution is the largest for spherical alkyl groups.

1.2 Electrolytes

1.2.1 Enthalpies of Solvation

Ionic solvation is influenced by a number of interactions. Firstly, the polarization interaction between the ionic charge and the surrounding solvent with dielectric constant ε. The corresponding enthalpy of solvation is given by the Born–Bjerrum equation

$$\Delta_s H_i = -\frac{(z_i e)^2 N_A}{2r_i}\left[1 - \frac{1}{\varepsilon} - \frac{T}{\varepsilon}\left(\frac{\partial \ln \varepsilon}{\partial T}\right)_P\right] \tag{3}$$

where N_A is Avogadro's number, $z_i e$ the charge of the ion and r_i its radius. This equation yields enthalpies of solvation which are too large, especially for small ions. Some of the attempts of obtain more realistic values apply an ionic radius enlarged with a rather arbitrary additive parameter δ, which is different for cations and anions. Taking into account the dielectric saturation of the solvent in the vicinity of the ion by modification of the value of the dielectric constant has no advantages as well, since the dependence of ε on the distance to the ion is not exactly known. Consequently, δ and ε become empirical factors which make it doubtful to use the Born–Bjerrum equation for the prediction of reliable enthalpies of solvation.

Better results are obtained with molecular solvation models which account for the interactions between the ionic charge and the dipoles and quadrupoles (or even higher multipoles) of the solvent molecules surrounding the ion and which include the lateral interactions between these solvent molecules. Usually these models also consider the effects of polarizabilities and those of dispersion and repulsion forces. The ion with one surrounding layer of solvent molecules is assumed to be a solvated complex with a radius large enough to represent the interaction with the remaining solvent by the Born–Bjerrum equation.

This molecular model approach explains also why the enthalpies of solvation of a cation and an anion with the same charge and radius are not equal. A change in the sign of the ionic charge leads to an inversion of a solvent molecule (with its dipole) in the solvated complex and the ion–dipole interaction will not change drastically. However, the orientation of the quadrupole of the solvent molecule is not affected by this inversion. Consequently, the energy of ion–quadrupole interaction changes sign if in a solvated complex the cation is substituted by an anion. The difference in the enthalpies of solvation of single-charged cations and anions with the same radius can be given by

$$\Delta_s H(M^+) - \Delta_s H(X^-) = \frac{A e \Theta}{R^3} - \frac{B \mu \Theta}{R^4} \tag{4}$$

in which μ and Θ are the dipole and quadrupole moments of the solvent molecule, e is the elementary charge, R the radius of the solvated complex ($R = r_i + 2r_s$, where r_i is the ionic radius and r_s the radius of the solvent molecule), while A and B are quantities dependent on the geometry of the solvated complex. Using this approach it is possible to divide the total enthalpies of solvation of electrolytes into the individual contributions of the ions.

Ionic enthalpies of solvation obtained in this way do not differ substantiallly between several solvents. Ion size is much more important. For Li^+ the values in different solvents lie around $-530\,kJ\,mol^{-1}$; for I^-, with a much larger radius, around $-300\,kJ\,mol^{-1}$. However, the uncertainties in the different contributions to the ionic enthalpy of solvation are considerable and hamper the analysis of specific effects in certain combinations of ion and solvent. To this end, enthalpies of transfer are much more useful.

1.2.2 Enthalpies of Transfer

Often studies on enthalpies of transfer of electrolytes apply procedures to divide $\Delta_{tr}H$ into contributions of the separate cation and anion. One of these procedures assumes equal enthalpies of transfer of the tetraphenylarsonium cation and the tetraphenylborate anion in every pair of solvents: $\Delta_{tr}H(Ph_4As^+) = \Delta_{tr}H(BPh_4^-)$.

In Table 3 ionic enthalpies of transfer based on this convention are given for the transfers PC $\longrightarrow$ DMSO in PC $\longrightarrow$ H$_2$O. Figure 2 shows that for alkali ions $\Delta_{tr}H^0(PC \longrightarrow DMSO)$ increases linearly with the radius of the cation. The same is true for R_4N^+ ions, but with a slope differing considerably from that of the alkali ions. The CH_2 increment in the values of $\Delta_{tr}H^0$ for R_4N^+ ions is $0.42\,kJ\,mol^{-1}$, comparable with a CH_2 increment of $0.36\,kJ\,mol^{-1}$ for alkanols in the same solvents. Apparently, the solvation of R_4N^+ ions depends on additive contributions of CH_2 groups.

The values for alkali ions in Fig. 2 are in accordance with the observation that DMSO as Lewis base is more basic than PC, while the acidity of the alkali ions as Lewis acids decreases from Li^+ to Cs^+. As a result $\Delta_{tr}H^0$ of Li^+ has the highest exothermic value, that of Cs^+ the lowest exothermic one.

Table 3. Ionic enthalpies of transfer $\Delta_{tr}H^0$ from propylenecarbonate to dimethylsulphoxide and water at 25°C (values in $kJ\,mol^{-1}$)

ion	PC $\longrightarrow$ DMSO	PC $\longrightarrow$ H$_2$O	ion	PC $\longrightarrow$ DMSO	PC $\longrightarrow$ H$_2$O
Li^+	-29.4	-3.0	Me_4N^+	$+1.0$	$+16.3$
Na^+	-17.5	$+10.2$	Et_4N^+	$+3.5$	-0.7
K^+	-13.0	$+21.9$	$n\text{-}Pr_4N^+$	$+5.1$	-11.7
Rb^+	-9.0	$+24.6$	$n\text{-}Bu_4N^+$	$+6.9$	-18.4
Cs^+	-5.5	$+26.8$	$n\text{-}Pen_4N^+$	$+8.5$	-20.0
Cl^-	-7.6	-26.4	$n\text{-}Hex_4N^+$	$+10.0$	-17.7
Br^-	-10.1	-13.6	$Phen_4As^+$	$+2.7$	$+14.6$
I^-	-9.5	$+3.3$	$BPhen_4^-$	$+2.7$	$+14.6$

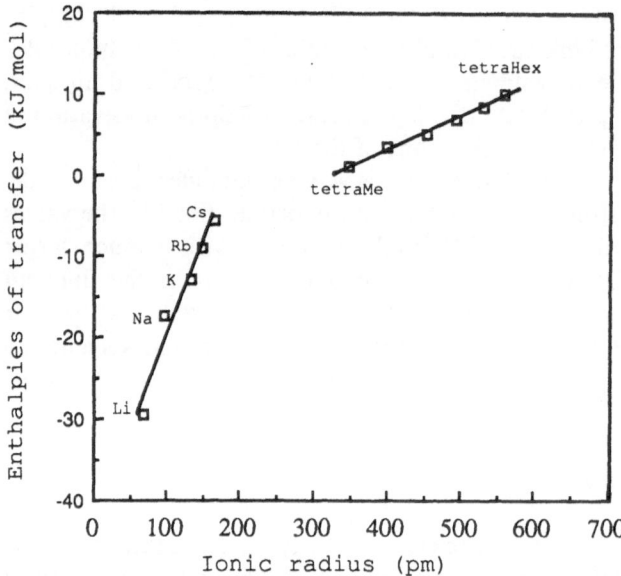

Fig. 15.2. Enthalpies of transfer of PC → DMSO for single-charged cations

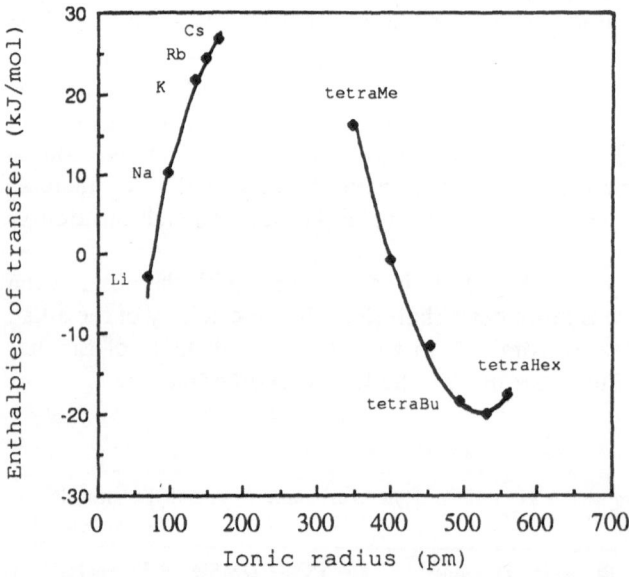

Fig. 15.3. Enthalpies of transfer of PC → water for single-charged cations

For transfers from PC to water, the variation of $\Delta_{tr}H^0$ with ionic radius of the alkali ions is comparable to that from PC to DMSO (see Fig. 3). However, the values of $\Delta_{tr}H^0$ are much more endothermic. Only $\Delta_{tr}H^0$ of Li^+ is negative. Apart from the difference in basicity between the two solvents, specific solvation effects of the alkali ions in water must be taken into account. The trend in

$\Delta_{tr}H^0$ going from Li^+ to Cs^+ corresponds to the change from the strongly hydrated Li^+ ion to the structure breaking Cs^+ ion.

The R_4N^+ ions show a dependence of $\Delta_{tr}H^0(PC \rightarrow H_2O)$ on ionic radius which is completely different, as well as in relation to the PC $\rightarrow$ DMSO transfers as to the PC $\rightarrow$ H_2O transfers of the alkali ions. As Fig. 3 shows, $\Delta_{tr}H^0(PC \rightarrow H_2O)$ becomes more exothermic as the ionic radius increases. Undoubtedly this is due to the peculiar hydration of R_4N^+ ions, the so-called hydrophobic hydration. It is assumed that this type of hydration is accompanied by a subtle change in the number and/or strength of the hydrogen bonds between the water molecules in the vicinity of the solute particle.

These considerations about ionic enthalpies of transfer have been made on basis of the extrathermodynamic assumption that $\Delta_{tr}H(Ph_4As^+) = \Delta_{tr}H(BPh_4^-)$. However, enthalpies of transfer of ions with phenyl groups, like Ph_4As^+, Ph_4P^+ and Ph_4Sb^+, appear to be strongly dependent on the nature of the central atom. This gives reason to cast doubt on the dividing convention as basis for physically realistic ionic enthalpies of transfer.

Fortunately, it is possible to avoid this problem by studying trends in the enthalpies of transfer of salts with a common ion. Table 4 gives the enthalpies of transfer of a number of alkali- and tetraalkylammonium bromides from the (reference) solvent DMF to the aprotic solvent DMSO and to the protic solvents NMF and water.

As can be inferred from this table, the values of $\Delta_{tr}H$ related to transfers between non-aqueous solvents have the same order of magnitude for all bromides. This is not true for transfers from DMF to H_2O, where clear differences occur between alkali bromides and tetraalkylammonium bromides. Values of $\Delta_{tr}H$ for the alkali bromides are in the range 30 to 35 kJ mol^{-1}, while values of $\Delta_{tr}H$ for R_4NBr deviate in an exothermic sense from this range. These deviations, which increase with the size of the alkyl groups, can be attributed to the exothermic contribution of the hydrophobic hydration of the R_4N^+ ions. From the data in Table 4 this contribution can be estimated as being approximately -55 kJ mol^{-1} in the case of Bu_4NBr.

Table 4. Enthalpies of transfer $\Delta_{tr}H^0$ of several bromides from dimethylformamide to dimethylsulphoxide, N-methylformamide and water at 25°C (values in kJ mol^{-1})

solute	DMF $\rightarrow$ DMSO	DMF $\rightarrow$ NMF	DMF $\rightarrow$ H$_2$O
LiBr	+6.0	+14.3	+28.7
NaBr	+6.0	+13.5	+30.3
KBr	+5.0	+12.8	+36.2
RbBr	+5.6	+13.6	+35.6
CsBr	+5.0	+13.5	+33.8
Me$_4$NBr	−3.6	+11.0	+8.2
Et$_4$NBr	+4.2	+11.7	−3.2
n-Pr$_4$NBr	+6.1	+10.9	−14.2
n-Bu$_4$NBr	+8.3	+10.0	−20.9

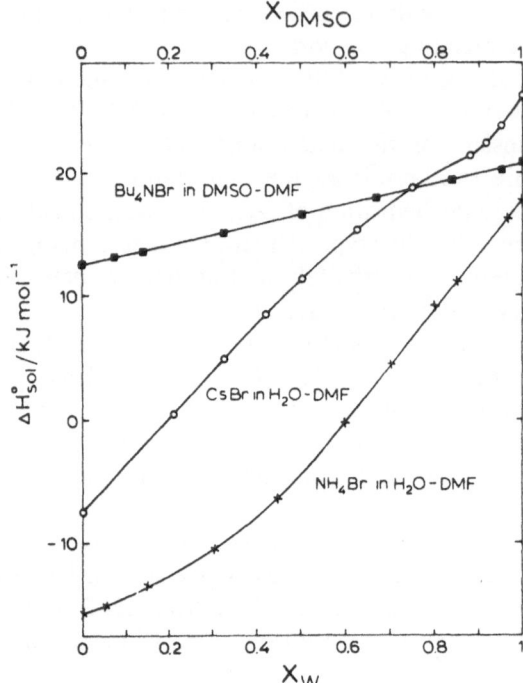

Fig. 15.4. Enthalpies of solution of some salts in mixtures of DMF and water

The different behaviour of alkali bromides and tetraalkylammonium bromides reveals itself even more clearly if we consider the changes in the enthalpies of solution of different bromides in mixtures of water and DMF covering the complete mole fraction range. Figure 4 shows that values of $\Delta_{sol}H$ of CsBr as well as NH_4Br change gradually in mixtures of DMF and water from the value in pure DMF to the value in pure H_2O. The changes are almost linear in the mole fraction of water, x_W. The same holds for $\Delta_{sol}H$ of Bu_4NBr in the non-aqueous mixtures DMF–DMSO and DMF–NMF. However, changes in $\Delta_{sol}H$ of Bu_4NBr in mixtures of DMF and water are completely different. Starting in pure DMF, the endothermic enthalpy of solution of Bu_4NBr changes, like that of CsBr, linearly with x_W to a considerably more endothermic value at $x_W = 0.7$, but at higher mole fractions the enthalpy of solution shifts strongly in an exothermic sense to end up with an exothermic value in pure water. That it is meaningful to attribute this behaviour to the peculiar solvation of alkyl groups in water and in mixtures with a high water content, is shown in Fig. 5 which compares enthalpies of solution of NH_4NBr and $BuNH_3Br$ in mixtures of DMF and water. At low water content the curves of both solutes are parallel, but at higher mole fractions of water the introduction of one butyl group affects the solvation considerably. It will be clear that in figures such as Fig. 5.4 and 5, only trends in $\Delta_{sol}H$ are important. They reflect directly the changes in solvation. Absolute values of $\Delta_{sol}H$ depend also on the lattice enthalpies of the crystalline salts. Therefore we consider the shapes of the

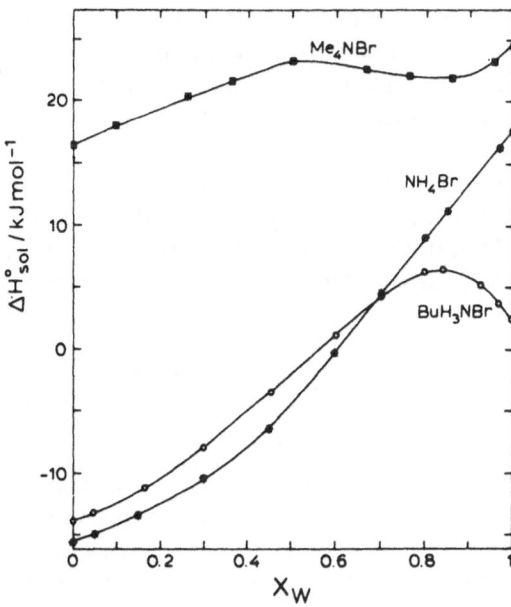

Fig. 15.5. Enthalpies of solution of NH_4NBr and $BuNH_3NBr$ in mixtures of DMF and water

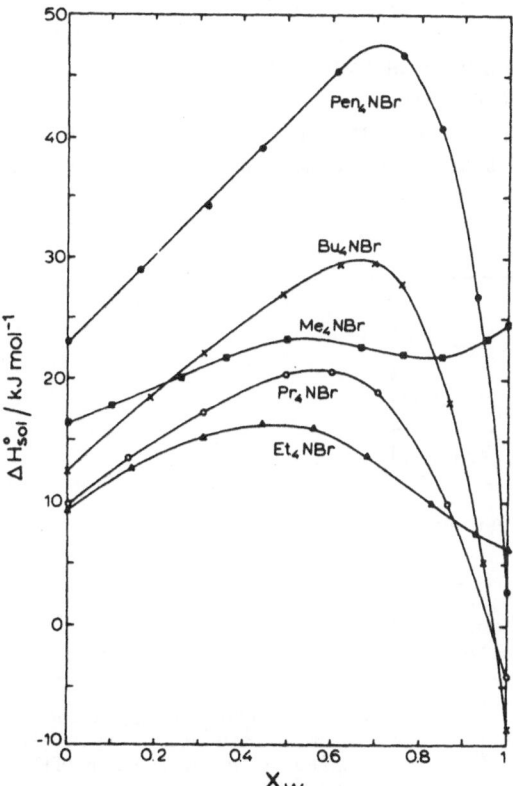

Fig. 15.6. Enthalpies of solution of tetra-alkylammoniumbromides in mixtures of DMF and water

different curves only and we do not pay attention to their position in the diagrams. This applies to Fig. 6, which compares the change in solvation of various tetraalkylammonium bromides and shows that the exothermic contribution of hydrophobic hydratation increases with the size of the alkyl groups.

The course of the enthalpies of solution of the larger tetraalkylammonium bromides can be described as a function of x_W with two parameters, the enthalpic contribution due to hydrophobic hydration in pure water, Hb(W), en n, a measure of the number of watermolecules involved in the hydrophobic hydration of the R_4N^+ ion, by means of:

$$\Delta_{sol}H(M) = x_W\Delta_{sol}H(W) + (1 - x_W)\Delta_{sol}H(DMF) + (x_W^n - x_W)Hb(W) \qquad (5)$$

where $\Delta_{sol}H(M)$ denotes the enthalpy of solution in the solvent mixture. The equation is based on a simple additive solvation model in which a clathrate-like cage of n water molecules envelopes each alkyl group. Values of Hb(W) vary from $-29\,kJ\,mol^{-1}$ for Et_4NBr to $-58\,kJ\,mol^{-1}$ for Pen_4NBr. In this model the influence of the non-aqueous component in the mixture is restricted to that of a "diluent" of the water structure. That will be approximately true if specific interactions between the cosolvent and water are small, as is apparently the case in aqueous mixtures of DMF. This view is substantiated by results in mixtures of water with other aprotic solvents. The course of the enthalpies of solution of these compounds in mixtures of water and DMSO or dimethyl-acetamide is closely similar to that in DMF.

Disturbing the hydrophobic character of the alkyl group by introducing terminal OH groups in the molecules largely influences the changes in solvation.

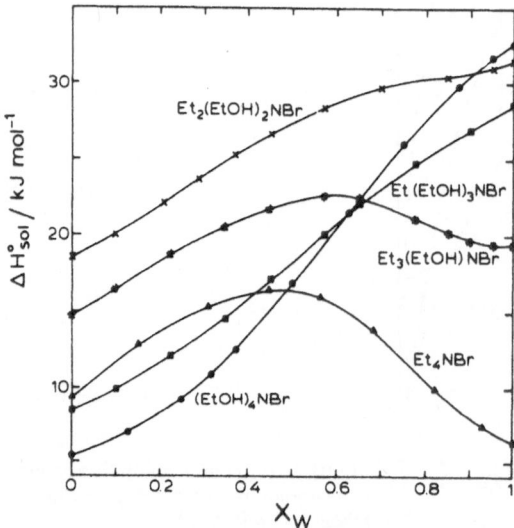

Fig. 15.7. Enthalpies of solution of hydroxy-substituted tetraalkylammonium bromides in mixtures of DMF and water

It is illustrated in Fig. 7 by some enthalpies of transfer of hydroxy substituted tetraalkylammonium compounds. The hydrophobic contribution in the results on $Et_3(EtOH)NBr$ is smaller than that for Et_4NBr and has disappeared completely with $(EtOH)_4NBr$. In fact $(EtOH)_4NBr$ shows a similar behaviour as CsBr.

From measurements at different temperatures it follows that the special hydration of tetraalkylammonium ions depends strongly on temperature. Its exothermic contribution decreases with increasing temperature, in accordance with the shift of number and/or strength of the hydrogen bonds in liquid water at rising temperatures.

2 The Energy of Solute–Solute Interactions

The chemical potential of a non-electrolyte B in solution can be given by

$$\mu_B = \mu_B^0 + RT[\ln c + B_2 c + B_3 c^2 + \cdots + B_n c^{n-1} + \cdots] \tag{6}$$

in which c is the concentration in mole per volume and B_n denote the virial coefficients of the chemical potential of solute B. This equation has its basis in the McMillan–Mayer theory of dilute solutions. The virial coefficients B_n are related to the forces of interaction in a group of n solute molecules influenced by the solvent. The coefficient B_2, related to the interactions between two solute molecules, is called the pairwise interaction coefficient and can be written as

$$B_2 = -\frac{1}{2} \int_0^\infty [\exp(-U/kT) - 1] 4\pi r^2 \, dr \tag{7}$$

where U is the potential of mean force of the two particles, dependent on their distance r and averaged over all relative orientations of the particles and over contributions of the solvent molecules. In a solution with one solute, B_2 is related to the interaction of two like molecules. If there are more solutes, also pairwise interactions between unlike molecules occur. Every pairwise interaction has its own B_2 coefficient.

For ionic solutions the equation for the chemical potential becomes more complicated, because of considerable Coulomb interactions between the ions even at very low concentrations. Since it is impossible to measure the chemical potentials of cations and anions seperately, the total chemical potential of the solute in an electrolyte solutions is expressed in a mean standard chemical potential by:

$$\mu_\pm = \mu_\pm^0 + RT[\ln c + A\sqrt{c} + Ec \ln c + B_2(\kappa)c + \cdots] \tag{8}$$

The terms in $\sqrt{c}$ and $c \ln c$ are the result of Coulomb interactions between the ions and can be calculated theoretically (e.g. with the Debye–Hückel theory) from the magnitudes of the ionic charges, the dielectric constant, and the temperature. For symmetrical electrolytes $(z_+ + z_- = 0)$ E equals zero. The second virial coefficient $B_2(\kappa)$ depends again on all forces between each pair of solute particles (i.e. ions) in the solution, but is now also related to the total concentration (or better ionic strength) through the Debye parameter κ.

By means of the usual thermodynamical relationships it is possible to derive expressions for quantities like enthalpy, entropy, heat capacities, etc., of the solutes from the equations for the chemical potentials.

2.1 Non-Electrolytes

The enthalpy of a solution of two solutes x and y in a solvent S can be related to the molalities m_x and m_y. To this end, the total enthalpy of the ternary system per kilogram of solvent, $H(m_x, m_y)$, is expressed as the enthalpy of 1 kg pure solvent, H_s^*, the standard partial molar enthalpies of both solutes, H_x^0 and H_y^0, and the pair-, triplet-, and higher order enthalpic interaction coefficients for the interactions between like and unlike solute molecules, $h_{xx}, h_{yy}, h_{xy}, h_{xxx}, h_{xxy}$, etc.:

$$H(m_x, m_y) = H_s^* + m_x H_x^0 + m_y H_y^0 + m_x^2 h_{xx} + m_y^2 h_{yy} + 2m_x m_y h_{xy}$$
$$+ m_x^3 h_{xxx} + m_y^3 h_{yyy} + 3m_x^2 m_y h_{xxy} + 3m_x m_y^2 h_{xyy} + \cdots \qquad (9)$$

For a binary system the corresponding relation is:

$$H(m_x) = H_s^* + m_x H_x^0 + m_x^2 h_{xx} + m_x^3 h_{xxx} + \cdots \qquad (10)$$

The enthalpic interaction coefficients are related to the coefficients in the McMillan–Mayer approach in a rather complicated way. They reflect the enthalpies of interaction of the solute particles and can be experimentally determined by calorimetric measurements of enthalpies of dilution, enthalpies of solution, and enthalpies of mixing.

The enthalpy of dilution of a solution of x in S refers to the process

$$\text{solution, molality } m_i + \text{solvent} \longrightarrow \text{solution, molality } m_f$$

or, schematically:

$$\begin{array}{c} N \, kg \, S \\ m_i N \, mol \, x \end{array} + N(m_i/m_f - 1) \, kg \, S \longrightarrow \begin{array}{c} (m_i/m_f) N \, kg \, S \\ m_i N \, mol \, x \end{array}$$

The measured change in enthalpy is

$$\Delta_{dil} H = N(m_i/m_f)[H_s^* + m_f H_x^0 + m_f^2 h_{xx} + m_f^3 h_{xxx} + \cdots]$$
$$- N(m_i/m_f - 1)H_s^* - N[H_s^* + m_i H_x^0 + m_i^2 h_{xx} + m_i^3 h_{xxx} + \cdots] \qquad (11)$$

Consequently, the molar enthalpy of dilution is given by:

$$\Delta_{dil}H_m(m_i \longrightarrow m_f) = (m_f - m_i)h_{xx} + (m_f^2 - m_i^2)h_{xxx} + \cdots \tag{12}$$

or

$$\frac{\Delta_{dil}H_m}{m_f - m_i} = h_{xx} + h_{xxx}(m_f + m_i) + h_{xxxx}(m_f^2 + m_f m_i + m_i^2) + \cdots \tag{13}$$

A plot of $\Delta_{dil}H_m/(m_f - m_i)$ vs. $(m_f + m_i)$ yields as intercept the value of h_{xx}, the limiting slope of the curve at $(m_f + m_i) \longrightarrow 0$ corresponds to h_{xxx}, while the curvature gives information on the values of the higher interaction coefficients. In many cases one or two interaction coefficients are sufficient to describe adequately the experimental results.

Enthalpic interaction coefficients of unlike solutes can be obtained from measurements of the enthalpy of solution of a pure substance x in a solution of substance y in the solvent S. The corresponding process is

$$m_x N \text{ mol pure } x + \begin{array}{c} N \text{ kg } S \\ m_y N \text{ mol } y \end{array} \longrightarrow \begin{array}{c} N \text{ kg } S \\ m_x N \text{ mol } x \\ m_y N \text{ mol } y \end{array}$$

The molar enthalpy of solution is given by:

$$\Delta_{sol}H_m(x \text{ in } S + y) = H_x^0 - H_x^* + m_x h_{xx} + m_x^2 h_{xxx} + 2m_y h_{xy} + 3m_x m_y h_{xxy} + 3m_y^2 h_{xyy} + \cdots \tag{14}$$

where H_x^* is the molar enthalpy of the pure compound x. At infinite dilution, the molality of x approaches zero and:

$$\Delta_{sol}H_m^\infty(x \text{ in } S + y) = H_x^0 - H_x^* + 2m_y h_{xy} + 3m_y^2 h_{xyy} + \cdots \tag{15}$$

$H_x^0 - H_x^*$ is equal to the (standard) enthalpy of solution of x in pure solvent S and h_{xy} can be obtained from:

$$\left[\frac{\partial \{\Delta_{sol}H_m^\infty(x \text{ in } S + y)\}}{\partial m_y} \right] = 2h_{xy}, \quad \text{if} \quad m_y \longrightarrow 0 \tag{16}$$

The molar enthalpy of transfer of solute x from pure solvent S to a solution of y in S can be written as:

$$\Delta_{tr}H_x(S \longrightarrow S + y) = 2m_y h_{xy} + 3m_y^2 h_{xyy} + 3m_x m_y h_{xxy} + \cdots \tag{17}$$

A similar relation holds for the transfer of solute y from $S \longrightarrow S + x$, so that

$$\frac{\Delta_{tr}H_x(S \longrightarrow S + y)}{m_y} = \frac{\Delta_{tr}H_y(S \longrightarrow S + x)}{m_x} = 2h_{xy} + 3m_y h_{xyy} + 3m_x h_{xxy} + \cdots \tag{18}$$

Enthalpic interaction coefficients of unlike molecules can be obtained also from measured enthalpies of mixing of a solution of x with a solution of y in the same solvent. From the scheme

$$\begin{array}{ccc} N_1 \text{ kg } S & N_2 \text{ kg } S & N_1 + N_2 \text{ kg } S \\ & + & \longrightarrow m_x N_1 \text{ mol } x \\ m_x N_1 \text{ mol } x & m_y N_2 \text{ mol } y & m_y N_2 \text{ mol } y \end{array}$$

it follows that

$$\Delta_{mix}H = \Delta_{dil}H(x) + \Delta_{dil}H(y) + 2m_x m_y \frac{N_1 N_2}{N_1 + N_2} h_{xy} + 3m_x^2 m_y \frac{N_1^2 N_2}{(N_1 + N_2)^2} h_{xxy}$$

$$+ 3m_x m_y^2 \frac{N_1 N_2^2}{(N_1 + N_2)^2} h_{xyy} + \cdots \tag{19}$$

in which $\Delta_{dil}H(x)$ and $\Delta_{dil}H(y)$ are the enthalpy changes if single solutions of the compounds x en y would be diluted from the same initial molality to the same final molality ($m_y = 0$ and $m_x = 0$, respectively).

Molecular interactions between solute particles in a particular solvent are very specific. The magnitude of the enthalpic interaction coefficients depends strongly on the nature of the solute molecules and on the properties of the solvent. Table 5 lists a number of enthalpic interaction coefficients in the solvent water, while Table 6 presents results in two non-aqueous solvents.

The results in water reflect the influence of the intermolecular interaction of the alkyl groups (hydrophobic interaction) with a positive contribution to h_{xx}. The cooperativity of this type of interaction is expressed by the positive values of h_{xxx}. Also the enthalpic interaction coefficients of unlike molecules with many or large alkyl groups are positive. However, pairwise interactions due to hydrogen bond formation or strong dipolar interactions, as with urea in water, lead to negative values of h_{xx}.

A completely different picture is given by the values of the enthalpic interaction coefficients in non-aqueous solvents. Most values are negative and alkyl groups give a negative contribution. In all cases the signs of h_{xx} and h_{xxx} are different. The values of h_{xx} are influenced by the possibility of hydrogen

Table 5. Enthalpic interaction coefficients of some solutes in water at 25°C

	h_{xx}	h_{xxx}		h_{xx}	h_{xxx}		h_{xx}	h_{xxx}
EtOH	243	65	NMF	272	—	glycol	362	—
n-PrOH	560	159	NMA	235	95	glucose	343	—
n-BuOH	1003	646	NMP	636	237	sucrose	577	—
t-BuOH	656	334	NBA	1477	1074	urea	−350	21

Units: h_{xx} in $J\,mol^{-1}\,(mol\,kg^{-1})^{-1}$; h_{xxx} in $J\,mol^{-1}\,(mol\,kg^{-1})^{-2}$. Abbreviations: NMF = N-methylformamide, NMA = N-methylacetamide, NMP = N-methylpropionamide, NBA = N-butylacetamide.

Table 6. Enthalpic interaction coefficients at 25 °C for several series of amides dissolved in N-methylformamide, NMF (protic) and N,N-dimethylformamide, DMF (aprotic)

	NMF		DMF	
	h_{xx}	h_{xxx}	h_{xx}	h_{xxx}
formamide	+95	−8	+119	—
propionamide	−31	+2	−286	+100
hexaanamide	−171	+16	−325	+49
N-methylacetamide	−1	−0.4	−124	+11
N-butylacetamide	−135	+10	−302	+36
N,N-diethylformamide	+97	−6	−24	+3
N,N-dibutylformamide	−15	—	−305	+37
N,N-diethylacetamide	+182	−11	−11	+1
N,N-dibutylacetamide	+55	−4	−342	+47

Units: h_{xx} in $J\,mol^{-1}\,(mol\,kg^{-1})^{-1}$; h_{xxx} in $J\,mol^{-1}\,(mol\,kg^{-1})^{-2}$

bond formation and the tendency of apolar groups to associate in a polar environment.

Only for series of related compounds it is possible to describe the values of h_{xx} and h_{xy} with a group interaction model. In such an approach, each solute molecule is considered to be composed of a limited number of molecular groups. The total interaction of two molecules is taken as the sum of the interactions of groups on one molecule with all groups on the other molecule and it is assumed that the group interactions do not interfere with each other. Then, the following equation results

$$h_{ij} = \sum_A \sum_B n_A^i n_B^j h_{AB} \tag{20}$$

in which h_{ij} is the enthalpic pairwise interaction coefficient of the molecules i and j, n_A^i the number of groups of type A in solute molecule i, n_B^j the number of groups of type B in solute molecule j and h_{AB} a parameter reflecting the A–B interaction. This additivity approach gives reasonable results and has some predictive power for series of related compounds in water. In non-aqueous solvents the results are much less satisfactory.

2.2 Electrolytes

The greater part of the enthalpy of an electrolyte solution is due to the Coulomb interactions between the ions. For a binary electrolyte this enthalpy can be described in the first approximation by an equation for the molar excess enthalpy derived from the theory of Debye and Hückel:

$$H_m^E = \frac{v/2|z_+z_-|A_H\sqrt{m}}{1 + B_H a\sqrt{m}} \tag{21}$$

In this equation v is the number of ions formed on complete dissociation of one molecule of the undissociated electrolyte, z_+ and z_- are the valencies of cation and anion, respectively. The parameters A_H and B_H depend on the nature of the solvent. The only quantity related to the type of ions in the solution, is the distance a, i.e. the closest approach of the cation and anion. Accurate experimental determinations of the excess enthalpies of electrolyte solutions in water show that this equation does not hold, since it does not allow for specific effects of the ions. This is most clearly demonstrated by measurements on the enthalpies of mixing of two aqueous electrolyte solutions with equal ionic strength. On mixing the Coulomb interactions do not change and consequently the Debye–Hückel contribution to the enthalpy of mixing will be zero. However, in many cases the observed enthalpy of mixing is not even small, showing that specific interactions of the ions can not be neglected. If one of the ions in both solutions is the same, say the anion, we have a solution of MX and one of NX. On mixing, the M...M and N...N interactions decrease by a simple dilution factor, while M...N interactions are formed. At a certain ionic strength this last interaction can be reflected by an enthalpic interaction coefficient h_{MN}. It appears that values of h_{MN} are only slightly dependent on the ionic strength, so that they can be assumed to represent the specific properties of the aqueous ions.

Solutes in water affect the liquid structure in their immediate environment. This region, often called the cosphere, can be divided into different zones schematically represented in Fig. 8. Zone I involves the primary solvation layer of the ion. Its structure is largely determined by strong ion-dipole interactions. In zone II the structure is determined by a competition between the directive forces of the electric field of the ion and the tendency of water molecules to form a structure as that in liquid water. Outside zone II, and hence outside the cosphere, the undisturbed water structure is present, characterized by the tendency of each water molecule to form hydrogen bonds with other water molecules. It is polarized by the electric field of the ion without any structural consequences. On basis of these views it is possible to distinguish between structure promoting ions, where zone I dominates, and structure breaking ions with properties which are largely determined by zone II. The first catagory comprises small and highly charged ions (Li^+, Na^+, F^-, Mg^{2+}, Ca^{2+}, Sr^{2+}, Ba^{2+}), the second large and singly charged ions (K^+, Rb^+, Cs^+, Cl^-, Br^-, I^-). For this last category the existence of a zone of type I has been questioned.

According to Young's rule, mixing of two ions of the same category in presence of a common counter ion gives rise to a positive enthalpic interaction coefficient. Table 7 presents a number of enthalpic interaction coefficients.

(Very) large enthalpic pair-interaction coefficients are the result of inter-actions with and between tetraalkylammonium ions. According to Young's rule the Me_4N^+ ion is a structure breaker. The Et_4N^+ ion is just on the border. In comparison with Me_4N^+ it is a structure maker, but with regard to the alkali ions it manifests itself as a structure breaker. The magnitude of the effects with R_4N^+ ions indicates that other factors are involved than those with the simple alkaline and halogen ions. Large R_4N^+ ions are hydrophobic structure promotors. The structure in their immediate vicinity is not determined by the centrosymmetrical electric field, but seems to be due to an enhanced interaction of water molecules ("iceberg formation" or better "hydrophobic hydration"). In

bulk

water

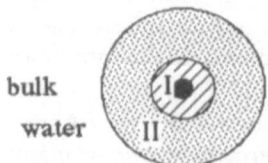

Fig. 15.8. The hydration model of Frank and Wen

Table 7. Enthalpic interaction coefficients of electrolytes with a common ion in water at 25°C

	h_{MN}		h_{MN}
$Na^+ \ldots K^+$	-159	$F^- \ldots Cl^-$	-94
$Li^+ \ldots Na^+$	$+343$	$F^- \ldots Br^-$	-70
$Li^+ \ldots K^+$	-259	$Cl^- \ldots Br^-$	$+13$
$Li^+ \ldots Me_4N^+$	-673	$F^- \ldots Ac^-$	$+45$
$Li^+ \ldots Et_4N^+$	-721	$Cl^- \ldots Ac^-$	-143
$Li^+ \ldots Pr_4N^+$	-2900	$Br^- \ldots Ac^-$	-135
$K^+ \ldots Me_4N^+$	$+497$		
$K^+ \ldots Et_4N^+$	$+492$		
$K^+ \ldots Pr_4N^+$	-1456	$Me_4N^+ \ldots Et_4N^+$	-208
$Cs^+ \ldots Me_4N^+$	$+343$	$Me_4N^+ \ldots Pr_4N^+$	-2567
$Cs^+ \ldots Et_4N^+$	$+306$	$Et_4N^+ \ldots Pr_4N^+$	-1290
$Cs^+ \ldots Pr_4N^+$	-1835		

Units: h_{MN} in $J\,mol^{-1}\,(mol\,kg^{-1})^{-1}$

that case Young's rule does not apply. However, it is possible to describe the enthalpic interaction coefficients in terms of "cosphere-overlap" (see Friedman). Mixing of a solution with Pr_4N^+ ions with another electrolyte decreases the overlap of the cospheres of the Pr_4N^+ ions and enables these ions to form a more complete hydrophobic cosphere. This is accompanied by a strengthening of hydrogen bonds and consequently it is an exothermic process. If it is assumed that the $Pr_4N^+ \ldots Pr_4N^+$ interactions dominate, this will result in negative enthalpic pairinteraction coefficients. The large exothermic enthalpies of dilution shown by R_4N salts with large alkyl groups are in accordance with the picture of a diminishing "cosphere-overlap" on diluting the solution.

3 Further Reading

Abraham M (1984) J Chem Soc Faraday Trans 1 80: 153
Bloemendal M (1985) Solute–solute interactions in non-aqueous solvents. Thesis, Vrije Universiteit, Amsterdam
Cassel RB, Wood RH (1974) J Phys Chem 78: 2460
Covington AK, Dickinson T (1973) Physical chemistry of organic solvent systems, Plenum, New York
Desnoyers JE, Perron G, Avédikian L, Morel JP (1976) J Solution Chem 5: 631
Desnoyers JE, Jolicoeur C (1983) In: Conway BE, Bockris JO'M, Yeager E (eds) Comprehensive treatise of electrochemistry, Plenum, New York, vol 5, p 1
De Visser C, Somsen G (1974) J Phys Chem 78: 1719
Frank HS, Wen WY (1957) Disc Faraday Soc 24: 133
Franks F, Reid DS (1973) In: Franks F (ed) Water, a comprehensive treatise, Plenum, New York, vol 2, p 323
Franks F, Pedley M, Reid DS (1975) J Chem Soc Faraday Trans 1, 80: 153
Friedman HL, Krishnan CV (1973) In: Franks F (ed) Water, a comprehensive treatise, Plenum, New York, vol 3, p 1

Heuvelsland WJM, De Visser C, Somsen G (1978) J Phys Chem 82: 29
Heuvelsland WJM (1980) Hydrophobic hydration in aqueous solvent mixtures. Thesis, Vrije Universiteit, Amsterdam
Krishnan CV, Friedman HL (1969) J Phys Chem 73: 3934
Rouw A, Somsen G (1981) J Chem Thermodynamics 13: 67
Rouw A, Somsen G (1982) J Chem Soc Faraday Trans 1, 78: 3397
Rouw AC (1982) Hydrophobic hydration of some organic compounds. Thesis, Vrije Universiteit, Amsterdam
Somsen G (1966) Rec Trav Chim Pays-Bas 85: 526

CHAPTER XVI
The Mobile Order Created by Hydrogen Bonds in Liquids

G.G. Siegel and P.L. Huyskens

In contrast with the rigid situation in mixed crystals, the neighbours of a given molecular group in a liquid constantly change in identity, distance and direction. The disorder in a liquid is a dynamic one. Nevertheless for the groups that form hydrogen bonds this change of environment occurs only during a limited fraction γ of the time. Furthermore during the complementary fraction of time $1 - \gamma$ where the groups are involved in H-bonding a correlation is established between the directions of given chemical bonds in the partners. H-bonding creates thus a kind of mobile order in liquids. As a consequence the g parameter of Kirkwood of hydrogen bonded pure liquids deviates in general markedly from one. This is not the case for strongly polar liquids where no H-bonds are present.

1 Molecular Structure in Crystals and in the Liquid State

"Structure" has a completely different meaning in a liquid than in a crystal and the concept of "quasi lattice" that is sometimes used for the thermodynamic study of liquids [1] has some nostalgic resonance but has nothing to do with the reality.

Let us compare the n-alkanes that are solid at room temperature (with more than 17 carbon atoms) with those that are liquid. In the crystals the position of each nucleus with respect to external axes is fixed within a few hundredths of an Angström corresponding to the thermal vibrations in the lattice [2]. In the liquids all molecules are perpetually moving with respect to each other. In the crystal all CH_3 groups are grouped in parallel bilayers separated from each

Intermolecular Forces
© Springer-Verlag Berlin Heidelberg 1991

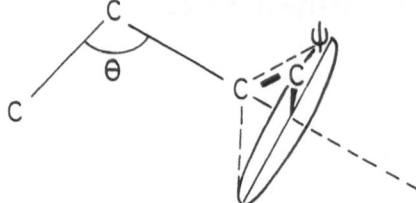

Fig. 1. Valence and azimuthal angles in paraffinic chains

other by the zig zag CH_2 chains. In these chains the C—C—C angles are all equal to 109.5° corresponding to an sp_3 hybridization of the atoms. Furthermore, in the forms stable at 25 °C, all carbon atoms of the chain lie in the same plane and this means that the azimuthal angle ψ describing (Fig. 1) the deviation of a third C—C bond from the plane of the two former bonds is uniformly equal to 0°. What remains from this structure in the liquid is the constancy of the C—C interdistance (1.54 Å) and of the valence angle. The azimuthal angles constantly fluctuate between 0° and 180° and this fluctuation is only sometimes partly limited by steric hindrances between neighbouring groups. The perpetually changing form of the molecule is that of a coil. In contrast to the crystals where the environment of the CH_3 groups is steadily formed by other CH_3 groups, there is a perpetual change in the contacts between the groups in the liquid state. An H atom of a CH_2 group will sometimes have other CH_2 groups as neighbours and sometimes CH_3 groups. In the course of time this atom will thus "see" the various groups of the molecules of the liquid defiling in front of it.

Each external atom of a molecule in the liquid phase has thus a time schedule for the various contacts.

In absence of any preference, this means in the case of "random contacts", this defiling of the various groups occurs prorata the fraction φ_j of the volume of the liquid that they occupy.

One can make a partition of the cohesion energy per unit volume, considering all the external atoms of a given group i and the fraction of time they spend in the contact with the atoms of the group j. This partition leads to the fraction f_{ij}.

One obtains for random contacts

$$f^0_{11} = \phi^2_1 \qquad f^0_{22} = \phi^2_2 \qquad f^0_{12} = 2\phi_1\phi_2$$

In the crystal, the groups have permanent contacts with frequencies

$$f^*_{11} = \phi_1 \qquad f^*_{22} = \phi_2 \qquad f^*_{12} = 0$$

Another aspect concerns the angles: when considering two groups temporarily in contact in the liquid the angle between two C—H directions of a given group is constant but the angle between two C—H directions in different groups constantly changes.

Thus, in the liquid, what remains constant are the interdistances and the angles governed by valence forces, whereas those determined by cohesion forces are spread over broader ranges. A direct consequence is that the maxima in the radial distribution functions detected by diffraction methods correspond to distances in chemical bonds and that the information given by these methods concerning the cohesion forces fails in accuracy.

In liquid alkanes, the "structure" in the sense of a constancy between narrow limits of interdistances and angles is only that governed by the chemical bonds.

2 Structuration of Liquids by H-Bonds. The Mobile Order

Specific interactions like H bonds are competitors of chemical bonds. To what extent do they participate in the structuration of liquids?

As a matter of fact, in liquids, H bonds bring about some additional constancy of distances and angles but, and that is essential, they do this *intermittently*. Actually H bonds in liquids, contrary to valence bonds, are regularly broken, their life times lying between 10^{-5} and 10^{-11} s [3].

Thus, H bonds in liquids do not lead, as do valence bonds or collegial lattice cohesion forces, to permanent constants with a constant distance between the nuclei. Actually, the H bond is broken during a fraction γ of the time and the contact is then interrupted. However, γ can be very small so that the contacts may be called "highly preferential".

In liquid n-butanol, for instance, the volume fraction ϕ_{OH} of the O—H group with respect to the whole molecule is 0.10. For random contacts an O—H proton would be in the vicinity of the OH group of a neighbouring molecule during 10% of the time. Actually the proton follows the oxygen atom of a neighbouring molecule in its walk through the solution, as a dog on a leash follows its master, for more than 95% of the time. *This confers to the H-bonded liquid a kind of order that is not encountered in crystals:the mobile order*. The consequence of this mobile order is that in pure butanol the frequency of contacts $f_{o...o}$ (0.095) between O—H groups is intermediate between the random value $f^0_{o...o} = 0.010$ and the value $f^*_{o...o} = 0.10$, encountered in the crystalline substance, but much nearer the latter.

H-bonding thus produces in liquids an additional structuration but his structuration is intermittent. Furthermore the spreading of the interdistance between the nuclei directly involved in the H-bond (O...O in butanol) is markedly larger than for valence bonds. This is reflected by the breadth of the peak v_σ corresponding to this vibration.

H bonding leads also to some constancy of angles. We have already said that, in liquid pentane for instance, there is no correlation between a C—H direction and the C—H directions of the neighbouring group in another molecule. In butanol however the angle between the directions of two neighbouring O—H groups lies mostly in the vicinity of the sp_3 angle of 109.5°

$$O—H...O\diagdown_H$$

This constancy of angles is of course less sharp in H bonds than in the case of valence angles. It may be perturbed by the competition between several H-bonds leading to a cyclic structure

Such structures are encountered in molar solutions of butanol in cyclohexane.

Other cohesion forces, as for instance the ones that lead to crystallization can also perturb the angles in H-bonds. The directions and characteristics of H-bonds in crystals are largely determined by the packing possibilities of the molecules in the lattice.

H-bonds in liquids lead to the intermittent formation of H-bonded chains as for instance

These chains are characterized like those in *n*-alkanes by the $R_{0...0}$ interdistance, the angle θ between two vicinal OH directions and the azimuthal angle ψ. The interdistances and the ψ angles are less fixed than in alkanes. Furthermore the proton may deviate from the line 0...0. Some ψ values may be prohibited because of the steric hindrances between the groups bonded to the oxygen atoms.

3 An Adequate Experimental Method for Studying the Mobile Order in Liquids: Dipolemetry

The larger variety of interdistances and angles in the temporary structuration brought about by H-bonds in liquids flattens the diffraction peaks and makes these methods less appropriate for the study of H-bonds in liquids.

A more interesting experimental study is that of the dipole moments. A method that is inapplicable to crystals where the moments cannot be oriented by an external field, becomes useful in the liquid phase where, due to the lability of the cohesion forces, the dipoles are orientable.

When one considers a given dipole $\vec{\mu}_1$ surrounded by $\mathcal{N}$ dipoles of the same value μ but with different orientations $\vec{\mu}_2$, $\vec{\mu}_3$, $\vec{\mu}_4$..., the Kirkwood parameter g_1 is defined as

$$g_1 = 1 + \sum_{j=2}^{j=\mathcal{N}} \cos\theta_{1j} = 1 + \sum_{j=2}^{j=\mathcal{N}} \mu_1\mu_j/\mu^2$$

where θ_{1j} is the angle between the dipole 1 and the dipole j. In a crystal these angles are fixed. In a liquid they change perpetually. In a liquid, due to this perpetual motion, the average value $\langle g_i \rangle$ observed for a given moment during a time long enough, is the same for all the dipoles i and equal to

$$g_{\text{liquids}} = 1 + \sum_{j \neq i} \langle \cos\theta_{ij} \rangle$$

When the orientations of the dipoles are independent of each other the mean value $\langle \cos\theta_{ij} \rangle$ is equal to zero and

$$g_{\text{liquid(random)}} = 1$$

For pure liquids, Kirkwood's correlation parameter can be determined from the dielectric constant ε^0, the refractive index n_0, the density ρ and the dipole moment μ^0 according to the equation that is derived from the theories of Onsager [4], Kirkwood [5] and Fröhlich [6].

$$g^0 = \left(\frac{9RT}{4\pi N}\right)\frac{(\varepsilon^0 - n^{02})(2\varepsilon^0 + n^{02})}{\mu^{02}(n^0 + 2)^2}\frac{M}{\rho^0}$$

M is the molar mass, the other symbols having their usual meaning[1]. M/ρ^0 is the molar volume $\bar{V}^0$ of the liquid.

μ^0 can be calculated from the contribution of group moments or determined from gas phase measurements where $g = 1$.

Experimental values of g^0 obtained in this way are given in Table 1 [7].

A striking difference appears between the polar liquids where, because of the lack of proton donor sites, no H-bonds can be formed, and the H-bonded liquids. In the first case the departures from the random values are rather moderate, never exceeding 25%. In contrast, in liquids with H-bonds the g parameters show strong deviations from unity.

---

[1] In principle n^0 is the "internal" refractive index that would be measured at the frequency of determination of ε^0 (usually $2 \times 10^6\,\text{s}^{-1}$) in absence of orientations of the dipoles. This value is not directly accessible by experiments and is usually replaced by n_D, the refractive index for the sodium line

Table 1. Kirkwood's correlation parameter g^0 of pure liquids at 25°C calculated from the molar volume $\bar{V}^0/cm^3mole^{-1}$, the dielectric constant ε^0, the refractive index n_D and the dipole moment $\mu/Debye$ of the free molecule

	μ^2	ε^0	n_D^2	$\bar{V}^0$	g^0
Polar liquids wwithout H-bonds					
Toluene	0.14	2.38	2.23	106.8	0.92
Triethylamine	0.57	2.38	1.96	139.8	0.93
Dichloromethane	2.53	9.00	2.02	64.3	1.20
Tetrahydrofurane	3.06	7.58	1.97	81.7	1.05
Cyclohexanone	8.40	18.25	2.10	104.0	1.23
Propionitrile	12.62	24.90	1.85	70.9	0.86
Dimethylformamide	14.47	36.71	2.04	77.0	1.14
Benzonitrile	16.00	25.20	2.32	102.6	0.80
H-bonded liquids					
Formic acid	1.98	58.50	1.87	37.8	7.14
Acetic acid	2.89	6.15	1.88	57.1	0.63
Butyric acid	2.86	2.96	1.95	110.0	0.33
Methanol	2.89	32.65	1.76	40.7	3.09
Ethanol	2.88	24.33	1.85	58.7	3.15
1-Propanol	2.55	20.10	1.91	75.1	3.58
1-Octanol	2.53	10.08	2.04	158.4	3.32
Water	3.38	78.54	1.77	18.1	2.84

Negative deviations are encountered in the carboxylic acids and are explained by the formation of cyclic dimers

$$R-C{<}^{O...H-O}_{O-H...O}{>}C-R$$

In the other cases the deviations are strongly positive.

In an H-bond between two O—H groups, the two O—H dipoles form an angle of 70.5°. If this angle is maintained for 95% of the time, the average value of $\cos \theta_{12}$ will be

$$\langle \cos \theta_{12} \rangle = 0.95 \cos 70.5 = 0.317$$

Suppose the O—H groups are involved in open chains with variable lengths. Each O—H group then participates in a chain with a mean number $\bar{n}_w$ of O—H groups. If for these chains the mean value $\langle \cos \psi \rangle = 0$, one calculates a g value of

$$g_{\bar{n}_w} = \left[\frac{1 + \cos \theta}{1 - \cos \theta} - \left(\frac{2\cos \theta}{\bar{n}_w} \right) \frac{1 - (\cos \theta)^{\bar{n}_w}}{(1 - \cos \theta)^2} \right]$$

where θ is the angle between two successive O—H directions. For long chains g tends to a limiting value

$$g_{lim} = \frac{1 + \cos\theta}{1 - \cos\theta}$$

For a tetrahedral angle $\theta = 70.5$ the limiting value should be g = 2. However, according to Table 1, the g^0 values of pure primary alcohols are around 3.45 ± 0.25. This was explained by the fact that, because of steric hindrances of the neighbouring groups, some rotamers around the H bonds are not formed and that the average value $\langle \cos\psi \rangle$ is positive. In this case the previous equation has to be modified to

$$g_{lim} = \left(\frac{1 + \cos\theta}{1 - \cos\theta}\right)\left(\frac{1 + \langle\cos\psi\rangle}{1 - \langle\cos\psi\rangle}\right)$$

where the azimuthal angle ψ is taken equal to zero in the "trans" position.

The mean value $\langle\cos\psi\rangle$ should thus be 0.45 in H-bonded chains of alcohols [8]. This can be explained by steric hindrances that prevent that angles ψ above a given value b should be reached

$$\int_0^b \cos\psi\, d\psi / b = 0.45$$

This yields a value of about $115°$ for b. The "forbidden" angle corresponds to that estimated by using a molecular model with the proper interdistances between the nuclei and the Van der Waals radii of the external atoms. The result is quite general and this steric hindrance effect was observed for all O—H...O bonds[2] [9]. This steric effect does not exist, in general, in OH...N bonds [10].

This is also predictable from molecular models.

When alcohol chains become shorter they show some tendency to cyclization. This can be shown by determining the evolution of the Kirkwood parameter upon dilution in an inert solvent (Fig. 2). For the tertiary alcohols this is already visible at high concentrations but for primary alcohols the effect only becomes very marked when the concentration drops below 6 mole per liter. Parameter g passes through a minimum of about 0.7 around 1 mole dm^{-3} indicating a maximum of cyclization at this concentration.

The g value of liquid water is 2.84 at $25°C$. This was first explained by Oster and Kirkwood [11] on the basis of the model of Bernal and Fowler [12]. According to this ice-like model each proton of the water molecule in the liquid is involved in equivalent H-bonds. Each oxygen atom is surrounded in a

---

[2] The effect does not exist in water

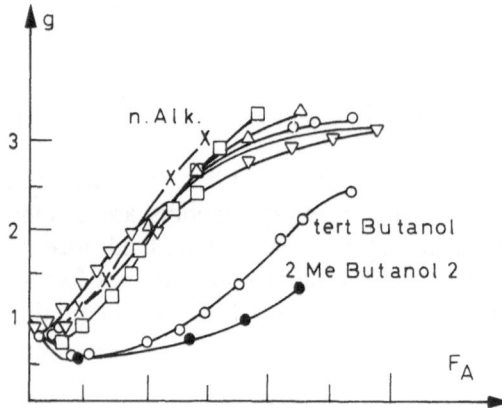

Fig. 2. Correlation factor g of Kirkwood as a function of the formal concentration $F_A/mol\,dm^{-3}$ of normal and tertiary alcohols at 25°C

tetrahedral way by four other oxygen atoms. In a regular tetrahedron the value of $\cos^2 \theta_{ij}$, for these four first neighbours would be equal to $\cos^2(109.5/2) = 0.33$. Assuming the other contributions to g vanish this leads to a g value of 2.33. In the wellknown model by Frank and Wen [13] there exists a dynamic equilibrium (flickering structure) between such "iceberg" water molecules and free ones. Again this model is marked by the nostalgic idea of recognizing in water the regularities observed in the crystal. Actually the equality of H-bonds in ice, and the strict tetrahedral angles between the oxygen atoms is imposed by the necessity of standardizing the cohesion forces in order to form the regular lattice. Such a necessity does not exist in liquid water and there is absolutely no reason to assume that the two hydrogen bonds in which the two protons are involved have the same strength. As a matter of fact there are experimental indications showing that they differ from each other [14].

In liquid water at room temperature both protons are in fact mostly involved in H-bonding. But, contrary to ice, both protons are also regularly disengaged during small fractions γ_1 and γ_2 of the time. When the first proton becomes free the H-bond to which the second proton belongs undergoes less constraint from the surroundings and becomes stronger. The probability γ_2 of breaking this bond will thus be markedly smaller than γ_1.

In a photograph of water within an observation time of 10^{-14} s some protons would be recognized "in the free state".

Most of them belong to chains of variable lengths. However two kinds of chains would be roughly distinguished: the strong ones and the weak ones. In the latter the O...O interdistances are larger, the bonds show important deviations from linearity and the bond angles θ deviate markedly from the tetrahedral value because the weaker bonds have to adapt themselves to the local constraints imposed by the stronger ones. These irregularities explain why in ice diffraction data give a constant number of four nearest neighbours at 2.76 Å whereas in liquid water the number of nearest neighbours fluctuates around 4.4 at a mean distance of 2.82 Å [15].

However it must be borne in mind that such a static description is not convenient when liquid water has to be considered during a time longer than 10^{-14} s. Actually each water molecule passes through all possible states that are found in the static description, the H-bonds constantly change owner, and the probability of vaporization depends on the ratios between the number of strong bonds and the number of water molecules, and the number of weak bonds and the number of water molecules.

4 References

1. See for instance Flory (1953) Principles of polymer chemistry, Cornell University Press, Ithaca p 497
2. Huyskens PL, Vandevyvere P, Siegel GG (1989) J Mol Struct 220: 555 (and references)
3. Kohler F (1974) In: Luck WAP (ed) Structure of water and aqueous solutions, Verlag Chemie, Weinheim, p 495 Hasted JB (1974) ibidem 377
4. Onsager L (1936) J Am Chem Soc 58: 1486
5. Kirkwood JG (1939) J Chem Phys 7: 911
6. Fröhlich H (1948) Trans Faraday Soc 44: 238
7. Huyskens PL, Siegel GG (1982) Croatica Chem Acta 55: 55
8. Huyskens PL, Cracco F (1960) Bull Soc Chim Belg 69: 422
9 Huyskens PL, Cleuren W, VanBrabant-Govaerts HM, Vuylsteke MA (1980) J Phys Chem 84: 2740
10. Huyskens PL, Cleuren W, Franz M, Vuylsteke A (1980) J Phys Chem 84: 2748
11. Oster G, Kirkwood JG (1943) J Chem Phys 11: 175
12. Bernal JD, Fowler RH (1933) J Chem Phys 1: 515
13. Frank HS, Wen WY (1957) Disc Faraday Soc 24: 133
14. Huyskens PL (1987) In: Kleeberg H (ed) Interactions of water in ionic and nonionic hydrates, Springer Berlin Heidelberg New York p 113
15. Marcus Y (1977) Introduction to liquid state chemistry, J Wiley, London

CHAPTER XVII
Hydrogen Bonding and Entropy

P.L. Huyskens and G.G. Siegel

A new general expression for the entropy S is given, based on the standard number $\langle W \rangle$ of "possibilities" of placing or moving for one molecule of a given kind. $\langle W \rangle$ is calculated in an independent way for all the different kinds of molecules present in the system. For a system containing N_A molecules A and N_B molecules of B, $S = k \ln \langle W_{oneA} \rangle^{N_A} \langle W_{oneB} \rangle^{N_B}$. This expression differs from that of Boltzmann but strictly agrees with that of Clausius, $dS = dq_{rev.}/T$. For the calculation of the entropy of placing in a liquid mixture, one has to use the equation $\langle W_{placingA} \rangle = x_A^{1/2} \Phi_A^{1/2}$ which combines the mole fraction x and the volume fraction Φ. The last term is related to the fact that in a liquid solution a given molecule has a larger volume at its disposal than its own one. During the fraction of time $(1 - \gamma_{Ah})$ during which an hydroxyl proton is involved in H-bonding it renounces to the possibility of visiting its mobile domain, DomA, and remains confined in a small part V_0 at the border of it. This creates in the liquid a kind of mobile order and a decrease of the entropy that is related to the ratio $V_0/\langle DomA \rangle$. The addition of an inert substance to the liquid increases $\langle DomA \rangle$, letting V_0 inaltered. This corresponds to a decrease of the entropy that constitutes the true origin of the hydrophobic effect. The quantitative treatment leads to an equation that, without the use of any adjustable constant, correctly predict the solubility of liquid alkanes in water.

1 A New General Expression for the Entropy [1]

The classical thermodynamic definition of the entropy S of a systems was initially given by R. Clausius. It is based on the amount of heat q absorbed by the system from the medium at a given absolute temperature T. In irreversible processes the amount of heat divided by the absolute temperature is always

Intermolecular Forces
© Springer-Verlag Berlin Heidelberg 1991

smaller than the given limiting value that characterizes the reversible process:

$$\frac{dq_{irr}}{T} < \frac{dq_{rev}}{T}.\tag{1}$$

The infinitesimal change of entropy dS, according to Clausius, is defined as

$$dS = dq_{rev}/T\tag{2}$$

and the second principle of thermodynamics can be written

$$\frac{dq}{T} - dS \leq 0\tag{3}$$

Another principle, known as the Nernst heat theorem, and also called the "Third Law" of thermodynamics, states that the entropy of all perfect and pure crystalline substances is zero at a temperature of absolute zero.

$$S_{T=0} = 0 \quad \text{(pure crystals)}.\tag{4}$$

At absolute zero all atoms are in the lowest level of energy. There is then no possibility of choice in pure crystals—either for the position or for the motions. The number W of possible arrangements for the individual particles is one in such a case.

It was Boltzmann who proposed, in 1895, to relate the entropy to the number of molecular arrangements W

$$S = k \ln W\tag{5}$$

with the Boltzmann constant k defined as the ratio between the gas constant and Avogadro's number

$$k = R/N.\tag{6}$$

Actually, most people associate the concept of entropy with a large number of individual particles. This idea is so widely prevalent that Eq. 5 may be considered in practice as an equivalent definition of the entropy.

However, in a recent paper [1], we have shown that the quantity S appearing in the Boltzmann equation is, in essence, not the same as the entropy S defined by Clausius. It is easy to show that the definition of Clausius is *extensive*: the amount of heat absorbed when, by means of semipermeable pistons, two gases are mixed together in a reversible way at a constant temperature is equal to

$$q_{rev} = RT(n_A \ln x_A^{-1} + n_B \ln x_B^{-1})\tag{7}$$

n_A and n_B being the number of moles and x_A, x_B the mole fractions. This amount is doubled when the number of moles of each kind is doubled.

However, the logarithm of the number of arrangements W_{exch} resulting from the possibility of exchange of the particles

$$\ln W_{exch} = \ln \frac{(n_A N + n_B N)!}{(n_B N)!(n_B N)!} \tag{8}$$

is not extensive. When n_A and n_B are multiplied by 2, $\ln W_{exch}$ is larger than twice the original value. Although the differences vanish when one uses Stirling's approximation, it is clear that $k \ln W_{exch}$ and S_{mixing} derived from the Clausius entropy are not strictly speaking the same.

Instead of ignoring this fundamental difference it is possible to adapt the Boltzmann equation so that it becomes strictly extensive.

For a pure substance in one phase we replace W by

$$\langle W_{one\,molecule} \rangle^{\mathcal{N}} \tag{9}$$

where $\langle W_{one\,molecule} \rangle$ is the average number of possibilities for one molecule and $\mathcal{N}$ the number of molecules. The average number of possibilities $\langle W_{one\,molecule} \rangle$ is independent of $\mathcal{N}$.

The definition of the entropy then becomes

$$S = k \ln \langle W_{one\,molecule} \rangle^{\mathcal{N}} \tag{10}$$

and this expression is rigorously extensive. It makes the entropy a quantity based on averages. This seems logical for a concept used for predictions. In this definition exceptional situations are not considered. The entropy is not related to the actual possibilities W of the system but to a *standardized* number $\langle W_{one\,molecule} \rangle^{\mathcal{N}}$. Let us examine this new definition in the following example. We consider a real gas containing N molecules in a volume V^0 under a pressure p where the Van der Waals corrections are still negligible. This gas is allowed to expand through a cock in an empty volume V' equal to V^0. When pressure equilibrium is reached half of the molecules have been transferred from V^0 to V'. At this moment most of the molecules still in V^0 will remain there for a very long time. Each molecule of the gas is indeed confined in a limited domain of space that moves only slowly. The average volume of this domain passes from V^0/N before the opening of the cock to $(V' + V^0)/N$ after. In calculating $\langle W^0_{one\,molecule} \rangle$ before, and $\langle W_{one\,molecule} \rangle$ after the opening of the cock we only consider the *average* possibilities and thus the average volumes of the domains. This average volume is multiplied by a factor $(V' + V^0)/V^0$. The corresponding increase of the entropy is thus given by

$$\Delta S_{\text{expansion}} = k \ln\left(\frac{\langle W \rangle}{\langle W^0 \rangle}\right)^N = k \ln\left(\frac{V' + V^0}{V^0}\right)^N$$

$$= nR \ln\left(\frac{V' + V^0}{V^0}\right) \tag{11}$$

This corresponds indeed to the Clausius value calculated from q_{rev}/T.

In this equation the system is considered in a standardized situation as if all molecules had the same volume at their disposal. In reality, fluctuations are possible and it may happen that more molecules are present in V' than in V^0 or inversely. However the new definition ignores these fluctuations and takes only average possibilities into account. As we have written "The clue for physicochemical predictions is the second law of thermodynamics. Now, we can only predict averages. And when we predict deviations from the average, they are in fact average deviations from the average. We cannot predict exceptions, we cannot predict miracles although exceptions and miracles may exist. In the history of Music, Salieri was "predictable", Mozart was not!"

Equation 10, which defines the entropy of a pure substance, seems not so fundamentally different from Boltzmann's Eq. (5).

However a drastic divergence appears when one considers systems with different components A, B, ... It makes sense to consider the average possibilities $\langle W_{\text{one molecule A}} \rangle$ but, also completely independently, the average possibilities $\langle W_{\text{one molecule B}} \rangle$.

Let us consider for instance a mixed crystal containing 100 atoms: 90 silver atoms and 10 gold atoms. The average number of places for one silver atom is clearly 10/9 and the average number of places for one gold atom is 10. These figures will not change when we take mixed crystals with 1000, 10000 or 10^{23} atoms.

The natural extension of Eq. 10 to systems with several components is thus

$$S = k \ln \langle W_{\text{one A}} \rangle^{\mathscr{N}_A} \langle W_{\text{one B}} \rangle^{\mathscr{N}_B} \cdots \tag{12}$$

This equation differs fundamentally from that of Boltzmann. Here the possibilities of the A molecules are considered completely independently of the possibilities of the B molecules. Both W of Boltzmann and $\langle W_{\text{one A}} \rangle^{\mathscr{N}_A} \times \langle W_{\text{one B}} \rangle^{\mathscr{N}_B}$ have a physical sense but it is not the same, W concerns the combined possibilities of A and B.

In a binary mixed crystal the separate average possibilities of placing one A and placing one B are given by

$$\langle W_{\text{placing one A}} \rangle = 1/x_A \tag{13}$$

$$\langle W_{\text{placing one B}} \rangle = 1/x_B. \tag{14}$$

The entropy of mixing deriving from the general Eq. 12 is then

$$\Delta S_{\text{mixing}} = kN[n_A \ln x_A^{-1} + n_B \ln x_B^{-1}]. \tag{15}$$

This expression is rigorous, when derived from Eq. 12. It is only an approximation obtained by using Stirling's formula, when it is derived from Eq. 8.

The new point of view we have adopted is thus to *consider the classical expression* (15) which is obviously extensive, *as being rigorous*. Although this will not affect the calculation of ΔS_{mix}, it has profound consequences for the real meaning of the entropy as it appears in the second law of thermodynamics.

2 Volume Correction in the Entropy of Mixing [2, 3]

When both molecules A and B have the same molar volume $\bar{V}_A$ and the same shape, the expression (15) which was derived here from a mixed crystal holds for all the other states of aggregations. This is widely accepted and also experimentally demonstrated. However the physical meaning may be different. Let us write this expression in the form

$$\Delta S_{mixing} = kN[n_A \ln \langle W_{placing\,one\,A} \rangle + n_B \ln \langle W_{placing\,one\,B} \rangle]. \qquad (16)$$

What we call $W_{placing\,one\,A}$ is the number of possibilities that could be obtained by exchanging one A and B molecules. But such an exchange is not realized.

The situation is quite different in the liquid mixtures. Here each molecule A occupies a domain, the volume of which is equal to

$$DomA = \frac{V}{n_A N} \qquad (17)$$

where V is the total volume of the system. But an inherent characteristic of this domain is that its localization always changes, because of the motions of the molecules in the liquid phase. The domain is thus not fixable with respect to external axes. It is also not orientable and has thus in principle spherical symmetry. Every thermodynamic treatment of the liquid phase has to take these characteristics into account and models based on a non-existing "pseudo lattice" have to be abandoned. DomA is thus a volume that characterizes the stationary wave function of each A molecule in the liquid. The entropy of mixing corresponds to the fact that the domain of A is shared by molecules of a different kind.

But here a subtle difficulty arises. When we add substance B to pure liquid A, not only possibilities of exchange between A and B are created but the volume of DomA is enlarged.

It should be noticed here that, in a *gas*, we can increase DomA without adding a foreign substances. This is not the case for a liquid.

A real fundamental question for ΔS_{mix} in a liquid is the following one: should we call this "entropy of exchange" (also called by Flory "combinatorial entropy") or "entropy of volume"?

The question remains unimportant as long as the molar volumes $\bar{V}_A$ and $\bar{V}_B$ are the same. As a matter of fact this is realized in gases as well as in mixed crystals. But, in liquid solutions, $\bar{V}_A$ generally differs from $\bar{V}_B$ (especially in polymer solutions).

If we now consider that ΔS_{mixing} refers to the possibilities of exchange, then we have to count the number of particles, irrespective of their volume and we find

$$\langle W_{placing\ one\ A} \rangle = 1/x_A. \tag{18}$$

On the contrary, if ΔS_{mix} is the consequence of the increase of the domain, we should write

$$\langle W_{placing\ one\ A} \rangle = 1/\phi_A \tag{19}$$

where ϕ_A is the valume fraction of A.

As a matter of fact Eq. 18 corresponds to the hypothesis upon which the classic Law of Raoult for the vapour pressure of an ideal binary liquid mixture is based. Equation 19 leads to an expression of ΔS_{mixing} that is known in the literature under the name "Flory-Huggins" and is generally used for polymer solutions [4].

We have proposed the adoption of an intermediate point of view, on the basis of the following considerations:

We have said that DomA is non-orientable with respect of external axes. But for each DomA there exists a direction in the solution in which the nearest molecule A of the same nature is encountered. Also this direction varies constantly. But, according to Einstein and Smoluchowski in this direction the A molecule is separated on average from the nearest one by

$$\sqrt{1/x_A} - 1 \tag{20}$$

molecules of B. This means that in this changing direction we can count $\sqrt{1/x_A}$ possibilities of exchange. This means that the introduction of B in pure A creates such possibilities of exchange and, besides, also increases the entropy of volume. Of course we may not count the possibilities twice. Therefore, according to these considerations, the number of choices for placing one A will be given by

$$\langle W_{placing\ one\ A} \rangle = (1/x_A)^{1/2}(1/\phi_A)^{1/2} \tag{21}$$

A general expression for all these equations giving the entropy of "placing" in binary mixtures is then

$$\Delta S_{placing} = R\alpha[n_A \ln \phi_A^{-1} + n_B \ln \phi_B^{-1}] + R(1-\alpha)[n_A \ln x_A^{-1} + n_B \ln x_B^{-1}] \tag{22}$$

With $\alpha = 1$ one obtains the Flory–Huggins expression, while $\alpha = 0$ gives the classical one and with $\alpha = 0.5$ one has the equation proposed by Huyskens and

Table 1. Partial molar entropy of mixing of the solute in the saturated solutions calculated with various values of α and experimental values [3]

Solute	Solvent	$\bar{V}_B$	$\bar{V}_S$	ϕ_{Bsat} exp	$(\partial \Delta S_{pl}/\partial n_B)$			
					$\alpha = 0$	$\alpha = 1$	$\alpha = 0.5$	exp
$n\text{-}C_{20}H_{42}$	$n\text{-}C_{15}H_{32}$	359.7	277.7	0.405	8.860	8.98	8.92	8.81
$n\text{-}C_{24}H_{50}$	$n\text{-}C_8H_{12}$	425.1	163.5	0.196	20.42	24.24	22.33	22.12
$n\text{-}C_{24}H_{50}$	$n\text{-}C_{10}H_{22}$	425.1	195.9	0.156	21.16	23.65	22.40	22.12
$n\text{-}C_{24}H_{50}$	$n\text{-}C_{12}H_{26}$	425.1	228.6	0.139	21.01	22.55	21.78	22.12
$n\text{-}C_{28}H_{58}$	$n\text{-}C_6H_{14}$	490.5	131.6	0.074	32.12	42.64	37.38	35.33
$n\text{-}C_{28}H_{58}$	$n\text{-}C_7H_{16}$	490.5	147.5	0.065	32.32	40.80	36.57	35.33
$n\text{-}C_{28}H_{58}$	$n\text{-}C_{10}H_{22}$	490.5	195.9	0.044	33.37	37.92	35.64	35.33
$n\text{-}C_{28}H_{58}$	$n\text{-}C_{12}H_{26}$	490.5	228.6	0.035	34.06	37.06	35.56	35.33
$n\text{-}C_{28}H_{58}$	$n\text{-}C_{16}H_{34}$	490.5	131.6	0.027	34.19	35.43	34.81	35.33
$n\text{-}C_{30}H_{62}$	$n\text{-}C_6H_{14}$	523.2	147.5	0.040	37.98	50.51	44.24	39.90
$n\text{-}C_{30}H_{62}$	$n\text{-}C_{10}H_{22}$	523.2	163.5	0.021	40.18	45.72	42.94	39.90
$n\text{-}C_{32}H_{66}$	$n\text{-}C_6H_{14}$	555.9	195.9	0.021	43.96	58.36	51.16	49.97
$n\text{-}C_{32}H_{66}$	$n\text{-}C_7H_{16}$	555.9	228.6	0.020	43.43	55.08	49.25	49.97
$n\text{-}C_{32}H_{66}$	$n\text{-}C_{10}H_{22}$	555.9	277.7	0.0102	46.73	53.24	49.99	49.97
$n\text{-}C_{32}H_{66}$	$n\text{-}C_{12}H_{26}$	555.9		0.0073	48.25	52.72	50.48	49.97
$n\text{-}C_{36}H_{24}$	$n\text{-}C_6H_{14}$	621.3		0.0045	57.80	75.72	66.76	62.02
$n\text{-}C_{36}H_{24}$	$n\text{-}C_7H_{12}$	621.3		0.0041	57.62	72.29	64.96	62.02
$n\text{-}C_{36}H_{24}$	$n\text{-}C_8H_{18}$	621.3		0.0039	57.19	69.30	63.24	62.02
$n\text{-}C_{36}H_{24}$	$n\text{-}C_{10}H_{22}$	621.3		0.0025	59.39	67.82	63.60	62.02
$n\text{-}C_{36}H_{24}$	$n\text{-}C_{12}H_{26}$	621.3		0.0019	60.39	66.35	63.37	62.02
$n\text{-}C_{36}H_{24}$	$n\text{-}C_{15}H_{32}$	621.3		0.0013	61.39	65.52	63.73	62.02

Haulait-Pirson. The partial derivative of this expression relative to n_B is

$$\left(\frac{\partial \Delta S_{placing}}{\partial n_B}\right)_{n_A} = -R\left[\ln \phi_B - \alpha\left(\frac{\bar{V}_B}{\bar{V}_A} - 1\right)\phi_A - (1-\alpha)\ln\left(\phi_B + \phi_A\frac{\bar{V}_B}{\bar{V}_A}\right)\right] \quad (23)$$

In the case of saturated solutions of solid n-alkanes in liquid n-alkanes this partial derivative multiplied by T is practically equal to the free energy of fluidization ΔG^0_{fluid} of the pure solid n-alkanes. As a matter of fact the heat of mixing of long alkanes in the liquid state can be neglected to a first approximation. ΔG^0_{fluid} is known from calorimetric measurements. This provides a tool for checking the various equation. In Table 1 the experimental value of the derivative at 25 °C obtained in this way is compared with the value calculated from the experimental ϕ_B at saturation, using the three values of the parameter α.

Table 1 clearly shows that the predicted values are markedly better with $\alpha = 0.5$. However the differences between the calculated values become only marked when $\bar{V}_B > 2\bar{V}_S$. This explains the success of the classical and the Flory–Huggins formulas.

3 The Entropy of the Mobile Order in H-Bonded Liquids and the Hydrophobic Effect. A Quantitative Treatment [5]

It is widely known that hydrocarbons are very slightly soluble in water. One can call this phenomenon "hydrophobicity". Several theories have been presented in the past to explain this poor solubility but, in general, they do not take the risk of making quantitative predictions. The theory of "ideal associated solutions" which neglects all effects not directly related to H-bonding, would predict a complete miscibility of pentane and water!

In general people are convinced that the hydrophobic effect is due to the necessity of breaking hydrogen bonds of water or of the alcohols in order to create the cavity in which the solute molecule can be placed. This is erroneous. Other people, especially those in biochemistry, think that there exists in the pure alkanes a special and obscure kind of cohesion force, "hydrophobic interactions" which should bind the alkane molecules together. For lower alkanes this is also erroneous.

These errors are easily demonstrated by the following experimental facts: alkanes, as shown in Table 2, are very poorly soluble in water but their solubilities vary only weakly with the temperature.

Such small changes of solubility with temperature demonstrate unambiguously that the dissolution does not require much energy. This is confirmed by calorimetric measurements. From the correlations of Somsen [6] relative to the hydration enthalpies, one deduces that the transfer of an alkane group CH_2 ($\bar{V}_{CH_2} = 16.4 \, cm^3 \, mol^{-1}$) from its own phase towards water needs only an energy expenditure of $2.3 \pm 0.3 \, kJ \, mol^{-1}$, only one tenth of that of the H-bonds in alcohols. This small amount can be accounted for by the changes in the dispersion forces.

This means that we do not have to break noticeable "hydrophobic interactions" in liquid pentane and that the number of O—H—O bonds that have to be broken in water upon introduction of the alkane also remains negligible. Now, it is clear that making a cavity in ice, the volume of which is five or six times larger than that corresponding to a water molecule, would involve the destruction of several H-bonds and therefore require an energy that should exceed fifty if not a hundred $kJ \, mol^{-1}$. The mobile H-bonds in the liquid adapt themselves clearly much better to the presence of the moving volume of the foreign substance.

Table 2. Solubilities in volume fractions ϕ_B of alkanes in water at various temperatures [5]

	5 °C	25 °C	40 °C	55 °C	99 °C
n-Pentane	6.6×10^{-5}	6.5×10^{-5}	6.7×10^{-5}	6.7×10^{-5}	11.1×10^{-5}
n-Hexane	2.5×10^{-5}	1.4×10^{-5}	1.5×10^{-5}	2.0×10^{-5}	3.4×10^{-5}
n-Heptane	3×10^{-6}	3.6×10^{-6}	3.9×10^{-6}	4.6×10^{-6}	8.2×10^{-6}
n-Octane	—	0.9×10^{-6}	—	—	1.6×10^{-6}

Similar considerations hold for the alcohols: ΔU_{res} practically does not change when going from ethanol ($58.7\,\text{cm}^3\,\text{mol}^{-1}$) to hexanol ($125.3\,\text{cm}^3\,\text{mol}^{-1}$) (see Chap. 1). This means that the addition of four additional CH_2 groups affects only in a very negligible way the proportion of O—H groups involved in H-bonding at a given moment. However, the low solubility of alkanes in water is an experimental fact. If the origin of the hydrophobicity is not of energetic nature, it must be caused by a reduction of the entropy upon introduction of the inert substance.

We have found this origin in the "mobile order" that characterizes H-bonded liquids. As said before, from a thermodynamic point of view, one of the most important characteristics of H-bonding in the *liquid* phase is the establishment of highly preferential contacts between given parts of the various molecules.

During the fraction γ_{Ah} of the time the O—H proton of an alcohol is free and during the fraction $1 - \gamma_{Ah}$ it occupies a small part v_0 of its domain, DomA, in the vicinity of one of the lone pairs of electrons of the oxygen atom of a neighbouring molecule. Neither v_0 nor DomA are fixed. Nevertheless the reduction of freedom for O—H group is clearly equal to the ratio v_0/DomA.

If γ_{Ah} would be entirely negligible, the effect on the molar entropy of the alcohol, brought about by the mobile order would be equal to

$$\overline{\Delta S}_{A\,(\text{mobile order})} = \overline{S}_{A\,(\text{with mob. order})} - \overline{S}_{A\,(\text{without m.o.})}$$
$$= R\ln\mathscr{R}_{\text{one molecule A}} = R\ln(v_0/\text{DomA}) \qquad (24)$$

$\mathscr{R}_{\text{one molecule A}}$ describes the reduction of freedom of one alcohol molecule in the liquid consecutive to the mobile order. As v_0 is much smaller than DomA, $\overline{\Delta S}_{A\,(\text{mobile order})}$ is negative.

When γ_{Ah} is not entirely negligible the expression should be corrected as

$$\overline{\Delta S}_{A\,(\text{mobile order})} = (1 - \gamma_{Ah})R\ln(v_0/\text{DomA})$$
$$- R(\gamma_{Ah}\ln\gamma_{Ah} + (1 - \gamma_{Ah})\ln(1 - \gamma_{Ah})) \qquad (25)$$

The last term corresponds to the possibilities of exchanging "free" and "bonded" A—H protons.

Let us first neglect the corrections to focus our attention on the physical basis of the relation. When we consider the series of the primary alcohols, we expect that v_0 remains constant but that DomA increases with the molecular weight. As a matter of fact, the differences of molar entropy between a primary alcohol and its homomorphous hydrocarbon is constant and positive in the gas phase whereas in the liquid phase it is negative and the absolute value increases with $\overline{V}_A^0$, the molar volume of the pure alcohol. This effect is clearly due to the mobile order that increases from ethanol to hexanol as shown in Table 3.

Table 3. Differences in the molar entropies $\overline{\Delta S^0}$ between the alcohols and the homomorphous hydrocarbons in the gas phase and in the liquid phase at 25 °C [5]

	$\overline{\Delta S^0_{gas}}/JK^{-1}\,mol^{-1}$	$\overline{\Delta S^0_{liquid}}/JK^{-1}\,mol^{-1}$
Ethanol	12.6	−27.7
1-Propanol	14.1	−33.4
1-Butanol	13.9	−36.8
1-Pentanol	14.1	−36.5
1-Hexanol	13.6	−39.0

When one goes from ethanol to hexanol, the absolute value of the negative entropy of the mobile order increases, because $DomA = \overline{V}_A/N$ increases whereas the confined volume v_0 remains constant.

Another way to increase DomA is, of course, to add pentane to ethanol, because in this case each alcohol molecule also has a larger domain at its disposal.

Thus, the addition of an inert substance to an alcohol (or to water) will increase the domain DomA of each molecule, and, as a consequence, the entropy of mobile order becomes more negative.

The chief reason for the hydrophobic effect is thus that the dissolution of a foreign substance in the perpetually moving molecular systems, increases the domain of each alcohol or water molecule and thus *extends the territory of the mobile order.*

A rigorous treatment of the mobile order considers not the entropy but the free energy. If the fraction of the time the A—H protons are "free" differs from one, the molar free energy of the pure liquid is given by

$$\overline{G}_A^{\cdot} = \overline{G}_A^{\cdot free} + RT \ln \gamma_{Ah}^{\cdot} \tag{26}$$

where $\overline{G}^{\cdot free}$ is the molar free energy in absence of H-bonding and mobile order. The effect of the latter on the chemical potential of the solute B is given by the equation

$$\left(\frac{\partial G}{\partial n_B}\right)_{mobile\ order} = n_A RT \left(\frac{\partial \ln \gamma_{Ah}}{\partial n_B}\right)_{n_A} \tag{27}$$

For alcohols γ_{Ah} depends on the formal concentration F_A of the alcohol according to the equation

$$1/\gamma_{Ah} = 1 + K_A F_A = 1 + K_A \phi_A/\overline{V}_A \tag{28}$$

where k_A is the association or insertion constant, and $\overline{V}_A$ the molar volume. This leads to the general equation

$$\left(\frac{\partial G}{\partial n_B}\right)_{mobile\ order} = \left[\frac{K_A \phi_A/\overline{V}_A}{1 + K_A \phi_A/\overline{V}_A}\right] RT \frac{\overline{V}_B}{\overline{V}_A} \phi_A \tag{29}$$

Table 4. Experimental and calculated solubilities of liquid alkanes in water at 25 °C (in volume fractions). Molar volume $\bar{V}_B$ of the liquid alkanes (in $cm^3 mol^{-1}$) [7]

Alkane	$\bar{V}_B$	$\phi_{B\,exp}$	$\phi_{B\,calc.}$
n-Pentane	116.1	6.17×10^{-5}	9.68×10^{-5}
n-Hexane	131.6	1.59×10^{-5}	2.83×10^{-5}
n-Heptane	147.5	4.17×10^{-6}	7.97×10^{-6}
n-Octane	163.5	9.16×10^{-7}	22.10×10^{-7}
n-Nonane	179.7	2.38×10^{-7}	6.01×10^{-7}
n-Decane	195.9	17.13×10^{-8}	16.3×10^{-8}
Cyclopentane	94.7	2.13×10^{-4}	5.20×10^{-4}
Methylcyclopentane	113.1	5.61×10^{-5}	12.30×10^{-5}
Propylcyclopentane	145.3	2.63×10^{-6}	9.50×10^{-6}
Cyclohexane	108.8	7.83×10^{-5}	17.2×10^{-5}
Methylcyclohexane	128.3	1.96×10^{-5}	3.68×10^{-5}
1-cis-2-dimethylcyclohexane	141.6	7.34×10^{-6}	12.80×10^{-6}
1-trans-2-dimethylcyclohexane	148.0	5.08×10^{-6}	7.66×10^{-6}
Cycloheptane	121.7	3.70×10^{-5}	6.21×10^{-5}
Cyclooctane	135.8	9.55×10^{-6}	20.30×10^{-6}
2-methylbutane	117.4	7.86×10^{-5}	8.73×10^{-5}
2-methylpentane	132.9	2.17×10^{-5}	2.55×10^{-5}
2-methylpentane	130.6	2.20×10^{-5}	3.07×10^{-5}
2-methylhexane	148.6	3.75×10^{-6}	7.30×10^{-6}
3-methylhexane	146.7	5.54×10^{-6}	8.50×10^{-6}
3-methylheptane	162.8	1.12×10^{-6}	2.57×10^{-6}
4-methyloctane	179.1	1.60×10^{-7}	6.31×10^{-7}
2,2-dimethylbutane	133.8	3.48×10^{-5}	2.38×10^{-5}
2,3-dimethylbutane	131.2	3.16×10^{-5}	2.92×10^{-5}
2,2-dimethylpentane	149.7	7.61×10^{-6}	6.68×10^{-6}
2,3-dimethylpentane	145.0	7.57×10^{-6}	9.73×10^{-6}
2,4-dimethylpentane	149.9	6.95×10^{-6}	6.58×10^{-6}
3,3,6-dimethylpentane	145.4	8.53×10^{-6}	9.43×10^{-6}
2,2,4-trimethylpentane	166.1	2.72×10^{-6}	1.79×10^{-6}
2,3,4-trimethylpentane	159.8	2.55×10^{-6}	2.98×10^{-6}
2,2,5-trimethylpentane	182.8	7.65×10^{-7}	4.68×10^{-7}

To a good approximation the factor between brackets, which corresponds to $-b_S$, is equal to $+1$.

$$b_S = \frac{-K_A \phi_A/\bar{V}_A}{1 + K_A \phi_A/\bar{V}_A} \simeq -1 \qquad (30)$$

One can use this equation to calculate the hydrophobic term at low values of the volume fraction of the alcohol.

In the case of water one has to take both protons into account and $b \simeq -2$.

At high dilution in water, the chemical potential of a dissolved substance is increased by an amount

$$\left(\frac{\partial G}{\partial n_B}\right)_{\substack{\text{mobile order} \\ \text{in water}}} = +2\frac{\bar{V}_B}{18}RT \qquad (31)$$

Neglecting the changes in the dispersion forces the chemical potential of a liquid alkane in water is thus given by

$$\mu_B = \mu_B^0 + RT\left[\ln \phi_B - 0.5\left(\frac{\bar{V}_B}{18} - 1\right) - 0.5\ln\frac{\bar{V}_B}{18} + 2\frac{\bar{V}_B}{18}\right] \qquad (32)$$

Assuming the chemical potential of the alkane in the organic phase in equilibrium is that of the pure alkane, the solubility of a liquid alkane in water should be given to a first approximation by

$$\phi_{\text{Bsat in water}} = \exp\left[0.5\left(\frac{\bar{V}_B}{18} - 1\right) + 0.5\ln\frac{\bar{V}_B}{18} - 2\frac{\bar{V}_B}{18}\right] \qquad (33)$$

The predicted values, using this equation that does not contain any adjustable parameter, are always of the correct order of magnitude (Table 4). This demonstrates the validity not only of the equation giving the effect of the mobile order but also of that of Huyskens and Haulait-Pirson used for the calculation of the entropy of mixing.

4 References

1. Huyskens PL, Siegel GG (1988) Bull Soc Chim Belg 97: 809
2. Huyskens PL, Haulait-Pirson MC (1985) J Mol Liq 31: 135
3. Huyskens PL, Siegel GG (1988) Bull Soc Chim Belg 97: 815
4. Flory PJ (1942) J Chem Phys 10: 51; Huggins ML (1942) J Phys Chem 46: 151; Hildebrand JH (1947) J Chem Phys 15: 225
5. Huyskens PL, Siegel GG (1988) Bull Soc Chim Belg 97:
6. Somsen G (1990) see Chap 15 of this book
7. Selected values from Ref. 5

CHAPTER XVIII
Specific Intermolecular Forces and the Permittivity and Conductivity of Solutions

J. Barthel

At all times electrolyte solutions were preferred systems for the study of specific intermolecular forces. There is no measurable concentration range in which electrolyte solutions show ideality. All interaction forces, due to ion–ion and ion–solvent interactions, are specific forces. They are in competition with each other as the solvation and association effects, and with the solvent–solvent interactions making up the solvent structure. Recent results from microwave measurements have shown and clarified the complex interplay of orientational, intramolecular, kinetic, H-bonding, diffusional, and migrational modes in electrolyte solutions. In this chapter solvent relaxation processes, ion-pair reorientation and formation kinetics, the re-establishment of electroneutrality, and electrolyte conductance at quasi-static frequencies are discusssed for protic and aprotic solvent systems, exemplifying the diversity of specific intermolecular forces in electrolyte solutions.

The Maxwell equations show that conductivity and permittivity of electrolyte solutions are complementary effects. They even imply one another, as shown in the theory of kinetic depolarization by the linear dependence of specific conductivity and the decrease in permittivity caused by the addition of electrolyte to a solvent to produce a solution of concentration c.

The linkage of conductivity and permittivity makes the joint discussion of these quantities desirable; it reveals new aspects and gives more insight in the structure and dynamics of electrolyte solutions.

Symbols and Abbreviations

I Fundamental Constants

N_A	:Avogadro constant $6.022137 \times 10^{23}\,mol^{-1}$
k	:Boltzmann constant $1.380658 \times 10^{-23}\,JK^{-1}$

Intermolecular Forces
© Springer-Verlag Berlin Heidelberg 1991

e	:elementary charge $1.602177 \times 10^{-19}\,C$
ε_0	:permittivity of vacuum $8.854188 \times 10^{-12}\,J^{-1}C^2m^{-1}$
μ_0	:permeability of vacuum $4\pi \times 10^{-7}\,Js^2C^{-2}m^{-1}$
c	:speed of light $2.997925 \times 10^8\,ms^{-1}$

II Superscripts and Subscripts
1 Superscripts

*	:complex quantity (field vectors $\vec{E}*$, $\vec{H}*$, $\vec{A}*$ or material properties $\varepsilon*, \sigma*, \eta*$), also
	:non-Coulombic part of the mean force potential (W_{ij}^*)
∞	:infinite dilution (Λ^∞)
$'$	:real part of a complex quantity (material properties $\varepsilon', \sigma', \eta'$), also
	:free ions (c', concentration; y', activity coefficient; etc), also
	:perturbed part of the correlation function (g'_{ij}), also
	:microscopic relaxation time (τ')
$''$	:imaginary part of a complex quantity (material properties $\varepsilon'', \sigma'', \eta''$)

2 Subscripts

A	:association
$_i$ or $_j$	:particles i or j (ions, molecules)
ij	:particle j in the vicinity of i
∞	:infinite frequency
$\pm$	:mean quantity (mean activity coefficient $y_\pm$)

III Symbols (# means an arbitrary subscript or superscript)

$\langle X \rangle$	:mean value of X
$a, a_\#$	:ion size parameter, see Fig. 14
$c, c_\#,$	:molarity [$mol\,dm^{-3}$]
$\vec{E}, \vec{E}_\#$	:electric field strength
f, f_i	:factor of Onsager's reaction field (of particle i)
f_p^{or}	:pulse response function of orientational polarization
F_p^{or}	:step response function of orientational polarization
g_{ij}, g_{ji}	:pair correlation functions
g_i	:weighting factor, see Eqs. (16)
ΔG_{298}^0	:Gibbs energy of a process at 298.15 K and 1 atm
ΔH_{298}^0	:enthalpy of a process at 298.15 K and 1 atm
i	:operator, $i^2 = -1$
$k_\#,$	:rate constant of a chemical reaction
$K, K_\#,$	:equilibrium constant of a chemical reaction
$m, m_\#,$	:molality [$mol(kg\ of\ solvent)^{-1}$]
$n_\#$	:refractive index
$\vec{P}, \vec{P}_\#$	:electric polarization

$\vec{r}, \vec{r}_{\#}, r, r_{\#}$:distance variable

$R, R_{\#}$:distance parameter, see Fig. 14

ΔS^0_{298} :entropy of a process at 298.15 K and 1 atm

t   :time variable

T   temperature [K]

T^0 :glass transition temperature

$u, u_{\#}$:ion mobility

$\vec{v}_{\#}$:particle velocity

$W_{\#},$:mean-force potential

$y_{\#},$:activity coefficient in the molar scale

$z_{\#}$:ion valency

α :degree of dissociation of an electrolyte, also

α_i :polarizability of particle i

ϑ :dielectric decrement, see Eqs. (26)

$\varepsilon, \varepsilon_{\#},$:relative permittivity

$\eta, \eta_{\#},$:generalized relative permittivity, see Eq. (4) also
          :dynamic viscosity

κ :Debye parameter, see Eq. (24b)

$\Lambda, \Lambda_{\#},$:equivalent conductance

$\mu, \mu_{\#}$:molecular dipole moment

ν :linear frequency of a wave

$\rho, \rho_{\#}$:particle density

$\sigma, \sigma_{\#}$:specific conductance

$\tau, \tau_{\#},$:relaxation time

ω :circular frequency of a wave

Remark: Symbols used only at single places with special meaning, which also may deviate from the meaning of the above list, are explained in the text.

1 Introduction

A liquid phase (pure chemical compound, mixture of nonelectrolytes, or electrolyte solution), when exposed to an electromagnetic field yields by its response function valuable information about the structure of the system, particle interactions, kinetic processes, and transport processes.

The phenomenological description of the interaction between electromagnetic fields and material systems is based on a system of Maxwell equations, field equations and material equations. Figure 1 shows these equations and gives particular expressions for different applications. The vectors $\vec{E}$ and $\vec{H}$ are the electric and magnetic field strengths; $\vec{D}$ is the dielectric displacement, and $\vec{B}$ is the magnetic flux density. $\delta\vec{D}/\delta t$ is the dielectric displacement current density resulting from a polarization of the material systems in an a.c. field, and $\vec{j}$ is the current density due to ionic conductivity.

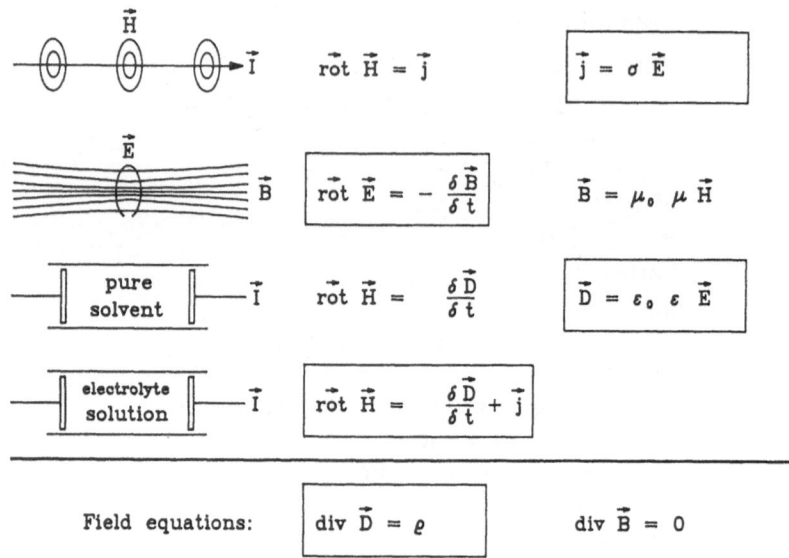

Fig. 1. The fundamental equations of electromagnetic fields. Maxwell equations for an electronic conductor ($\vec{H}, \vec{E}$: 1st and 2nd line), a non-conducting liquid ($\vec{H}, \vec{E}$: 3rd and 2nd line), and an electrolyte solution ($\vec{H}, \vec{E}$: 4th and 2nd line); material equations defining specific conductance σ, relative permeability μ, and relative permittivity ε; field equations defining the sources of field lines of dielectric displacement $\vec{D}$ and magnetic flux density $\vec{B}$. For further explanations see text

The Maxwell equations express the two fundamental laws of electrodynamics:

1. Ampère's law: Every electric current of density $\vec{j}$, and/or displacement current $\delta\vec{D}/\delta t$, through an arbitrary open surface in the space is accompanied by a magnetic field of field strength $\vec{H}$ in the boundary curve of this surface.
2. Faraday's law: Every flow, variable with time, of magnetic flux density $\vec{B}$ through an arbitrary open surface in the space induces an electric field of field strength $\vec{E}$ around the boundary of this surface.

The electric ($\vec{D}, \vec{E}$) and magnetic ($\vec{B}, \vec{H}$) field vectors are related by material equations where ε_0 and μ_0 are the permittivity and permeability of the vacuum; ε and μ, are the relative permittivity and relative permeability of the medium. A third material equation is Ohm's law relating current density $\vec{j}$ and electric field strength $\vec{E}$ via electric conductivity σ. The field equations show that single charges are the sources of the electric field $\vec{D}$, whereas the magnetic field $\vec{B}$ has no sources. In this paper only the framed equations will be used.

From Legendre's vectorial identity when applied to an electromagnetic field

$$\overrightarrow{\text{rot rot }} \vec{A}^* = \text{grad div } \vec{A}^* - \nabla^2\vec{A}^* \tag{1a}$$

and the equations in Fig. 1, follows the wave equation

$$\nabla^2\vec{A}_0^* + k^{*2}\vec{A}_0^* = 0 \tag{1b}$$

for the propagation of a monochromatic electromagnetic wave of circular frequency $\omega(\omega = 2\pi\nu, \nu = $ linear frequency) in the investigated medium.

Vector $\vec{A}^*, \vec{A}^* = \vec{A}_0^* \exp(i\omega t)$, stands for the periodically changing electric $(\vec{E}^*)$ and magnetic $(\vec{H}^*)$ field strength vectors, perpendicular one to the other.

It is common use to characterize physical quantities underlying attenuation and phase shift in dissipating systems with the help of complex (superscript: asterisk) quantities.

From the Maxwell equations it follows that

$$k^* = k_0 \left[\varepsilon^* - i\frac{\sigma^*}{\varepsilon_0\omega} \right]^{1/2}; \quad k_0 = \frac{\omega}{c} \tag{2a, b}$$

In Eqs. (2) k_0 is the propagation coefficient of the electromagnetic wave of circular frequency ω in vacuo, ε^* is the complex relative permittivity of the medium, σ^* is the complex specific electric conductivity (only occurs for electrically conducting media), and c is the speed of light. The dependence of k^* on the permeability μ^* of the medium is not included in Eq. (2a), since it is negligible for all solutions studied here.

$$\varepsilon^* = \varepsilon' - i\varepsilon''; \quad \sigma^* = \sigma' + i\sigma'' \tag{3a, b}$$

The complex permittivity ε^* and conductivity σ^* are frequency-dependent quantities, Strictly, in the application of Maxwell theory, they cannot be separately measured. The only measurable quantity at frequency ω is the "generalized permittivity" η^*. This depends on the attenuation and phase shift of the electromagnetic wave travelling through the dissipative liquid phase.

$$\eta^* = \varepsilon^* - i\frac{\sigma^*}{\omega\varepsilon_0}; \quad \eta' = \varepsilon' + \frac{\sigma'}{\omega\varepsilon_0}; \quad \eta'' = \varepsilon'' + \frac{\sigma''}{\omega\varepsilon_0} \tag{4a, b, c}$$

2 The Complex Permittivity

The generalized complex permittivity of electrically non-conductive systems $(\sigma^* = 0)$ is reduced to the permittivity of the medium, ε^*. The electric polarization $\vec{P}$ with its macroscopic and microscopic definitions links the relative permittivity to the molecular quantities $\vec{\mu}$ (dipole moment) and α (polarizability) of the investigated system

$$\text{macroscopic:} \quad \vec{P} = \varepsilon_0(\varepsilon - 1)\vec{E} \tag{5}$$

$$\text{microscopic:} \quad \langle\vec{P}\rangle = \langle\vec{P}_\mu\rangle + \langle\vec{P}_\alpha\rangle \tag{6}$$

$\langle\vec{P}_\mu\rangle$ is the sum of the permanent mean dipole moments per unit of volume,

$\langle \vec{P}_\mu \rangle = \sum \langle \vec{\mu}_k \rangle$, and $\langle \vec{P}_\alpha \rangle$ is that of the dipole moments induced by the electric field $\vec{E}_k$ at position of the polarizable particle k, $\langle \vec{P}_\alpha \rangle = \sum \alpha_k \vec{E}_k$.

In a mixture of non-associated polar compounds Y_i (dipole molecules of dipole moment μ_i and polarizability α_i), present at particle densities ρ_i (particles/unit of volume) in quasi-static fields (field of low frequency, e.g. for small molecules $\nu < 10^6$ Hz), the static permittivity ε and the molecular quantities μ_i, α_i and f_i (f_i = factor of the Onsager reaction field) are linked by the Onsager equation, see Ref. [1]

$$\frac{(\varepsilon - 1)(2\varepsilon + 1)}{3\varepsilon} = \frac{1}{\varepsilon_0} \sum_i \frac{\rho_i}{1 - f_i \alpha_i} \left[\alpha_i + \frac{1}{3kT} \frac{\mu_i^2}{1 - f_i \alpha_i} \right] \tag{7}$$

Equation (7) was extended by Kirkwood [2] to include the case of one-component auto-associating systems, and by Winkelmann and Quitsch [3] to include associating multicomponent systems.

The application of a quasi-static field $\vec{E}$ to molecular systems produces an equilibrium polarization $\langle \vec{P}^{eq} \rangle$ as the response. The dipole molecules follow the polarity changes of the electromagnetic field without time lag; hence, no energy dissipation takes place.

With increasing frequency of the applied field situations are reached in which the polarity change of the field causes significant variation of the field strength within periods characteristic of molecular motions (dipole molecule orientation, atomic and electronic polarization) or reactions (such as hydrogen-bond formation etc.). Polarization then lags behind the electric field and energy is dissipated in the system. This effect is called dielectric dispersion. Dielectric dispersion is represented by the frequency dependence of the real part ε' of the complex permittivity $\varepsilon^*[\varepsilon' = \varepsilon'(\nu)$, dispersion curve]; the frequency dependence of its imaginary part ε'' yields the absorption curve [$\varepsilon'' = \varepsilon''(\nu)$].

The lag of polarization behind the electric field is characterized by the "loss-angle" δ; the dissipated energy per unit volume and time is $\dot{W}$

$$\mathrm{tg}\delta = \frac{\varepsilon''(\omega)}{\varepsilon'(\omega)}; \quad \dot{W} = \frac{1}{2} E_0^2 \omega \varepsilon_0 \varepsilon''(\omega) \tag{8a, b}$$

In a schematic way Fig. 2 shows the typical behaviour of the frequency dependence of the real (ε') and imaginary (ε'') parts of the complex permittivity of non-conducting simple liquids such as water, methanol, acetonitrile, etc. The first plateau (0 to 10^8 Hz) is the range of static permittivity ($\varepsilon'(\nu) = \varepsilon'(0) = \varepsilon$; $\varepsilon''(\nu) = 0$). The steep slope which follows it is the frequency region where the dipole moment orientation lags behind the applied field ($\nu \sim 10^9$ Hz) and finally cannot follow at all the polarity change of the field. This relaxation process, where $\varepsilon(\nu)$ decreases from $\varepsilon'(0) = \varepsilon$ to $\varepsilon'(\infty) = \varepsilon_\infty$ reduces the orientational part of the polarization from its equilibrium value at permittivity ε to zero at ε_∞. The quantity ($\varepsilon - \varepsilon_\infty$) is called dispersion amplitude. In contrast, the induced

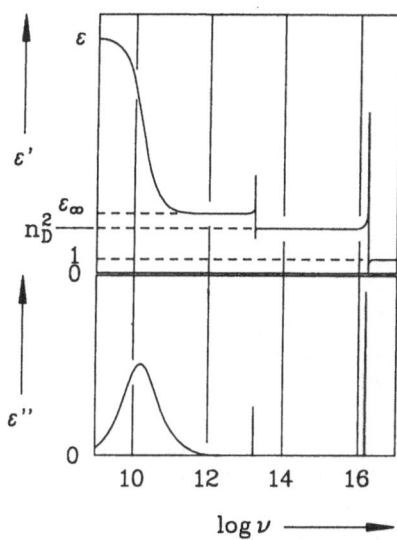

Fig. 2. Schematic representation of the dispersion curve ε' and the absorption curve ε'' of a polar liquid. For explanation see text

polarization $\langle \vec{P}_\alpha \rangle$ fluctuation is very fast and is not affected in the relaxation region of molecular reorientation. At the plateau of ε_∞ the total polarization $\vec{P}$ equals the induced polarization. This fact is used experimentally for the separate determination of $\vec{P}_\mu$ and $\vec{P}_\alpha$

$$\langle \vec{P}_\mu \rangle = \varepsilon_0(\varepsilon - \varepsilon_\infty)\vec{E}; \quad \langle \vec{P}_\alpha \rangle = \varepsilon_0(\varepsilon_\infty - 1)\vec{E} \qquad (9a, b)$$

Resonant transitions in the IR to UV regions are due to intramolecular atomic and electronic displacements. Figure 2 illustrates two such resonant transitions, one in the IR region where ε_∞ decreases to n_D^2 ($= \varepsilon'$ at the frequency of the sodium D-line), and the other in the UV-region where n_D^2 decreases to 1. This result can be used to split the total induced polarization P_α into its atomic electronic components

$$\langle \vec{P}_\alpha \rangle = \langle \vec{P}_\alpha \rangle_{at} + \langle \vec{P}_\alpha \rangle_{el}; \quad \langle \vec{P}_\alpha \rangle_{at} = \varepsilon_0(\varepsilon_\infty - n_D^2)\vec{E} \qquad (10a, b)$$

The information from Fig. 2 is incomplete because it does not contain the energy dissipation from kinetic or transport processes. The corresponding modes will be discussed in the following sections.

Some data are given to illustrate the preceding discussion on dielectric dispersion and absorption as it is sketched in Fig. 2. The point of inflexion of the dispersion curve is situated at the same frequency as the maximum of the absorption curve, this frequency being the reciprocal relaxation time τ of the bulk reorientation process [4]: τ (water) $= 8.3$ ps; τ (methanol) $= 51$ ps, τ (ethanol) $= 162$ ps; τ (acetonitrile) $= 3.5$ ps; τ (N,N-dimethylformamide) $= 11$ ps; τ (formamide) $= 37$ ps, τ (N-methylformamide) $= 123$ ps.

In contrast to the frequency domain study of Fig. 2, where ε' and ε'' were observed over a large frequency range as functions of the frequency of perturbing

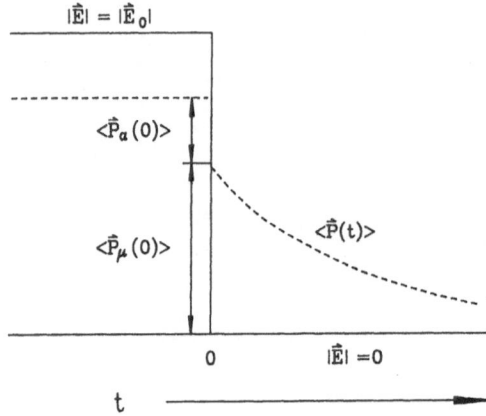

Fig. 3. Time domain method. Electric field jump at time $t = 0$ and step-response function. The induced polarization $\langle \vec{P}_\alpha \rangle$ breaks down at $t = 0$ without time lag; orientational polarization $\langle \vec{P}_\mu \rangle$ decreases monotonically, controlled by a step-response function $F_p^{or}(t)$ of relaxational type

monochromatic electromagnetic waves, a time domain study looks at the time-dependent response of polarization on a perturbation. Figure 3 illustrates how a homogeneous medium, initially exposed to an electric field of field strength $\vec{E}_0$ returns to equilibrium when the field is switched off at time $t = 0$. The induced polarization $\langle \vec{P}_\alpha \rangle$ breaks down without a time lag, whereas the orientational polarization $\langle \vec{P}_\mu \rangle$ decreases monotonously to its final value

$$\langle \vec{P}(0)\vec{P}(t) \rangle = \langle \vec{P}(0)\vec{P}(0) \rangle F_p^{or}(t) \tag{11}$$

$F_p^{or}(t)$ is the step-response function of the orienational polarization $\langle \vec{P}_\mu \rangle$.

The step-response function $F_p^{or}(t)$ is of the relaxational type; $F_p^{or}(0) = 1$; $F_p^{or}(\infty) = 0$

The simplest type of a step-response function is obtained for a one-component system in which dipole reorientation is controlled by a kinetic first order law

$$-\frac{d}{dt} \langle \vec{P}(0)\vec{P}(t) \rangle = k^{or} \langle \vec{P}(0)\vec{P}(t) \rangle; \quad k^{or} = \frac{1}{\tau} \tag{12a, b}$$

The rate constant k^{or} of the reorientation process is the reciprocal relaxation time τ. From Eq. (12a) follows

$$F_p^{or}(t) = \exp\left(-\frac{t}{\tau}\right) \tag{12c}$$

The conversion from the time to the frequency domain is made via consideration of a sequence of infinitely small steps of the field strength $\vec{E}$ at subsequent times t' controlled by the pulse-response function of the orientational polarization

$$f_p^{or}(t') = -\frac{dF_p^{or}(t-t')}{d(t-t')} \tag{13}$$

which is the time derivative of the step-response function F_p^{or}.

The application of a harmonically changing electric field $E(t)$, $E(t) = E_0 \exp(i\omega t)$, that is, a monochromatic electromagnetic wave of circular frequency ω, yields the relation [5]

$$\langle \vec{P}_\mu(\omega, t) \rangle = \varepsilon_0(\varepsilon - \varepsilon_\infty)\vec{E}(t) \int_0^\infty e^{i\omega t'} f_p^{or}(t')dt' \tag{14a}$$

$$= \varepsilon_0(\varepsilon - \varepsilon_\infty)\vec{E}(t)L_{i\omega}[f_p^{or}(t')]$$

where $L_{i\omega}[f_p^{or}(t')]$ is the Laplace transform of $f_p^{or}(t')$. Equation (14a) yields the frequency dependence of permittivity

$$\varepsilon^*(\omega) = \varepsilon'(\omega) - i\varepsilon''(\omega)$$

$$= \varepsilon_\infty + (\varepsilon - \varepsilon_\infty)L_{i\omega}[f_p^{or}(t)] \tag{14b}$$

In the case of the kinetic first order process of reorientation, yielding the step-response function given by Eq. (12c), the pulse response function and its Laplace transform are given by the relations

$$f_p^{or}(t) = \frac{1}{\tau}\exp\left(-\frac{t}{\tau}\right); \quad L_{i\omega}[f_p^{or}(t)] = \frac{1}{1 + i\omega\tau} \tag{15a, b}$$

and Eq. (14b) can be written

$$\varepsilon^*(\omega) = \varepsilon_\infty + (\varepsilon - \varepsilon_\infty)\frac{1 - i\omega\tau}{1 + \omega^2\tau^2} \tag{15c}$$

$$\varepsilon'(\omega) = \varepsilon_\infty + (\varepsilon - \varepsilon_\infty)\frac{1}{1 + \omega^2\tau^2} \tag{15d}$$

$$\varepsilon''(\omega) = (\varepsilon - \varepsilon_\infty)\frac{\omega\tau}{1 + \omega^2\tau^2} \tag{15e}$$

$$\left(\varepsilon' - \frac{\varepsilon + \varepsilon_\infty}{2}\right)^2 + (\varepsilon'')^2 = \left(\frac{\varepsilon - \varepsilon_\infty}{2}\right)^2 \tag{15f}$$

Equations (15d), (15e) and (15f) are respectively the projections of the three-dimensional curve $\varepsilon^*(\omega)$ in the (ε', ω)-plane (the dispersion curve shown in Fig. 2), in the (ε'', ω)-plane (the absorption curve shown in Fig. 2), and in the $(\varepsilon'', \varepsilon')$-plane (the Argand diagram, also called Cole–Cole plot). According to Eq. (15f) the Argand diagram is a semicircle of radius $[(\varepsilon - \varepsilon_\infty)/2]$.

Figure 4 shows all the curves described by Eqs. (15c) to (15f) for water in a three-dimensional diagram. They were obtained by a multivariate analysis of

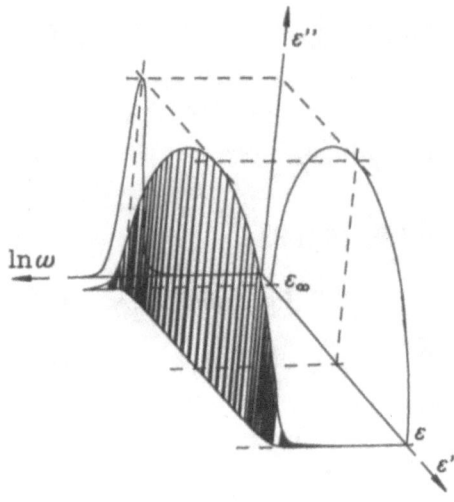

Fig. 4. Complex permittivity ε^* of water at 25°C in the frequency range 0.9 to 40 GHz and its projections in the (ε', ω)-plane (dispersion curve), (ε'', ω)-plane (absorption curve), and (ε', ε'')-plane (Argand diagram). From Ref. [6]

the data from measurements on pure water in the frequency region 0.9 to 40 GHz ($\varepsilon = 78.5$, $\varepsilon_\infty = 5.3$, $\tau = 8.3$ ps) [6, 7].

The relaxation time τ is the reciprocal of the frequency at the inflexion point of the dispersion curve, the maximum of the absorption curve, or the apex of the Argand diagram. Generally, more than one relaxation process occurs in a system. Each one has its characteristic time constant (relaxation time). These can be observed when a sufficiently large frequency range is covered. Figure 4 shows only a limited section from 0.9 to 40 GHz of the complete water diagram. The corresponding relaxation process is the reorientation process which would re-establish with a rate constant of τ^{-1} the undisturbed (equilibrium) structure of water when the perturbing field is switched off. In a complete diagram it is followed at higher frequencies by the relaxation process of hydrogen bond formation with a relaxation time of about 1 ps [8].

A system with more than one relaxation process can be analysed using a discontinuous relaxation time distribution

$$f_p^{or}(t) = \sum \frac{g_j}{\tau_j} \exp\left(-\frac{t}{\tau_j}\right); \quad g_j = \frac{\varepsilon_j - \varepsilon_{j\infty}}{\varepsilon - \varepsilon_\infty}; \quad \sum g_j = 1 \qquad (16a,b,c)$$

if the relaxation times of the processes are significantly different. The complex permittivity is given as

$$\varepsilon^*(\omega) = \varepsilon_\infty + (\varepsilon - \varepsilon_\infty) \sum_{j=1}^{n} \frac{g_j}{1 + i\omega\tau_j} \qquad (16d)$$

Here ε' decreases in n steps from ε to ε_∞, and every step is a relaxation process of dispersion amplitude $(\varepsilon_j - \varepsilon_{j\infty})$ and relaxation time τ_j. Consequently, the

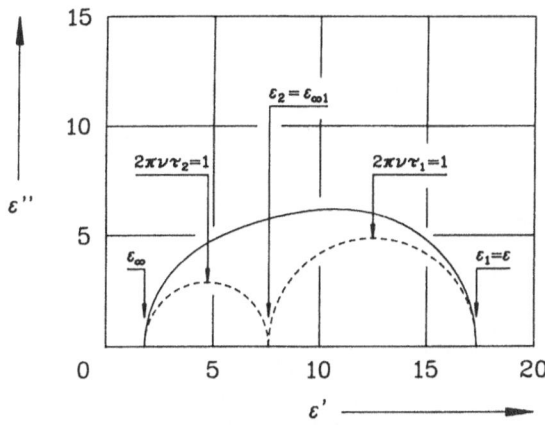

Fig. 5. Full line: Argand diagram of a mixture of propylene carbonate and dimethoxyethane (mole fraction of PC $x_{PC} = 0.20$). *Dashed line*: Relaxation of propylene carbonate (relaxation time $\tau_1 = 22$ ps) and dimethoxyethane (relaxation time $\tau_2 = 4.7$ ps) as obtained by multivariate data analysis from the *continuous line* [14]

dispersion curve has the corresponding sequence of inflection points, the absorption curve has n maxima, and the Argand diagram contains n semicircles, see Fig. 5 in which the symbols of Eqs. (16) are explained.

Relaxation models with continuous relaxation time distributions $G(\ln \tau)$ are based on pulse response functions of the type of

$$f_p^{or} = \int_{\tau=0}^{\infty} \frac{G(\ln \tau)}{\tau} \exp\left(-\frac{t}{\tau}\right) d(\ln \tau); \quad \int_{\tau=0}^{\infty} G(\ln \tau) d(\ln \tau) = 1 \qquad (17a, b)$$

yielding the permittivity function

$$\varepsilon^*(\omega) = \varepsilon_{\infty} + (\varepsilon - \varepsilon_{\infty}) \int_{\tau=0}^{\infty} \frac{G(\ln \tau)}{1 + i\omega\tau} d(\ln \tau) \qquad (17c)$$

Various empirical formulae are used for the description of continuous relaxation time distributions; examples are the equations of Cole–Cole [9], Kirkwood–Fuoss [10], Davidson–Cole [11], Fröhlich [12], and Havriliak–Negami [13]. A survey is given in Ref. [5].

Diagrams with several relaxation processes occur for mixtures of polar compounds. The Argand diagram for a propylene carbonate-dimethoxyethane mixture in Fig. 5 is an example [14]. They also are found for pure polar liquids provided a sufficiently large frequency range is scanned. For example, methanol shows three relaxation processes [8]: (1) a structural one at $\tau_1 = 51$ ps, (2) a H-bonding one at $\tau_3 = 1.2$ ps, like water, and a further process ($\tau_2 = 7.1$ ps) which is attributed to the rotation of free OH-groups of the methanol chain ends. In water a relaxation time corresponding to this process ought also to exist; however, it cannot be separated because the relaxation time τ_2 doesn't differ sufficiently from the bulk relaxation time, $\tau_1 = 8.3$ ps.

Apolar compounds, like CCl_4 or C_6H_6 do not yield relaxation steps and have no relaxation times. In binary mixtures with polar compounds they reduce

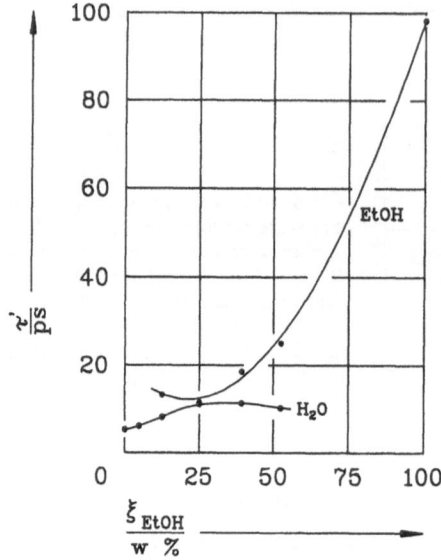

Fig. 6. Microscopic relaxation times τ' (Eq. 19) of ethanol water mixtures as a function of alcohol weight per cent ξ_{EtOH} at 25°C. Only the two bulk relaxation times τ_1 (ethanol) and τ_2 (water) are shown. Rotation of end-standing OH-groups and H-bonding are suppressed for the sake of clearness. From Ref. [7]

the dispersion amplitudes of the polar compounds because of volume effects. Up to rather high concentrations H-bonding of protic polar compounds is not affected by apolar substances.

In mixtures of protic substances a complex pattern of relaxation processes is found. For instance, water–ethanol mixtures from 0.9 to 40 GHz yield various relaxation times: ethanol (τ_1 bulk), water (τ_2 bulk), rotation of OH, and H-bonding [7]. At 25w% (weight %) of ethanol τ_1 and τ_2 are equal, see Fig. 6. The 25w% mixture shows various specific properties, such as a minimum compressibility, a maximum excess enthalpy of mixing, a maximum Kirkwood correlation factor g_{ij} (i = water, j = ethanol), and a maximum of the enthalpy of transfer of ions from aqueous solution to the mixed solvent system water–ethanol.

Relaxation times obtained from high frequency permittivity measurements are macroscopic relaxation times concerning the orientation of the polarization vector, see Eqs. (12); they do not provide information on the reorientation of dipolar molecules; that is controlled by the correlation function

$$\varphi(t) = \frac{\langle \vec{\mu}(0)\vec{\mu}(t) \rangle}{\langle \vec{\mu}(0)\vec{\mu}(0) \rangle} \tag{18}$$

which has a microscopic relaxation time τ'.

Actually there is no general theory which connects the macroscopic (τ) and microscopic (τ') relaxation times. In a special case, the relaxation of a rotational diffusion process of not associating molecules can be approximated with the help of the Powless–Glarum equation [15, 16]

$$\tau' = \frac{3\varepsilon}{2\varepsilon + \varepsilon_\infty} \tau \tag{19}$$

and linked via τ' to the Stokes–Einstein–Debye equation, see Ref. [17]

$$\tau' = \frac{3V_m f}{kT} \eta \tag{20}$$

relating the microscopic relaxation time τ' to the dynamic vicosity η of the medium in which the dipole molecule rotates. V_m is the molar volume of the rotating molecule and f is a friction factor.

3 Electrolyte Solutions

In the context of Eqs. (2) it has been stressed that electric conductivity and permittivity generally cannot be determined separately. However, at very low frequencies (as $\nu \to 0$) the energy loss in an electrically conducting liquid phase per unit volume and time is completely due to the charge transport in the solution.

$$\dot{W} = \tfrac{1}{2} E_0^2 \omega \varepsilon_0 \eta''(\omega) = \tfrac{1}{2} E_0^2 \sigma \tag{21}$$

The charge transport is controlled by Ohm's law

$$\vec{j} = \sigma \vec{E}; \quad \sigma = e \sum \rho_k |z_k| u_k; \quad \sigma = 10^3 \Lambda_c \tag{22a, b, c}$$

where e is the charge of proton, and ρ_k, z_k, and u_k are the density, valency, and mobility of the ionic species k in the solution. Equation (22c) defines the molar conductance Λ at molarity c [mol dm^{-3}] of the electrolyte solution.

At very low concentration the molar conductance of a completely dissociated 1, 1 electrolyte ($C^{z+} A^{z-}$) is given by the limiting law [18, 19]

$$\Lambda = \Lambda^\infty - S\sqrt{c} \tag{23}$$

where Λ and Λ^∞ are the molar conductances at molarities c and zero, respectively. The limiting slope S is calculable (see Table 2).

The variation of molar conductance with frequency, the so-called Debye–Falkenhagen effect, is very small. The dispersion of conductance of a 1.1 electrolyte in the concentration range of the Debye–Onsager limiting law yields the characteristic time constant τ^{el} for disturbed electroneutrality to return to its equilibrium after switching off the perturbing electric field

$$\tau^{el} = \frac{2}{\kappa^2 (u_+ + u_-)kT}; \quad \kappa^2 = \frac{e^2(\rho_+ z_+^2 + \rho_- z_-^2)}{\varepsilon_0 \varepsilon kT} \tag{24a, b}$$

τ^{el} is known as the Debye relaxation time [20, 41]; κ is the Debye parameter.

Investigations on the high frequency permittivity of electrolyte solutions neglect the frequency dependence of conductance; σ^* in Eqs. (2) to (4) is replaced by the conductivity measured at zero or quasi-static frequencies ($\sigma' = \sigma; \sigma'' = 0$). The generalized permittivity, Eq. (4), then takes the simple form

$$\eta^* = \varepsilon^* - i\frac{\sigma}{\omega\varepsilon_0}; \quad \eta' = \varepsilon'; \quad \eta'' = \varepsilon'' + \frac{\sigma}{\omega\varepsilon_0} \qquad (25a, b, c)$$

Polyelectrolytes and biochemical systems generally do not permit this simplification.

Figure 7 shows the Argand diagrams η'' vs ε' and ε'' vs ε' of pure water (curve 1) and a 0.7931 M aqueous solution of KCl (curve 2: η'' vs ε', curve 3: ε'' vs ε'). The diagram ε'' vs ε' was obtained from the measurements of η'' for KCl at various frequencies ω after substraction of the conductivity term, $\sigma/\omega\varepsilon_0$ in Eq. (25c), where σ is the separately measured specific conductivity at quasi static frequencies (200 to 10 000 Hz). For the non-conducting solvent water is $\eta'' = \varepsilon''$ (curve 1). Curve 2 shows clearly that $\eta'' \to \infty$ for $\omega \to 0$, according to Eq. (25c); the broken line is the asymptote due to conductivity [21].

Potassium chloride is completely dissociated in aqueous solution. The "free" ions of this example, K^+ and Cl^-, have no dipole moment which could be orientated in an electric field. Therefore, they do not yield particular dispersion amplitudes and relaxation times. However, the reduction of the solvent dispersion amplitude by the ions is distinctly larger than the volume effect of apolar compounds. The very high electric charge densities at the ion surfaces and possible specific effects (hydrogen bonds of anions, electron interactions with cations etc.) obstruct the orientation of the adjacent solvent molecules in the applied electric field. Solvent molecules partially orientated by the ions in competition with the external field give rise to a concentration-dependent

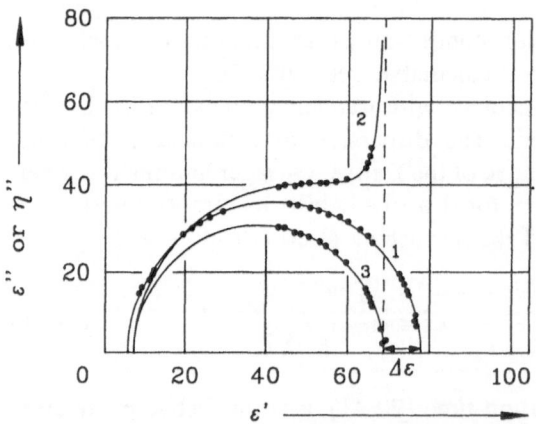

Fig. 7. Argand diagram of a completely dissociated electrolyte and comparison with the diagram of its pure solvent. Solvent: water at 25°C (*curve 1*), electrolyte 0.7931 M KCl (*curve 2*: before, *curve 3*: after conductivity correction); $\Delta\varepsilon$: decrease of solvent permittivity caused by the ions, see Eq. (26a)

dielectric decrease of the solvent permittivity

$$\Delta\varepsilon = \varepsilon_{solv} - \vartheta c + \beta c^{3/2} \tag{26a}$$

The dielectric decrement

$$\vartheta = \lim_{c\to 0}\left[\varepsilon_{solv} - \varepsilon_{sol}(c)\right] \tag{26b}$$

where ε_{solv} and $\varepsilon_{sol}(c)$ are the static permittivities (as $v \to 0$) of the pure solvent, and the solvent in the electrolyte solution at molarity c, characterizes the fraction of solvent molecules which are "irrotationally bound" to ions (dielectric saturation), so that they cannot participate in the relaxation process of the solvent. The number ϑ is an ion- and solvent-specific quantity. ϑ varies with the charge density at the ion surface: $Li > Na > K > Rb > Cs$ [21], see also Fig. 8.

There are several theories which calculate solvation numbers with the aid of this dielectric decrement [22–25]. As well as this "static" solvation effect of electrolyte solutions there is also a kinetic depolarization effect [26–29] which will be discussed in the chapter on electrolyte conductance.

In contrast to the free ions in electrolyte solutions, the ion pairs behave as dipoles if their life time is comparable to the period of polarity change of the external field—or longer. Solutions with partially associated electrolytes exhibit at least three processes. In Fig. 9 this characteristic behaviour is exemplified for acetonitrile solutions of Bu_4NBr [21] which exhibit

1. a high frequency relaxation process—dispersion amplitude $(\varepsilon_2 - \varepsilon_\infty)$, relaxation time τ_2—for reorientation of the solvent;
2. a low frequency relaxation process—dispersion amplitude $(\varepsilon_1 - \varepsilon_{\infty 1})$, relaxation time τ_1—for reorientation of the ion pairs $[Bu_4N^+Br^-]^0$;
3. a transport process—the conductivity of the electrolyte solution $[\eta' = \varepsilon; \eta'' \to \infty$ according to Eq. (25c)].

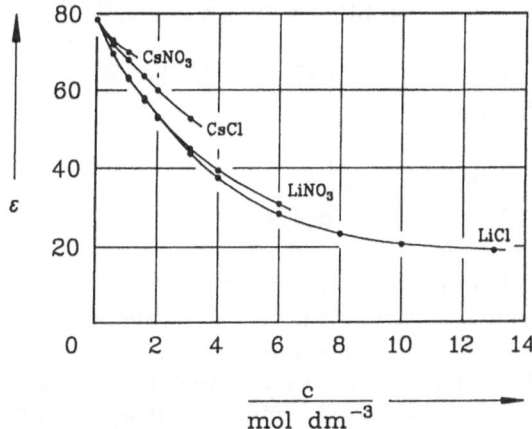

Fig. 8. Permittivity of aqueous lithium and cesium salt solutions at 25°C up to saturation concentrations

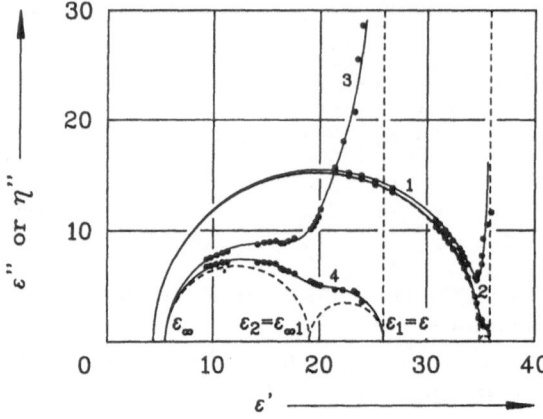

Fig. 9. Argand diagram of Bu₄NBr solutions of acetonitrile at 25°C. *Continuous lines:* measured data (ε', η''): *curve 1* (0.058 M), *curve 3* (1.384 M); after substraction of the conductivity contribution $(\varepsilon', \varepsilon'')$: *curve 2* (0.058 M), curve 4 (1.384 M). *Dashed lines:* relaxation regions of solvent and ion pairs

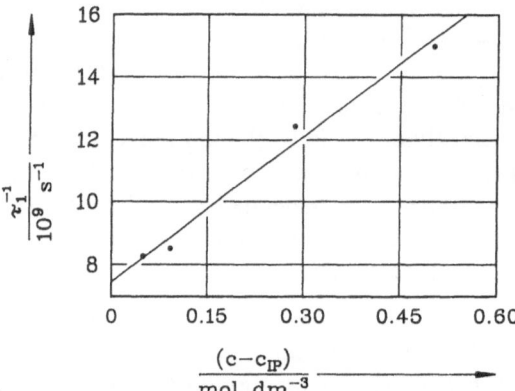

Fig. 10. Dependence of the ion-pair relaxation time on the concentration of free ions in acetonitrile solutions of Bu₄NBr at 25°C. The Argand diagrams at two concentrations of this organic solvent system are given in Fig. 9. Free ion concentration $(c - c_{IP})$ is calculated from the dispersion amplitude $(\varepsilon_1 - \varepsilon_{\infty 1})$. For results see Table 1

The dielectric decrease of the electrolyte solution $\Delta\varepsilon$, not shown in Fig. 10, is $\Delta\varepsilon = \varepsilon_{solv} - \varepsilon_2(c)$, $\varepsilon_2(c)$ being the permittivity of the solvent of the electrolyte solution at concentration c. The permittivity of the electrolyte solution is $\varepsilon_1(c)$. At low electrolyte concentration, solution permittivity $\varepsilon_1(c)$ is larger than the permittivity of the pure solvent, $\varepsilon_1(c) > \varepsilon_{solv}$.

After appropriate rearrangement [30, 31] of Eq. (7) to yield the dispersion amplitude of the ion–pair relaxation process

$$\varepsilon_1 - \varepsilon_{\infty 1} = \frac{3\varepsilon_1}{\varepsilon_0 kT(2\varepsilon_1 + 1)} \cdot \frac{\mu_{IP}^2 \rho_{IP}}{(1 - \alpha_{IP} f_{IP})^2}$$

this equation may be used to calculate the ion-pair concentration c_{IP} (via the ion-pair density ρ_{IP}) from the dispersion amplitude $(\varepsilon_1 - \varepsilon_{\infty 1})$ of the ion-pair relaxation process. To achieve this, a model is made representing the ion pair as a prolate ellipsoid; the length of the ellipsoid axis is obtained either from the configuration of cation and anion directly in contact (contact ion pair), or with the inclusion of a solvent molecule (solvent-shared ion pair). The ion charges and the centre-to-centre distance of the ions in the ellipsoid permit the calculation of the dipole moment μ_{IP}. The ion polarizabilities yield the ion-pair polarizability α_{IP}, and the reaction field factor f_{IP} of the prolate ellipsoid then can be calculated as usual [32, 33]. Knowledge of the ion-pair concentration c_{IP} at various electrolyte concentrations c permits the calculation of the concentration-dependent equilibrium constant of ion-pair formation, K_c

$$C^{z+} + A^{z-} \underset{k_2}{\overset{k_1}{\rightleftharpoons}} [C^{z+} A^{z-}]^o; \quad K_c = \frac{k_1}{k_2} = \frac{c_{IP}^{eq}}{c_+^{eq} c_-^{eq}} \qquad (27a, b)$$

and the thermodynamic association constant K_A by extrapolation of K_c to zero electrolyte concentration. In Table 1 association constants obtained from microwave permittivity studies are compared with those obtained from conductivity and heat of dilution measurements.

The inspection of the concentration-dependence of the microscopic relaxation time τ_1' [calculated from τ_1 using Eq. (19)] for the rotational diffusion process of the ion pair shows that τ_1' decreases with increasing viscosity of the electrolyte solution and thus appears to be inconsistent with Eq. (20). The inconsistency disappears when taking into account that the dipole does not only rotate as a rigid dipole molecule; a kinetic mode related to the reaction rate of the chemical reaction, Eq. (27a), is superimposed on the rotational mode.

At first order perturbations $\delta c_i(\delta c_+ = \delta c_- = -\delta c_{IP})$ in the vicinity of equilibrium the chemical reaction follows the time law

$$\frac{d[c_{IP}^{eq} + \delta c_{IP}]}{dt} = k_1 [c_+^{eq} + \delta c_{IP}][c_-^{eq} + \delta c_{IP}] - k_2 [c_{IP} - \delta c_{IP}] \qquad (28a, b)$$

$$0 = k_1 c_+^{eq} c_-^{eq} - k_2 c_{IP}^{eq}$$

Table 1. Rate constants k_1 and k_2 of the ion-pair formation and dissociation reactions, association constants K_A, and microscopic relaxation times τ_{IP}' of ion-pair reorientation from high frequency permittivity measurements[a]

Electrolyte Solution	$k_1 \times 10^{-9}$ dm^3(mol s)$^{-1}$	$k_2 \times 10^{-7}$ s^{-1}	K_A/dm^{3}mol^{-1} Amel.[b]	Lit.	τ_{IP}'	τ_{SED}' [e]
Bu$_4$NBr/CH$_3$CN	7.8 ± 0.7	50 ± 10	17 ± 5	20^c	143 ± 8	175
CdSO$_4$/H$_2$O	1.8 ± 0.4	0.7 ± 0.2	280 ± 50	$245^c, 239^d$	136 ± 8	146
MgSO$_4$/H$_2$O	1.80 ± 0.01	1.03 ± 0.01	150 ± 40	$156^c 161^d$	136 ± 1	130

[a] From Ref. [6]: [b] calculated from the concentration dependence of the dispersion amplitude: [c] from conductance measurements: [d] calorimetric (heat of dilution) measurements:; [e] Stokes–Einstein–Debye relaxation time (Eq. 20)

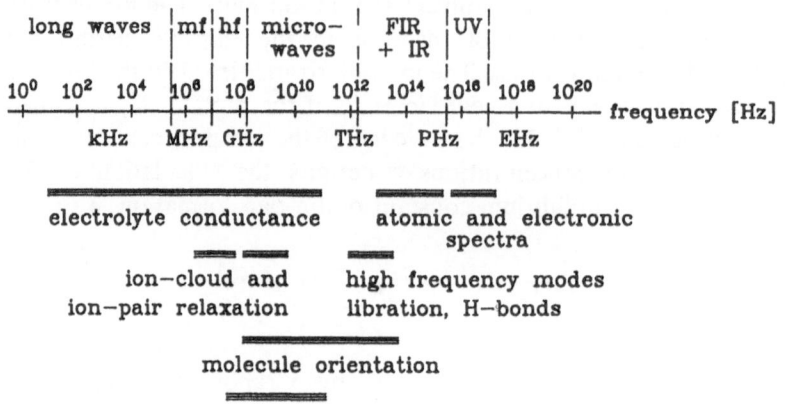

Fig. 11. Relaxation modes and resonant transition contributing to the generalized permittivity of solutions and their characteristic frequency ranges

Retaining only the first order terms yields $[c_+^{eq} = c_-^{eq} = c - c_{IP}^{eq}]$

$$d\frac{(\delta c_{IP})}{dt} = [2k_1(c - c_{IP}^{eq}) + k_2]\delta c_{IP} = \frac{1}{\tau^{kin}}\delta c_{IP} \tag{28c}$$

The relaxation time of this kinetic mode is

$$\tau_{IP}^{kin} = \frac{1}{k_2 + 2k_1(c - c_{IP}^{eq})} \tag{29}$$

and superimposition of the processes yields the observed relaxation time τ_{IP} (mes.)

$$\frac{1}{\tau_{IP}(mes.)} = \frac{1}{\tau_{IP}^{or}} + \frac{1}{\tau_{IP}^{kin}} = \frac{1}{\tau_{IP}^{or}} + k_2 + 2k_1(c - c_{IP}^{eq}) \tag{30}$$

The linear plot of τ_{IP}^{-1}(mes.) vs $(c - c_{IP}^{eq})$ yields the quantities τ_{IP}^{or} and the rate constants k_1 and k_2 of the ultra-rapid reactions of ion-pair formation and decomposition. An example is given in Fig. 10. The data for some electrolyte solutions are given in Table 1; after transformation of τ_{IP}^{or} to microscopic relaxation times a satisfactory agreement is found with Stokes–Einstein–Debye relaxation times.

To round off the discussion on solvent and solution permittivities Fig. 11 shows the phenomena and the spectral ranges on which information is given in the preceding chapters.

4 Phenomenology of Electrolyte Conductivity

The conductivity of electrolyte solutions is controlled by ion–ion and ion–solvent molecule interactions in perturbed equilibria which are so complex that a statistical–mechanical treatment of electrolyte conductivity actually is possible only for low electrolyte concentrations. Before a brief discussion of the statistical theory of conductivity is given in the following chapter, important results of empirical investigations, including highly concentrated solutions, will be presented.

Specific conductivity σ is related to molar conductance Λ by its definition, see Eq. (22c) [34, 35].

$$\sigma = f\Lambda m \tag{31a}$$

The total variation of specific conductance $d\sigma$ at increase dm of molality is given by the relation

$$d\sigma = f(\Lambda dm + md\Lambda) \tag{31b}$$

where Λ is the equivalent conductance of the solution, m is the molality of the electrolyte, and f is a dimensional factor. The addition of ions to the solution $(dm > 0)$ increases the conductivity through the term Λdm of Eq. (31b). However, this is opposed to the simultaneously occurring conductivity decrease caused by the hindered ion movement (decreasing ion mobility) at increasing charge density $(d\Lambda < 0)$, see Eq. (39) and subsequent text.

Consequently, the specific conductivity σ is seen to pass through a maximum $(d\sigma = 0)$; $\sigma = \sigma_{max}$ at a molality $m = m_{max}$. Figure 12 illustrates these maxima

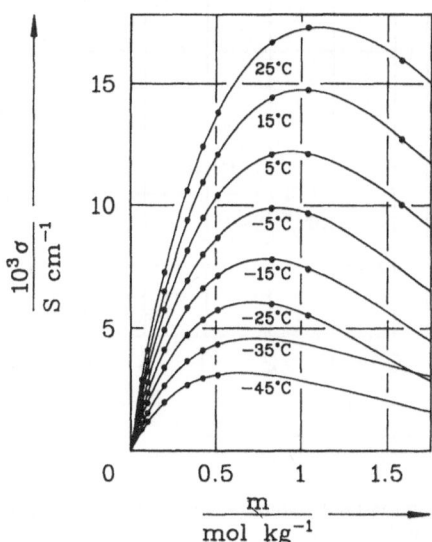

Fig. 12. Specific conductance σ as a function of molality m and temperature. Solutions of LiAsF$_6$ in propylene carbonate-dimethoxyethane mixtures (weight fraction of propylene carbonate 32 w%). From Ref. [36].

for LiAsF$_6$ solutions in mixtures of propylene carbonate and dimethoxyethane (32w% of PC) at various temperatures [36].

The height of the maximum σ_{max} and its position m_{max} are correlated so that the larger is m_{max} the higher is σ_{max}. Whilst this statement is obvious from the results shown in Fig. 12, it is generally valid and has led to establish Figs. 13 correlating σ_{max} and m_{max} for a variety of salts in propylene carbonate and methanol [35]. The arrangement of the salts according to increasing σ_{max} (or m_{max}) in the aprotic solvent propylene carbonate shows that the salts appear to be ordered according to the descending sequence of their Stokes' radii. In contrast, the protic solvent methanol produces a sequence in which the tetraalkylammonium salts are found distinctly below the alkali metal salts. The difference in these specific ion–solvent interactions underlying the plots of Fig. 13 will be discussed in a subsequent chapter.

However, another important conclusion can be drawn from Fig. 13. The $\sigma_{max} - m_{max}$-correlation leads us to the assumption that there must be an energy barrier for every solvent which mainly depends on the properties of the solvent and on the temperature. Indeed, at concentration m_{max} the solutions of different salts in a given solvent have nearly equal activation energies $E(\sigma)$ of charge transport, $E^{(\sigma)} = - R[d \ln \sigma / d(1/T)]$; this is independent of the position and height of the maximum [37, 38].

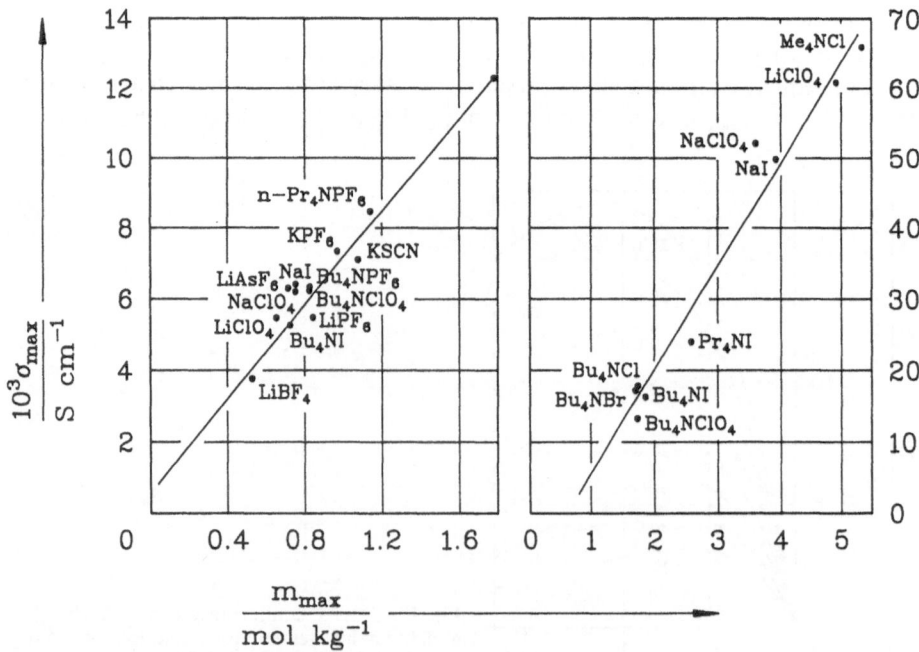

Fig. 13. Correlation of σ_{max} vs m_{max} for alkali metal and tetraalkylammonium salts in a: propylene carbonate (left); and b: methanol (right) at 25°C. From Ref. [35]

Another corresponding state exists for the infinite dilute solution. Here again solutions of various salts in a given solvent have almost equal activation energies [35].

The analysis of the measured conductivities at high salt concentration can be advantageously carried out with the help of the Vogel–Fulcher–Tamann equation

$$\sigma_m(T) = A_m^{(\sigma)} \exp\left[\frac{-B_m^{(\sigma)}}{R[T - T^0(m)]}\right] \tag{32}$$

which yields temperature-independent parameters $A_m^{(\sigma)}$, $B_m^{(\sigma)}$ and $T^0(m)$ at every molality m of electrolyte. $A_m^{(\sigma)}$ is a parameter of the type of (αm + higher terms), $T^0(m)$ is the glass transition temperature of the binary solution at electrolyte molality m, and $B_m^{(\sigma)}$ is related to the activation energy of conductivity $E^{(\sigma)}(m, T)$

$$T^0(m) = T(0) + am + bm^2 \tag{33a}$$

$$E^{(\sigma)}(m, T) = B_m^{(\sigma)}\frac{T^2}{[T - T^0(m)]^2} \tag{33b}$$

An equation similar to Eq. (32), with parameters ($A^{(\varphi)}$, $B^{(\varphi)}$ and T^0), holds for fluidity $\varphi(\varphi = \eta^{-1}$ where η is the viscosity of the liquid).

Conductivity measurements on the 13 almost completely dissociated electrolytes in propylene carbonate of Fig. 13a between $-45°C$ and $25°C$ and viscosity measurements on pure propylene carbonate at the same temperatures lead to [35]

1. an equal glass transition temperature $T^0(0)$ for the pure solvent, as obtained by extrapolation of $T^0(m)$ for every solution, $\lim_{m \to 0} T^0(m) = (151.6 \pm 0.9)$ K.
2. an equal glass transition temperature $T^0(0)$ from conductivity and viscosity measurements, viscosity: $T^0(0) = (152.1 \pm 0.7)$ K
3. equal coefficients $B_0^{(\sigma)}[(3350 \pm 50)$ J mol^{-1}] and $B^{(\varphi)}[(3420 \pm 50)$ J mol^{-1}]

This example clearly shows that the transport activation energies at infinite dilution are equal for all the electrolyte solutions investigated and equal to that of viscous flow of the solvent.

5 Some Aspects of the Statistical–Mechanical Theory of Conductance

The theory of ionic conductivity is presented in several text books and monographs to which the interested reader is referred [39–43]. No notable progress in the field has been made during the last decade. In this chapter only a few aspects are presented which are helpful to understand the fundamentals of the conductance equations of completely dissociated and partially associated electrolyte solutions.

The theory of ionic conductivity normally begins with Onsager's continuity equation for stationary ionic movements

$$\text{div}_i[g_{ij}\vec{v}_{ij}] + \text{div}_j[g_{ij}\vec{v}_{ji}] = 0 \tag{34}$$

where g_{ij} is the pair-correlation function of the ions i and j and $\vec{v}_{ij}$ is the relative velocity of the ion j with regard to the moving ion i, see Fig. 14. The models of the electrolyte solution used are Hamiltonian models at Mayer–McMillan level.

The time-dependent pair-correlation functions of Eq. (34) $g_{ij}(\vec{r}_1, \vec{r}_2; t)$ take into account the perturbation of the spherical symmetry of the equilibrium pair-correlation function $g_{ij}^0(r)[r = |\vec{r}_1 - \vec{r}_2| = |\vec{r}_{12}|]$ with the help of a first order perturbation function $g_{ij}'(\vec{r}_1, \vec{r}_2; t)$ of axial symmetry

$$g_{ij}(\vec{r}_1, \vec{r}_2; t) = g_{ij}^0(r) + g_{ij}'(\vec{r}_1, \vec{r}_2; t) \tag{35}$$

The symmetry axis is directed along the electric field $\vec{E}$ which produces an increased local concentration of unlike charged ions in front of, and a decreased local concentration of ions behind the moving ion i. The perturbation function fulfils the conditions

$$g_{ii}' = g_{jj}' = 0; \quad \lim_{\vec{r}_{ij} \to \infty} g_{ij} = 0 \tag{36a, b}$$

A further boundary condition prevents two moving ions, i and j, from approaching each other within distances less than R_{ij}

$$\left[(\vec{v}_{ij} - \vec{v}_{ij}) \frac{\vec{r}_{ij}}{r_{ij}} \right]_{R_{ij}} = 0 \tag{36c}$$

The distance R_{ij} of closest approach of the ions depends on the ion-ion and ion–solvent interactions.

Mayer–McMillan models use potential energies based on mean force potentials at infinite dilution of the electrolyte

$$W_{ij}^\infty(r) = \frac{e^2 z_i z_j}{4\pi\varepsilon_0 \varepsilon kT} \cdot \frac{1}{r} + W_{ij}^*(r) \tag{37}$$

The first term on the right side of Eq. (37) is the Coulomb potential of the ions i and j in a homogeneous medium (solvent) of relative permittivity ε. The second term is the mean force potential of the non-Coulombic forces which includes

Table 2. Coefficients of the conductance equation for binary symmetrical electrolytes $C^{z+}A^{z-}$ $(z_+ = |z_-|).$[a,b]

$S = S_1\Lambda^\infty + S_2$:	$S_1 = 0.82043 \cdot 10^6 z^3(\varepsilon T)^{-3/2}$
	$S_2 = 82.484 \cdot 10^{-5} z^2(\varepsilon T)^{-1/2}\eta^{-1}$
$E = E_1\Lambda^\infty - 2E_2$:	$E_1 = 6.7749 \cdot 10^{12} z^6(\varepsilon T)^{-3}$
	$E_2 = 0.99750 \cdot 10^3 z^5(\varepsilon T)^{-2}\eta^{-1}$
$J = J_1\Lambda^\infty + J_2$:	$J_1 = 2E_1'[2b^{-1} + 2b^{-2} - b^{-3} + 0.9074 + \ln(\varkappa R/c^{1/2})]$
	$J_2 = E_2'[35b^{-1}/3 + 2b^{-2} - 2.0689 - 4\ln(\varkappa R/c^{1/2})]$
$G = G_1\Lambda^\infty + G_2$:	$G_1 = \varkappa c^{-1/2} R E_1'(0.6094 + 4.4748b^{-1} + 3.8284b^{-2})$
	$G_2 = \varkappa c^{-1/2} R E_2'(-1.3693 + 34b^{-1}/3 - 2b^{-2})$

$\varkappa/c^{1/2} = 0.50290 \cdot 10^{12} z(\varepsilon T)^{-1/2}$; $b = 16.709 \cdot 10^{-6} z^2(\varepsilon T)^{-1}R^{-1}$:

[a] From Ref. [33]; [b] $R_{ij} = R$

the specific molecular interaction forces

$$W_{ij}^*(r) = \begin{cases} \infty & \text{if} \quad r \leq a_{ij} \\ W_{ij}^* & \text{if} \quad a_{ij} \leq r \leq R_{ij} \\ 0 & \text{if} \quad r \geq R_{ij} \end{cases} \tag{38}$$

In Eqs. (38) a_{ij} is the contact distance of the ions i and j, $a_{ij} = a_i + a_j$, R_{ij} is the distance of closest approach (see Eq. (36c)). Distance parameters and mean force potential will be discussed in the following chapter.

Several approximations are needed in order to produce a conductance equation which is valid at least at low electrolyte concentrations

$$\Lambda = \Lambda^\infty - \Lambda^{el} - \Lambda^{rel} \tag{39}$$

In Eq. (39) Λ and Λ^∞ are the molar conductance at concentrations c and zero, respectively. At infinite dilution molar conductance is not hindered by ion–ion interaction and hence exhibits its maximum value. The term Λ^{el} (the electrophoretic term) contains the contributions which diminish the conductance because of the hydrodynamic coupling of the ions; Λ^{rel} (the relaxation term) takes into account the ion–ion interactions which result from the perturbed equilibrium distribution. The frequency dependence of molar conductance, which yields the relaxation time τ^{el}, Eqs. (24), is completely reflected in Λ^{rel}. Equation (23) is the limiting law of Eq. (39).

Truncated series developments of Λ^{el} and Λ^{rel} up to the $c^{2/3}$ terms (2nd virial coefficient approximation) yield the relation

$$\Lambda = \Lambda^\infty - Sc^{1/2} + Ec\log c + J(R)c - G(R)c^{3/2} + \cdots \tag{40}$$

which is valid for completely dissociated symmetrical electrolytes. Table 2 contains the coefficients $S, E, J(R), G(R)$ of Eq. (40) for a symmetrical electrolyte $C^{z+}A^{z-}(z_+ = |z_-|)$ in a solvent of viscosity η and relative permittivity ε. Only the coefficients $J(R)$ and $G(R)$ depend on the distance parameter R, and hence insight in the specific interactions of the electrolyte solution is limited to these terms.

6 The Chemical Model of Electrolyte Solutions

The chemical model of electrolyte solutions aims at a Hamiltonian model at Mayer–McMillan level which is as near as possible to factual chemical reality. To achieve this, the model underlying the calculation of the pair-correlation functions $g_{ij}^0(r)$ subdivides the space around an ion i into three regions as illustrated in Fig. 14 [44, 45].

1. $r \leqq a_{ij}$, the region within which no pair configuration is possible because $a_{ij} = a_i + a_j$ is the contact distance of the ions i and j
2. $a_{ij} \leqq r \leqq R_{ij}$, the region within which a paired state of oppositely-charged ions, the so-called ion pair, suppresses long-range interactions with other ions in the solution. In dilute solutions the occupation of this region by ions of the same sign, or by more than two ions, can usually be neglected.
3. $r \geqq R_{ij}$, the region of long-range ion–ion Coulombic interactions.

Table 3 shows the mean force potentials for a solution of the electrolyte compound $C^{z^+}A^{z^-}$.

The undisturbed pair-correlation functions $g_i^{0(1)}(r), g_i^{0(2)}(r), g_i^{0(3)}(r)$ for the regions (1), (2), (3) are obtained from the mean force potentials $W_{ij}^{(1)}(r)$,

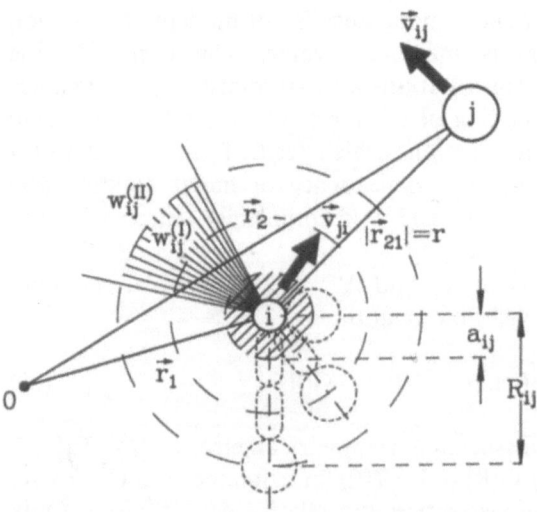

Fig. 14. The chemical model of electrolyte solutions. 0: observer; i, j; ions i and j at mutual distance $\vec{r}_{21}$ and distances $\vec{r}_1$ and $\vec{r}_2$ from the observer. Special positions (contact, separation by one or two orientated solvent molecules) are sketched with broken lines; a_{ij}, R_{ij}: distance parameters; W_{ij}: mean force potentials; $\vec{v}_{ij}$ and $\vec{v}_{ji}$: relative velocities of the ions i and j

Table 3. Mean force potentials of the chemical model in the range of low to moderate electrolyte concentrations[a]

Region	Potential $W_{ij}(r)$
$r \leq a_{ij}$	∞
$a_{ij} \leq r \leq R_{ij}$	$\dfrac{e^2 z_i z_j}{4\pi\varepsilon_0\varepsilon}\dfrac{1}{r} - \dfrac{e^2 z_i z_j}{4\pi\varepsilon_0\varepsilon}\dfrac{\varkappa}{1+\varkappa R_{ij}} + W_{ij}^*$
$r \geq R_{ij}$	$\dfrac{e^2 z_i z_j}{4\pi\varepsilon_0\varepsilon}\dfrac{\exp[\varkappa(R_{ij}-r)]}{1+\varkappa R_{ij}}$

[a] From Ref. [45]; $\varkappa$ is the Debye parameter, for its definition see Eq. (24b)

$W_{ij}^{(2)}(r)$, $W_{ij}^{(3)}(r)$ of Table 3.

$$g_{ij}^{0(i)}(r) = \exp\left[-\frac{W_{ij}^{(i)}(r)}{kT}\right]; \quad W_{ij}^{(i)}(r) = W_{ij}^{el(i)}(r) + W_{ij}^{*(i)}(r) \qquad (41a, b)$$

The link between thermodynamic excess quantities of an electrolyte solution and the chemical model is the integral

$$\int_0^\infty r^2 g_{ij}^0 dr = \int_0^{a_{ij}} r^2 g_{ij}^{0(1)} dr + \int_{a_{ij}}^{R_{ij}} r^2 g_{ij}^{0(2)} dr + \int_{R_{ij}}^\infty r^2 g_{ij}^{0(3)} dr \qquad (42)$$

The first integral on the right hand side of Eq. (42) is equal to zero $[W_{ij}^{(1)}(r) = \infty]$. The second yields the concentration-dependent equilibrium constant of ion-pair formation (configuration of $++$ and $--$ may be neglected at low concentrations)

$$K_c = 4000\pi N_A \exp\left[-\frac{2\varkappa q}{1+\varkappa R_{+-}}\right] \int_{a_{+-}}^{R_{+-}} r^2 \exp\left[\frac{e^2|z_+z_-|}{4\pi\varepsilon_0\varepsilon kT}\cdot\frac{1}{r} - \frac{W_{+-}^*}{kT}\right] dr; \qquad (43)$$

The ion–ion interactions of "free" ions are described by the mean activity coefficient [44, 45].

$$y'_\pm = \exp\left[-\frac{\varkappa q}{1+\varkappa R_{+-}}\right]; q = \frac{e^2|z_+z_-|}{8\pi\varepsilon_0\varepsilon kT} \qquad (44a, b)$$

From Eqs. (43) and (44) the thermodynamic association constant K_A

$$K_A = \frac{K_c}{y'^2_\pm} = 4000\pi N_A \int_{a_{+-}}^{R_{+-}} r^2 \exp\left[\frac{2q}{r} - \frac{W_{+-}^*}{kT}\right] dr \qquad (45a)$$

and the degree of dissociation α can then be calculated

$$K_A = \frac{(1-\alpha)}{\alpha^2 c}\frac{1}{y'^2_\pm} \qquad (45b)$$

It follows that all thermodynamic properties of an electrolyte solution are controlled by the same association constant. A great number of precise measurements confirm this statement [45].

In the framework of the chemical model the activity coefficient of the partially associated electrolyte is given by the relation

$$y_\pm = \alpha y'_\pm \tag{45c}$$

The use of the chemical model in the theory of electrolyte conductance requires two presuppositions:

1. ion pairs are electrically neutral particles and cannot transport electric charges
2. the perturbation function $g'_{ij}(\vec{r}_1, \vec{r}_2; t)$ equals zero in the region $a_{ij} \leq r \leq R_{ij}$

From the first of these presuppositions the system of equations controlling the conductance of partially associated electrolytes follows as

$$\Lambda = \alpha[\Lambda^\infty - S(\alpha c)^{1/2} + E\alpha c \log(\alpha c) + J(R)\alpha c - G(R)(\alpha c)^{3/2}] \tag{46a}$$

$$K_A = \frac{(1-\alpha)}{\alpha^2 c y'^2_\pm}; \quad y'_\pm = \exp\left[\frac{-\kappa q}{1 + \kappa R}\right] \tag{46b, c}$$

Equations (46) are reduced to Eq. (40) if $\alpha = 1$.

The analysis of conductance data using Eqs. (46) yields the quantities Λ^∞, R and K_A (or α).

From the second presupposition it follows that association constants obtained from conductance measurements and those from thermodynamic properties must be equal. This statement can also be confirmed by various examples [45, 54].

Temperature- and pressure-dependent conductance data yield the temperature- and pressure-dependence of the association constants

$$\left(\frac{\delta \ln K_A}{\delta t}\right)_p = \frac{\Delta H^0_A}{RT^2}; \quad \left(\frac{\delta \ln K_A}{\delta p}\right)_T = -\frac{\Delta V_A}{RT} \tag{47a, b}$$

from which the enthalpies of ion-pair formation, ΔH^0_A, and the change in molar volume, ΔV_A, in the ion-pair formation process can be determined. ΔH^0_A is also available from calorimetric measurements of heats of dilution, and ΔV_A is available from measurements of the pressure-dependence of the solution density. The agreement of ΔH^0_A and ΔV_A data determined by thermodynamic methods and conductance is excellent [45, 46, 54].

The agreement of the association constants (which reflect the short-range part of Coulombic and the specific non-Coulombic forces) determined by different methods (mw-spectroscopy, transport processes, thermodynamic properties) illustrates the reliability of the chemical model. In contrast to the

Table 4. Gibbs energies, enthalpies, and entropies of the association process of 1.1 electrolytes in ethanol[a,b]

$$\Delta G_A^0 = - RT \ln K_A$$

Electrolyte	$\dfrac{\Delta G_{298}^0}{J\,mol^{-1}}$	$\dfrac{\Delta S_{298}^0}{J K^{-1}\,mol^{-1}}$	$\dfrac{\Delta H_{298}^0}{J\,mol^{-1}}$
NaI	− 10200	67.0⎱ −18.2[c]	9700
NaBPh$_4$	− 9450	48.8⎰	5100
KI	− 11200	75.0	11100
KSCN	− 11000	74.3	11200
CsI	− 12300	74.3	9800
Pr$_4$NBr	− 12150	51.8	3300
Pr$_4$NI	− 12900	50.4	2100
Pr$_4$NClO$_4$	− 14100	49.8	1800
i-Am$_3$BuNI	− 13100	49.5⎱ −16.4[c]	1700
i-Am$_3$BuNBPh$_4$	− 13800	33.1⎰	−3900

[a] From Ref. [46]; [b] K_A according to Eq. (45); [c] almost equal entropy decrease for the anion change I^- to BPh_4^-

attempts of the older electrolyte theories, unambiguous systematic reproduction of the solution properties is possible if the specific interaction forces are included in the short-range part of the pair-correlation functions (in this case in the association constant).

In particular, the temperature-dependence of association gives valuable information about the structure of electrolyte solutions, as seen in Table 4. Here the change in Gibbs energies ΔG_{298}^0 for ion-pair formation at 298 K of ethanol solutions does not differ too greatly—a normal pattern for 1.1 electrolytes in almost all solvents. In contrast, the values of ΔS_{298}^0 and ΔH_{298}^0 obtained from the temperature dependence of ΔG_{298}^0 subdivide the electrolytes into classes. Alkali metal salts with solvated cations and anions exhibit entropies of ion-pair formation which are about $20\,J mol^{-1} K^{-1}$ larger than those of tetraalkylammonium salts with solvated anions. In both cases also anion classes can be recognized; the exchange of the solvated anions against the unsolvated tetraphenylborate anion is accompanied with an almost equal change in entropy ($18.2\,J mol^{-1}$ for the sodium and $16.4\,J mol^{-1} K^{-1}$ for the i—Am$_3$BuN salt).

The temperature dependence of Gibbs' energy, $(\delta\Delta G^0/\delta T)_p$, yielding the pattern of Table 4 is also observed in other protic solvents and reveals markedly different behaviour between the tetraalkylammonium and alkali metal salts in MeOH, EtOH and PrOH [47]. Significant minima in the temperature range $-45 > \Theta/°C \leqq +25$ are observed for the tetraalkylammonium salts (Fig. 15 shows an example) whereas no minima occur in this temperature range for the alkali metal salt solutions. The position of the minimum depends on the anion and is independent of the cation as shown in Table 5.

Another convincing example of specific interactions in electrolyte solutions is provided by solutions of the $Bu_{4-n}H_n$-picrates in nitrobenzene. Here only Bu_4NPi is strongly dissociated ($K_A \sim 0$), whilst the other cations, apt to form

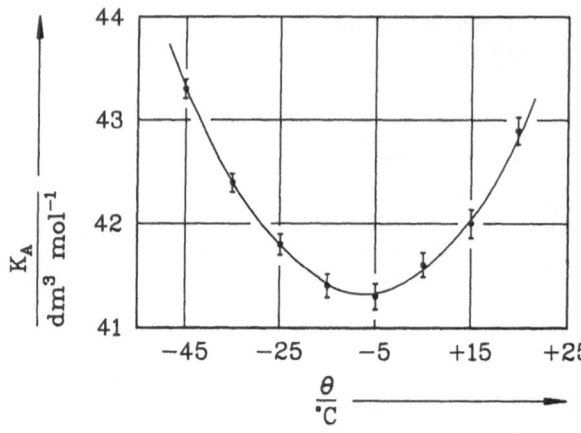

Fig. 15. Temperature dependence of the association constant of Pr_4NI in methanol. Error bars are shown to prove the significance of the minimum. From Ref. [47]

Table 5. Cation-independent temperature at the minimum of ion association of tetraalkylammonium salts[a]

MEOH:	CL^-	$< SCN^-$	$< BR^-$	$< I^-$	$< CLO_4^-$	
	$-70°C$	$-40°C$	$-20°C$	$-5°C$	$+15°C$	
ETOH:			BR^-	$< I^-$	$< CLO_4^-$	$< B(C_6H_5)_4$
			$-20°C$	$0°C$	$+15°C$	$+45°C$
PROH:				I^-	$< CLO_4^-$	$< B(C_6H_5)_4^-$
				$-10°C$	$+15°C$	$+45°C$

[a] From Ref. [47]

H-bonds with the anion, are strongly associated ($526 < K_A/mol\,dm^{-3} < 685$) [48].

Specific interactions are responsible for the minimum of association of potassium alcoholates in the series Li-, Na-, K-, Rb-, Cs-OAlk (Alk = CH_3, C_2H_5, C_3H_7) in methanol, ethanol, and propanol resulting from the superposition of two effects. On the one hand there is a volume effect which increases from Li to Cs, whilst on the other the charge density on the cation (which effectively decreases from Li to Cs) causes the switching of H-bonds from the adjacent amphiprotic solvent to the anion [49].

Important contributions about the relationship between specific interactions and conductance were made by Gilkerson [50–52] and Huyskens and co-workers [53]. The addition of cation- and/or anion-solvating agent L to the electrolyte solution, where ion pairs C^+A^- of the electrolyte exist, produces a sequence of competing equilibria, e.g.

$$C^+ + L \rightleftharpoons CL^+; \quad CL^+ + L \rightleftharpoons CL_2^+; \quad A^- + L \rightleftharpoons AL^-;$$

$$[C^+A^-] + L \rightleftharpoons [C^+LA^-]$$

with equilibrium constants K_1^+, K_2^+, K_1^-, K_1^{IP}. These can be obtained from conductance measurements. This procedure finds practical interest also when the problem of trace water (here $L = H_2O$) in nonaqueous solvents is concerned.

Almost all conductance measurements on mixed solvent systems are good examples for ion and solvent specific interactions since they rarely fulfil the predictions based on solvent permittivity. Specific short-range interactions again can be given as the reason for the failures here, cf. Ref. [49].

To terminate the series of examples evidencing the important role of specific intermolecular forces on dielectric and transport properties we revert to the kinetic depolarization effect relating the dielectric decrement with specific conductivity. Kinetic depolarization results from the interaction of solvent molecules with (solvated) ions moving in the electric field $\vec{E}$. According to the theory, the decrease of the static permittivity, $\Delta\varepsilon = \Delta\varepsilon_{solv} - \varepsilon_{sol}(c)$, is proportional to the specific conductivity, σ.

$$\Delta\varepsilon = k \cdot \sigma; \quad k = p\frac{\varepsilon_{solv} - \varepsilon_{\infty solv}}{\varepsilon_{solv}} \cdot \frac{\tau_{solv}}{\varepsilon_0} \tag{48a, b}$$

The constant of proportionality k is solely determined by the dielectric data of the solvent and the friction factor p characterizing hydrodynamic boundary conditions: $p = 1$ for sticking, $p = 2/3$ for slipping movements of the ions.

Linear functions $\Delta\varepsilon$ vs σ are usually found [4]. In general, the experimental depolarization factor k_{exp} exceeds the value predicted by Eqs. (48). To overcome this difficulty, specific ion–solvent interactions are considered yielding solvation numbers (corrected numbers of irrotationally bound solvent molecules) based on the superposition of kinetic depolarization and dielectric saturation. Generally, the result is reasonable for aprotic but not convincing for protic solvents, see Ref. [4] and quoted literature.

7 References

1. Böttcher CJF (1973) Dielectrics in static fields, 2nd edn. Elsevier, Amsterdam (Theory of electric polarization, vol 1)
2. Kirkwood JG (1939) J Chem Phys 7: 911
3. Winkelmann J, Quitzsch K (1976) Z Phys Chem (Leipzig) 257: 678
4. Barthel J, Buchner R (1986) Pure Appl Chem 58: 1077
5. Böttcher CJF, Bordewijk P (1978) Dielectric in time-dependent fields, 2nd edn. Elsevier, Amsterdam (Theory of Electric Polarization, vol 2)
6. Barthel J, Bachhuber K, Buchner R (1988) In: Moreau M, Turq P (eds) Chemical reactivity in liquids. Plenum, New York, p 55
7. Barthel J, Buchner R, Steger H (1989) Wiss Z Tech Hochsch Chem. "Carl Schorlemmer" Leuna-Merseburg 31: 409
8. Barthel J, Bachhuber K, Buchner, R, Hetzenauer H (1990) Chem Phys Lett 165: 369
9. Cole KS, Cole RH (1941) J Chem Phys 9: 341
10. Kirkwood JG, Fuoss RM (1941) J Chem Phys 9: 329
11. Davidson DW, Cole RH (1950) J Chem Phys 18: 1417
12. Fröhlich H (1958) Theory of dielectrics, 2nd edn. Oxford University Press, Oxford
13. Havriliak S, Negami S (1966) J Polym Sci C14: 99

14. Barthel J, Feuerlein F (1986) Z Phys Chem NF 148: 157
15. Powles JG (1953) J Chem Phys 21: 633
16. Glarum SH (1960) J Chem Phys 33: 1371
17. Dole JC, Kivelson D, Schwartz RN (1981) J Phys Chem 85: 2169
18. Onsager L (1927) Phys Z 28: 277
19. Onsager L (1926) Phys Z 27: 388
20. Falkenhagen H, Kelbg G (1959) in: Bockris J O'M (ed) Modern aspects of electrochemistry, no 2. Butterworth, London, p 1
21. Unpublished data from the author's laboratory
22. Lestrade J-C, Badiali JP, Cachet H (1975) In: Davies M (ed) Dielectric and related molecular properties, vol 2. The Chemical Society, London, p. 106
23. Badiali JP, Cachet H, Lestrade J-C (1981) Pure Appl Chem 53: 1383
24. Kaatze U (1983) Z Phys Chem NF 135: 51
25. Giese K, Kaatze U, Pottel R (1970) J Phys Chem 74: 3718
26. Hubbard JB, Onsager L (1977) J Chem Phys 67: 4850
27. Hubbard JB (1978) J Chem Phys 68: 1649
28. Hubbard JB, Colonomos P, Wolynes PG (1979) J Chem Phys 71: 2652
29. Felderhof BU (1984) Mol Phys 51: 801
30. Cavell EAS, Knight PC, Sheik MA (1972) J Chem Soc Faraday Trans 67: 2225
31. Cavell EAS, Knight PC (1972) J Chem Soc Faraday Trans II 68: 765
32. Scholte TG (1949) Physica 15: 437
33. Barthel J, Buchner R, Wittmann H-J (1984) Z Phys Chem NF 139: 23
34. Barthel J, Wachter R, Gores H-J (1979) In: Conway BE, Bockris JO'M (eds) Modern aspects of electrochemistry, vol 13. Plenum, New York, p 1
35. Barthel J (1985) Pure Appl Chem 57: 355
36. Barthel J, Gores H-J (1985) Pure Appl Chem 57: 1071
37. Gores H-J, Barthel J (1980) J Solution Chem 9: 939
38. Barthel J, Gores H-J, Schmeer G (1979) Ber Bunsenges Phys Chem 81: 911
39. Robinson RA, Stokes RH (1959) Electrolyte solutions, 2nd edn. Butterworths, London
40. Fuoss RM, Accascina F (1959) Electric conductance. Interscience Publishers, New York
41. Falkenhagen H (1971) Theorie der Elektrolyte. Hirzel, Stuttgart
42. Fernandez-Prini R (1973) In: Covington AK, Dickinson T (eds) Physical chemistry of organic solvent systems. Plenum, London, New York, p 525
43. Justice J-C (1983) In: Conway BE, Bockris JO'M, Yeager E (eds) Comprehensive treatise of electrochemistry, vol 5. Plenum, London, p 223
44. Barthel J (1979) Ber Bunsenges Phys Chem 83: 252
45. Barthel J, Gores H-J, Schmeer G, Wachter R (1983) Top Curr Chem 111: 33
46. Barthel J, Neueder R, Feuerlein F, Straßer F, Iberl L (1986) J Solution Chem 12: 449
47. Barthel J, Krell M, Iberl L, Feuerlein F (1986) J Electroanal. Interfacial. Electrochem. 214: 485
48. Witschonke CR, Kraus CA (1947) J Am Chem Soc 69: 2472
49. Barthel J (1976) Ionen in nichtwäßrigen Lösungen. Steinkopff, Darmstadt
50. Ralph EK, Gilkerson WR (1964) J Am Chem Soc 95: 8551
51. Ezell JB, Gilkerson WR (1968) J Phys Chem 72: 144
52. Aitken HW, Gilkerson WR (1973) J Am Chem Soc 95: 8551
53. Macau J, Lamberts L, Huyskens P (1971) Bull Soc Chim Fr (no 7): 2387
54. Barthel J (1985) In: Glaeser PS (ed) The role of data in scientific progress, Elsevier, Amsterdam, p 337

CHAPTER XIX
The Role of Hydrogen Bonds in Biochemistry

Y. Engelborghs

Most biochemical reactions occur in water. H-bridges present among the water molecules are responsible for the hydrophobic effect, i.e. the preferential association of molecules that are not able to form H-bonds with water. This effect is a very important driving force for the folding of biopolymers and the formation of membranes.

H-bonds are also responsible for the specificity of the structures formed. In proteins particular H-bonds are responsible for the secondary structures, e.g. α-helix, β-strands, β-turns. In fact, the whole structure can be considered as an H-bond network. The physical properties of polysaccharides and the double-strand structure of nucleic acids also depend on the presence of many H-bonds.

Due to their short lifetime and their steric restrictions, H-bonds are very important in biological recognition and catalysis, which has to be dynamic and very specific.

Intermolecular Forces
© Springer-Verlag Berlin Heidelberg 1991

1 Introduction

Hydrogen bonds are very abundant in biological systems. Certainly this must be a reflection of the relevance of their intrinsic properties for biology. Although they are relatively weak compared to a covalent C—C bond, a large number of them can contribute considerably to stability. Their intrinsic restrictions on stoichiometry and geometry make them very useful in the process of recognition, which usually has to be very specific, yet dynamic and reversible.

Of course, hydrogen bonds are also very important in the determination of the properties of water, in which most biochemistry is taking place. In ice, the water molecules form a crystalline lattice by making regular H-bridges. In liquid water, many hydrogen bridges are broken, but many are still present. Their presence leads to the very high boiling point of water. Of course, they do not form a static lattice anymore, but a very dynamic one, where every water molecule interacts with others whose identity continuously changes. Molecules that are able to make H-bridges with water dissolve rather easily. However, molecules that cannot make hydrogen bridges with water molecules, stick together. This phenomenon is called the hydrophobic effect. The driving force for this association is of entropic nature, and it is not due to the specific interactions among the hydrophobic molecules themselves. The classical picture is that the water molecules which are in contact with a hydrophobic cavity, reorient themselves and regenerate as many new hydrogen bridges as had been broken. In doing so they increase the local order above that of bulk water. When hydrophobic molecules associate they reduce their contact with water and this is accompanied by an increase of the entropy. (For a detailed discussion of the hydrophobic effect see Ref. [1]. Another view on the nature of the entropy effects is found in Ref. [2]).

The structure of biopolymers is determined by the properties of their building blocks with respect to water. Proteins are built from 20 different amino acids, some of which can readily form H-bridges, others are very hydrophobic. The native structure of proteins is the one where hydrophobic contacts, Van der Waals contacts, steric constraints of H-bonds, electrostatic interactions and interactions with solvent molecules are optimized within the degrees of freedom offered by the rotations around the covalent bonds. Clearly hydrophobic contacts will largely be found in the interior of proteins. Extraction of hydrophobic contacts from water will be a major driving force for the folding of proteins to their native conformation.

The same hydrophobic effect is the driving force for the association of detergent molecules into micelles. Detergent molecules have a hydrophobic tail and a hydrophilic head. The tails stick together in the micelles.

Due to the hydrophobic effect, most hydrophobic amino acid side chains are buried in the interior of protein molecules. Kauzmann previously compared the interior of a protein with an oil droplet [3]. However, H-bridge-forming groups are also present in the interior and they are all involved in bridges. Their steric constraints leads to the formation of very specific structures. This is born out by the results of X-ray crystallography. In this way H-bonds play an important role in the stability and especially the specificity of biological structures. This will be discussed in some more detail for the different classes of biological macromolecules known.

The dynamic character of H-bonds makes them very suitable to participate in biological functions. The functions involve the frequent formation and breakage of bonds. A number of examples will be given where even a very small number of H-bonds plays a very important role in biological dynamics.

2 Structural Role of H-Bridges in Biopolymers

2.1 Polysaccharides

The building blocks of polysaccharides are sugar molecules, a great variety of structures of which are known [4]. Here we compare α- and β-glucose. The only difference is the position of the —OH group on atom C-1 (Fig. 1). The polymerization of these sugars leads to two polymers with largely different physical properties. α-Glucose polymerizes into helical structures with H-bridges within the chain. In this way it forms amylose or starch, which is a white powder in dry state. It is the energy reserve, e.g., in potatoes. However, β-glucose polymerizes into cellulose which forms strong fibers by lateral H-bridges and which is the polymeric building block of wood. This enormous difference between the physical properties of the two polymers is due to the different types of H-bridge networks that can be made in the polymer, thanks to the different orientation of the OH-group in the monomer. (The strength of wood is also determined by the presence of chemical crosslinkers such as lignine.)

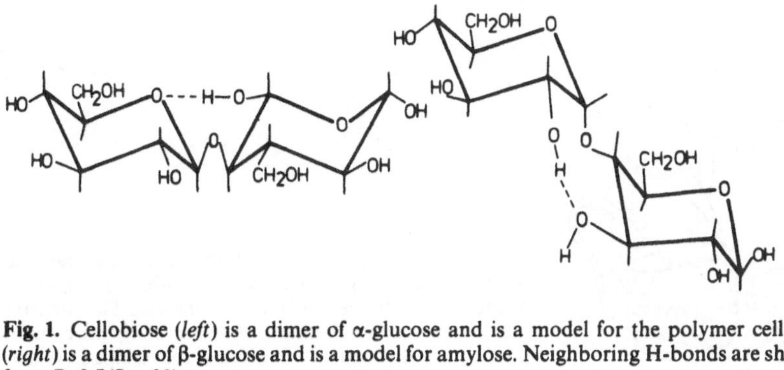

Fig. 1. Cellobiose (*left*) is a dimer of α-glucose and is a model for the polymer cellulose. Maltose (*right*) is a dimer of β-glucose and is a model for amylose. Neighboring H-bonds are shown (Redrawn from Ref. [4] p 38)

2.2 Proteins

Proteins are polypeptides built from 20 naturally occurring amino acids. Their sequence is determined by the gene from which they are derived. Nature has evolved a complex machinery for protein synthesis that shows a remarkable accuracy. However, no machinery was developed to fold the nascent polypeptide chain into an active protein. The folding of the chain into the so-called native structure is a spontaneous process. In proteins, stretches are present where subsequent amino acids are folded according to a recurring pattern: the so-called secondary structures (primary = sequence; tertiary = 3-D structure; quaternary = subunit association). Drawings of these structures can be found in all textbooks of biochemistry. The include the well-known α-helix structure, as described by Pauling and Corey in 1931 [5]. The latter structure is characterized by the formation of H-bridges between $>$C$=$O of amino acid i and HN$<$ of amino acid i + 3. The structure is a right-handed helix. As such it allows for having the bulky side chains on the outer surface. It is clear that the presence of this secondary structure proves the importance of H-bridges within a protein. In the unfolded chain all the H-bridge-forming groups interact with water. Maximal stability of the folded state clearly implies the maximal formation of new H-bridges in the interior of the protein.

Other regular structures are found such as the parallel and antiparallel β-sheets, again characterized by the formation of H-bridges between the amide hydrogen and carbonyl oxygen of the backbone, this time, however, of distant groups. Several strands together form sheets, which are usually curved, and a cylindrical structure can be found in so-called β-barrels, e.g., in the enzyme triose-phosphate isomerase (Fig. 2) [4]. Here, the different strands of the β-sheet are connected at the outside by α-helices.

The β-turns constitute a final element of secondary structure. These are the sharp bends found at the surface which allow the backbone to fold back in order to make a compact structure. A comparison of these turns shows that they are stabilized by one or a few H-bonds. H-bonds are not restricted to secondary structures alone. In fact, the data of the protein structures analyzed

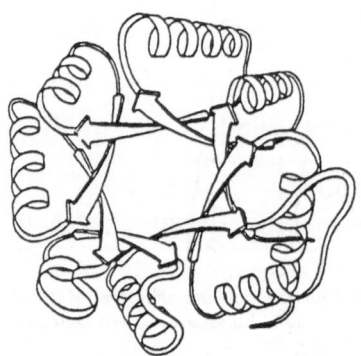

Fig. 2. Spacial distribution of the α-helices and the β-strands (*arrows*) folded into a β-barrel structure of the enzyme triose phosphate isomerase. The β-barrel has a structure reminiscent of the cooling tower of a power station. A top view is given (From Ringe D, (1989) Nature 339:658)

by X-ray crystallography prove that all the possibilities to make H-bridges in the interior of proteins are exploited. A protein can be considered being a three-dimensional network of H-bonds. A detailed quantitative description of the different types of H-bonds found in proteins is found in Ref. [6].

2.3 Nucleic Acids

Our genetic information is stored in genes. From the chemical point of view, these are nucleic acids, i.e., polymers of nucleotide bases. They form helical structures stabilized by the stacking of the bases on top of each other [8]. They

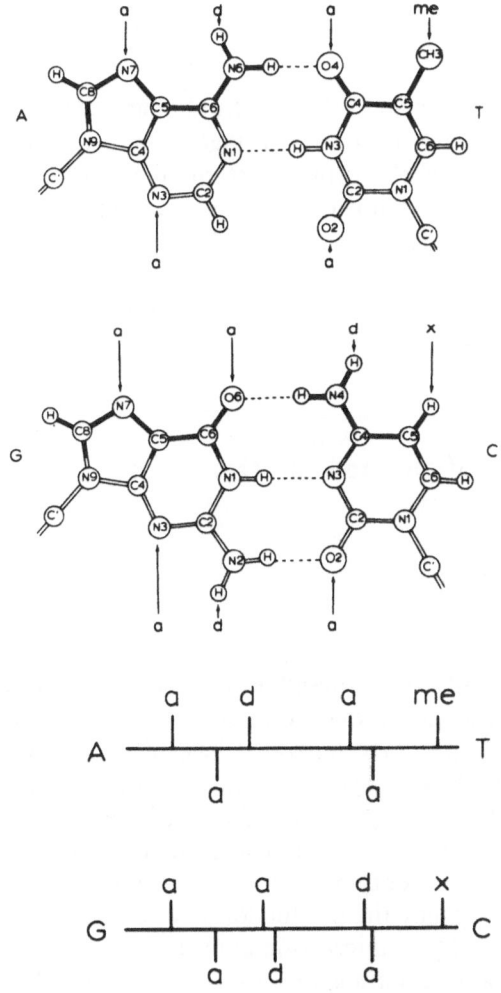

Fig. 3. (a) Pattern of H-bridges between the base pairs AT and GC in DNA. Note that the two pairs have the same geometry of their connection to the sugars and so to the backbone. Remaining possibilities for the formation of H-bridges between the paired bases and proteins are indicated by the arrows with (a) for an H-bond acceptor and (d) for a donor. These sites are exposed in the grooves of the double helix. (b) Stick figure representation of the two base pairs, clearly indicating the different patterns obtained. Groups protruding into the major groove are drawn above the horizontal line (Redrawn from Ref. [9] p 306)

associate to make double strands. The association of the two strands is based on the complementarity of G and C forming three H-bridges and A and T forming only two H-bridges. As Fig. 3a shows, the attachment of the bases to the phosphodiester backbone is independent of the nature of the base. This complementarity immediately explains the early observation of Chargaff that the occurrence in DNA of A(G) equals that of T(C). This structure, with the two strands storing the same information, is the ideal solution to a demand for high fidelity, while extensive copying is going on. Very good complementarity is necessary for accuracy and has to be combined with a relative easiness for separating the strands. These properties are reminiscent of the properties of H-bridges themselves, and it should not surprise us that this type of chemical bonds is quite abundant in nature.

2.4 Membranes

The association of phospholipids is partly due to their hydrophobic character, and therefore due to the tendency of water to form H-bridges. The geometry of the hydrophobic molecules plays an important role in determining the shape of the association product formed. Detergent molecules, for instance, prefer the formation of micelles because they are rather conical. Phospholipids, however, can make large vesicles of planar or only slightly curved bilayers. The hydrophobic interior layers form a strong barrier against the transport of hydrophilic molecules and protects the chemistry inside the cell.

3 The Role of H-Bridges in Specificity and the Dynamics of Molecular Interactions

3.1 Protein–Nucleic Acid Interactions

Nucleic acids constitute the substrate of a large manifold of molecular interactions. When cells multiply, the nucleic acids that store genetic information have to be replicated. Therefore, strands need to be separated. The differentiation of various cells in an organism implies the selective expression of certain genes. This means that certain genes become active, are transcribed into messenger-RNA, which is then translated into proteins. Bacteria can grow on different media. In general, they produce only the enzymes necessary for the transformation of the available materials, whereby the production of other enzymes is switched off. All these phenomena imply the interaction of proteins (enzymes) with nucleic acids. A number of these interactions are not dependent on the sequence of the nucleic acids, and arise largely out of electrostatic interactions with the chain of negative charges at the phosphodiester backbone. These proteins are known to glide over the surface of the nucleic acids [9]. However, the specific expression of certain genes—while others being suppressed—implies the interaction of proteins with specific sites on nucleic acids. The same is true for restriction enzymes. The latter are enzymes which split the chain at very specific sites. This sequence-specific recognition is based on interactions with the bases in the minor and major groove on the surface of the nucleic acids. The bases give donors and acceptors the possibility to interact. Each base pair forms a specific pattern of donors and acceptors (represented in the so-called stick figures, see Fig. 3b). With a small number of base pairs, a very specific geometric pattern of donors and acceptors can be formed. Proteins find their target sequence on a nucleic acid remarkably fast, despite the low occurrence of these sequences. The mechanism generally accepted for the explanation of these high speeds, is that DNA-binding proteins, when they encounter a nucleic acid by diffusion, stick to the phosphodiester backbone by the electrostatic interactions and glide over their surface. This random motion allows for the rapid scanning of the whole sequence. During this rapid scanning, the target sequence has to be recognized by their H-bridge interactions. Here again, the highly dynamic character of H-bonds is of crucial importance. A nice compilation of the combinations of H-bond possibilities between amino acid side chains and nucleic acid bases was published by Hélène and Lancelot [10].

3.2 Specificity During Catalysis

In this chapter we want to find out whether a single H-bridge can make an important difference. We therefore take a closer look at enzyme catalysis. All biological reactions are catalyzed by proteins called enzymes. Their catalytic power is much higher than that of their industrial counterparts, even though they work in very moderate conditions of temperature and pressure. Next to their amazing efficiency, enzymes also show a remarkable specificity, i.e., they can make a distinction between a number of very related chemical substances. This specificity implies that the enzyme is able to recognize the different chemical molecules, in other words is able to interact with them and to maximally exploit the differences between their structures.

Specificity is pushed to an extreme in these processes where a very high accuracy is demanded, e.g., in all the steps of protein synthesis. Accuracy of this complex process depends on the specificity of each of the enzymes involved. Here we describe a particular step in more detail, i.e., the coupling of the amino acid valine to its corresponding val-tRNA, by the enzyme val-tRNA synthetase [11]. This enzyme has to discriminate the very related amino acid threonine. A study of the error frequency shows that the latter is extremely low, i.e., of the order of 1 threonine added for 10^6 valine molecules. This extreme specificity must be based on the exploitation of all the differences between the two amino acids. However, if we compare threonine and valine, the only difference is the presence of an —OH group in threonine and a —CH_3 group in valine. The threonine-specific site certainly exploits the possibility to form two H-bridges with the threonine-OH, whereas the valine-specific site must be hydrophobic on the same spot. How did nature go about to achieve such a high accuracy? The answer lies in the presence of an editing function. The tRNA synthetases have a second site specific for the most frequent mistake, where hydrolysis of the wrong adduct occurs (Fig. 4):

$$Val + ATP + E_{Val} \longrightarrow E_{Val} \cdot Val\text{-}AMP + PPi$$

$$E_{Val} \cdot Val\text{-}AMP + tRNA^{Val} \longrightarrow E_{Val} + Val\text{-}tRNA^{Val}$$

but if the wrong adduct would have been formed, hydrolysis occurs at the

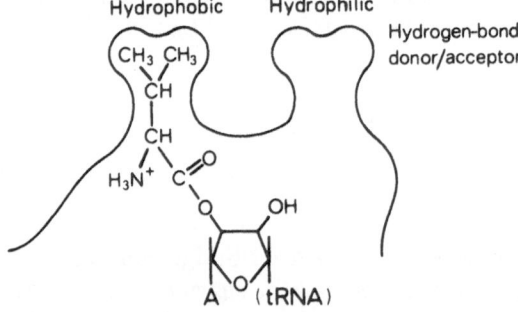

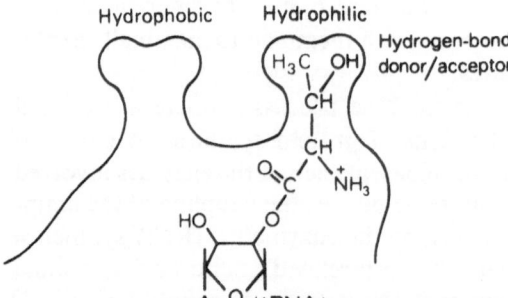

Fig. 4. Schematic representation of the active site of Val-tRNA^{Val}-synthetase, with a synthesis site where Val is preferred and a hydrolysis site where the most probable mistake (threonine) can be hydrolyzed. Between synthesis and hydrolysis a translocation takes place, probably by $2' \longrightarrow 3'$ hydroxyl acyl transfer (From Ref. [11] p 356)

second site:

$$E_{Val} \cdot Thr\text{-}AMP + tRNA^{Val} \longrightarrow E_{Val} + Thr + AMP + tRNA^{Val}$$

For this editing process the specificity is again determined by the difference between the —OH group of threonine and the —CH_3 group of valine. Obviously, the overall specificity is the product of the two. Whenever high accuracy is necessary, nature has developed analogous types of editing mechanisms.

3.3 H-Bridges and the Rate of Catalysis

Enzymes essentially speed up rates of reactions by stabilizing the transition state of that particular reaction. Again the question arises whether a single H-bond can be important. The modern technique of protein engineering can help us to find the answer. As an example, we take again the enzyme that links the correct amino acid to its corresponding t-RNA, in this case tyrosine-tRNA

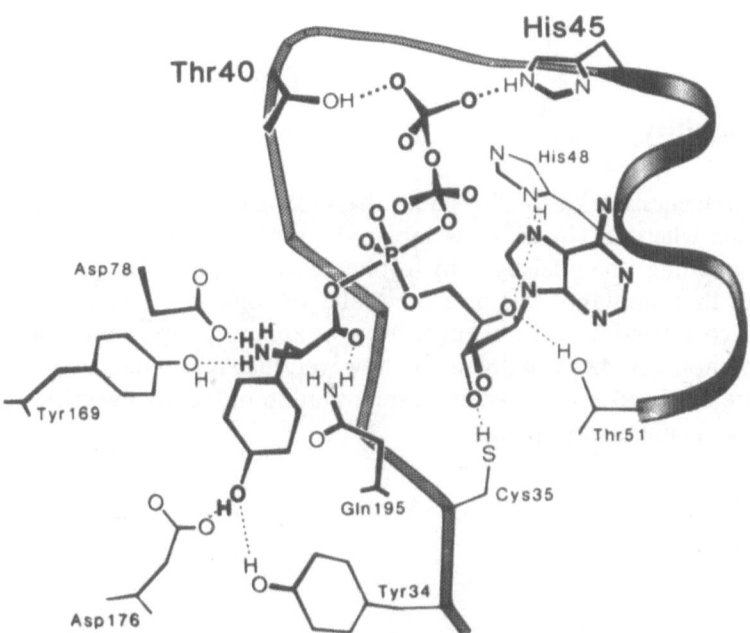

Fig. 5. Representation of the active site of the enzyme Tyr-tRNA^{Tyr}-synthetase with the substrates in the hypothetical transition state conformation (Phosphate is pentavalent). Site-directed mutagenesis allowed for the identification of those residues that only influence the catalytic rate. These are supposed to be only involved in the binding to the transition state conformation (From Ref. [12] p 322)

synthetase. The first step in the reaction is an activation of the amino acid by its reaction with ATP:

$$Tyr + ATP + E \longrightarrow E \cdot Tyr \cdot ATP$$

$$E \cdot Tyr \cdot ATP \longrightarrow E \cdot Tyr\text{-}AMP + PPi$$

In the group of Fersht [12, 13], site-directed mutagenesis was applied to replace an amino acid Thr-40 by an Ala, which makes it impossible to form H-bridges with the threonine-OH group. The enzymatic parameters of this mutant were studied in great detail. These studies clearly show that the binding of ATP or of Tyr are not influenced by this mutation. However, the rate of the synthesis is greatly reduced, as well as the binding of the reaction product Tyr-AMP. This proves that the —OH group of threonine stabilizes the transition state and the product of the reaction, by way of H-bond formation. This stabilization of the transition state causes a very important increase in the reaction rate. Similar results are obtained when residues like His-45 are replaced by Ala. Some residues, however, play a role in the binding of the substrate in the "ground state" as well as in the transition state. On the basis of such an analysis of mutants, the structure of the transition state could be deduced and fitted into the structure of the enzyme as obtained from X-ray analysis of a product complex (Fig. 5).

4 Thermal Stability

Site-directed mutagenesis has also allowed for the making of a series of mutants of, e.g., lysozyme where a Thr-157 was replaced by Ala, so as to replace a group with H-bonding potential by a hydrophobic group [14]. The thermal denaturation of these mutants was measured, and the differences in thermal stability can be correlated with the changes brought about by these mutations. X-ray crystallographic studies usually reveal, however, that very often compensating effects are observed, so that an exact quantitation of the involvement of a single H-bridge is not always possible.

5 Conclusion

Hydrogen bonds have a number of properties that make them exceptionally suited to contribute to biological interactions, where stability has to be combined with specificity and dynamic character. By additivity, they play an important role in the stability of all biopolymers, i.e., proteins, nucleic acids, and polysaccharides. The necessary complementarity of the donor and the acceptor groups allows for the construction of extremely specific recognition patterns that are used in the sequence-specific interaction of certain proteins with nucleic acids.

Even a single H-bridge can play an important role where specificity is pushed to its limits, for instance, in enzyme catalysis and superspecificity.

6 References

1. Tanford C (1980) The hydrophobic effect: Formation of micelles and biological membranes, Wiley, New York
2. Huyskens P, Siegel G (1991) In: Huyskens P, Huyskens-Zeegers Th, Luck W (eds) Intermolecular forces: An introduction to modern methods and results, Springer-Verlag, Berlin Heidelberg New York Tokyo
3. Kauzmann W (1959) Adv Protein Chem 14: 1
4. Rees DA (1977) Polysaccharide Shapes, Chapman and Hall, London (Outlines Studies in Biology)
5. Pauling L, Corey RB (1951) Proc Natl Acad Sci, USA 37: 235
6. Banner DW, Bloomer HC, Petsko GA, Philips DC, Pogson CJ, Wilson AI (1975) Nature 255: 609
7. Baker EN, Hubbard RE (1984) Progr Biophys Mol Biol 44: 97
8. Saenger W (1983) Principles of nucleic acid structure Springer-Verlag, New York Berlin Heidelberg Tokyo
9. Von Hippel PH, Bear DG (1983) In: Sund H, Veeger C (eds) Mobility and Recognition in Cell Biology. De Gruyter, Berlin, p 291
10. Hélène C, Lancelot G (1982) Progr Biophys Molec Biol 39: 1
11. Fersht AR (1985) Enzyme structure and mechanism, 2nd edn, Freeman & Co, New York, p 354
12. Fersht AR, Leatherbarrow RJ, Wells TNC (1986) Trends Biol Sci 11: 321
13. Fersht AR, Shi JP, Knill-Jones J, Lowe DM, Wilkinson AJ, Blow DM, Brick P, Carter P, Waye MMY, Winter G (1985) Nature 314: 235
14. Alber T, Dao-pin S, Wilson K, Wozniak JA, Cook SP, Matthews BW (1987) Nature 330: 41

CHAPTER XX
Hydrogen Bonds in Crystals

G.S.D. King

Hydrogen bonds are one of the weaker types of force holding atoms together in the crystalline state. The possibilities and limitations of X-ray and neutron diffraction methods for their characterisation are introduced.

A short section on the role of hydrogen bonds in inorganic crystals is followed by a more thorough discussion of their function in organic and biological molecular crystals. The most common types of bond ($-O-H \cdots O \langle$, $-O-H \cdots N \langle$ — and $\rangle N-H \cdots O \langle$) are mentioned and particular attention is paid to the last of these because of their importance in proteins and nucleic acids.

Examples are given of statistical studies of molecular geometry, using data from the Cambridge Structural Database, to study the relation between geometry and electronegativity. The role of hydrogen bonds in determining the structure of nucleic acids and in the different functions of ribonucleic acids and deoxyribonucleic acids is discussed. Finally, the existence of weak $-\rangle C-H \cdots O \langle$ bonds is demonstrated.

1 Introduction

Crystals are held in the ordered state by various interatomic forces which may be strong—covalent or ionic bonds, weak—hydrogen bonds—or Van der Waals forces. While some crystals are held together exclusively by one sort of

Intermolecular Forces
© Springer-Verlag Berlin Heidelberg 1991

Table 1. Melting point and weakest bond

Substance	Weakest bond	Melting point (°C)
Hydrogen, H_2	Van der Waals	−259
Benzene, C_6H_6	Van der Waals	3
Phenol, C_6H_5OH	Van der Waals + H-bond	100
Glucose, $C_6H_{12}O_6$	H-bond	110
Sodium chloride, NaCl	Ionic	801
Diamond, C	Covalent	>3500

interactive force, in most cases several types of bonding play a role and the weakest of these determines the physical cohesion of the crystal, which becomes evident in such properties as hardness and melting point.

In Table 1 we give the melting point and type of the weakest bond for a number of substances.

Van der Waals interaction is due to initial polarisation of the molecular electron clouds leading to weak electrostatic forces. Hydrogen bond formation involves the sharing of the proton of a hydrogen atom attached to a donor atom with an acceptor atom.

Inorganic crystals, in particular minerals, which contain hydrogen bonds, are more reactive and, in nature, more readily undergo weathering than crystals with only ionic and covalent bonds. In organic molecular crystals, hydrogen bonds will be formed whenever there are suitable —OH or $>$NH groups available.

We shall use the word "acceptor" to describe the atom that accepts a hydrogen bond from a hydrogen atom covalently bonded to a third atom—the "donor". It should be noted that the acceptor must have a free electron pair and can also be an electron donor.

We shall first discuss the methods used to observe hydrogen bonds in crystals. A few examples will be given of hydrogen bonds in inorganic crystals and we shall conclude with a comprehensive study of the different types of hydrogen bonds found in organic and biological crystals.

2 The Observation of Hydrogen Bonds in Crystals

The positions of atoms in the unit cell of a crystal can be determined by X-ray diffraction methods with an accuracy which increases with the number of electrons in the atom. This means that hydrogen atom positions can only be determined with a relatively low accuracy. However, diffraction measurements at low temperature (90–120 K), where the thermal vibration of the atoms is reduced, give a good accuracy. In contrast, the use of neutron diffraction, where the scattering power of an atom is not dependent on atomic numbers, gives accurate values for hydrogen atom positions and even better values for deuterium.

It is important to note that X-rays are scattered by electrons but not by atomic nuclei. That electron cloud surrounding a hydrogen nucleus will be displaced towards the atom to which it is bonded. Neutrons are scattered by atomic nuclei and hardly at all by electrons so that a neutron scattering density distribution gives the positions of the nuclei. Thus the distance between the centre of mass of a hydrogen atom electron cloud and the centre of mass of the adjoining atom (the X-ray bond length) will be less than that between the nuclei (the neutron bond length).

Modern X-ray crystal structure determinations are sufficiently accurate for the determination of hydrogen bond topology (except for compounds containing heavy atoms where the contribution of the hydrogen atoms to the total scattering is negligible). Accurate structure determinations will usually give a good measure of hydrogen bond distances. A comparative study of $>$N—H...O bonds in crystals that have been investigated both by X-ray and by neutron diffraction is discussed below.

3 Hydrogen Bonds in Inorganic Crystals

In inorganic crystals, hydrogen bonds involve either hydroxyl groups or free water molecules and their function is closely linked to the double role of water:

1. increasing the effective ionic radius of a small cation by forming a coordination polyhedron of roughly the same size as that of the oxyanion in the structure and thereby allowing efficient packing. Thus, the cation in copper sulphate is $Cu(H_2O)_4^{2+}$ which is comparable in size to the SO_4^{2-} anion.
2. as a space filler between cation and anion. The presence of the hydrogen atoms gives the possibility of a crystal's being be held together by other than electro-static forces.

A simple examples of the role of hydrogen bonds in inorganic crystals is to be seen in the structure of gypsum $CaSO_4 \cdot 2H_2O$, in which the hydrogen atom positions have been determined by neutron diffraction studies [1]. In this structure there are double layers of calcium and sulphate ions in which each calcium atom is bonded to four sulphate ions in the same layer and to two sulphates in the second layer.

Each calcium atom is also linked to two water oxygens so that its coordination is raised to eight. Each water molecule forms two hydrogen bonds to sulphates in adjacent double layers. These hydrogen bonds are the only links between the double layers. They are strong enough to hold the structure together but weak enough to allow easy cleavage between the double layers.

The physical properties of the clay minerals, many of which have the ability to expand by taking up large amounts of water (or hydroxylated organic compounds) are due to hydrogen bond formation. A single layer consists of an infinite hexagonal sheet of Si_4O_{10} units linked by its free oxygen atoms to an aluminium hydroxide sheet. The hydroxyls of this sheet can form hydrogen bonds directly to the doubly bonded oxygens of the silicate layer in the unhydrated clays with interlayer spacing of 7.3 Å or via a series of water molecules to the silicate layer in the hydrated clays where the interlayer spacing can be as much as 10 Å.

A general discussion of the structures of clay minerals has been published [2].

4 Hydrogen Bonds in Organic Crystals

The principal types of hydrogen bond are:

$$-O-H...O\langle$$
$$-O-H...N\langle\ -$$
$$\rangle N-H...O\langle$$

and

$$\rangle N-H...N\langle\ -$$

although a hydrogen atom attached to sulphur may also act as an electron acceptor. The molecular packing in a crystal is determined by the possibility of hydrogen bond formation; all hydroxyl, amine and imino hydrogens will form hydrogen bonds unless prevented by steric hindrance. The geometry of the bond depends on the location of the free electron pair of the acceptor atom. If this is an sp^3 atom, the hydrogen atom will lie out of the plane of the acceptor and its covalent bonds. If the acceptor is an sp^2 atom, such as a carbonyl oxygen, the hydrogen atom will be in or near the plane of the carbonyl group.

4.1 —O—H...O⟨ Bonds

Some of the most frequently encountered —OH...O⟨ bonds are those between carboxyl groups which often form centrosymmetrical dimers:

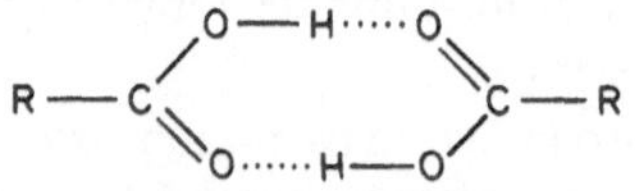

but can also form infinite chains as in the case of acid salts such as:

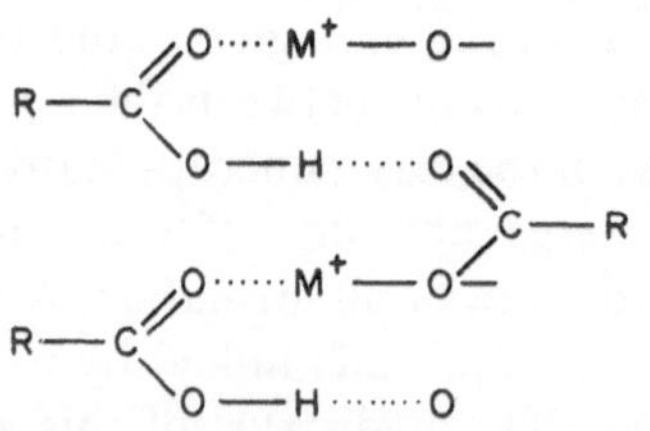

In this case the hydrogen bond appears to be symmetrical although this symmetry may be due to disorder or to a time-average of the two asymmetrical structures.

The second important class of —O—H...O< bonds is from a hydroxyl hydrogen to a hydroxyl or an ether oxygen as in sugars. In all these bonds the O...O distance is 2.75 ± 0.25 Å except in some intramolecular bonds where steric factors force the oxygen atoms to approach each other more closely.

4.2 —O—H...N< — Bonds

These bonds are found when it is sterically difficult to form —O—H...O< bonds, or when packing considerations make their formation more favourable. Thus, in nicotinic acid [3] the molecules are linked in chains by hydrogen bonds from the acid hydroxyl of one molecule to the pyridine nitrogen of the next.

4.3 >N—H...O< Bonds

>N—H...O< bonds have been more thoroughly studied than any other type because of their function in the living cell (proteins and nucleic acids). A study [4] of 57 hydrogen bonds in 26 compounds for which both X-ray and neutron diffraction measurements are available has shown that the neutron N—H distance is very close to 1.03 Å and that, by adjusting the hydrogen atom coordinates from the X-ray structure determination so as to give this N—H distance, it is possible to obtain corrected values for the H...O distance and the N—H...O angle which agree very well with the neutron values in all X-ray results.

Statistical studies of the conformation of 1500 >N—H...O=C< hydrogen bonds from nearly 900 crystal structure determinations [5, 6] show that there is a strong tendency for the hydrogen atom to lie near a free electron pair of the acceptor oxygen atom. In half the cases studied the H...O direction is within 30° of the >C=O plane and its projection in this plane makes an angle of $125 \pm 15°$ with the C=O direction. Deviations from these values are greater for intramolecular bonds where steric hindrance is involved. For intermolecular bonds, modal values of 1.85 and 2.85 Å were found for H...O and N...O, respectively, with the N—H...O angle greater than 155°.

4.3.1 >N—H...O=C< Bonds in Amino Acids and Proteins

In the early 1950s, many crystal structure determinations of amino acids were carried out. The information on molecular dimensions and hydrogen bonding led to the suggestion [7] that the α-helix and the β-pleated sheet were probable stable conformations of polypeptide chains in protein molecules. Both of these structures are stabilished by hydrogen bonds. In the α-helix the chain is twisted into a spiral with 3.7 amino acid residues per turn. Each peptide in the chain is linked via its imino group to the carbonyl of the third peptide in sequence from it and via its carbonyl to the imino groups of the third peptide preceding it. The β-pleated sheet consists of flat polypeptide ribbons each linked by peptide–peptide hydrogen bonds to a neighbouring ribbon which may be either parallel or antiparallel with the first ribbon. In this way a more-or-less flat sheet is built up. The first definite confirmation of the existence of the α-helix came with the determination of the crystal structure of sperm whale myoglobin [8].

The α-helix and β-pleated sheet are the principal secondary structures of proteins (the primary structure being the amino acid sequence) although other well-defined structures are found as are bends which usually arise from the incorporation of proline or hydroxyproline in the chain. There are also regions of apparently irregular conformation which often form the active site of the protein molecule at which reactants are brought together. The arrangement of secondary regions gives the tertiary structure which is designed according to the function of the molecules in the living cell.

4.3.2 $\rangle$N—H...O$\langle$ Bonds in Nucleosides, Nucleotides and Nucleic acids

Much attention has been given to the geometry of individual nucleosides. In particular, the effect of protonation has been studied by Taylor and Kennard [9]. They found that the principal differences between protonated and un-protonated structures are to be found in the bond angles rather than in bond lengths and derived functions which could be used to discriminate between the two species. Thus, for cytosines they postulate the discriminator

$$W = 12.38*C2\text{-}N3\text{-}C4 + 10.44*N1\text{-}C2\text{-}O2 - 2784.7 \text{ (angles in degrees)}.$$

W is positive for cytosine bearing a hydrogen atom on N-3 (**1** and **2**) and negative for those without (resonance forms **3** and **4**).

We have calculated W for all cytosines in the September 1989 Cambridge Structural Database [10] with a quoted mean standard deviation for a carbon–

1   **2**

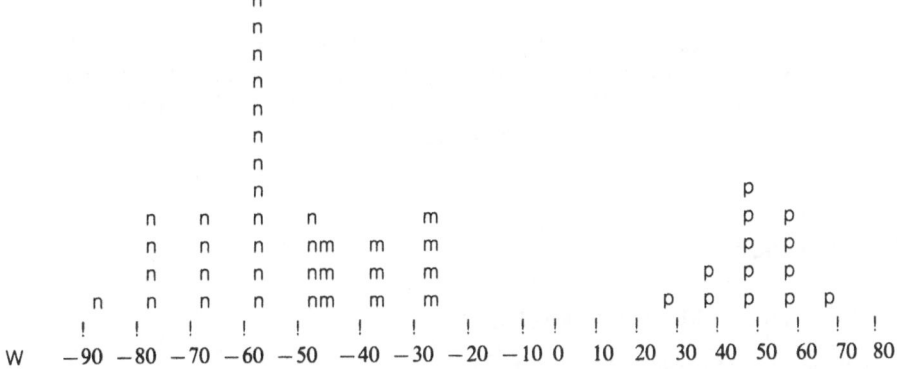

carbon bond of 0.01 Å or less and a crystallographic R-factor not exceeding 0.05. In 40 crystal structures we found 49 individual cytosines of which 14 are protonated and bear a hydrogen atom at N-3 (p in the histogram below), 9 have N-3 linked to a transition-metal atom (m) and the remaining 26 have N-3 with a free electron pair (n). This confirms the earlier results of Taylor and Kennard. We also find that compounds with hydrogen bonds between N-3 and nitrogen have lower W values than do those with hydrogen bonds between N-3 and oxygen. Thus, the lowest W values are found for compound with $N_3 \ldots H-N$ bonds and the highest for those with $N_3-H \ldots O$ bonds. The unprotonated compounds have the formal structure **5**. There are no examples of structures **2** and **4** with a formal positive charge at N-4 although these resonance forms may make a small contribution to the effective structure. A charge density study by combined X-ray and neutron diffraction at low temperature would be necessary for a precise assessment of the contributions of the different forms to the molecular structure.

Distribution of W for neutral (n), protonated (p) and N-3-metal (m) cytosines:

```
 n
 n
 n
 n
 n
 n
 n
 n                                                          p
       n   n   n    n        m                          p   p
       n   n   n    nm   m   m                          p   p
       n   n   n    nm   m   m                      p   p   p
   n   n   n   n    nm   m   m                  p   p   p   p   p
   !   !   !   !    !    !   !    !    !   !    !   !   !   !   !   !   !
W  -90 -80 -70 -60 -50 -40 -30 -20 -10 0   10  20  30  40  50  60  70  80
```

The study of the crystal structures of purines and pyrimidines, in particular of the hydrogen bonds that can be formed between them, led to the postulation [11] of the double helix structure of deoxyribonucleic acids (DNA), a structure that could explain the X-ray diffraction pattern obtained by R.E. Franklin and R.G. Gosling [12]. The two polynucleotide chains are linked by hydrogen bonds between a purine of one chain and a pyrimidine of the other. The geometry of these molecules allows adenine to pair only with cytosine, and guanine only with thymine. The base pairs form the steps of a spiral staircase while the sugar phosphate chains form the handrail. There are no free hydrogen donor groups on the outside of the helix which therefore has no external specificity which would allow it to react with selected molecules. DNA therefore has a passive role in the living cell, that of an information carrier. Ribonucleic acids (RNA), on the other hand, have a free hydroxyl group at the 2′-position of the ribose. This enables them to form hydrogen bonds with other molecules and to play an active role in decoding the information in DNA (messenger-RNA) and in controlling the amino acid sequence during protein synthesis (transfer-RNA).

4.4 — $>$C—H...O Bonds

R. Taylor and O. Kennard [13] have made a statistical study of 113 compounds of which the crystal strucure had been determined by neutron diffraction so that the hydrogen atom positions were known with great accuracy. They showed that in many cases the H...Y distance (rHY) is significantly less than the sum of the Van der Waals radii of hydrogen (VH) and of Y (VY). The difference $d = VH + VY - rHY$ had an average value of 0·03 Å for the 661 contacts in these structures. This confirmed that the values of VH and VY were reasonable. Nevertheless, there were 46 contacts with $d > 0·3$ Å of which 42 were C—H...O contacts. Most of these short distances occur when the carbon atom is adjacent to a nuetral or positively charged nitrogen atom. 21 of the 64 C—H next to a positively charged nitrogen form contacts with an oxygen atom where the C—H direction lies within 30° of a free electron pair of an oxygen atom at least 0·3 Å closer than the sum of the Van der Waals radii.

This work suggests that H...O attractions ca lead to the formation of C—H...O bonds when the carbon atom is next to a neutral or positively charged nitrogen atom and when there are no hydroxyl, amino or imino groups which would preferentially act as hydrogen donors.

5 References

1. Atoji M, Rundle RE (1958) J Chem Phys 29: 1306
2. Brindley GW, Brown G (Eds) (1980) Crystal structures of clay minerals and their X-ray identification Mineralogical Society, London
3. Wright WB, King GSD (1953) Acta Cryst 6: 305
4. Taylor R, Kennard O (1983) Acta Cryst B39: 133
5. Taylor R, Kennard O, Versichel W (1983) JACS 105: 5761
6. Taylor R, Kennard O, Versichel W (1984) Acta Cryst B40: 280
7. Pauling L, Corey RB, Branson HR (1951) Proc Nat Acad Sci WAsh 27: 205
8. Kendrew JC, Bodo G, Dintzis HM, Parrish RG, Wyckoff H, Phillips DC (1958) Nature 181: 662
9. Taylor R, Kennard O (1982) J Mol Struct 78: 1
10. Allen FH, Kennard O, Taylor R (1983) Acc Chem Res 16: 146
11. Watson JD, Crick FHC (1953) Nature 171: 737
12. Franklin RE, Gosling RG (1953) Nature 171: 740
13. Taylor R, Kennard O (1982) JACS 104: 5063

CHAPTER XXI
Role of Intermolecular Interactions in Chromatographic Separations

J. Ceulemans

A survey is presented of column chromatography, with special emphasis on intermolecular interactions that are at the origin of selectivity in chromatographic separations. In gas chromatography, differences in electronic dispersion, dipole-induced dipole and dipole-dipole interaction as well as differences in hydrogen bonding, E.D.A.-complexation and coordination-compound formation of the stationary phase (liquid or solid) with solutes to be separated may all contribute to selectivity. In addition, differences in the size and shape of solute molecules may also contribute to selectivity in special cases. The characterization of stationary liquids according to their potential for interaction is discussed. Numerous examples are presented to illustrate the role of intermolecular interactions in analytical separations by gas chromatography. Differences in both non-specific and specific intermolecular interactions between solutes to be separated and the stationary as well as the mobile phase may contribute to selectivity in liquid chromatography. In addition, separations may be achieved on the basis of differences in ionic equilibria as well as on the basis of differences in the size and shape of solute molecules, depending on the chromatographic technique employed. A survey of liquid chromatography techniques is given. A number of examples are presented to illustrate the role of intermolecular interactions in liquid chromatography separations.

Intermolecular Forces
© Springer-Verlag Berlin Heidelberg 1991

1 Introduction

Chromatography (literally "color-writing" from the Greek) derives its name from the experiments of the Russian botanist, MS Tswett, who performed interesting scientific research using column chromatography at the beginning of this century. In a paper published in 1906 he describes the separation of pigments of green leaves on a calcium carbonate adsorbent column. After rinsing with petroleum ether, the components of the dye mixture formed zones of different colors on the column. This observation gave the method its name. In the course of time the meaning of the word "chromatography" has considerably expanded, and today the method has but little to do with "separation according to color". Chromatography is a collective name for all separation processes in which the separation of the components of a mixture is effected by differences in their partition between a stationary phase and a mobile phase that flows over it. Chromatographic separations may be performed in columns (column chromatography), on paper (paper chromatography), and on a thin layer of material coated on a glass, plastic, or aluminum plate (thin-layer chromatography). For the sake of brevity, only column chromatography is discussed in this article.

Depending on the physical state of the mobile phase, a division is made between gas and liquid chromatography. Within each of these divisions, subdivisions based on the stationary phase are designated. Thus, gas–solid chromatography (GSC) involves a column packed with an adsorbent or a capillary column in which the inner surface is coated with a thin adsorbent layer, whereas gas–liquid chromatography (GLC) involves a column in which the (effective) stationary phase is a liquid. Similarly, liquid chromatography in columns can be liquid–solid chromatography (LSC) or liquid–liquid chromatography (LLC). The boundary between gas and liquid chromatography has become somewhat diffuse in recent years by the introduction of supercritical fluid chromatography. Similarly, the introduction of bonded phase chromatography has made the borderline between GSC and GLC and between LSC and LLC less distinctly defined.

2 General Principle

Chromatographic separations depend on differences in the partitioning of the components of a mixture between a stationary phase and a mobile phase that flows over it. The capacity ratio k' is defined as:

$$k' = n_s/n_m \tag{1}$$

where n_s and n_m are the number of moles of individual solutes in the stationary phase and mobile phase, respectively. If a solute did spend no time at all in the stationary phase, i.e., $k' = 0$, its retention time t_R would merely be the time taken for the mobile phase to traverse the column. Let this time be t_d, the dead time. In the case of $k' > 0$, for every unit of time that an average molecule of an individual solute spends in the mobile phase, it spends k' units of time in the stationary phase. The average molecule must necessarily spend a time equal to t_d in the mobile phase in order to traverse the column. The time spent in the stationary phase therefore equals:

$$t'_R = t_d \cdot k' \tag{2}$$

This is the net or dead-time adjusted retention time. The total time t_R required for an average molecule of an individual solute to traverse the column is given by:

$$t_R = t_d + t_d \cdot k' = t_d(1 + k') \tag{3}$$

The separation of the components of a mixture is possible when these components are characterized by different k' values in the analysis. Such differences may result from differences in interaction with the stationary phase in gas chromatography and from differences in interaction with both the mobile phase and stationary phase in liquid chromatography. Quite in general, chromatographic separations may result from differences in:

1. molecular size and shape
2. electronic dispersion
3. dipole-induced dipole interactions
4. dipole–dipole interactions
5. hydrogen bonding
6. EDA complexation
7. coordination-compound formation
8. ionic equilibria

The effective separation of the components of a mixture depends both on the selectivity and on the efficiency of the column. Selectivity is measured by the separation factor or relative retention r_{21}, which may be obtained from retention data of two compounds to be separated by:

$$r_{21} = t'_{R.2}/t'_{R,1} = k'_2/k'_1 \tag{4}$$

Column efficiency is measured by the number of theoretical plates N:

$$N = 5.54 \, (t_R/W_{1/2})^2 \tag{5}$$

and the effective plate number N':

$$N' = 5.54(t_R'/W_{1/2})^2 \tag{6}$$

where $W_{1/2}$ is the peak width at half peak height. The resolution R_s of two compounds to be separated, which is defined as:

$$R_s = 2\frac{t_{R,2} - t_{R,1}}{W_1 + W_2} \tag{7}$$

where W is the peak width on the base line between the tangents of the peak, is related to these parameters by:

$$R_s = \frac{\sqrt{N'}}{4}\frac{r_{21} - 1}{r_{21}}. \tag{8}$$

Unfortunately, effective plate numbers are not constant within a chromatographic analysis, but depend on the value of the capacity ratio, resulting in a complex relation between resolution, selectivity, and efficiency. In the (quite common) case that a (quasi-)linear relation between peak width and retention time is observed, a straightforward description of the relation between resolution, selectivity, and efficiency is given by:

$$R_s = \sqrt{N_{inf}}\frac{r_{21} - 1}{2(r_{21} + 1) - 4t_q/t_{R,1}'} \tag{9}$$

where N_{inf} is the number of theoretical plates at infinite capacity and $t_{R,1}'$ is the net retention time of the least sorbed of two compounds to be separated [1, 2]. The time quality factor t_q in this equation is defined as:

$$t_q = t_c - t_d \tag{10}$$

where t_c is the intercept on the time axis of the (quasi-)linear relation between peak width and retention time. This parameter, which is always negative, is a measure of the decrease of N' and deterioration of column resolving power with decreasing retention within an analysis.

Quite in general, column efficiency is higher in gas chromatography than it is in liquid chromatography. Selectivities achievable in liquid chromatography, on the other hand, surpass those obtainable by gas chromatography. Since this article is about the role of intermolecular interactions in chromatographic separations, it mainly deals with selectivity. Column efficiency is only discussed sporadically.

3 Gas Chromatography

Since in gas chromatography the mobile phase is a gas, interaction of solute molecules with this mobile phase is in general negligible compared to interaction with the stationary phase. As a consequence, processes in gas chromatography are somewhat analogous to evaporation and sublimation. Energy and enthalpy are lower in the stationary phase, but entropy is higher in the mobile phase, so that solute molecules distribute between these two phases. When this distribution is different for the components of a mixture, then they are separated in the chromatographic process.

In the discussion of the role of intermolecular interactions in gas chromatographic separation, different retention measures will be used. The retention time as such is rather inadequate because of its dependence on experimental conditions; (column temperature, flow rate, amount of stationary phase). Only if different compounds or different classes of compounds are analyzed on the same column under exactly the same conditions, can meaningful comparisons be made. One way of circumventing the problem is to relate the retention data of the compounds investigated to those of one or several reference substances. Relative retention as defined above is such a retention measure. A more sophisticated measure is the retention index I, which is defined as:

$$I = 100\, n_C + 100 \frac{\log t'_{R,i} - \log t'_{R,nc}}{\log t'_{R,nc+1} - \log t'_{R,nc}} \tag{11}$$

where $t'_{R,i}$ is the net retention time of the species of interest i; $t'_{R,nc}$ is the net retention time of the n-alkane eluted next before it; $t'_{R,nc+1}$ is the net retention time of the n-alkane eluted next after it, and n_C is the carbon number of the n-alkane of retention $t_{R,nc}$. Both relative retention and the retention index still depend to some extent on column temperature.

An absolute retention measure that is of interest for the discussion of the role of intermolecular interactions is the specific retention volume V_g, which is defined as:

$$V_g = \frac{j\dot{V}_0(t_R - t_d)}{W_s} \times \frac{273.16}{T_c} \tag{12}$$

where j is the pressure drop correction factor or compressibility factor, $\dot{V}_0$ is the volume flow rate of carrier gas at the outlet, W_s is the weight of stationary liquid in the column, and T_c is the absolute column temperature.

Unless indicated otherwise, graphs inserted for illustrative purposes were constructed on the basis of retention data taken from a general compilation work, the Handbook of Chromatography [3].

3.1 Gas–Solid Chromatography

In gas–solid chromatography the separation is dependent on adsorption effects. The adsorption of vapors onto surfaces can be classified into two main classes: (i) physical adsorption, in which the vapors are held on the surface of the adsorbent by intermolecular interactions, and (ii) chemisorption, in which the vapors form a chemical bond with the surface. In general, physical adsorption is relatively weak and the bonds between vapor and surface are formed and broken easily and rapidly. Chemisorption, on the other hand, usually requires considerable activation energy and thus tends to occur at higher temperatures. Furthermore, adsorption and desorption may be slow. Gas chromatography, therefore, uses physical adsorption almost exclusively.

Separation in gas–solid chromatography may be due to different processes, dependent on the nature of the stationary phase and of the compounds to be separated. Different adsorbents may be used.

(i) Molecular sieves are zeolites, in which the water of crystallization is driven off by the action of heat, so that a network of empty holes remains. These holes are of molecular proportions, and of a uniform diameter which is determined by the crystal structure of the parent zeolite. Molecular sieves are specific adsorbents for those molecules which are of such a size and shape that they can enter the holes. As a consequence, chromatographic separations on the basis of molecular size and shape may be performed on these materials. The most characteristic use of molecular sieves in gas chromatography is provided by molecular sieve 5Å, in which the holes are such that they can contain linear chains of methylene groups, but no branched chains. As a result, they can be used as subtractors for removing n-alkyl compounds from branched compounds. This not only works for alkanes, but also for other classes of chemical compounds, such as alkenes and alkanols.

(ii) Graphitized carbon black, which is obtained by heating charcoal to 2500–3000°C, forms a nonspecific adsorbent that is both apolar and non-complexing. Electronic dispersion is dominant in its interaction with vapors in all cases and it is the sole interaction mechanism for apolar solutes. Relations between retention and structure resemble those for solutes on apolar non-complexing stationary liquids (see below).

(iii) Microporous aromatic hydrocarbon polymer beads are marketed under the names of Porapak, Polypak, Phasepak, and various grades of Chromosorb. On this material, dipole-induced-dipole interactions are much more important for polar solutes than on graphitized carbon black.

(iv) Silica gel, an amorphous form of hydrated silica, is a powerful adsorbent of large specific surface area. The adsorptive properties of silica gel depend on the presence of surface silanol (Si—OH) groups. Thus, beside electronic dispersion, dipole induction, dipole–dipole interaction, and hydrogen-bond formation can play an important role in the separation process. Retention properties resemble those of stationary liquids containing hydroxyl groups (see below).

3.2 Gas–Liquid Chromatography

Gas–liquid chromatography was introduced in 1952 by James and Martin [4], following an earlier suggestion made by Martin and Synge [5]. The fact that a great variety of liquids with different properties can be used as the stationary phase, resulting in packings suitable for the most diverse separation tasks, greatly contributed to the wide-spread use of gas–liquid chromatography. The stationary liquid may be coated onto an inert solid support and then packed into the column (packed column). A great variety of solids has been applied as support, from high surface area adsorbents to nonporous glass beads. The great majority of work, however, has been done using either kieselguhr (i.e., diatomaceous earth, e.g., Celite and Chromosorb W) or crushed firebrick. An alternative way to hold the stationary liquid in the column is to spread it as a thin film onto the internal wall of a narrow bore tube. Such columns are called capillary or open tubular columns. The columns are referred to as wall-coated open tubular (WCOT) columns, if the walls of the tubing are used without the addition of any material that could be considered as solid support. If the walls are first coated with a layer of fine particulate support material, which is then coated with the stationary phase, then columns are referred to as support-coated open tubular (SCOT) columns. The efficiency of open tubular columns considerably exceeds that of packed columns and increases with decreasing internal diameter.

3.2.1 Separation on Apolar Noncomplexing Columns

Retention of apolar noncomplexing solutes on these columns occurs solely through electronic dispersion. Since cohesion of such solutes is also solely due to dispersion forces, correlations of retention data with boiling points are to be expected. There are, however, many systematic small deviations from such correlations. For example, branched alkanes tend to have smaller retentions than their boiling point suggests (see Fig. 1). Cycloalkanes, on the other hand, tend to be retained more. Unsaturation has no important effect on retention, as is evident from Fig. 2 in which the retention index of n-alkenes and aromatic

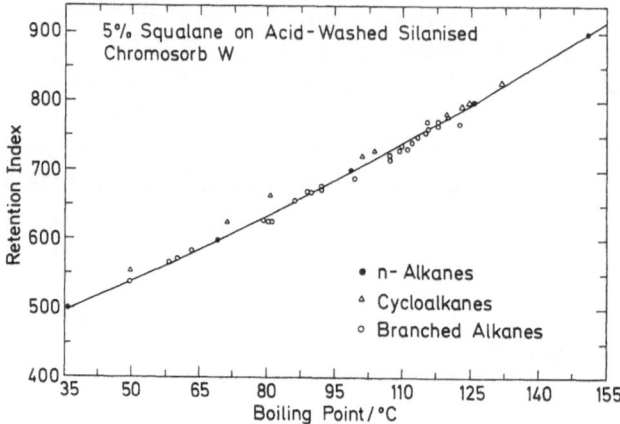

Fig. 1. Correlation of the retention index of n-alkanes, branched alkanes and cycloalkanes on squalane with boiling point data

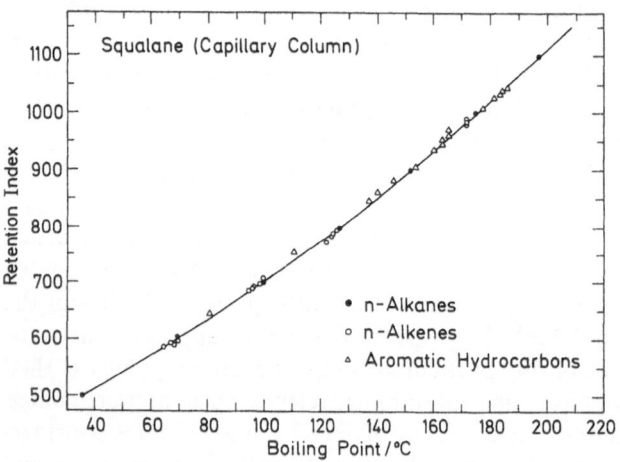

Fig. 2. Correlation of the retention index of n-alkanes, n-alkenes and aromatic hydrocarbons on squalane with boiling point data

hydrocarbons is compared with that of n-alkanes on squalane (2,6,10,15,19,23-hexamethyltetracosane).

Retention of polar and complexing solutes is much less than expected from their boiling point, because dipole–dipole interactions and/or complexation, which contribute to cohesion in these solutes, are absent in the chromatographic separation process. This is illustrated in Fig. 3, in which the retention of alkanes, esters, ketones, and alcohols on squalane is compared. Among these compounds, alcohols, in which both dipole–dipole interaction and hydrogen-bond formation contribute to cohesion, are retained least when boiling point is taken as a

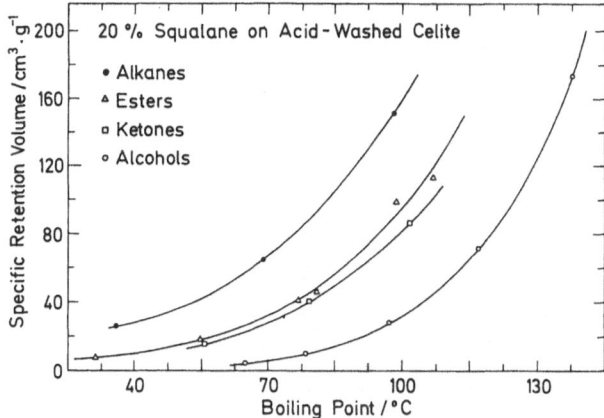

Fig. 3. Comparison of the retention characteristics of alkanes, esters, ketones, and alcohols on squalane, taking normal boiling points as reference criterion

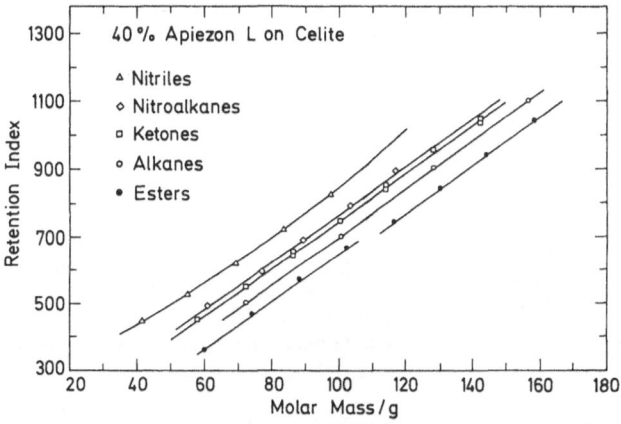

Fig. 4. Comparison of the retention characteristics of nitriles, nitroalkanes, ketones, esters, and alkanes on Apiezon L, taking molar mass as reference criterion

reference criterion. Polar solutes can induce a dipole moment in an apolar stationary liquid and dipole-induced-dipole interactions can therefore contribute to the separation process. This is evident from Fig. 4, in which the retention of nitriles, nitroalkanes, ketones, esters, and alkanes on Apiezon L is compared. Nitriles, nitroalkanes, and ketones are more retained than alkanes when molar mass is taken as a reference criterion. Esters are retained less, which can be attributed to smaller per-weight electronic dispersion as a result of the presence of two (moderately bulky) oxygen atoms. When halogen atoms are present in the molecule, decreased per-weight electronic dispersion has a very pronounced effect, as is evident from Figs. 5 and 6. The effect increases with increased atomic weight of the halogen atom. Decreased per-weight electronic dispersion is also

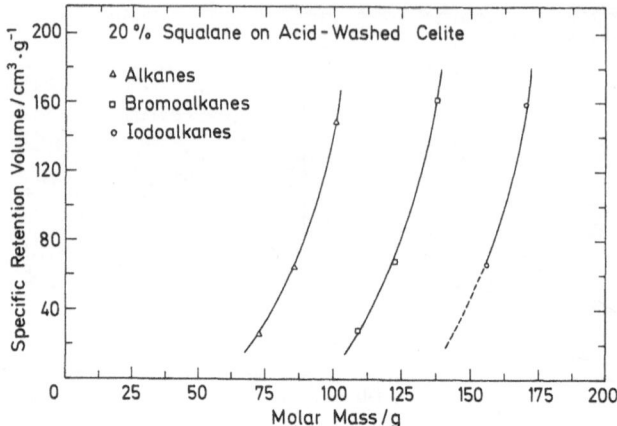

Fig. 5. Comparison of the retention characteristics of alkanes, bromoalkanes, and iodoalkanes on squalane, taking molar mass as reference criterion

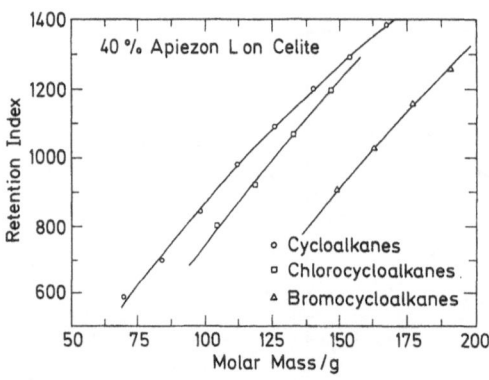

Fig. 6. Comparison of the retention characteristics of cycloalkanes, chlorocycloalkanes, and bromocycloalkanes on Apiezon L, taking molar mass as reference criterion

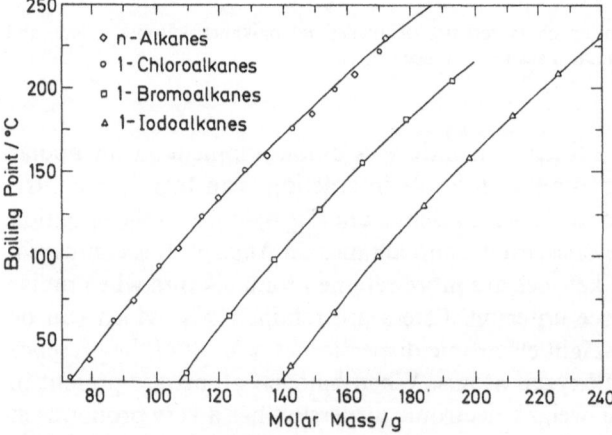

Fig. 7. Comparison of boiling point data of alkanes and haloalkanes, taking molar mass as reference criterion

clearly observable from boiling point data; (see Fig. 7). For 1-chloroalkanes, dipole–dipole interactions compensate for the decreased per-weight electronic dispersion with respect to that of alkanes, but bromoalkanes and especially iodoalkanes have much lower boiling points than alkanes when molar mass is taken as the reference criterion.

3.2.2 Separation on Polar and/or Hydrogen-Bonding Columns

The specific retention volume of saturated hydrocarbons is less on polar and/or complexing columns than it is on apolar noncomplexing ones. Dipole–dipole interactions must decrease and/or hydrogen bonds must be broken in order to allow an alkane to enter a polar and/or complexing stationary liquid, and these interactions are not replaced by suitably strong new interactions. The larger the proportion of polar and complexing groups in the stationary liquid, the smaller is the retention of a saturated hydrocarbon. For "dilute" stationary liquids, i.e., liquids in which the majority of the molar volume is occupied by apolar noncomplexing groups, the decrease in retention is moderate and such stationary liquids are easily usable for analysis of saturated hydrocarbons. With "concentrated" polar and/or complexing stationary liquids, in which saturated hydrocarbons are sparingly soluble, the decrease in the retention of such compounds may be very great and their chromatography is either hard or impossible in them.

Unsaturation in compounds to be separated has an important effect on their retention on polar columns, as a result of dipole-induced-dipole interactions. This is illustrated in Fig. 8, in which retention indices of cycloalkanes, cyclo-alkenes, and alkylbenzenes on Emulphor-O are compared. Dipole-induced-dipole interactions do not compensate fully, however, for the decrease in dipole–dipole interactions and/or hydrogen bonding, which occurs when the unsaturated solute enters the stationary liquid. This explains, for instance, why the specific

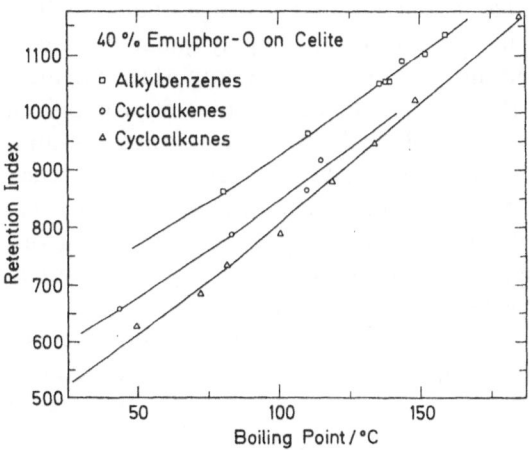

Fig. 8. Comparison of the retention characteristics of cycloalkanes, cyclo-alkenes, and alkylbenzenes on Emul-phor-O, taking normal boiling points as reference criterion

Table 1. Specific retention volume of different compounds on n-hexadecane, palmitonitrile and n-hexadecanol

Compound	Molar mass (g)	Boiling point (°C)	Dipole moment (Debye)	Specific retention volume[a] ($cm^3 g^{-1}$)		
				n-Hexadecane	Palmitonitrile	n-Hexadecanol
1-Pentene	70.13	29.2[b]	0.34	83	63	32
Propionitrile	55.08	97.2	3.60	87	521	99
t-Butanol	74.12	82.2	1.67	92	390	320

[a] From Ref. [6]; [b] boiling point at 740 mm Hg

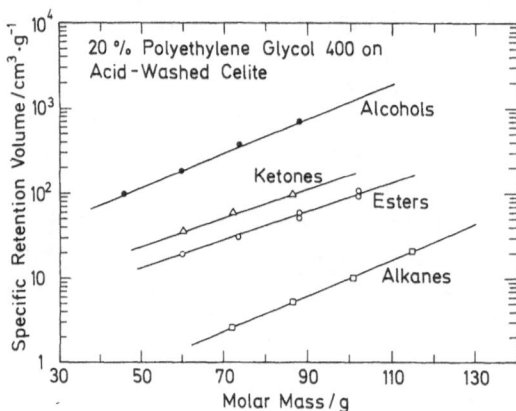

Fig. 9. Comparison of the retention characteristics of alkanes, esters, ketones, and alcohols on polyethylene glycol, taking molar mass as reference criterion

retention volume of 1-pentene in palmitonitrile and n-hexadecanol is lower than it is in n-hexadecane; (see Table 1). The effect is most pronounced in n-hexadecanol, in which both dipole–dipole interaction and hydrogen bonding contribute to cohesion.

Polar solutes and solutes capable of forming hydrogen bonds are further retained as a result of dipole–dipole interaction and hydrogen bonding. This is illustrated by retention data in Table 1. In the highly polar stationary liquid palmitonitrile, dipole–dipole interaction is a very important retention mechanism for polar solutes and the specific retention volume of propionitrile thus exceeds that of the less-polar t-butanol. In the strongly hydrogen-bonding stationary liquid n-hexadecanol, on the other hand, the retention is inversed with t-butanol having more than three times the retention of propionitrile. The contribution of dipole–dipole interaction and hydrogen bonding is further illustrated in Fig. 9, in which specific retention volumes of alkanes, esters, ketones, and alcohols on polyethylene glycol are compared. Ketones and esters are more retained than alkanes, when molar mass is taken as the reference criterion, mainly as a result of dipole–dipole interaction. Alcohols are retained much more than the other three classes of compounds, as a result of hydrogen bonding with an ether function of the polyether polyethylene glycol:

$$R$$
$$|$$
$$O$$
$$|$$
$$H$$
$$\vdots$$
$$-CH_2-O-CH_2-O-CH_2-O-CH_2-$$

Within an homologous series, retention increases with molar mass for all classes of compounds, as a result of increased electronic dispersion per molecule.

3.2.3 Separation on the Basis of Coordination-Compound Formation

The association of a metal ion with other ions or molecules called ligands produces a composite complex compound generally called a coordination compound. The complexes that are best known are those of the transition metals. Because of the availability of d-orbitals for bonding, the transition-metal complexes are of the highest stability and of the greatest number.

Like ammonia, amines may form coordination compounds with transition-metal ions. The stability of the complexes decreases in the following order: $NH_3 > CH_3NH_2 > (CH_3)_2NH > (CH_3)_3N$. On the basis of this, separations of amines using metal salts of higher fatty acids as stationary phase have been performed. Results given in Table 2, in which retention data of amines on Apiezon L and manganese stearate columns are compared, clearly show the importance of complexation to the separation process. For a specific class of compounds, retention increases with molar mass on both columns. On the apolar noncomplexing Apiezon L column, an increase is also observed on going from primary to secondary and tertiary amines. On the manganese stearate column the relation is inversed, however, which is in line with the stability of the respective complexes.

Coordination compounds in which an olefin is bonded to a metal ion are also observed. Such olefin complexes allow very specific chromatographic

Table 2. Relative retention[a,b] of amines on Apiezon L and manganese stearate

Amine	Apiezon L	Manganese Stearate
Pr^nNH_2	0.173	3.655
Bu^nNH_2	0.215	6.032
Pr^n_2NH	0.276	2.134
Bu^n_2NH	0.758	4.871
Pr^n_3N	0.619	0.649
Bu^n_3N	2.382	3.342

[a] Relative to retention of mesitylene; [b] from Ref. [7]

Table 3. Relative retention[a,b] of different C_4 hydrocarbons on a hydrocarbon mixture obtained by oligomerization of *iso*-butene (*iso*-butene trimer) and on a saturated solution of silver nitrate in ethylene glycol

Compound	*iso*-Butene trimer	Saturated solution of $AgNO_3$ in ethylene glycol
iso-Butane	5.6	0.7
n-Butane	7.5	0.7
Butadiene	6.7	10.0
1-Butene	6.7	6.2
iso-Butene	6.7	3.2
cis-2-Butene	8.0	1.7
trans-2-Butene	8.9	5.5

[a] Relative to retention of ethylene; [b] from Ref. [8]

separations of olefins. In general, silver nitrate is used dissolved in some non-volatile polar liquid, of which ethylene glycol, propylene glycol, and benzyl cyanide are most common. Saturated hydrocarbons behave as if the silver nitrate were not there. Thus, when ethylene glycol is used as solvent for the silver nitrate, the retention of alkanes is very low, whereas olefins are specifically retained. This is illustrated in Table 3, in which the relative retention of a number of C_4 hydrocarbons on a hydrocarbon mixture formed by oligomerization of *iso*-butene and on a saturated solution of silver nitrate in ethylene glycol is compared. When benzyl cyanide is used as solvent, alkanes are more strongly retained so that simple mixtures of alkanes and alkenes can be analyzed in a single run.

3.2.4 Characterization of Liquid Stationary Phases

Liquid stationary phases may be characterized, as far as their polarity and their ability to form hydrogen bonds and other types of complexes is concerned, by comparing retention data (preferably retention indices) of suitable test substances on these phases with those on an apolar non-complexing stationary liquid. One advantage of such comparisons is that anomalies due to differences in per-weight electronic dispersion resulting from bulky atoms in solute molecules cancel out, as is evident from Fig. 10, in which retention indices of cycloalkanes, chloro- and bromocycloalkanes, and cycloalkanones on Apiezon L and Emulphor-O are compared. Chloro- and bromocycloalkanes, which have about the same group dipole moment, fall on the same line, despite the presence of bulky chlorine and bromine atoms in these molecules.

Parallel straight lines are generally obtained when retention indices of different classes of compounds on two liquid stationary phases are compared, as exemplified by Figs. 11 and 12. The results shown clearly indicate that dipolar interactions (respectively dipole-induced-dipole and dipole–dipole) are very

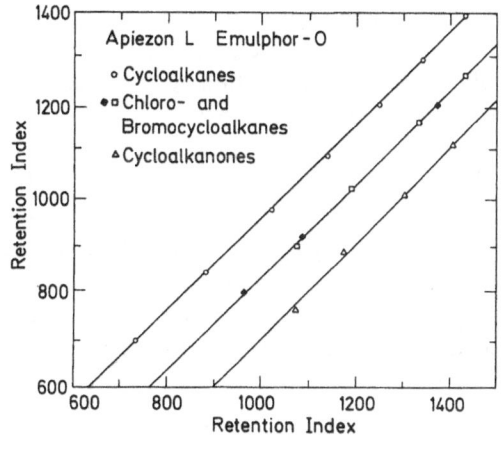

Fig. 10. Retention index of a number of cycloalkanes, chloro- and bromocyclo-alkanes and cycloalkanones on Apiezon L as a function of their retention index on Emulphor-O

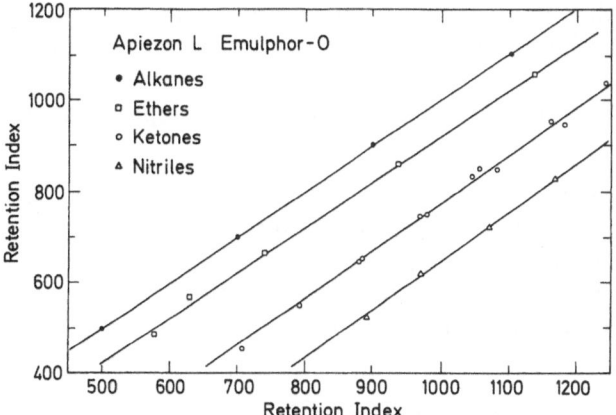

Fig. 11. Retention index of a number of alkanes, ethers, ketones, and nitriles on Apiezon L as a function of their retention index on Emulphor-O

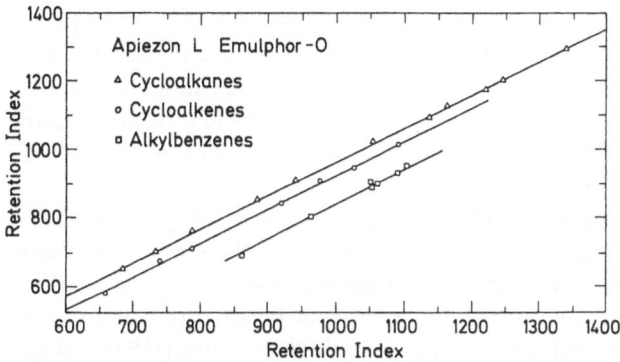

Fig. 12. Retention index of a number of cycloalkanes, cycloalkenes, and alkylbenzenes on Apiezon L as a function of their retention index on Emulphor-O

important on Emulphor-O, since shifts are in line with respectively the polarizability and the group dipole moment of the different classes of compounds
investigated. For the sake of simplicity, comparisons are limited in practice to
only a few test substances and columns are characterized by:

$$\Delta I = I^P - I^P_{squalane} \tag{13}$$

where I^P is the retention index for probe P on the stationary phase to be
characterized, and $I^P_{squalane}$ is the retention index for the same test substance on
the apolar noncomplexing reference phase squalane. ($\Delta I = 0$ for n-alkanes).
These are the Rohrschneider–McReynolds phase constants. As test solutes,
Rohrschneider [9] proposed the use of:

 (i) benzene (dipole induction and weak proton acceptor capability),
 (ii) nitromethane (strongly polar, with some weak proton-acceptor capability),
 (iii) 2-butanone (polar, with proton-acceptor but no proton-donor capabilities),
 (iv) ethanol (polar, with both proton-donor and proton-acceptor capabilities),
     and
 (v) pyridine (polar with strong proton-acceptor but no proton-donor
     capabilities).

McReynolds [10] replaced three of these test substances by similar but less
volatile compounds for practical reasons and increased the number of test substances to ten. The choice of these test substances appears somewhat arbitrary
and both a reduction in the number and a change in the identity of the test
solutes has been recommended by different investigators.

4 Liquid Chromatography

In liquid chromatography the mobile phase is a liquid, whereas the stationary
phase may be a solid (LSC) or a liquid (LLC). As a consequence, the sample
can interact selectively with both the mobile phase and the stationary phase.
Moreover, the nature and extent of interaction with the mobile phase can easily
be adapted to the requirements of a particular separation by a change in chemical
composition of this phase. Hence, liquid chromatography is a more versatile
technique than gas chromatography and can often achieve more difficult
separations.

The efficiency of classical liquid chromatography, in which gravity flow is
used, is rather low. By a reduction in the particle size of the stationary phase-
materials used, higher efficiencies may be obtained. The small size of these
particles leads to a considerable resistance to solvent flow, so that the mobile
phase has to be pumped through the column under high pressure; (HPLC: High-
Performance Liquid Chromatography). The stationary phases used are called
microparticulate column packings and are commonly uniform porous silica

particles with nominal diameters of 3, 5, or 10 µm. Typically, the column is 10 to 50 cm long and has an internal diameter of 2 to 5 mm.

In addition to the two main types of liquid chromatography, a number of special techniques exist, e.g., ion-exchange chromatography, ion-pair chromatography, affinity chromatography, and size-exclusion chromatography.

More and more the trend in liquid chromatography is to use so-called bonded phases, in which the stationary phase is chemically bonded to the solid support. Bonded phases are prepared by reacting the surface silanol groups of silica gel with a chlorosilane. With monofunctional silanes each molecule of the silylating agent can react with only one silanol group:

$$\text{Si}-\text{OH} + \text{Cl}-\underset{\underset{R}{|}}{\overset{\overset{R}{|}}{\text{Si}}}-R' \longrightarrow \text{Si}-\text{O}-\underset{\underset{R}{|}}{\overset{\overset{R}{|}}{\text{Si}}}-R' + \text{HCl} \tag{14}$$

More complicated surface structures can be produced by changing the functionality of the silylating agent and the conditions under which the reaction is carried out. A crosslinked polymeric layer can be produced at the silica surface by the use of di- or trichlorosilanes in the presence of moisture. The simple structure is to be preferred, however, as it is better defined and easier to manufacture reproducibly than the more complicated surface structures. The R groups in Eq. (14) are usually methyl groups. The nature of R' can be apolar (e.g., octyl, octadecyl, phenyl), polar (e.g., $-(CH_2)_3NH_2$, $-(CH_2)_3CN$), or ionizable (e.g., sulfonic acid, quaternary ammonium). The introduction of ionizable groups produces bonded phases with ion-exchange properties. The range of functional groups that can be bonded to silica is very wide, and for specialized applications (e.g., the separation of chiral compounds) some fairly exotic bonded phases are available.

4.1 Liquid–Solid Chromatography

In liquid–solid chromatography, separations are based on differences in adsorption on a solid stationary phase. The extent of this adsorption depends on: (i) interactions of different components of a mixture with the solid stationary phase, (ii) interactions of the eluent with the solid stationary phase, and (iii) interactions with the eluent of the different components of a mixture. This may be represented by the following scheme.

$$\begin{array}{ccc} & \text{Solid stationary phase} & \\ \swarrow & & \searrow \\ \text{Compounds} & \longleftrightarrow & \text{Liquid mobile} \\ \text{to be separated} & & \text{phase (eluent)} \end{array}$$

Dissolved solute molecules X must compete with mobile phase molecules M for a place on the adsorbent surface:

$$X + M_{ads} \rightleftharpoons X_{ads} + M. \tag{15}$$

Thus, in order to obtain reasonable retention and good separation, interaction of the eluent with the solid stationary phase should not be too extensive.

Different adsorbents may be used, but silica gel is by far the most studied and utilized. The adsorption sites on the surface of silica gel are silanol (Si—OH) groups. These can be present as isolated groups, or can be hydrogen-bonded to one another. These two types of silanol groups have different adsorptive strength. The strength of the interaction between the adsorbent and eluent molecules increases as the polarity and hydrogen-bonding capacity of the eluent increases. Thus, maximal adsorption and retention of solute molecules is obtained with apolar noncomplexing eluents. Desorption can be caused by gradual addition of a polar and/or hydrogen-bonding substance to the mobile phase. (In liquid chromatography, the composition of the mobile phase can be made to change in a predetermined way during the separation, which is called gradient elution.) In a typical analysis, hexane or 2,2,4-trimethylpentane would be used as base eluent, with a gradual increase of the concentration of acetonitrile or some other polar and/or hydrogen-bonding substance during the analysis.

4.2 Liquid–Liquid Chromatography

Liquid–liquid chromatography uses a liquid stationary phase coated onto a finely divided inert solid support. Different interactions are important, either for the separation process or for the stability of the column: (i) interaction of the liquid stationary phase with the solid support, (ii) interaction of the liquid mobile phase with the liquid stationary phase, (iii) interaction of the compounds to be separated with the liquid stationary phase, and (iv) interaction of the compounds to be separated with the liquid mobile phase. This may be represented by the following scheme:

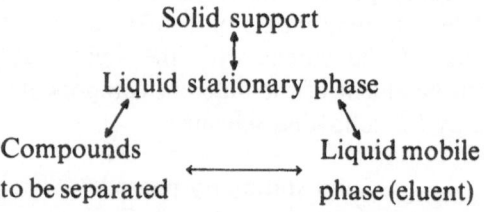

The interaction of the liquid stationary phase with the solid support does not play a role in the separation process, but is of prime importance for the stability of the column. Clearly, the interaction with the solid support should be much more important than with the eluent in order for the column to be stable. Thus, normal practice in this regard would be to use a polar support material, a polar stationary liquid, and an apolar mobile liquid. This is called normal-phase liquid chromatography. Reverse-phase liquid chromatography, on the other hand, uses an apolar stationary liquid and a polar mobile phase. As a result of the introduction of bonded phases, the reverse-phase mode has become normal practice, however. Indeed, the majority of work in HPLC at the moment is done using bonded phases in which octadecyl groups are attached to the surface of silica particles.

4.3 Optimization of Selectivity in Liquid Chromatography

Selectivity in liquid chromatography is controlled and optimized mainly by the nature and composition of the mobile phase. Even with diverse sample types, the same stationary phase can often be used for the separation. Since reverse-phase liquid chromatography has become common practice, the discussion will be limited to this mode.

In the reverse-phase mode (apolar stationary phase) interaction of compounds to be separated with the stationary phase is mainly through electronic dispersion; (dipole induction and in special cases hydrogen bonding can also contribute to retention). As mobile phase a polar and/or complexing liquid is used, resulting in dipole induction, dipole–dipole interaction, and/or hydrogen bonding with the compounds to be separated. Such interactions lower retention, as is evident from Fig. 13, in which the retention of alkanes, alkenes, and alkylbenzenes on squalene is compared, using acetonitrile as eluent. Alkylbenzenes are retained

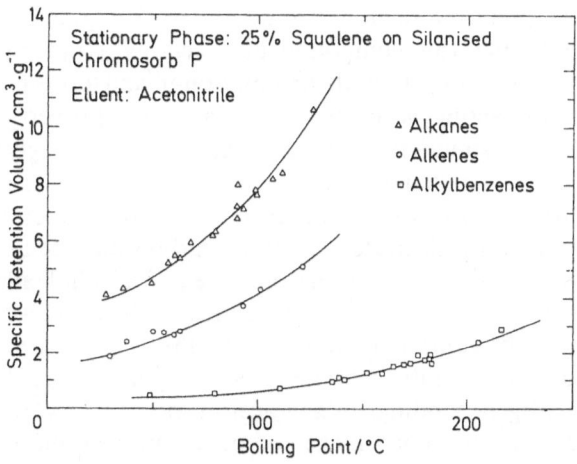

Fig. 13. Comparison of the retention of alkanes, alkenes and alkylbenzenes on squalene with acetonitrile as eluent, taking normal boiling points as reference criterion

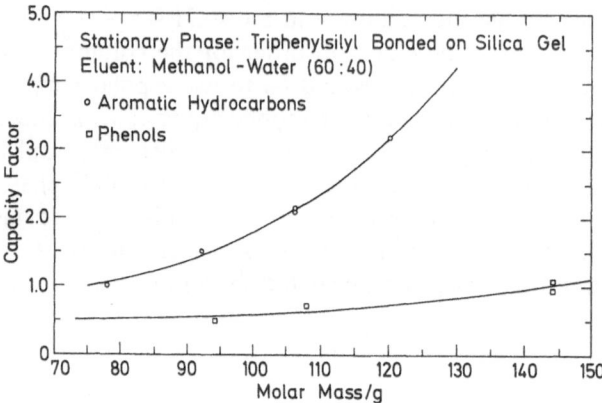

Fig. 14. Comparison of the retention of aromatic hydrocarbons and phenols on triphenylsilyl bonded on silica gel with methanol-water as eluent, taking molar mass as reference criterion. Graph based on retention data taken from Ref. [11]

less than alkenes and much less than alkanes, as a result of stronger dipole induction by the highly polar mobile phase. A further illustration is provided by retention data of aromatic hydrocarbons and phenols on triphenylsilyl bonded on silica gel, shown in Fig. 14, in which a mixture of methanol and water is used as eluent. As a result of dipole–dipole interaction and hydrogen bonding with the mobile phase, phenols elute much more rapidly than aromatic hydrocarbons.

Selectivities obtainable with liquid chromatography exceed those of gas chromatography. This mainly stems from the fact that not only pure liquids but also mixtures of liquids may be used as mobile phase and that the composition of such mixtures may be changed in a predetermined way during the analysis. As a result of this, it is possible to really fine-tune the interaction of the mobile phase with the compounds to be separated. In reverse-phase liquid chromatography, acetonitrile, tetrahydrofuran, and methanol are favored liquids in this regard. With acetonitrile polar interactions are dominant. The polarity of tetrahydrofuran is much less, but this compound also acts as a proton acceptor. Methanol finally is polar and has both proton-donor and proton-acceptor capabilities. Solutes interacting according to a particular property become less retained relative to other solutes interacting not at all or less strongly when that property is emphasized by a change in chemical composition of the mobile phase. Results taken from supercritical fluid chromatography experiments, shown in Fig. 15, may serve to illustrate this. With carbon dioxide as base eluent, the retention of toluene, nitrobenzene, and benzyl alcohol all decrease with increased concentration of methanol. The extent of the decrease differs according to the nature of the compound, however, and is in the logical order: toluene (dipole-induced-dipole) < nitrobenzene (dipole–dipole) < benzyl alcohol (dipole–dipole and hydrogen bonding). When acetonitrile is added to the mobile phase, the decrease in retention of toluene and nitrobenzene is more

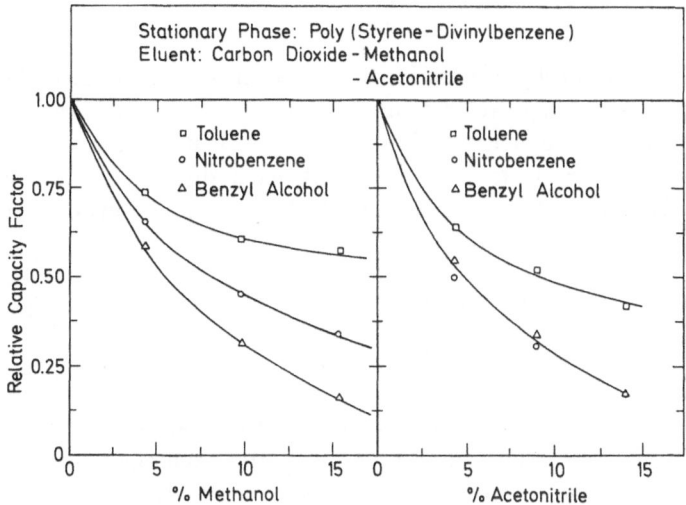

Fig. 15. Effect of the addition of methanol and acetonitrile to the mobile phase on the retention of toluene, nitrobenzene, and benzyl alcohol on a poly(styrene-divinylbenzene) column using carbon dioxide as base eluent. Graph based on retention data taken from Ref. [12]

pronounced, in line with the higher dipole moment of acetonitrile with respect to methanol. Such an effect is not observed for benzyl alcohol, however, because acetonitrile has little tendency to form hydrogen bonds. For a proper understanding, it should finally be remarked that with polar and/or complexing mobile phase mixtures an increase in polarity and complexing capability in many instances leads to an *increase* in retention, because solutes become more excluded from the mobile phase by mutual interactions of the mobile phase molecules.

4.4 Ion-Exchange and Ion-Pair Chromatography

For the chromatography of ionic solutes, both ion-exchange and ion-pair chromatography can be utilized. Ion-pair chromatography has the advantage over ion-exchange chromatography, of using the same bonded stationary phases that are used for the separation of neutral compounds. Weak acids or bases may also be separated on such conventional columns by simple suppression of ionization through adjustment of the pH, a method referred to as ion-suppression chromatography. In this method, samples may be chromatographed on a reverse-phase column (e.g., C-18) using methanol or acetonitrile plus a buffer solution as the mobile phase.

4.4.1 Ion-Exchange Chromatography

In ion-exchange chromatography, separation is based on ionic equilibria. As stationary phase an ion-exchange resin is used, the surface of which carries positive or negative charges to give ion-exchange sites (R^+ or R^-). Counterions of opposite charge (Y^+ or Y^-) are associated with each site and these can exchange with similarly charged ions in the mobile phase. Sample ions (X^- or X^+) may thus exchange with these counterions:

$$R^+Y^- + X^- \rightleftharpoons R^+X^- + Y^- \tag{16}$$

$$R^-Y^+ + X^+ \rightleftharpoons R^-X^+ + Y^+ \tag{17}$$

The retention depends on the strength of interaction of the sample ions with the exchange sites. Ions that react only weakly with the exchange sites are poorly retained whilst ions with strong interactions elute slowly. As a result, different ionic solutes in a sample are separated in the process.

Both strong and weak ion-exchange resins may be used for chromatographic separations. Strong ion-exchange resins contain strong acid or base groups that are fully ionized over a wide pH range. Strong cation-exchange resins contain sulfonic acid exchange sites that are fully ionized above about pH 2. Strong anion-exchange resins contain quaternary ammonium exchange sites that are fully ionized up to about pH 10. Weak cation- and anion-exchange resins, on the other hand, contain respectively carboxylic acid and amino groups. These resins have a higher exchange capacity, but they are ionized only over a restricted pH range. Ion-exchange materials used for chromatographic separations may be microparticulate resins based on styrene-divinylbenzene copolymers or bonded phases based on microparticulate silica. Weak cation-exchange resins are not available based on silica, because they are only fully ionized at high pH where silica is appreciably soluble.

Ligand-exchange chromatography is a related technique, whereby as stationary phase an ion-exchange resin is used, on which a suitable metal ion is sorbed, the counterion again being displaced. Different ligands may be separated on such columns as a result of differences in the stability of the coordination compounds formed; the higher the stability, the greater the retention. Alternatively, the metal ion may be added to the mobile phase resulting in inversed retention characteristics.

4.4.2 Ion-Pair Chromatography

The method of ion-pair chromatography is derived from the field of solvent extraction. An ionized compound (A^+_{aq}) can be extracted from water into an organic solvent by using a suitable counterion (B^-_{aq}) to form an ion-pair:

$$A^+_{aq} + B^-_{aq} \rightleftharpoons (A^+B^-)_{org}. \tag{18}$$

The ion-pair behaves as a nonionic polar molecule, soluble in organic solvents. The method has been extended to chromatography, allowing analysis of ions on conventional chromatographic columns. In ion-pair chromatography, reverse-phase separation is used on an octadecyl-bonded column, with the ion-pairing reagent added to the mobile phase. The ion pairs are effectively separated as neutral polar molecules. For cations, typical ion-pairing reagents are alkyl sulfonic acids, e.g., pentane, hexane, heptane, or octane sulfonic acid. For anions, tetrabutylammonium salts are frequently used. A typical example is the separation of inorganic ions, e.g., Cl^-, Br^-, SO_4^{2-}, on conventional reverse-phase columns by ion-pair formation. For instance:

$$(C_4H_9)_4N^+ + Cl^- \rightleftharpoons (C_4H_9)_4N^+Cl^- \tag{19}$$

The mechanism in this method of separation is still open to some debate. The simplest model assumes that separation occurs by partitioning of the neutral ion-pairs between the mobile phase and the octadecyl-bonded phase. This mechanism cannot explain all the experimental results, and there is no doubt that it is a considerable oversimplification. As a matter of fact, separation would frequently be expected to be rather poor, since the bulk of the neutral ion-pair is often provided by the ion-pairing reagent (see for instance Eq. (19)). It seems reasonable to expect that in such cases much greater separative power may be obtained by playing on the equilibrium constant of the ion-pair-formation reaction, since the ions to be separated interact little with the stationary phase, whereas the interaction of the ion pairs formed is very substantial (but not necessarily very different for the different ionic solutes involved). An alternative mode of separation can be provided by the fact that the octadecyl-bonded phase may become loaded to some extent with the ion-pairing reagent as a result of the organic substituents it contains. This produces stationary charged sites at which ion-exchange can take place.

4.5 Affinity Chromatography

Affinity chromatography is a highly selective method, in which separation results from the specific interaction of protein molecules with suitable ligands. Such interactions are often referred to as "lock-and-key" processes, depending as they do on the correct spatial configuration of the substituent groups. The ligand exhibiting a specific binding affinity for the protein is covalently bonded to a gel matrix; this constitutes the stationary phase. The sample molecules are added in a suitably buffered mobile phase, resulting in the bonding and thus retention of those components that have a specific affinity for the ligand. Other components are eluted from the column. Subsequently, the specific interaction between the protein molecules and the ligand is weakened by a change in the composition or pH of the mobile phase, so that the protein is released and eluted from the column.

4.6 Size-Exclusion Chromatography

In size-exclusion chromatography, alternatively termed gel-permeation, gel-filtration, and molecular-sieve chromatography, separation is due to differences in the size and shape of the molecules to be separated. Size-exclusion chromatography is carried out on a porous stationary phase, which may be based on either silica or on a polymer gel. Large molecules are totally excluded from the pore structure of the gel and thus elute very rapidly. All such components of a mixture elute as a single peak and are not separated from one another; (total exclusion limit). Very small molecules (e.g., the solvent molecules) can penetrate into even the smallest pores and therefore take the longest time to elute from the column. Again, all such components of a mixture are not separated from one another but elute as a single peak (total permeation limit). Molecules of intermediate size are able to diffuse into some of the pores to an extent dependent on their size and shape and can consequently be separated. Provided that exclusion is the only separation mechanism (i.e., no specific adsorption, partition, or ion-exchange), the entire sample elutes between the total exclusion and total permeation limit.

The technique of size-exclusion chromatography is used for characterization of monomeric and polymeric species with molar masses above about 400. It is thus particularly suited to separations of natural and manufactured polymeric materials. Stationary phases used in the technique are microparticulate materials consisting of styrene-divinylbenzene copolymers or silica. These are available in a range of pore sizes for the separation in different molar mass ranges. The degree of separation achievable is somewhat limited, however, since it is found as a general rule that molar mass differences of at least 10% are required before any separation can be achieved.

5 References

1. Ceulemans J (1984) J Chromatogr Sci 22: 296
2. Ceulemans J (1986) J Chromatogr Sci 24: 147
3. Zweig G, Sherma J (eds) (1972) Handbook of Chromatography, vol 1, CRC Press, Cleveland
4. James AT, Martin AJP (1952) Biochem J 50: 679
5. Martin AJP, Synge RLM (1941) Biochem J 35: 1358
6. Littlewood AB (1970) Gas Chromatography, 2nd edn, Academic Press, New York, p 103

7. Barber DW, Phillips CSG, Tusa GF, Verdin A (1959) J Chem Soc 18
8. Bradford BW, Harvey D, Chalkley DE (1955) J Inst Petrol 41: 80
9. (a) Rohrschneider L (1966) J Chromatogr 22: 6; (b) Rohrschneider L (1967) Adv Chromatogr 4: 333
10. McReynolds WR (1970) J Chromatogr Sci 8: 685
11. Szabo G, Csato E, Offenmüller K, Dévai M, Borbély-Kuszmann A, Liptai G (1988) Chromatographia 26: 255
12. Smith RM, Marsin Sanagi M (1988) Chromatographia 26: 77

Subject Index